Harmonic Analysis on Finite Groups

CAMBRIDGE STUDIES IN ADVANCED MATHEMATICS 108

HARMONIC ANALYSIS ON FINITE GROUPS

Line up a deck of 52 cards on a table. Randomly choose two cards and switch them. How many switches are needed in order to mix up the deck?

Starting from a few concrete problems, such as the random walk on the discrete circle and the Ehrenfest diffusion model, this book develops the necessary tools for the asymptotic analysis of these processes. This detailed study culminates with the case-by-case analysis of the cut-off phenomenon discovered by Persi Diaconis.

This self-contained text is ideal for students and researchers working in the areas of representation theory, group theory, harmonic analysis and Markov chains. Its topics range from the basic theory needed for students new to this area, to advanced topics such as Gelfand pairs, harmonics on posets and q-analogs, the complete analysis of the random matchings, and the representation theory of the symmetric group.

CAMBRIDGE STUDIES IN ADVANCED MATHEMATICS

All the titles listed below can be obtained from good booksellers or from Cambridge University Press. For a complete series listing visit: http://www.cambridge.org/series/sSeries.asp?code=CSAM

Harmonic Analysis on Finite Groups
Representation Theory, Gelfand Pairs and Markov Chains

TULLIO CECCHERINI-SILBERSTEIN
Università del Sannio, Benevento

FABIO SCARABOTTI
Università di Roma "La Sapienza", Rome

FILIPPO TOLLI
Università Roma Tre, Rome

CAMBRIDGE
UNIVERSITY PRESS

CAMBRIDGE UNIVERSITY PRESS

Cambridge, New York, Melbourne, Madrid, Cape Town, Singapore, São Paulo, Delhi

Cambridge University Press
The Edinburgh Building, Cambridge CB2 8RU, UK

Published in the United States of America by Cambridge University Press, New York

www.cambridge.org
Information on this title: www.cambridge.org/9780521883368

First published 2008

Printed in the United Kingdom at the University Press, Cambridge

A catalogue record for this publication is available from the British Library

ISBN 978-0-521-88336-8 hardback

To Katiuscia, Giacomo, and Tommaso

To my parents and Cristina

To my Mom, Rossella, and Stefania

Contents

Preface

In September 2003 we started writing a research expository paper on "Finite Gelfand pairs and their applications to probability and statistics" [43] for the proceedings of a conference held in Batumi (Georgia). After a preliminary version of that paper had been circulated, we received several emails of appreciation and encouragement from experts in the field. In particular, Persi Diaconis suggested that we expand that paper to a monograph on Gelfand pairs. In his famous 1988 monograph "Group representations in probability and statistics" [55] there is a short treatement of the theory of Gelfand pairs but, to his and our knowledge, no book entirely dedicated to Gelfand pairs was ever written. We thus started to expand the paper, including some background material to make the book self-contained, and adding some topics closely related to the kernel of the monograph. As the "close relation" is in some sense inductive, we pushed our treatement much further than what Persi was probably expecting. In all cases, we believe that our monograph is in some sense unique as it assembles, for the first time, the various topics that appear in it.

The book that came out is a course in finite harmonic analysis. It is completely self-contained (it only requires very basic rudiments of group theory and of linear algebra). There is also a large number of exercises (with solutions or generous hints) which constitute complements and/or further developments of the topics treated.

For this reason it can be used for a course addressed to both advanced undergraduates and to graduate students in pure mathematics as well as in probability and statistics. On the other hand, due to its completeness, it can also serve as a reference for mature researchers.

It presents a very general treatment of the theory of finite Gelfand pairs and their applications to Markov chains with emphasis on the cutoff phenomenon discovered by Persi Diaconis.

The book by Audrey Terras [220], which bears a similar title, is in some sense orthogonal to our monograph, both in style and contents. For instance, we do not treat applications to number theory, while Terras does not treat the representation theory of the symmetric group.

We present six basic examples of diffusion processes, namely the random walk on the circle, the Ehrenfest and the Bernoulli–Laplace models of diffusion, a Markov chain on the ultrametric space, random transpositions and random matchings. Each of these examples bears its own peculiarity and needs specific tools of an algebraic/harmonic-analytic/probabilistic nature in order to analyze the asymptotic behavior of the corresponding process.

These tools, which we therefore develop in a very self-contained presentation are: spectral graph theory and reversible Markov chains, Fourier analysis on finite abelian groups, representation theory and Fourier analysis of finite groups, finite Gelfand pairs and their spherical functions, and representation theory of the symmetric group. We also present a detailed account of the (distance-regular) graph theoretic approach to spherical functions and on the use of finite posets.

All this said, one can use this monograph as a textbook for at least three different courses on:

(i) **Finite Markov chains** (an elementary introduction oriented to the cutoff phenomenon): Chapters 1 and 2, parts of Chapters 5 and 6, and Appendix 1 (the discrete trigonometric transforms).

(ii) **Finite Gelfand pairs** (and applications to probability): Chapters 1–8 (if applications to probability are not included, then one may omit Chapters 1 and 2 and parts of the other chapters).

(iii) **Representation theory of finite groups** (possibly with applications to probability): Chapters 3, 4 (partially), 9, 10 and 11.

This book would never have been written without the encouragement and suggestions of Persi Diaconis. We thank him with deepest gratitude.

We are also grateful to Alessandro Figà Talamanca who first introduced us to Gelfand pairs and to the work of Diaconis.

We express our gratitude to Philippe Bougerol, Philippe Delsarte, Charles Dunkl, Adriano Garsia, Rostislav I. Grigorchuk, Gerard Letac,

Arun Ram, Jan Saxl and Wolfgang Woess for their interest in our work and their encouragement.

We also acknowledge, with warmest thanks, the most precious careful reading of some parts of the book by Reza Bourquin and Pierre de la Harpe who pointed out several inaccuracies and suggested several changes and improvements on our expositions.

We finally express our deep gratitude to David Tranah, Peter Thompson, Bethan Jones and Mike Nugent from Cambridge University Press for their constant and kindest help at all the stages of the editing process.

Roma, 14 February 2007   TCS, FS and FT

Part I

Preliminaries, Examples and Motivations

1
Finite Markov chains

1.1 Preliminaries and notation

Let X be a finite set and denote by $L(X) = \{f : X \to \mathbb{C}\}$ the vector space of all complex-valued functions defined on X. Clearly $\dim L(X) = |X|$, where $|\cdot|$ denotes cardinality.

For $x \in X$ we denote by δ_x the *Dirac function* centered at x, that is

$$\delta_x(y) = \begin{cases} 1 & \text{if } y = x \\ 0 & \text{if } y \neq x. \end{cases}$$

The set $\{\delta_x : x \in X\}$ is a natural basis for $L(X)$ and if $f \in L(X)$ then $f = \sum_{x \in X} f(x)\delta_x$.

The space $L(X)$ is endowed with the scalar product defined by setting

$$\langle f_1, f_2 \rangle = \sum_{x \in X} f_1(x)\overline{f_2(x)}$$

for $f_1, f_2 \in L(X)$, and we set $\|f\|^2 = \langle f, f \rangle$. Note that the basis $\{\delta_x : x \in X\}$ is orthonormal with respect to $\langle \cdot, \cdot \rangle$. Sometimes we shall write $\langle \cdot, \cdot \rangle_{L(X)}$ to emphasize the space where the scalar product is defined if other spaces are also considered.

If $Y \subseteq X$, the symbol $\mathbf{1}_Y$ denotes the *characteristic function* of Y:

$$\mathbf{1}_Y(x) = \begin{cases} 1 & \text{if } x \in Y \\ 0 & \text{if } x \notin Y; \end{cases}$$

in particular, if $Y = X$ we write $\mathbf{1}$ instead of $\mathbf{1}_X$.

For $Y_1, Y_2, \ldots, Y_m \subseteq X$ we write $X = Y_1 \coprod Y_2 \coprod \cdots \coprod Y_m$ to indicate that the Y_j's constitute a *partition* of X, that is $X = Y_1 \cup Y_2 \cup \cdots \cup Y_m$ and $Y_i \cap Y_j = \emptyset$ whenever $i \neq j$. In other words the symbol $\coprod$ denotes a *disjoint union*. In particular, if we write $Y \coprod Y'$ we implicitly assume that $Y \cap Y' = \emptyset$.

3

If $A : L(X) \to L(X)$ is a linear operator, setting $a(x, y) = [A\delta_y](x)$ for all $x, y \in X$, we have that

$$[Af](x) = \sum_{y \in X} a(x, y) f(y) \tag{1.1}$$

for all $f \in L(X)$ and we say that the *matrix* $a = (a(x, y))_{x,y \in X}$, *indexed* by X, *represents* the operator A.

If the linear operators $A_1, A_2 : L(X) \to L(X)$ are represented by the matrices a_1 and a_2, respectively, then the composition $A_1 \circ A_2$ is represented by the corresponding product of matrices $a = a_1 \cdot a_2$ that is

$$a(x, y) = \sum_{z \in X} a_1(x, z) a_2(z, y).$$

For $k \in \mathbb{N}$ we denote by $a^k = \left(a^{(k)}(x, y)\right)_{x,y \in X}$ the product of k copies of a, namely

$$a^{(k)}(x, y) = \sum_{z \in X} a^{(k-1)}(x, z) a(z, y).$$

We remark that (1.1) can be also interpreted as the product of the matrix a with the column vector $f = (f(x))_{x \in X}$.

Given a matrix a and a column or, respectively, a row vector f, we denote by a^T and by f^T the transposed matrix (i.e. $a^T(x, y) = a(y, x)$ for all $x, y \in X$) and the row, respectively column transposed vector. This way we also denote by $f^T A$ the function given by

$$[f^T A](y) = \sum_{x \in X} f(x) a(x, y). \tag{1.2}$$

With our notation, the identity operator is represented by the identity matrix which may be expressed as $I = (\delta_x(y))_{x,y \in X}$. If X is a set of cardinality $|X| = n$ and $k \leq n$, then a k-subset of X is a subset $A \subseteq X$ such that $|A| = k$.

If $v_1, v_2, \ldots, v_m$ are vectors in a vector space V, then $\langle v_1, v_2, \ldots, v_m \rangle$ will denote their linear span.

1.2 Four basic examples

This section is an informal description of four examples of finite diffusion processes. Their common feature is that their structure is rich in symmetries so that one can treat them by methods and techniques from finite harmonic analysis.

The main scope of this book is to study these examples in full detail and present all the mathematical background.

Example 1.2.1 (The simple random walk on the discrete circle) Let C_n denote a regular n-gon. We number the vertices consecutively from 0 to $n - 1$: we regard these numbers as elements of $\mathbb{Z}/n\mathbb{Z}$, i.e. mod n. This way, there is an edge connecting the jth with the $(j + 1)$st for all $j = 0, 1, \ldots, n - 1$. This is the *discrete circle*.

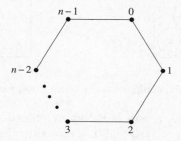

Figure 1.1. The discrete circle C_n

Suppose that a person is randomly moving on the vertices of C_n according to the following rule. The time is discrete $(t = 0, 1, 2, \ldots)$ and at each instant of time there is a move.

At the beginning, that is at time $t = 0$, the random walker is in 0.

Suppose that at time t he is at position j. Then he tosses a fair coin and he moves from j to either $j + 1$ or to $j - 1$ if the result is a head or a tail, respectively. In other words, given that at time t he is in j, at time $t + 1$ he can be at position $j + 1$ or $j - 1$ with the same probability $1/2$.

Figure 1.2. The equiprobable moves $j \mapsto j - 1$ and $j \mapsto j + 1$

Example 1.2.2 (The Ehrenfest diffusion model) The Ehrenfest diffusion model consists of two urns numbered $0, 1$ and n balls, numbered $1, 2, \ldots, n$. A *configuration* is a placement of the balls into the urns. Therefore there are 2^n configurations and each of them can be identified with the subset A of $\{1, 2, \ldots, n\}$ corresponding to the set of balls contained in the urn 0. Note that there is no ordering inside the urns.

The *initial configuration* corresponds to the situation when all balls are inside urn 0.

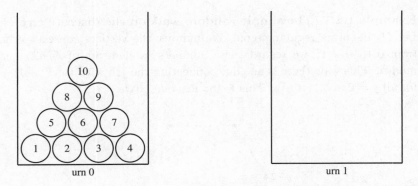

Figure 1.3. The initial configuration for the Ehrenfest urn model ($n = 10$)

Then, at each step, a ball is randomly chosen (each ball might be chosen with probability $1/n$) and it is moved to the other urn.

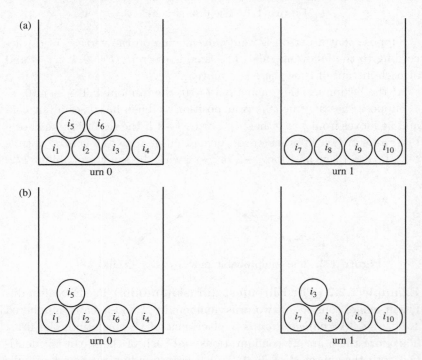

Figure 1.4. (a) A configuration at time t. (b) The configuration at time $t+1$ if the chosen ball is i_3

Denoting by Q_n the set of all subsets of $\{1, 2, \ldots, n\}$, we can describe this random process as follows. If we are in a configuration $A \in Q_n$ then, at the next step, we are in a new configuration $B \in Q_n$ of the form $A \coprod \{j\}$ for some $j \notin A$ or $A \setminus \{j'\}$ for some $j' \in A$ and each of these configurations is chosen with probability $1/n$ (actually, to avoid a parity problem, in our study of this model we shall make a slight change in the definition of the mixing process).

Example 1.2.3 (The Bernoulli–Laplace diffusion model) The Bernoulli–Laplace diffusion model consists of two urns numbered $0, 1$ and $2n$ balls, numbered $1, 2, \ldots, 2n$. A *configuration* is a placement of the balls into the two urns, n balls each. Therefore there are $\binom{2n}{n}$ configurations, each of them can be identified with an n-subset A of $\{1, 2, \ldots, 2n\}$ corresponding to the set of balls contained in the urn 0. The *initial configuration* corresponds to the situation when the balls contained in urn 0 are $1, 2, \ldots, n$ (clearly the balls $n + 1, n + 2, \ldots, 2n$ are contained in the urn 1).

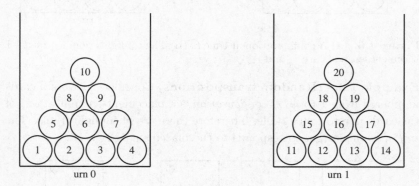

Figure 1.5. The initial configuration for the Bernoulli–Laplace urn model $(n = 10)$

Then at each step two balls are randomly chosen, one in urn 0, the other one in urn 1 and switched.

Denoting by Ω_n the set of all n-subsets of $\{1, 2, \ldots, 2n\}$, we can describe this random process as follows. If we are in a configuration $A \in \Omega_n$ then at the next step we are in a new configuration $B \in \Omega_n$ of the form $A \coprod \{i\} \setminus \{j\}$ for some $i \notin A$ and $j \in A$ and each of these configurations is chosen with probability $1/n^2$.

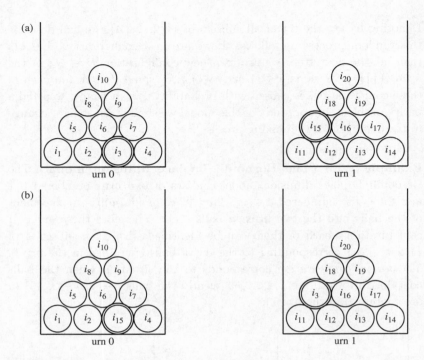

Figure 1.6. (a) A configuration at time t. (b) The configuration at time $t+1$ if the chosen balls are i_3 and i_{15}

Example 1.2.4 (Random transpositions) Consider a deck of n cards numbered from 1 to n. A *configuration* is a placement (permutation) of the cards in a row on a table. Therefore there are $n!$ configurations. The *initial configuration* corresponds to the placement $1, 2, \ldots, n$.

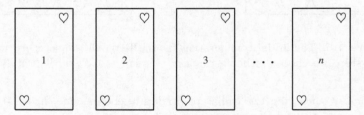

Figure 1.7. The initial configuration of the deck

Then at each step both the left and the right hand independently choose a random card. Note that the possibility is not excluded that the same card is chosen by both hands: such an event may occur with probability $1/n$.

If we have chosen two *distinct* cards we switch them, otherwise we leave the configuration unchanged.

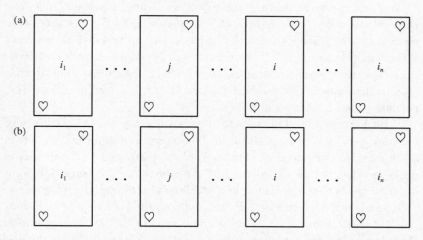

Figure 1.8. (a) A card configuration at time t. (b) The card configuration at time $t+1$ if the chosen cards are $i \neq j$

In other words, denoting by $T_n = \{\{i, j\} : i, j \in \{1, 2, \ldots, n\}, i \neq j\}$ the set of all unordered pairs of cards, with probability $1/n$ we leave the configuration unchanged, while, with probability $(n-1)/n$ we randomly pick one of the $n(n-1)/2$ elements in T_n. Altogether, at each step each element $t \in T_n$ can be chosen with probability $2/n^2$.

The following description is equivalent. Denote by S_n the *symmetric group* of degree n and by $T_n = \{g \in S_n : \exists i \neq j$ s.t. $g(k) = k, \forall k \neq i, j$ and $g(i) = j\}$ the set of all *transpositions*. Denote by $1_{S_n} \in S_n$ the identity element $(1_{S_n}(k) = k, \ \forall k \in \{1, 2, \ldots, n\})$. Each configuration, say $\mathbf{i} = (i_1, i_2, \ldots, i_n)$ corresponds to the element $g_\mathbf{i} \in S_n$ defined by $g_\mathbf{i}(j) = i_j$ for all $j = 1, 2, \ldots, n$. We start at the identity 1_{S_n} and, at each step, we choose either a random element $t \in T_n$ with probability $2/n^2$ or the identity $t = 1_{S_n}$ with probability $1/n$. This way, if at the kth step we are in the configuration $\mathbf{i}$ then, at the $(k+1)$st step we are in the configuration $t \cdot \mathbf{i}$ corresponding to the group element $t g_\mathbf{i}$.

In each of the above examples, we have described a diffusion model with a deterministic initial configuration. For the discrete circle the random walker is at position (vertex) 0; in the Ehrenfest model all balls are in urn 0 (Figure 1.3); in the Bernoulli-Laplace model the balls $1, 2, \ldots, n$ are in urn 0 and the balls $n+1, n+2, \ldots, 2n$ in urn 1 (Figure 1.5);

finally for the random transpositions the cards $1, 2, \ldots, n$ are placed in increasing order (Figure 1.7).

For each model we address the following natural question: *how long (how many steps) does it take to mix things up?* For instance, how many random transpositions (Example 1.2.4) are needed to generate a random permutation? It is indeed clear, even at an intuitive level, that repeated random transpositions do mix the cards, that is that after sufficiently many random transpositions, all configurations (i.e. permutations) become equiprobable.

In 1981 Diaconis and Shahshahani gave a striking answer to the above question: $\frac{1}{2}n \log n$ transposition are necessary and sufficient to generate a random permutation. Moreover they discovered that this model presents what is now called a *cutoff phenomenon*: the transition from *order* to *chaos* is concentrated in a small neighborhood of $\frac{1}{2}n \log n$.

Subsequently Diaconis and Shahshahani studied the models in Examples 1.2.1, 1.2.2 and 1.2.3. They showed that in the last two examples there is a cutoff phenomenon after $k = \frac{1}{4}n \log n$ steps, while for the discrete circle, the position of the random walker becomes random after $k = n^2$ steps, but no cutoff phenomenon occurs.

We shall first treat Examples 1.2.1 and 1.2.2 because they are significantly simpler (they only require elementary finite abelian harmonic analysis) and together they provide instances of absence and presence of the cutoff phenomenon. They will be discussed in the next chapter, while in the present one we continue by providing some probabilistic background.

Example 1.2.3 will be discussed in the second part of the book. It is a paradigmatic, but sufficiently simple, example of a problem on a homogeneous space on which there is a treatable set of spherical functions.

Example 1.2.4, which, historically, was the first to be analyzed, requires a deep knowledge of the representation theory of the symmetric group, a beautiful subject in mathematics on its own, which is a cornerstone both in pure and in applied mathematics (see, for instance, the book of Sternberg [212] for the applications to physics).

1.3 Markov chains

In this section we give the formal definition of a Markov chain in a simplified setting. We begin with some basic notions of probability theory.

A *probability measure* (or *distribution*) on a finite set X is a function $\nu : X \to [0, 1]$ such that $\sum_{x \in X} \nu(x) = 1$. It is called *strict* if $\nu(x) > 0$

for all $x \in X$. For a subset $A \subseteq X$ the quantity $\nu(A) = \sum_{x \in A} \nu(x)$ is the probability of A. Clearly $\nu(A \cup B) = \nu(A) + \nu(B) - \nu(A \cap B)$ for all $A, B \subseteq X$, $\nu(\emptyset) = 0$ and $\nu(X) = 1$.

The subsets of X are usually called the *events*. We also say that (X, ν) is a *finite measure space*.

Example 1.3.1 Let h, k be two positive integers and $X = \{r, b\}$ be a two-set. Define $\nu : X \to [0, 1]$ by setting $\nu(r) = \frac{h}{h+k}$ and $\nu(b) = \frac{k}{h+k}$. It is immediate to check that (X, ν) is a probability space. It can be used to modelize a single drawing from an urn containing h red balls and k blue balls: $\nu(r)$ (resp. $\nu(b)$) is the probability that the chosen ball is red (resp. blue).

Given two events $A, B \subseteq X$, with $\nu(A) > 0$, the *conditional probability* of B assuming A is

$$\nu(B|A) = \frac{\nu(A \cap B)}{\nu(A)}.$$

It expresses the probability of the event B given that the event A has occurred.

The following properties are obvious:

- $\nu(B|A) = 1$ if $A \subseteq B$;
- $\nu(\emptyset|A) = 0$;
- $\nu(B_1 \cup B_2|A) = \nu(B_1|A) + \nu(B_2|A) - \nu(B_1 \cap B_2|A)$;

for all $A, B, B_1, B_2 \subseteq X$ with $\nu(A) > 0$, and clearly $\nu(\cdot|A)$ is a probability measure on A.

Moreover, given elements $A_1, A_2, \ldots, A_n \subseteq X$ with $\nu(A_1 \cap A_2 \cap \cdots \cap A_{n-1}) > 0$, we have the so-called *Bayes sequential formula*

$$\nu(A_1 \cap A_2 \cap \cdots \cap A_n) = \nu(A_1)\nu(A_2|A_1)\nu(A_3|A_1 \cap A_2) \qquad (1.3)$$
$$\cdots \nu(A_n|A_1 \cap A_2 \cap \cdots \cap A_{n-1}).$$

Indeed:

$$\nu(A_1 \cap A_2 \cap \cdots \cap A_n) = \nu(A_n|A_1 \cap \cdots \cap A_{n-1})\nu(A_1 \cap \cdots \cap A_{n-1})$$
$$= \nu(A_n|A_1 \cap \cdots \cap A_{n-1})\nu(A_{n-1}|A_1 \cap \cdots \cap A_{n-2})$$
$$\cdot \nu(A_1 \cap \cdots \cap A_{n-2})$$
$$= \cdots =$$
$$= \nu(A_n|A_1 \cap \cdots \cap A_{n-1})\nu(A_{n-1}|A_1 \cap \cdots \cap A_{n-2})$$
$$\cdot \nu(A_{n-2}|A_1 \cap \cdots \cap A_{n-3}) \cdots \nu(A_2|A_1)\nu(A_1).$$

Two events $A, B \subseteq X$ are *independent* if

$$\nu(A \cap B) = \nu(A)\nu(B).$$

Note that, if $\nu(A), \nu(B) > 0$, then condition above is equivalent to

$$\nu(B|A) = \nu(B) \text{ and } \nu(A|B) = \nu(A).$$

Example 1.3.2 (Choice with replacement) Consider the four elements set $X = \{rr, rb, br, bb\}$ and h, k two positive integers. Define $\nu : X \to [0, 1]$ by setting

$$\nu(rr) = \frac{h^2}{(h+k)^2}, \quad \nu(rb) = \nu(br) = \frac{hk}{(h+k)^2} \text{ and } \nu(bb) = \frac{k^2}{(h+k)^2}.$$

It is immediate to check that (X, ν) is a probability space. Set $A = \{rb, rr\}$ and $B = \{br, rr\}$. Then

$$\nu(A \cap B) = \nu(A)\nu(B) \equiv \frac{h^2}{(h+k)^2},$$

in other words the events A and B are independent.

This probability space described above serves as a model for the following process. Consider an urn with h red balls and k blue balls. A choice *with replacement* consists in picking up two balls from the urn subject to the restriction that the second ball is chosen only after the first ball is put back into the urn. Then the set of all four possible outcomes of the extraction can be identified with the set X: for instance we associate with the event that the first ball chosen is red and the second one is blue the element $rb \in X$. This way the sets A and B correspond to the result of the *first* chosen ball and the *second* chosen ball to be red, respectively. Moreover, taking into account Example 1.3.1, we may observe that $\nu(rb)$ is given by the product of the probability to choose a red ball with the probability to choose a blue ball.

Example 1.3.3 (Choice without replacement) In the preceding model, suppose the second ball is chosen *without replacement* of the first ball.

As before $X = \{rr, rb, br, bb\}$ denotes the set of all four possible outcomes of this extraction and ν, the associated probability on X, is now defined by

$$\nu(rr) = \frac{h(h-1)}{(h+k)(h+k-1)}, \quad \nu(rb) = \nu(br) = \frac{hk}{(h+k)(h+k-1)}$$

and

$$\nu(bb) = \frac{k(k-1)}{(h+k)(h+k-1)}.$$

With A and B as in the previous example, we have

$$\nu(B|A) = \frac{h-1}{h+k-1} \neq \frac{h}{h+k} = \nu(B)$$

and A and B are no longer independent.

Let X and Y be two finite sets and μ a probability measure on Y. A function $\xi : Y \to X$ is called a *random variable*. For $x \in X$, we set $\mu\{\xi = x\} = \mu(\{y \in Y : \xi(y) = x\})$.

Given random variables $\xi_1, \xi_2, \ldots, \xi_n : Y \to X$, we say that they are (*collectively*) *independent* if

$$\mu\{\xi_1 = x_1, \xi_2 = x_2, \cdots, \xi_n = x_n\} = \mu\{\xi_1 = x_1\}\mu\{\xi_2 = x_2\}$$
$$\cdots \mu\{\xi_n = x_n\}$$

where, for $x_1, x_2, \ldots, x_n \in X$, $\mu\{\xi_1 = x_1, \xi_2 = x_2, \cdots, \xi_n = x_n\} = \mu(\{y \in Y : \xi_i(y) = x_i, i = 1, 2, \ldots, n\})$.

Example 1.3.4 (The Bernoulli scheme) Let ν be a probability measure on a finite set X. Denote by $X^{(n)} = X \times X \times \cdots \times X$ (n times) the nth cartesian product of X and by $\mu = \nu^{(n)}$ the corresponding nth fold product measure defined by

$$\nu^{(n)}(x_1, x_2, \ldots, x_n) = \nu(x_1)\nu(x_2) \cdots \nu(x_n).$$

If $\xi_j : X^{(n)} \to X$ denotes the jth projection, namely $\xi_j(x) = x_j$ for all $x = (x_1, x_2, \ldots, x_n)$, then the random variables $\xi_1, \xi_2, \ldots, \xi_n$ are independent.

Indeed

$$\nu^{(n)}\{\xi_1 = \overline{x_1}, \xi_2 = \overline{x_2}, \cdots, \xi_n = \overline{x_n}\} = \nu^{(n)}(\overline{x_1}, \overline{x_2}, \cdots, \overline{x_n})$$
$$= \nu(\overline{x_1})\nu(\overline{x_2}) \cdots \nu(\overline{x_n})$$

and

$$\nu^{(n)}\{\xi_j = \overline{x_j}\} = \sum_{\substack{x_1, x_2, \ldots, x_{j-1}, \\ x_{j+1}, \ldots, x_n \in X}} \nu(x_1)\nu(x_2) \cdots \nu(x_{j-1})\nu(\overline{x_j})\nu(x_{j+1}) \cdots \nu(x_n)$$
$$= \nu(\overline{x_j})$$

so that

$$\nu^{(n)}(\xi_1 = \overline{x_1})\nu^{(n)}(\xi_2 = \overline{x_2})\cdots\nu^{(n)}(\xi_n = \overline{x_n}) = \nu(\overline{x_1})\nu(\overline{x_2})\cdots\nu(\overline{x_n}).$$

The Bernoulli scheme is the model of an experiment in which we choose n times an element of X according to the same probability distribution ν on X; see also Example 1.3.2. Usually, in the Bernoulli scheme, one considers infinite sequences $\xi_1, \xi_2, \ldots, \xi_n, \ldots$ of random variables; to treat this infinite setting one needs the full strength of measure theory ($X^{(n)}$ is then replaced by $X^{(\infty)}$ the set of all infinite sequences in X), [27].

A *stochastic matrix* $P = (p(x,y))_{x,y \in X}$ on X is a real valued matrix indexed by X such that

- $p(x,y) \geq 0$ for all $x, y \in X$
- $\sum_{y \in X} p(x,y) = 1$ for all $x \in X$.

If $\xi_1, \xi_2, \ldots, \xi_n : Y \to X$ are random variables and $\mu\{\xi_1 = x_1, \xi_2 = x_2, \ldots, \xi_{n-1} = x_{n-1}\} > 0$, we set

$$\mu(\xi_n = x_n | \xi_1 = x_1, \xi_2 = x_2, \ldots, \xi_{n-1} = x_{n-1})$$
$$= \frac{\mu\{\xi_1 = x_1, \xi_2 = x_2, \ldots, \xi_n = x_n\}}{\mu\{\xi_1 = x_1, \xi_2 = x_2, \ldots, \xi_{n-1} = x_{n-1}\}}.$$

Definition 1.3.5 (Markov chain) Let X be a finite set, ν a probability distribution on X and P a stochastic matrix on X. A (homogeneous) *Markov chain* with *state space* X, *initial distribution* ν and *transition matrix* P is a finite sequence $\xi_0, \xi_1, \ldots, \xi_n : Y \to X$ of random variables, where (Y, μ) is a finite measure space, satisfying the following conditions:

$$\mu\{\xi_0 = x\} = \nu(x)$$
$$\mu(\xi_{k+1} = x_{k+1} | \xi_0 = x_0, \xi_1 = x_1, \ldots, \xi_k = x_k) = p(x_k, x_{k+1})$$

for all $k = 0, 1, \ldots, n-1$ and $x, x_1, x_2, \ldots, x_n \in X$ such that $\mu\{\xi_0 = x_0, \xi_1 = x_1, \ldots, \xi_k = x_k\} > 0$.

When the initial distribution ν equals a Dirac measure δ_x one says that the Markov chain *starts (deterministically)* at x.

Theorem 1.3.6 *A sequence* $\xi_0, \xi_1, \ldots, \xi_n : Y \to X$ *is a Markov chain with initial probability distribution* ν *and transition matrix* P *if and only if for all* $x_0, x_1, \ldots, x_n \in X$ *one has*

$$\mu\{\xi_0 = x_0, \xi_1 = x_1, \ldots, \xi_n = x_n\} = \nu(x_0)p(x_0, x_1)p(x_1, x_2)$$
$$\cdots p(x_{n-1}, x_n). \qquad (1.4)$$

In particular,

$$\mu\{\xi_k = x\} = \sum_{x_0, x_1, \ldots, x_{k-1} \in X} \nu(x_0)p(x_0, x_1)p(x_1, x_2) \cdots p(x_{k-1}, x),$$

that is $\mu\{\xi_k = x\} = \nu^{(k)}(x)$, *where* $\left(\nu^{(k)}\right)^T = \nu^T P^k$ *(as in (1.2)) for* $k = 0, 1, \ldots, n$ *and*

$$\mu(\xi_{k+1} = x_{k+1}|\xi_0 = x_0, \xi_1 = x_1, \ldots, \xi_k = x_k) = \mu(\xi_{k+1} = x_{k+1}|\xi_k = x_k).$$

The last condition expresses the fact that if the present ξ_k *is known, then the* future ξ_{k+1} *does not depend on the past* $\xi_0, \xi_1, \ldots, \xi_{k-1}$.

Proof Suppose that $\xi_0, \xi_1, \ldots, \xi_n : Y \to X$ is a Markov chain with initial probability distribution ν and transition matrix P. Then by the Bayes sequential formula (1.3) we have

$$\mu\{\xi_0 = x_0, \xi_1 = x_1, \ldots, \xi_n = x_n\}$$
$$= \mu\{\xi_0 = x_0\}\mu(\xi_1 = x_1|\xi_0 = x_0)$$
$$\cdot \mu(\xi_2 = x_2|\xi_0 = x_0, \xi_1 = x_1)$$
$$\cdots \mu(\xi_n = x_n|\xi_0 = x_0, \xi_1 = x_1, \ldots, \xi_{n-1} = x_{n-1})$$
$$= \nu(x_0)p(x_0, x_1)p(x_1, x_2) \cdots p(x_{n-1}, x_n).$$

Conversely, if (1.4) holds, then summing over $x_n \in X$, as P is stochastic, we get

$$\mu\{\xi_0 = x_0, \xi_1 = x_1, \ldots, \xi_{n-1} = x_{n-1}\} = \nu(x_0)p(x_0, x_1)p(x_1, x_2)$$
$$\cdots p(x_{n-2}, x_{n-1}).$$

Continuing this way, one gets

$$\mu\{\xi_0 = x_0, \xi_1 = x_1, \ldots, \xi_k = x_k\} = \nu(x_0)p(x_0, x_1)p(x_1, x_2) \cdots p(x_{k-1}, x_k)$$

for all $k = 0, 1, \ldots, n$.

Therefore $\mu\{\xi_0 = x_0\} = \nu(x_0)$ and, for $k = 0, 1, \ldots, n-1$,

$$\mu(\xi_{k+1} = x_{k+1}|\xi_0 = x_0, \xi_1 = x_1, \ldots, \xi_k = x_k)$$
$$= \frac{\mu\{\xi_0 = x_0, \xi_1 = x_1, \ldots, \xi_{k+1} = x_{k+1}\}}{\mu\{\xi_0 = x_0, \xi_1 = x_1, \ldots, \xi_k = x_k\}}$$
$$= p(x_k, x_{k+1})$$

and the theorem is proved. $\qquad \square$

Remark 1.3.7 The term homogeneous in Definition 1.3.5 refers to the fact that in

$$\mu(\xi_{k+1} = x_{k+1} | \xi_k = x_k) = p(x_k, x_{k+1})$$

the matrix P is the same for all k's. In the nonhomogeneous case, the matrix P varies: for each k there is a transition matrix P_k. In the following we shall only consider homogeneous Markov chains.

Example 1.3.8 This is the basic measure theoretic construction in the theory of Markov chains (we describe it only in our simplified setting).

Let (X, ν) be a finite measure space and P be a stochastic matrix indexed by X.

Take $Y = X^{(n+1)} = \{(x_0, x_1, \ldots, x_n) : x_j \in X\}$ and set

$$\mu(x_0, x_1, \ldots, x_n) = \nu(x_0) p(x_0, x_1) p(x_1, x_2) \cdots p(x_{n-1}, x_n).$$

It is easy to show that μ is a probability measure on Y. Denoting, as in Example 1.3.4, by $\xi_j : Y \to X$ the projection relative to the jth component, we have

$$
\begin{aligned}
\mu\{\xi_0 = \overline{x}\} &= \mu\{(x_0, x_1, \ldots, x_n) : x_0 = \overline{x}\} \\
&= \sum_{x_1, x_2, \ldots, x_n \in X} \nu(\overline{x}) p(\overline{x}, x_1) p(x_1, x_2) \cdots p(x_{n-1}, x_n) \\
&= \nu(\overline{x})
\end{aligned}
$$

and

$$
\begin{aligned}
\mu(\xi_{k+1} = \overline{x} | \xi_k = \overline{x}) &= \frac{\mu\{(x_0, x_1, \ldots, x_n) : x_k = \overline{x}, x_{k+1} = \overline{x}\}}{\mu\{(x_0, x_1, \ldots, x_n) : x_k = \overline{x}\}} \\
&= \frac{\sum_{\substack{x_0, x_1, \ldots, x_{k-1} \\ x_{k+2}, \ldots, x_n \in X}} [\nu(x_0) p(x_0, x_1) \cdots p(x_{k-1}, \overline{x}) \cdot p(\overline{x}, \overline{x'}) \cdot}{\sum_{\substack{x_0, x_1, \ldots, x_{k-1} \\ x_{k+1}, \ldots, x_n \in X}} [\nu(x_0) p(x_0, x_1) \cdots p(x_{k-1}, \overline{x}) \cdot p(\overline{x}, x_{k+1}) \cdot} \\
&\qquad \frac{\cdot \, p(\overline{x'}, x_{k+2}) \cdots p(x_{n-1}, x_n)]}{p(x_{k+1}, x_{k+2}) \cdots p(x_{n-1}, x_n)]} \\
&= \frac{\left(\sum_{x_0, x_1, \ldots, x_{k-1} \in X} \nu(x_0) p(x_0, x_1) \cdots p(x_{k-1}, \overline{x})\right) p(\overline{x}, \overline{x'})}{\sum_{x_0, x_1, \ldots, x_{k-1} \in X} \nu(x_0) p(x_0, x_1) \cdots p(x_{k-1}, \overline{x})} \\
&= p(\overline{x}, \overline{x'}).
\end{aligned}
$$

Usually, as we alluded to above, in the definition of a Markov chain and in the construction in Example 1.3.8, one considers infinite sequences

$\xi_0, \xi_1, \ldots, \xi_n, \ldots$ and $(x_0, x_1, \ldots, x_n, \ldots)$; see, for a detailed account on measure theory, Shiryaev's book [199] and, for more on Markov chains, the books by Billingsley [27] and Stroock [215]. For the point of view of ergodic theory and dynamical systems we refer to Petersen's monograph [177]

In this book we are mainly interested in the asymptotic behavior of the distributions $\nu^{(k)}$'s as $k \to \infty$. The simplified approach will largely suffice in this setting.

A more intuitive description of a Markov chain is the following. Time is discrete, namely $t = 0, 1, 2, \ldots$. At time $t = 0$ we choose a point in X according to the initial distribution ν (i.e. $x \in X$ is chosen with probability $v(x)$). Then, if at time t we are at point $x \in X$, at time $t+1$ we move to $y \in X$ with probability $p(x, y)$.

This way, $\nu(x_0)p(x_0, x_1)p(x_1, x_2) \cdots p(x_{n-1}, x_n)$ represents the probability of the path $(x_0, x_1, \ldots, x_n) \in X^{(n)}$ among all paths of the same length n.

Setting $\nu^{(0)} = \nu$ and, for all $k = 1, 2, \ldots$,

$$
\begin{aligned}
\nu^{(k)}(x) &= \sum_{x_0, x_1, \ldots, x_{k-1} \in X} \nu(x_0)p(x_0, x_1)p(x_1, x_2) \cdots p(x_{k-1}, x) \\
&= \sum_{x_0 \in X} \nu(x_0)p^{(k)}(x_0, x),
\end{aligned}
$$

then $\mu\{\xi_k = x\} = \nu^{(k)}(x)$, that is $\nu^{(k)}$ is the distribution probability after k steps (at time $t = k$): $\nu^{(k)}$ represents the probability of being in state x after k steps (at time $t = k$).

Remark 1.3.9 In the following we will identify a Markov chain with an initial probability distribution ν and a transition matrix P on a finite space X. We also say that $\{\nu^{(k)}\}_{k \in \mathbb{N}}$ is the *sequence* of probability distributions of the Markov chain.

For the four basic examples in Section 1.2 we have:

1. $X = C_n$, $\nu = \delta_0$, $p(j, j+1) = \frac{1}{2} = p(j+1, j)$ and $p(i, j) = 0$, otherwise, for $i, j = 0, 1, \ldots, n-1$.
2. $X = Q_n$, $\nu = \delta_{A_0}$, where $A_0 = \{1, 2, \ldots, n\}$, $p(A, B) = \frac{1}{n}$ if either $A \subset B$ and $|B| = |A| + 1$, or $B \subset A$ and $|A| = |B| + 1$, and $p(A, B) = 0$ otherwise.
3. $X = \Omega_n$, $\nu = \delta_{A_0}$, where $A_0 = \{1, 2, \ldots, n\}$, $p(A, B) = \frac{1}{n^2}$ if $|A \cap B| = n - 1$ and $p(A, B) = 0$ otherwise.

4. $X = S_n$, $\nu = \delta_1$, where 1 denotes the unit element of S_n,

$$p(g,h) = \begin{cases} \frac{2}{n^2} & \text{if } hg^{-1} \in T_n \\ \frac{1}{n} & \text{if } g = h \\ 0 & \text{otherwise,} \end{cases}$$

where T_n denotes the set of all transpositions in S_n.

1.4 Convergence to equilibrium

Let $P = (p(x,y))_{x,y \in X}$ be a stochastic matrix. A *stationary distribution* for P is a probability measure π on X such that

$$\pi^T = \pi^T P$$

that is $\pi(y) = \sum_{x \in X} \pi(y) p(x,y)$ for all $y \in X$.

We say that P is *ergodic* if there exists $n_0 \in \mathbb{N}$ such that

$$p^{(n_0)}(x,y) > 0 \qquad \forall x, y \in X. \tag{1.5}$$

The following is known as "Markov ergodic theorem" or "convergence to equilibrium theorem" [99, 199]. It shows that under the hypothesis of ergodicity, the chain has a stationary distribution which is also the limit of the transition probabilities, that is, it shows that "things really mix up".

Theorem 1.4.1 *A stochastic matrix P is ergodic if and only if there exists a strict probability distribution π on X such that*

$$\lim_{n \to \infty} p^{(n)}(x,y) = \pi(y) \qquad \forall x, y \in X.$$

In other words, the above limits exist, they are independent of x and form a strict probability distribution. Moreover, if P is ergodic, then such a π is the unique stationary distribution for P.

Proof Suppose that P is ergodic and set $\varepsilon = \min_{x,y \in X} p^{(n_0)}(x,y)$, so that, in virtue of (1.5), one has $0 < \varepsilon < 1$. Set

$$M^{(n)}(y) = \max_{x \in X} p^{(n)}(x,y), \qquad m^{(n)}(y) = \min_{x \in X} p^{(n)}(x,y).$$

It is clear that $M^{(n)}(y) \geq m^{(n)}(y)$. As $p^{(n+1)}(x,y) = \sum_{z \in X} p(x,z) \cdot p^{(n)}(z,y)$, one has that

$$m^{(n)}(y) = \sum_{z \in X} p(x,z)m^{(n)}(y) \leq p^{(n+1)}(x,y) \leq \sum_{z \in X} p(x,z)M^{(n)}(y)$$
$$= M^{(n)}(y)$$

so that

$$M^{(n+1)}(y) \leq M^{(n)}(y) \quad \text{and} \quad m^{(n+1)}(y) \geq m^{(n)}(y).$$

This shows that $\{M^{(n)}(y)\}_{n \in \mathbb{N}}$ and $\{m^{(n)}(y)\}_{n \in \mathbb{N}}$ are, for all $y \in X$, bounded monotone sequences and therefore they converge. Call $\pi_1(y)$ and $\pi_2(y)$ their limits, respectively. For any $r \geq 0$ we have

$$p^{(n_0+r)}(x,y) = \sum_{z \in X} p^{(n_0)}(x,z)p^{(r)}(z,y)$$
$$= \sum_{z \in X} [p^{(n_0)}(x,z) - \varepsilon p^{(r)}(y,z)]p^{(r)}(z,y)$$
$$\qquad + \varepsilon \sum_{z \in X} p^{(r)}(y,z)p^{(r)}(z,y)$$
$$= \sum_{z \in X} [p^{(n_0)}(x,z) - \varepsilon p^{(r)}(y,z)]p^{(r)}(z,y) + \varepsilon p^{(2r)}(y,y)$$
$$\geq_{(*)} m^{(r)}(y) \sum_{z \in X} [p^{(n_0)}(x,z) - \varepsilon p^{(r)}(y,z)] + \varepsilon p^{(2r)}(y,y)$$
$$= (1-\varepsilon)m^{(r)}(y) + \varepsilon p^{(2r)}(y,y),$$

where $\geq_{(*)}$ follows from

$$p^{(n_0)}(x,z) - \varepsilon p^{(r)}(y,z) \geq p^{(n_0)}(x,z)[1 - p^{(r)}(y,z)] \geq 0.$$

It follows that

$$m^{(n_0+r)}(y) \geq (1-\varepsilon)m^{(r)}(y) + \varepsilon p^{(2r)}(y,y). \qquad (1.6)$$

Analogously, one can prove that

$$M^{(n_0+r)}(y) \leq M^{(r)}(y)(1-\varepsilon) + \varepsilon p^{(2r)}(y,y). \qquad (1.7)$$

From (1.6) and (1.7) we get

$$M^{(n_0+r)}(y) - m^{(n_0+r)}(y) \leq (1-\varepsilon)[M^{(r)}(y) - m^{(r)}(y)]$$

and iterating

$$M^{(kn_0+r)}(y) - m^{(kn_0+r)}(y) \leq (1-\varepsilon)^k[M^{(r)}(y) - m^{(r)}(y)] \to 0 \qquad (1.8)$$

as k goes to infinity. Therefore we have

$$\pi_1(y) = \lim_{n\to\infty} M^{(n)}(y) = \lim_{n\to\infty} m^{(n)}(y) = \pi_2(y);$$

call $\pi(y)$ this common value.

Then from (1.8) we have, with $n = kn_0 + r$ and $0 \leq r < n_0$

$$|p^{(n)}(x,y) - \pi(y)| \leq M^{(n)}(y) - m^{(n)}(y) \leq (1 - \varepsilon)^{[n/n_0]-1} \qquad (1.9)$$

and therefore $\lim_{n\to\infty} p^{(n)}(x,y) = \pi(y)$ for all $y \in X$.

Also, by taking the limit for $n \to \infty$ in

$$p^{(n+1)}(x,y) = \sum_{z\in X} p^{(n)}(x,z)p(z,y)$$

we get

$$\pi(y) = \sum_{z\in X} \pi(z)p(z,y)$$

so that π is stationary for P.

Moreover, as for all n we have $\sum_{y\in X} p^{(n)}(x,y) = 1$, we analogously deduce that $\sum_{y\in X} \pi(y) = 1$; finally as $m^{(n)}(y) \geq m^{(n_0)}(y) \geq \varepsilon > 0$ for $n \geq n_0$, we also have that π is strict.

Conversely, suppose that $\lim_{n\to\infty} p^{(n)}(x,y) = \pi(y)$ for all x and $y \in X$, with π a strict probability distribution. Then, for every x,y there exists $n_0(x,y)$ such that $p^{(n)}(x,y) > 0$ for all $n \geq n_0(x,y)$. Taking $n_0 = \max_{x,y\in X} n_0(x,y)$ we get that $p^{(n)}(x,y) > 0$ for all $n \geq n_0$ and therefore P is ergodic.

To prove uniqueness, suppose that π' is another stationary distribution for an ergodic P. Then

$$\pi'(y) = \sum_{x\in X} \pi'(x)p^{(n)}(x,y) \to \sum_{x\in X} \pi'(x)\pi(y) = \pi(y)$$

as $n \to \infty$, showing that indeed $\pi' = \pi$. □

The significance of this theorem is clear: after sufficiently many steps, the probability of being in y almost equals $\pi(y)$ and it is independent of the initial distribution.

From the proof of the previous theorem we can also extract the following corollary.

Corollary 1.4.2 *Let P be an ergodic stochastic matrix and π its stationary distribution. Then there exist $A > 0$, $0 < c < 1$ and $n_1 \in \mathbb{N}$ such that, for all $n \geq n_1$,*

$$\max_{x,y\in X} |p^{(n)}(x,y) - \pi(y)| \leq A(1 - c)^n.$$

Proof This is an immediate consequence of (1.9). Indeed, if $n = n_0 k + r$ with $0 \leq r < n_0$, then $\frac{n}{n_0} = k + \frac{r}{n_0}$ and

$$(1 - \varepsilon)^{[n/n_0] - 1} = (1 - \varepsilon)^{k-1} = (1 - \varepsilon)^{\frac{n}{n_0} - \frac{r}{n_0} - 1}$$
$$= (1 - \varepsilon)^{-\frac{r}{n_0} - 1}[(1 - \varepsilon)^{\frac{1}{n_0}}]^n$$

so that we can take $A = (1 - \varepsilon)^{-2}$, $1 - c = (1 - \varepsilon)^{\frac{1}{n_0}}$ and $n_1 = n_0$. $\square$

Exercise 1.4.3 Let p be the transition probability kernel for the simple random walk on the discrete circle C_{2m+1}. Using the estimate in Corollary 1.4.2 show that

$$\max_{y \in X} |p^{(k)}(x, y) - \pi(y)| \leq 2 e^{-\frac{k}{m 2^{2m+1}}}$$

for $k \geq m$ where $\pi(x) = \frac{1}{2m+1}$.

We do not go further into the general theory of finite Markov chains. In the following sections we concentrate on a special class of Markov chains that have a good spectral theory. For nice expositions of the general theory we refer to the monographs by Behrends [18], Bremaud [32], Durrett [81], Häggström [116], Kemeny and Snell [142], Seneta [195], Shiryaev [199] and Stroock [215].

1.5 Reversible Markov chains

Definition 1.5.1 Given a stochastic matrix P, we say that P is *reversible* if there exists a strict probability measure π on X such that

$$\pi(x) p(x, y) = \pi(y) p(y, x) \qquad (1.10)$$

for all $x, y \in X$. We also say that P and π are in *detailed balance*.

The following characterization of reversibility was kindly indicated to us by Pierre de la Harpe (it corresponds to Exercise 5.8.1 in [20]).

Proposition 1.5.2 *A stochastic matrix $P = (p(x, y))_{x, y \in X}$ is reversible if and only if the following two conditions hold:*

(i) *if $p(x, y) \neq 0$ then $p(y, x) \neq 0$, for all $x, y \in X$;*
(ii) *for any $n \in \mathbb{N}$ and for any choice of $x_0, x_1, \ldots, x_n \in X$ one has*

$$p(x_0, x_1) \cdots p(x_{n-1}, x_n) p(x_n, x_0) = p(x_0, x_n) p(x_n, x_{n-1})$$
$$\cdots p(x_1, x_0). \qquad (1.11)$$

Proof We first observe that the condition (1.10) defining reversibility is equivalent to

$$p(y, x) = \frac{\pi(x)}{\pi(y)} p(x, y) \tag{1.12}$$

for all $x, y \in X$, which, in case $p(x, y) \neq 0$ becomes

$$\frac{p(y, x)}{p(x, y)} = \frac{\pi(x)}{\pi(y)}. \tag{1.13}$$

Suppose that P is reversible and that π is a strict probability on X in detailed balance with P. Then (i) follows immediately from (1.12).

To prove (ii), by (i) we can suppose that in (1.11) one has $p(x_i, x_{i\pm1}) \neq 0$ for all $i = 0, 1, \ldots, n \mod n + 1$. Then (1.11) is equivalent to

$$\frac{p(x_0, x_1)}{p(x_1, x_0)} \frac{p(x_1, x_2)}{p(x_2, x_1)} \cdots \frac{p(x_{n-1}, x_n)}{p(x_n, x_{n-1})} \frac{p(x_n, x_0)}{p(x_0, x_n)} = 1.$$

which, by virtue of (1.13), becomes

$$\frac{\pi(x_1)}{\pi(x_0)} \frac{\pi(x_2)}{\pi(x_1)} \cdots \frac{\pi(x_n)}{\pi(x_{n-1})} \frac{\pi(x_0)}{\pi(x_n)} = 1 \tag{1.14}$$

which is trivially satisfied.

Conversely, suppose that P satisfies conditions (i) and (ii). Consider the relation $\sim$ in X defined by setting $x \sim y$ if either $x = y$ or there exist $x_0, x_1, \ldots, x_n \in X$ such that $x_0 = x$, $x_n = y$ and $p(x_i, x_{i+1}) \neq 0$ for $i = 0, 1, \ldots, n - 1$. Note that by (i) this is an equivalence relation. Let then $X = X^1 \coprod X^2 \coprod \cdots \coprod X^N$ denote the corresponding partition of X induced by $\sim$.

For $i \in \{1, 2, \ldots, N\}$ fix $x_0^i \in X^i$ and set $X_0^i = \{x_0^i\}$,

$$X_1^i = \{x_1 \in X^i \setminus X_0^i : p(x_0^i, x_1) \neq 0\}$$

and, more generally,

$$X_{n+1}^i = \{x_{n+1} \in X^i \setminus (X_0^i \coprod X_1^i \coprod \cdots \coprod X_n^i) :$$
$$\exists x_n \in X_n^i \text{ s.t. } p(x_{n+1}, x_n) \neq 0\}.$$

Then $X^i = X_0^i \coprod X_1^i \coprod \cdots \coprod X_{N_i}^i$ for some $N_i \in \mathbb{N}$. Set $\overline{\pi}(x_0^i) = 1$,

$$\overline{\pi}(x_1) = \frac{p(x_0^i, x_1)}{p(x_1, x_0^i)}$$

for all $x_1 \in X_1^i$ and, more generally, for all $x_n \in X_n^i$, $n \leq N_i$,

$$\overline{\pi}(x_n) = \frac{p(x_0^i, x_1)}{p(x_1, x_0^i)} \frac{p(x_1, x_2)}{p(x_2, x_1)} \cdots \frac{p(x_{n-2}, x_{n-1})}{p(x_{n-1}, x_{n-2})} \frac{p(x_{n-1}, x_n)}{p(x_n, x_{n-1})} \tag{1.15}$$

where $x_j \in X_j^i$ and $x_j \sim x_{j+1}$ for all $j = 1, 2, \ldots, n-1$. Note that by virtue of (1.14) the definition of (1.15) is well posed (it does not depend on the particular choice of the chain $x_0^i \sim x_1 \sim x_2 \sim \cdots \sim x_{n-1} \sim x_n$).

Set $c = \sum_{x \in X} \overline{\pi}(x)$ and define $\pi = \frac{1}{c}\overline{\pi}$. By construction one has $\pi(x) > 0$ for all $x \in X$ and $\sum_{x \in X} \pi(x) = 1$. In other words, π is a strict probability on X. We are only left to show that P and π are in detailed balance.

Let $x, y \in X$. If $p(x, y) = 0$ then, recalling (i), (1.10) is trivially satisfied. Otherwise, there exist $0 \leq i \leq N$ and $0 \leq n \leq N_i - 1$ such that $x, y \in X^i$ and, up to exchanging x with y, we may suppose that $x \in X_n^i, y \in X_{n+1}^i$. But then, if $x_n = x$, $x_{n+1} = y$, $x_j \in X_j^i$ and $x_j \sim x_{j+1}$ for all $j = 1, 2, \ldots, n-1$,

$$
\begin{aligned}
\pi(x)p(x, y) &= \frac{1}{c} \frac{p(x_0^i, x_1)}{p(x_1, x_0^i)} \frac{p(x_1, x_2)}{p(x_2, x_1)} \cdots \frac{p(x_{n-1}, x_n)}{p(x_n, x_{n-1})} p(x_n, x_{n+1}) \\
&= \frac{1}{c} \frac{p(x_0^i, x_1)}{p(x_1, x_0^i)} \frac{p(x_1, x_2)}{p(x_2, x_1)} \\
&\quad \cdots \frac{p(x_{n-1}, x_n)}{p(x_n, x_{n-1})} \frac{p(x_n, x_{n+1})}{p(x_{n+1}, x_n)} p(x_{n+1}, x_n) \\
&= \pi(y)p(y, x)
\end{aligned}
$$

and this ends the proof. $\qquad\square$

Set $\langle f_1, f_2 \rangle_\pi = \sum_{x \in X} f_1(x)\overline{f_2(x)}\pi(x)$ and define a linear operator $P : L(X) \to L(X)$ by setting $(Pf)(x) = \sum_{y \in X} p(x, y)f(y)$ for all $f_1, f_2, f \in L(X)$.

Proposition 1.5.3 *P and π are in detailed balance if and only if P is self-adjoint with respect to the scalar product $\langle \cdot, \cdot \rangle_\pi$.*

Proof Let $f_1, f_2 \in L(X)$. Then, supposing that P and π are in detailed balance, we have

$$
\begin{aligned}
\langle Pf_1, f_2 \rangle_\pi &= \sum_{x \in X} \left[\sum_{y \in X} p(x, y)f_1(y) \right] \overline{f_2(x)}\pi(x) \\
&= \sum_{x \in X} \sum_{y \in X} \pi(x)p(x, y)f_1(y)\overline{f_2(x)} \\
&= \sum_{x \in X} \sum_{y \in X} \pi(y)p(y, x)f_1(y)\overline{f_2(x)} \\
&= \langle f_1, Pf_2 \rangle_\pi.
\end{aligned}
$$

Conversely, supposing that P is selfadjoint, one easily checks that, for all $x, y \in X$,

$$\pi(x)p(x,y) = \langle P\delta_y, \delta_x \rangle_\pi = \langle \delta_y, P\delta_x \rangle_\pi = \pi(y)p(y,x).$$

$\square$

Remark 1.5.4 Suppose now that P and π are in detailed balance, so that, by the above proposition, P is selfadjoint. From classical linear algebra, we deduce that the real matrix P can be diagonalized over the *reals*, that is there exist *real* valued eigenvalues $\{\lambda_x : x \in X\}$ and a *real* matrix $U = (u(x,y))_{x,y \in X}$ such that

$$\begin{cases} \sum_{y \in X} p(x,y)u(y,z) = u(x,z)\lambda_z \\ \sum_{x \in X} u(x,y)u(x,z)\pi(x) = \delta_y(z). \end{cases} \tag{1.16}$$

Note that we use the set X to parameterize the eigenvalues only to simplify the notation.

The first equation in (1.16) tells us that each column of U (that is, each function $y \mapsto u(y,z)$) is an eigenvector of P; the second one tells us that these columns are orthogonal with respect to $\langle \cdot, \cdot \rangle_\pi$.

Setting $\Delta = (\lambda_x \delta_x(y))_{x,y \in X}$ (the diagonal matrix of the eigenvalues) and $D = (\pi(x)\delta_x(y))_{x,y \in X}$ (the diagonal matrix of coefficients of π) and denoting by $I = (\delta_x(y))_{x,y \in X}$ the identity matrix, (1.16) may be rewritten in the form

$$\begin{cases} PU = U\Delta \\ U^T DU = I. \end{cases} \tag{1.17}$$

The following proposition explains how one obtains the powers of the transition matrix P using its spectral analysis in (1.16) and (1.17).

Proposition 1.5.5 $p^{(n)}(x,y) = \pi(y) \sum_{z \in X} u(x,z)\lambda_z^n u(y,z).$

Proof From the orthogonality relations (namely the second equation in (1.17)) we can deduce the following *dual orthogonality relations*: $U^T D$ is the inverse of U and therefore $UU^T D = I$, that is,

$$\sum_{y \in X} u(x,y)u(z,y) = \frac{1}{\pi(z)}\delta_z(x). \tag{1.18}$$

Then from $PU = U\Delta$ we get $P = U\Delta U^T D$, that is,

$$p(x,y) = \pi(y) \sum_{z \in X} u(x,z)\lambda_z u(y,z). \tag{1.19}$$

Iterating this argument we have

$$P^n = U \Delta^n U^T D, \tag{1.20}$$

which coincides with the formula in the statement. □

Example 1.5.6 (Two state Markov chains) Suppose that $X = \{0, 1\}$. Then a stochastic matrix on X is necessarily of the form

$$P = \begin{pmatrix} p(0,0) & p(0,1) \\ p(1,0) & p(1,1) \end{pmatrix} = \begin{pmatrix} 1-p & p \\ q & 1-q \end{pmatrix}$$

with $0 \le p, q \le 1$. This transition matrix is reversible if and only if it is trivial or $0 < p, q$: indeed in the second case it is easy to check that P is in detailed balance with the distribution π given by $\pi(0) = \frac{q}{p+q}$ and $\pi(1) = \frac{p}{p+q}$.

It is easy to check that P is ergodic if and only if $0 < p, q$ and $\min\{p, q\} < 1$ and that this is in turn equivalent to the conditions $0 < p, q$ and $|1 - p - q| < 1$. Setting

$$U = \begin{pmatrix} 1 & \sqrt{\frac{p}{q}} \\ 1 & -\sqrt{\frac{q}{p}} \end{pmatrix}$$

then U is the matrix whose columns are the eigenvectors of P and are orthonormal with respect to the scalar product $\langle \cdot, \cdot \rangle_\pi$ (see (1.16)) which in the present case is given by

$$\langle (f(0), f(1)), (g(0), g(1)) \rangle_\pi = \frac{f(0)g(0)q + f(1)g(1)p}{p+q}.$$

The corresponding eigenvalues are $\lambda_0 = 1$ and $\lambda_1 = 1 - p - q$. Then, with $\Delta = \begin{pmatrix} 1 & 0 \\ 0 & 1-p-q \end{pmatrix}$ and $D = \begin{pmatrix} \frac{q}{p+q} & 0 \\ 0 & \frac{p}{p+q} \end{pmatrix}$, the formula $P^k = U \Delta^k U^T D$ (cf. (1.20)), becomes

$$P^k = \begin{pmatrix} \frac{q}{p+q} + \frac{p}{p+q}(1-p-q)^k & \frac{p}{p+q} - \frac{p}{p+q}(1-p-q)^k \\ \frac{q}{p+q} - \frac{q}{p+q}(1-p-q)^k & \frac{p}{p+q} + \frac{q}{p+q}(1-p-q)^k \end{pmatrix}. \tag{1.21}$$

We end this section by giving the first elementary property of the spectrum of P (note that actually this property does not require reversibility).

Proposition 1.5.7 1 *is always an eigenvalue of P. Moreover, if λ is another eigenvalue then $|\lambda| \le 1$.*

Proof It is obvious that $P\mathbf{1} = \mathbf{1}$, showing that 1 is an eigenvalue (we recall that $\mathbf{1}$ is the characteristic function of X; see Section 1.1). Suppose now that $Pf = \lambda f$ with f nontrivial. Choose $x \in X$ such that $|f(x)| \geq |f(y)|$ for all $y \in X$ (X is finite). Then

$$|\lambda f(x)| = |\sum_{y \in X} p(x,y)f(y)| \leq \sum_{y \in X} p(x,y)|f(y)|$$
$$\leq |f(x)| \sum_{y \in X} p(x,y) = |f(x)|$$

and therefore $|\lambda| \leq 1$. □

In the following two sections we shall explore the geometrical structure underlying a reversible Markov chain in order to derive other properties of its spectrum.

1.6 Graphs

A *simple, undirected graph* with *loops* is a couple $\mathcal{G} = (X, E)$ where X is a set of *vertices* and E is a family of unordered pairs $\{x, y\}$ of elements of X possibly collapsing to a singleton (namely when $x = y$), which is called the set of *edges*. In other words if Ω denotes the family of all $2-$subsets of X, then E might be identified with a subset of $\Omega \coprod X$. Usually we simply say that X is a graph with edge set E. If $\{x, y\} \in E$, x and y are called *neighbors* or *adjacent* and we write $x \sim y$. We also say that the vertices x and y are *incident* with the edge $\{x, y\}$ and that $\{x, y\}$ *joins* x and y. Moreover if $x \sim x$ we say that $\{x\}$ is a *loop*.

The *degree* of a vertex x is the number edges that are incident to x; in symbols $\deg x = |\{y \in X : y \sim x\}|$. When $\deg x = d$ is constant and the graph has no loops we say that the graph is *regular* of *degree d*.

Example 1.6.1 Figure 1.9 represents the graph with vertex set $V = \{x, y, z\}$ and edge set $E = \{\{x, y\}, \{y, z\}, \{z\}\}$. Observe that $\deg x = 1$, $\deg y = \deg z = 2$.

Given a graph $\mathcal{G} = (X, E)$, a *path* in $\mathcal{G}$ is a sequence $p = (x_0, x_1, \ldots, x_n)$ of vertices such that $x_i \sim x_{i+1}$ for all $i = 0, 1, \ldots, n-1$. The vertices x_0 and x_n are called the *initial* and *terminal* vertices of p, respectively, and one says that p connects them. If $x_0 = x_n$ one says that the path is *closed*. Moreover the *length* of the path p is the nonnegative

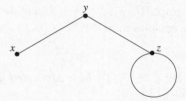

Figure 1.9.

number $|p| = n$. The *inverse* of a path $p = (x_0, x_1, \ldots, x_n)$ is the path $p^{-1} = (x_n, x_{n-1}, \ldots, x_1, x_0)$; note that $|p^{-1}| = |p|$. Given two paths $p = (x_0, x_1, \ldots, x_n)$ and $p' = (y_0, y_1, \ldots, y_m)$ with $x_n = y_0$ we define their *composition* as the path $p \cdot p' = (x_0, x_1, \ldots, x_n = y_0, y_1, \ldots, y_m)$; note that $|p \cdot p'| = |p| + |p'|$.

For $x, y \in X$ say that $x \cong y$ if there exists a path connecting them: clearly $\cong$ is an equivalence relation. The equivalence classes are called the *connected components* of $\mathcal{G}$. $\mathcal{G}$ is *connected* if there exists a unique connected component, in other words if given two vertices in X there exists a path connecting them; in this case, the *geodesic distance* $d(x, y)$ of two vertices is the minimal length among the paths connecting x and y.

A graph $\mathcal{G} = (X, E)$ is *bipartite* if there exists a nontrivial partition $X = X_0 \coprod X_1$ of the set of vertices such that $E \subseteq \{\{x_0, x_1\} : x_0 \in X_0, x_1 \in X_1\}$, that is every edge joins a vertex in X_0 with a vertex in X_1. Note that a bipartite graph has no loops and that, if $\mathcal{G}$ is connected, then the partition of the set of vertices is unique.

Example 1.6.2 Figure 1.10 shows the bipartite graph $\mathcal{G} = (X, E)$ with vertex set $X = X_0 \coprod X_1$, where $X_0 = \{x, y\}$ and $X_1 = \{u, v, z\}$, and edge set $E = \{\{x, u\}, \{x, v\}, \{y, v\}, \{y, z\}\}$.

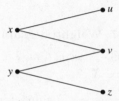

Figure 1.10.

Exercise 1.6.3 A graph $\mathcal{G} = (X, E)$ is bipartite if and only if it is *bicolorable*, i.e. there exists a map $\phi : X \to \{0, 1\}$ such that $x \sim y$ infers $\phi(x) \neq \phi(y)$.

Proposition 1.6.4 *Let $\mathcal{G} = (X, E)$ be a connected graph. The following conditions are equivalent.*

(i) $\mathcal{G}$ *is bipartite;*

(ii) *every closed path in $\mathcal{G}$ has even length;*

(iii) *there exists $x_0 \in X$ such that every closed path containing x_0 has even length;*

(iv) *given $x, y \in X$, then for all paths p connecting x and y one has $|p| \equiv d(x, y) \mod 2$, that is $|p| - d(x, y)$ is even.*

Proof Suppose that $\mathcal{G}$ is bipartite and let $X = X_0 \coprod X_1$ the corresponding partition of the vertices. Let $p = (x_0, x_1, \ldots, x_n)$ be a closed path. Without loss of generality we may suppose that $x_0 \in X_0$. But then $x_1 \in X_1$, $x_2 \in X_0$ and, more generally, $x_{2i} \in X_0$ and $x_{2i+1} \in X_1$ for all i's. Since $x_n = x_0$ we necessarily have that $n = |p|$ is even. This shows (i) $\Rightarrow$ (ii).

(ii) $\Rightarrow$ (iii) is trivial.

To show that (iii) $\Rightarrow$ (i) define $X_0 = \{x \in X : d(x, x_0) \text{ is even}\}$ and $X_1 = \{x \in X : d(x, x_0) \text{ is odd}\}$. We are left to show that if $y, y' \in X_0$ then $y \nsim y'$. Indeed if $p = (x_0, x_1, \ldots, x_{2n} = y)$ and $p' = (x_0, x'_1, \ldots, x'_{2m} = y')$ are two paths connecting x_0 with y and y', respectively, then, if $y \sim y'$ and setting $q = (y, y')$ we have that the composed path $p \cdot q \cdot (p')^{-1}$ is a closed path of odd length containing x_0, contradicting the hypothesis. The same arguments hold for X_1.

(iv) $\Leftrightarrow$ (ii): let p' be a minimal path connecting x and y, so that $|p'| = d(x, y)$; let p be another path connecting x and y. Then $|p'| + |p| = |p'| + |p^{-1}| = |p' \cdot p^{-1}| \equiv 0 \mod 2$ as $p' \cdot p^{-1}$ is a closed path. This completes the proof. $\square$

1.7 Weighted graphs

Let $\mathcal{G} = (X, E)$ be a graph. A *weight* on $\mathcal{G}$ is a function $w : X \times X \to [0, +\infty)$ such that for $x, y \in X$

$$\begin{cases} w(x, y) = w(y, x) & \text{(symmetry)} \\ w(x, y) > 0 & \text{if and only if } x \sim y. \end{cases}$$

In other words, w is a function that assigns to each edge $\{x, y\} \in E$ a positive number, namely $w(x, y)$. We then say that $\mathcal{G} = (X, E, w)$ is a *weighted graph*.

With a weight w we associate a stochastic matrix $P = (p(x, y))_{x, y \in X}$ on X by setting

$$p(x, y) = \frac{w(x, y)}{W(x)}$$

where $W(x) = \sum_{z \in X} w(x, z)$.

The corresponding Markov chain (whatever the initial distribution is) is called a *random walk* on the weighted graph $\mathcal{G}$.

P is reversible; it is in detailed balance with the distribution π defined by

$$\pi(x) = \frac{W(x)}{W}$$

where $W = \sum_{z \in X} W(z)$.

Indeed, for all $x, y \in X$ we have

$$\pi(x)p(x, y) = \frac{W(x)}{W} \cdot \frac{w(x, y)}{W(x)} = \frac{w(x, y)}{W} = \frac{w(y, x)}{W} = \pi(y)p(y, x).$$

Also note that if X does not have *isolated* vertices (that is vertices that are not incident to any other vertex), then $\pi(x) > 0$ for all $x \in X$. Conversely, suppose that X is a set and P a transition matrix on X in detailed balance with a strict probability measure π. Setting $w(x, y) = \pi(x)p(x, y)$ we have that $\pi(x)p(x, y) = \pi(y)p(y, x)$ implies $w(x, y) = w(y, x)$ and taking $E = \{\{x, y\} : w(x, y) > 0\}$ we have that $\mathcal{G} = (X, E, w)$ is a weighted graph.

This shows that reversible Markov chains are essentially the same thing as random walks on weighted graphs.

Exercise 1.7.1 Show that the two state Markov chain in Example 1.5.6 may be obtained by considering the weight function given by $w(0, 0) = \frac{q(1-p)}{p+q}$, $w(0, 1) = w(1, 0) = \frac{pq}{p+q}$ and $w(1, 1) = \frac{p(1-q)}{p+q}$ on the graph in Figure 1.11.

Figure 1.11.

Remark 1.7.2 Every graph $\mathcal{G} = (X, E)$ can be given the *trivial weight* $w(x, y) = 1$ if $x \sim y$ and $w(x, y) = 0$, otherwise; in this case, we refer to $\mathcal{G}$ as to an *unweighted graph*.

In what follows $\mathcal{G} = (X, E, w)$ is a *finite* weighted graph and P the associated transition matrix. We denote by $\sigma(P) = \{\lambda \in \mathbb{C} : \det(P - \lambda I) = 0\}$ (in other words, the set of all eigenvalues of P) the *spectrum* of P. The *multiplicity* of an eigenvalue $\lambda \in \sigma(P)$ of P equals the dimension of the eigenspace $V_\lambda = \{f \in L(X) : Pf = \lambda f\}$. We recall (see (1.16)) that each V_λ has a basis consisting of real valued functions. Note that one can also write $(Pf)(x) = \sum_{\substack{y \in X: \\ y \sim x}} p(x, y) f(y)$.

Proposition 1.7.3 *Let $\mathcal{G} = (X, E, w)$ be a finite weighted graph. The multiplicity of the eigenvalue 1 of the transition matrix P equals the number of connected components of $\mathcal{G}$. In particular, the multiplicity of 1 is one if and only if $\mathcal{G}$ is connected.*

Proof It is obvious that if f is constant in each connected component, then $Pf = f$. Conversely, suppose that $Pf = f$ with $f \in L(X)$ non-identically zero and real valued. Let $X_0 \subset X$ be a connected component of $\mathcal{G}$ and denote by $x_0 \in X_0$ a maximum point for $|f|$ in X_0, i.e. $|f(x_0)| \geq |f(y)|$ for all $y \in X_0$; we may suppose, up to passing to $-f$, that $f(x_0) \geq 0$. Then $f(x_0) = \sum_{y \in X_0} p(x_0, y) f(y)$ and as $\sum_{y \in X_0} p(x_0, y) = 1$ we have $\sum_{y \in X_0} p(x_0, y)[f(x_0) - f(y)] = 0$. Since $p(x_0, y) \geq 0$ and $f(x_0) \geq f(y)$ for all $y \in X_0$, we deduce that $f(y) = f(x_0)$ for all $y \sim x_0$. Let now $z \in X_0$; then, by definition, there exists a path $p = (x_0, x_1, \ldots, x_n = z)$ connecting x_0 to z. In the previous step we have established that $f(x_1) = f(x_0) \geq f(y)$ for all $y \in X_0$ so that we can iterate the same argument to show that $f(x_0) = f(x_1) = f(x_2) = \cdots = f(x_{n-1}) = f(x_n) = f(z)$ i.e. f is constant in X_0. This shows that f is constant on the connected components of X and this ends the proof. $\qquad\square$

Note that connectedness is a structural property of $\mathcal{G}$ and therefore the multiplicity of the eigenvalue 1 of P does not depend on the specific (positive) values of the weight w.

The following proposition provides another example of a structural (geometrical) property that reflects on the spectral theory of the graph independently on the weight function.

Proposition 1.7.4 *Let $\mathcal{G} = (X, E, w)$ be a finite connected weighted graph and denote by P the corresponding transition matrix. Then the following are equivalent:*

(i) $\mathcal{G}$ *is bipartite;*

(ii) *the spectrum $\sigma(P)$ is symmetric: $\lambda \in \sigma(P)$ if and only if $-\lambda \in \sigma(P)$;*

(iii) $-1 \in \sigma(P)$.

Proof Suppose that $\mathcal{G}$ is bipartite with $X = X_0 \coprod X_1$ and that $Pf = \lambda f$. Set $\widetilde{f}(x) = (-1)^j f(x)$ for $x \in X_j$, $j = 0, 1$. Then, for $x \in X_j$ we have:

$$
\begin{aligned}
[P\widetilde{f}](x) &= \sum_{y:y \sim x} p(x,y)\widetilde{f}(y) \\
&= (-1)^{j+1} \sum_{y:y \sim x} p(x,y)f(y) \\
&= (-1)^{j+1} \lambda f(x) \\
&= -\lambda \widetilde{f}(x).
\end{aligned}
$$

We have shown that is λ is an eigenvalue for f, then $-\lambda$ is an eigenvalue for $\widetilde{f}$; this gives the implication (i) $\Rightarrow$ (ii).

(ii) $\Rightarrow$ (iii) is obvious.

(iii) $\Rightarrow$ (i): suppose that $Pf = -f$ with f nontrivial and real valued. Denote by $x_0 \in X$ a point of maximum for $|f|$; then, up to switching f to $-f$, we may suppose that $f(x_0) > 0$. We then have that $-f(x_0) = \sum_{y:y \sim x_0} p(x_0, y)f(y)$ infers $\sum_{y:y \sim x_0} p(x_0, y)[f(x_0) + f(y)] = 0$ and since $f(x_0) + f(y) \geq 0$ we deduce $f(y) = -f(x_0)$ for all $y \sim x_0$. Set $X_j = \{y \in X : f(y) = (-1)^j f(x_0)\}$ for $j = 0, 1$. Arguing as in the proof of the previous proposition, we see that $X = X_0 \coprod X_1$: indeed X is connected and if $p = (x_0, x_1, \ldots, x_m)$ is a path, then $f(x_j) = (-1)^j f(x_0)$. Finally if $y \sim z$ we clearly have $f(y) = -f(z)$ so that X is bipartite. This ends the proof. $\qquad\square$

Remark 1.7.5 We observe that for $x, y \in X$ and $n \in \mathbb{N}$, one has

$$
p^{(n)}(x,y) = \sum p(x_0, x_1)p(x_1, x_2) \cdots p(x_{n-1}, x_n) \geq 0
$$

where the sum runs over all paths $p = (x_0, x_1, \ldots, x_n)$ of length n connecting $x_0 = x$ to $x_n = y$: clearly this sum vanishes if and only if such paths of length n do not exist. Moreover if $p^{(n_0)}(x,y) > 0$ for all $x, y \in X$ the same holds for all $n \geq n_0$: indeed $p^{(n_0+1)}(x,y) = \sum_{z \in X} p(x,z)p^{(n_0)}(z,y) > 0$ and induction applies.

Theorem 1.7.6 *Let $\mathcal{G} = (X, E)$ be a finite graph. Then the following conditions are equivalent.*

(i) *$\mathcal{G}$ is connected and not bipartite;*

(ii) *for every weight function on X, the associated transition matrix P is ergodic.*

Proof (ii) $\Rightarrow$ (i): Suppose that for a weight function on X, the associated transition matrix P is ergodic. This means that there exists $n_0 > 0$ such that

$$p^{(n_0)}(x, y) > 0 \quad \text{for all } x, y \in X. \tag{1.22}$$

Then X is connected: indeed, from (1.22) and Remark 1.7.5, we deduce that every two vertices x and y are connected by a path of length n_0.

Analogously, X is not bipartite: for $n > n_0$ one has

$$p^{(n)}(x, y) = \sum_{z \in X} p^{(n-n_0)}(x, z) p^{(n_0)}(z, y) > 0$$

showing that there exist both even and odd paths connecting x and y (compare with Proposition 1.6.4).

Conversely, suppose that X is connected and not bipartite. Denote by $\delta = \max_{x,y \in X} d(x, y)$ the diameter of X. From Proposition 1.6.4 we know that for each $x \in X$ there exists a closed path of odd length starting and ending at x, call it $p(x)$. Set $2M + 1 = \max_{x \in X} |p(x)|$. If $z \sim x$, then, for all n, the path $q_{2n} = (x, z, x, z, \ldots, x, z, x)$ has length $2n$; thus the path p_m which equals q_m if m is even and equals the composition $p(x) \cdot q_{2t}$ for $m = |p(x)| + 2t$ odd, yields a closed path of any length $m \geq 2M$ starting and ending at x.

This implies that for all $n \geq 2M + \delta$ and any $x, y \in X$ there exists a path from x to y of length exactly n: indeed, denoting by $p_{x,y}$ a minimal path connecting x to y, so that $|p_{x,y}| = d(x, y) \leq \delta$, setting $m = n - d(x, y) \geq 2M$ we have that $p_m \cdot p_{x,y}$ (where p_m is as above) is a path of length n connecting x to y.

In virtue of the Remark 1.7.5, we conclude that $p^{(n)}(x, y) > 0$ for all $x, y \in X$ and $n \geq 2M + \delta$, that is P is ergodic. $\qquad\square$

We note that if P is ergodic and in detailed balance with π, then its stationary distribution coincides with π. Indeed summing over $x \in X$ in $\pi(x)p(x, y) = \pi(y)p(y, x)$ we have $\sum_{x \in X} \pi(x)p(x, y) = \pi(y)$ showing that π is indeed the stationary distribution with respect to P (uniqueness follows from ergodicity of P).

We now present another proof of the Markov ergodic theorem (Theorem 1.4.1) in the case of reversibility for the Markov chain.

Theorem 1.7.7 *Let $\mathcal{G} = (X, E, w)$ be a connected not bipartite weighted graph. Let P denote the associated stochastic matrix which is in detailed balance with the distribution π. Then*

$$\lim_{n \to \infty} p^{(n)}(x, y) = \pi(y)$$

for all $x, y \in X$.

Proof Recalling the spectral decomposition in Proposition 1.5.5 we have

$$p^{(n)}(x, y) = \pi(y) \sum_{z \in X} u(x, z) \lambda_z^n u(y, z). \tag{1.23}$$

Since X is connected, by Proposition 1.7.3 the eigenvalue 1 has multiplicity one; therefore there exists z_0 such that $\lambda_z < \lambda_{z_0} = 1$ for all $z \neq z_0$. Moreover $u(x, z_0) = 1$ for all $x \in X$. Since X is not bipartite, $\lambda_z > -1$ for all $z \in X$. Therefore (1.23) may be rewritten in the form

$$p^{(n)}(x, y) = \pi(y) + \pi(y) \sum_{z \in X \setminus \{z_0\}} u(x, z) \lambda_z^n u(y, z) \tag{1.24}$$

and therefore $p^{(n)}(x, y) \to \pi(y)$ for $n \to \infty$ as, for all $z \neq z_0$, one has $-1 < \lambda_z < 1$. $\qquad\square$

In other words, a reversible transition matrix P is ergodic if and only if the eigenvalue $\lambda = 1$ has multiplicity *one* and $\lambda = -1$ is *not* an eigenvalue; moreover, these conditions are equivalent to the convergence to equilibrium.

Definition 1.7.8 The quantity $\lambda^* = \max \{|\lambda_z| : z \in X \setminus \{z_0\}\}$ is called the *the second largest eigenvalue modulus*. It is clear that λ^* gives rise in (1.24) to the slowest term that tends to 0.

1.8 Simple random walks

Let $\mathcal{G} = (X, E)$ be a finite graph. If we take on X the trivial weight function as in Remark 1.7.2 then the associated Markov chain is called the *simple random walk* on X. That is, if we are in x we move to a neighbor y with probability $\frac{1}{\deg x}$. The associated transition matrix P is given by

$$p(x, y) = \begin{cases} \frac{1}{\deg x} & \text{if } x \sim y \\ 0 & \text{otherwise} \end{cases}$$

for all x, $y \in X$, and it is in detailed balance with the measure

$$\pi(x) = \frac{\deg x}{\sum_{y \in X} \deg y}.$$

Example 1.8.1 The *complete graph K_n* on n vertices is the graph whose vertex set K_n has cardinality n and whose edge set is $E = \{\{x, y\} : x, y \in K_n, \ x \neq y\}$: that is two vertices are connected if and only if they are distinct. K_n is regular: each vertex has degree $n - 1$.

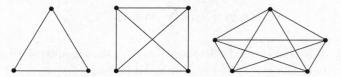

Figure 1.12. The complete graphs K_3, K_4 and K_5

The transition matrix P of the simple random walk on K_n is given by

$$p(x, y) = \begin{cases} \frac{1}{n-1} & \text{if } x \neq y \\ 0 & \text{if } x = y \end{cases}$$

and it is in detailed balance with the uniform measure $\pi(x) = \frac{1}{n}$ for all $x \in K_n$. The graph is connected and it is bipartite if and only if $n = 2$. Therefore the simple random walk is ergodic for all $n \geq 3$. Moreover, setting $W_0 = \{\text{constant functions on } K_n\}$ and $W_1 = \{f : K_n \to \mathbb{C} : \sum_{y \in K_n} f(y) = 0\}$ we have:

$$f \in W_0 \Rightarrow \sum_{y \in K_n} p(x, y) f(y) = f(x)$$

$$f \in W_1 \Rightarrow \sum_{y \in K_n} p(x, y) f(y) = \frac{1}{n-1} \sum_{\substack{y \in K_n \\ y \neq x}} f(y)$$

$$= \frac{1}{n-1} \sum_{y \in K_n} f(y) - \frac{1}{n-1} f(x) = -\frac{1}{n-1} f(x).$$

Therefore

- W_0 is an eigenspace of P corresponding to the eigenvalue 1 whose multiplicity is equal to $\dim W_0 = 1$.
- W_1 is an eigenspace of P corresponding to the eigenvalue $-\frac{1}{n-1}$ whose multiplicity is equal to $\dim W_1 = n - 1$.

Example 1.8.2 (The discrete circle) Example 1.2.1 is also an example of a simple random walk on un unweighted graph, that is on C_n. Note that C_n is connected and, as it consists of a single closed path, it is bipartite if and only if n is even. Therefore the random walk is ergodic if and only if n is odd. Its spectrum will be computed in the next chapter.

Example 1.8.3 (The Ehrenfest diffusion model) The Ehrenfest diffusion model in Example 1.2.2 may be seen as the simple random walk on the graph whose vertex set is Q_n, the set of all subsets of $\{1, 2, \ldots, n\}$, and edge set is $\{\{A, B\} : |A \triangle B| = |A \backslash B| + |B \backslash A| = 1\}$. In other words, we join A and B when there exists $j \in \{1, 2, \ldots, n\}$ such that $A = B \coprod \{j\}$ or $B = A \coprod \{j\}$.

We can give another description of the Eherenfest diffusion model. Define $\{0, 1\}^n = \{(a_1, \ldots, a_n) : a_k \in \{0, 1\},\ k = 1, \ldots, n\}$ the set of all binary n-tuples. Then there is a natural bijection

$$
\begin{aligned}
Q_n &\leftrightarrow \{0, 1\}^n \\
A &\leftrightarrow (a_1, \ldots, a_n) : a_k = \left\{ \begin{array}{l} 1 \text{ if } k \in A \\ 0 \text{ if } k \notin A. \end{array} \right.
\end{aligned}
$$

The Ehrenfenst urn model, using $\{0, 1\}^n$ as the state space, has the following description: at each time choose a random coordinate, say a_k, and change it (if it is 0, it becomes 1 and vice versa). In the corresponding graph on $\{0, 1\}^n$, two vertices $(a_1, \ldots, a_n)$ and $(b_1, \ldots, b_n)$ are connected if and only if $|\{k : a_k \neq b_k\}| = 1$.

In what follows, we will use indifferently Q_n and $\{0, 1\}^n$ to describe the state space of the Eherenfest urn model; we will usually write Q_n but we will have in mind the identification $Q_n \leftrightarrow \{0, 1\}^n$

For $n = 1$ the graph is simply a segment:

Figure 1.13(a)

For $n = 2$ we have a square (it coincides with C_4):

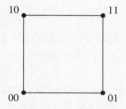

Figure 1.13(b)

For $n = 3$ we have a cube:

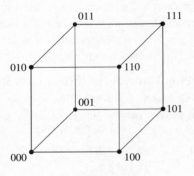

Figure 1.13(c)

Therefore Q_n is also called the n-dimensional *hypercube*.

The graph Q_n is connected and always bipartite: if we set

$$X_0 = \{A \in Q_n : |A| \text{ is even}\} \qquad X_1 = \{A \in Q_n : |A| \text{ is odd}\}$$

then it is clear that two vertices in X_0 (or in X_1) cannot be connected. Therefore the simple random walk on Q_n is not ergodic. To obviate this fact, one can consider the following variation of the simple random walk:

$$p(x, y) = \begin{cases} \frac{1}{n+1} & \text{if } x \sim y \\ \frac{1}{n+1} & \text{if } x = y \\ 0 & \text{otherwise.} \end{cases} \qquad (1.25)$$

That is we add a loop to each vertex in Q_n. This is a simple trick that may be used to deal with random walks on bipartite graphs. The simple random walk on Q_n (without loops) has a simple description in terms of the urn model: at each stage choose randomly a ball with probability $\frac{1}{n}$ and move it to the other urn. The Markov chain in (1.25) is a simple modification: we allow the model to remain in its current state with probability $\frac{1}{n+1}$. The spectrum of this random walk will be computed in the next chapter.

Exercise 1.8.4 (1) Show that the Bernoulli–Laplace diffusion model is a simple random walk on an (unweighted) graph and that it is always ergodic.

(2) Show that the random transpositions model is a random walk on a weighted graph with a loop at every vertex and in which the weight function takes only two values (a value on the loops, the other value on

the non-loops), that is it is obtained by a simple random walk by adding a loop at every vertex with the same weight.

The spectrum of the Bernoulli–Laplace diffusion model will be computed in Chapter 6, while the spectrum of the random transposition model in Section 10.7.

1.9 Basic probabilistic inequalities

In this section, we want to give some notions and tools for a quantitative study of the convergence to equilibrium distribution for an ergodic reversible Markov chain. First of all, we introduce a natural distance between two probability measures; if μ and ν are probability measures on X their *total variation distance* is given by

$$\|\mu - \nu\|_{TV} = \max_{A \subseteq X} \left| \sum_{x \in A} (\mu(x) - \nu(x)) \right| \equiv \max_{A \subseteq X} |\mu(A) - \nu(A)|.$$

It is easy to show that if we denote by $\|\mu - \nu\|_{L^1(X)}$ the standard $L^1(X)$-distance i.e. $\|\mu - \nu\|_{L^1(X)} = \sum_x |\mu(x) - \nu(x)|$ we have that

$$\|\mu - \nu\|_{TV} = \frac{1}{2}\|\mu - \nu\|_{L^1(X)}. \tag{1.26}$$

Indeed let B be a subset of X such that $\|\mu - \nu\|_{TV} = \left| \sum_{x \in B} (\mu(x) - \nu(x)) \right|$. Then it is clear that $\mu - \nu$ does not change sign on B and takes the opposite sign on the complement B^c of B. Suppose without loss of generality that $\mu - \nu$ is nonnegative on B. Then, using the fact that μ, ν are probability distributions, we have

$$\sum_{x \in B} |\mu(x) - \nu(x)| = \sum_{x \in B} (\mu(x) - \nu(x))$$

$$= 1 - \sum_{x \in B^c} \mu(x) - 1 + \sum_{x \in B^c} \nu(x)$$

$$= \sum_{x \in B^c} (\nu(x) - \mu(x))$$

$$= \sum_{x \in B^c} |\mu(x) - \nu(x)|$$

which gives

$$\|\mu - \nu\|_{L^1(X)} = \sum_{x \in B} |\mu(x) - \nu(x)| + \sum_{x \in B^c} |\mu(x) - \nu(x)| = 2\|\mu - \nu\|_{TV}.$$

It follows that

$$\|\mu - \nu\|_{TV} = \frac{1}{2} \max_{\phi} \left| \sum_{x \in X} (\mu(x) - \nu(x))\phi(x) \right| \qquad (1.27)$$

where the maximum is over all $\phi : X \to \mathbb{C}$ such that $|\phi(x)| \leq 1$ for all $x \in X$. Indeed, the maximum is achieved by taking

$$\phi(x) = \begin{cases} \frac{|\mu(x)-\nu(x)|}{\mu(x)-\nu(x)} & \text{if } \mu(x) \neq \nu(x) \\ 0 & \text{otherwise.} \end{cases}$$

Let P be a reversible transition matrix, in detailed balance with the strict probability measure π.

For $x \in X$ set $\nu_x^{(k)}(y) = p^{(k)}(x,y)$; in other words, $\nu_x^{(k)}$ is the distribution probability after k steps if the initial distribution is δ_x.

Introduce the scalar product

$$\langle f_1, f_2 \rangle_{\frac{1}{\pi}} = \sum_{x \in X} f_1(x)\overline{f_2(x)} \frac{1}{\pi(x)}$$

with the norm $\|f\|_{\frac{1}{\pi}}^2 = \langle f, f \rangle_{\frac{1}{\pi}}$. Then we have:

Proposition 1.9.1 *Suppose that P is an ergodic reversible Markov chain in detailed balance with π. With the notation from Theorem 1.7.7, for every $x \in X$, we have*

$$\|\nu_x^{(n)} - \pi\|_{\frac{1}{\pi}}^2 = \sum_{\substack{z \in X \\ z \neq z_0}} \lambda_z^{2n} u(x,z)^2.$$

Proof We recall that $\lambda_{z_0} = 1$ (see the proof of Theorem 1.7.7). Then,

$$\|\nu_x^{(n)} - \pi\|_{\frac{1}{\pi}}^2 = \sum_{y \in X} \left[p^{(n)}(x,y) - \pi(y) \right]^2 \frac{1}{\pi(y)}$$

$$\text{from (1.24)} = \sum_{y \in X} \pi(y) \sum_{\substack{z_1, z_2 \in X \\ z_1, z_2 \neq z_0}} u(x,z_1)u(x,z_2)\lambda_{z_1}^n \lambda_{z_2}^n u(y,z_1)u(y,z_2)$$

$$= \sum_{\substack{z_1, z_2 \in X \\ z_1, z_2 \neq z_0}} u(x,z_1)u(x,z_2)\lambda_{z_1}^n \lambda_{z_2}^n \sum_{y \in X} \pi(y)u(y,z_1)u(y,z_2)$$

from $(1.16) = \displaystyle\sum_{\substack{z_1,z_2 \in X \\ z_1,z_2 \neq z_0}} u(x,z_1)u(x,z_2)\lambda_{z_1}^n \lambda_{z_2}^n \delta_{z_1}(z_2)$

$$= \sum_{\substack{z \in X \\ z \neq z_0}} u(x,z)^2 \lambda_z^{2n}.$$

$\square$

The following corollary is our first version of the celebrated Diaconis–Shahshahani upper bound lemma.

Corollary 1.9.2 (Upper bound lemma)

$$\|\nu_x^{(n)} - \pi\|_{TV}^2 \leq \frac{1}{4} \sum_{\substack{z \in X \\ z \neq z_0}} \lambda_z^{2n} u(x,z)^2.$$

Proof From (1.26) we have

$$\|\nu_x^{(n)} - \pi\|_{TV}^2 = \frac{1}{4}\|\nu_x^{(n)} - \pi\|_{L^1(X)}^2$$

$$= \frac{1}{4}\left\{ \sum_{y \in X} \left[\left|p^{(n)}(x,y) - \pi(y)\right| \frac{1}{\sqrt{\pi(y)}} \right] \sqrt{\pi(y)} \right\}^2$$

by Cauchy–Schwarz $\leq \dfrac{1}{4}\|\nu_x^{(n)} - \pi\|_{\frac{1}{\pi}}^2 \displaystyle\sum_{y \in X} \pi(y)$

$$= \frac{1}{4} \sum_{\substack{z \in X \\ z \neq z_0}} \lambda_z^{2n} u(x,z)^2.$$

$\square$

In what follows we show that Corollary 1.9.2 is a good tool to bound $\|\nu_x^{(n)} - \pi\|_{TV}^2$. Indeed, we will present several examples where also a lower bound for $\|\nu_x^{(n)} - \pi\|_{TV}^2$ is available and the two estimates agree.

Example 1.9.3 For the two state Markov chain (Example 1.5.6) the upper bound lemma gives:

$$\|\nu_0^{(n)} - \pi\|_{TV}^2 \leq \frac{1}{4}(1 - p - q)^{2n}\frac{p}{q}. \tag{1.28}$$

By the explicit formula 1.21, $\|\nu_0^{(n)} - \pi\|_{TV}^2 = (\frac{p}{p+q})^2(1 - p - q)^{2n}$ and so equality holds in (1.21) if and only if $p = q$.

A looser bound, that uses only the second largest eigenvalue modulus, namely $\lambda^* = \max\{|\lambda_z| : z \in X \setminus \{z_0\}\}$ (see Definition 1.7.8), is the following.

Corollary 1.9.4

$$\|\nu_x^{(n)} - \pi\|_{TV}^2 \leq \frac{1 - \pi(x)}{4\pi(x)}|\lambda^*|^{2n}.$$

Proof

$$
\begin{aligned}
\|\nu_x^{(n)} - \pi\|_{TV}^2 &\leq \frac{1}{4} \sum_{z \in X \setminus \{z_0\}} \lambda_z^{2n} u(x,z)^2 \\
&\leq \frac{1}{4}|\lambda^*|^{2n} \sum_{z \in X \setminus \{z_0\}} u(x,z)^2 \\
&= \frac{1}{4}|\lambda^*|^{2n}\left(\frac{1}{\pi(x)} - u(x,z_0)^2\right) \\
&= \frac{1}{4}|\lambda^*|^{2n}\left(\frac{1}{\pi(x)} - 1\right) \\
&= \frac{|\lambda^*|^{2n}}{4}\frac{1 - \pi(x)}{\pi(x)},
\end{aligned}
$$

where the first inequality follows from Corollary 1.9.2 and the first equality from (1.18). $\square$

The method in the proof of the previous corollary is an example of the so-called *analytic techniques* that will be not treated in our book. There is a growing literature on this subject; for instance Saloff-Coste's lecture notes [184], the book in preparation by Aldous and Fill [2], and the monographs by Stroock [215], Behrends [18] and Bremaud [32]. A quick introduction is also available in [60].

We end this section fixing some notation and recalling two inequalities that will be used to obtain lower bound estimates. Let X be a finite set, μ a probability measure on X and f a real valued function defined on X. The *mean value* of f respect to μ is

$$E_\mu(f) = \sum_{x \in X} f(x)\mu(x)$$

and the *variance* of f respect to μ is

$$\mathrm{Var}_\mu(f) = E_\mu\left((f - E_\mu(f))^2\right) \equiv E_\mu(f^2) - E_\mu(f)^2.$$

Proposition 1.9.5 (Markov inequality)

$$\mu\{x \in X : |f(x)| \geq \alpha\} \leq \frac{1}{\alpha} E_\mu(|f|), \quad \forall \alpha > 0.$$

Proof

$$\frac{1}{\alpha} E_\mu(|f|) = \frac{1}{\alpha} \sum_{\substack{x \in X: \\ |f(x)| \geq \alpha}} |f(x)|\mu(x) + \frac{1}{\alpha} \sum_{\substack{x \in X: \\ |f(x)| < \alpha}} |f(x)|\mu(x)$$

$$\geq \frac{1}{\alpha} \sum_{\substack{x \in X: \\ |f(x)| \geq \alpha}} |f(x)|\mu(x)$$

$$\geq \sum_{\substack{x \in X: \\ |f(x)| \geq \alpha}} \mu(x)$$

$$= \mu\{x \in X : |f(x)| \geq \alpha\}.$$

□

Corollary 1.9.6 (Chebyshev inequality)

$$\mu\{x \in X : |f(x) - E_\mu(f)| \geq \alpha\} \leq \frac{\text{Var}_\mu(f)}{\alpha^2}, \quad \forall \alpha > 0.$$

Proof

$$\mu\{|f - E_\mu(f)| \geq \alpha\} = \mu\{|f - E_\mu(f)|^2 \geq \alpha^2\}$$

$$\leq \frac{1}{\alpha^2} E_\mu\left(|f - E_\mu(f)|^2\right)$$

$$= \frac{\text{Var}_\mu(f)}{\alpha^2}.$$

□

1.10 Lumpable Markov chains

Let X be a finite space and $P = (p(x,y))_{x,y \in X}$ a stochastic matrix indexed by X. In this section we study a condition under which from a Markov chain with transition probability P we can construct another Markov chain with a smaller state space.

Suppose that $\mathcal{P}$ is a partition of X, that is $X = \coprod_{A \in \mathcal{P}} A$. Let ν be a probability distribution on X and consider the sequence of probability distributions on X

$$\begin{cases} \nu^{(0)} \equiv \nu \\ \nu^{(k)}(y) = \sum_{x \in X} \nu^{(k-1)}(x) p(x,y) & k = 1, 2, \ldots \end{cases}$$

associated with the Markov chain with initial distribution ν and transition probability P.

We consider the sequence $\{\nu^{(k)}|_{\mathcal{P}}\}_{k \in \mathbb{N}}$ of probability distributions on the set $\mathcal{P}$ obtained by restricting $\nu^{(k)}$ to $\mathcal{P}$: for any $A \in \mathcal{P}$, the value of $\nu^{(k)}|_{\mathcal{P}}$ on A equals $\nu^{(k)}(A) \equiv \sum_{x \in A} \nu^{(k)}(x)$.

Definition 1.10.1 We say that the stochastic matrix P is *lumpable* with respect to the partition $\mathcal{P}$ if, for any initial distribution ν, the sequence $\nu^{(k)}|_{\mathcal{P}}$ is the sequence of probability distributions of a Markov chain (the *lumped chain*) on $\mathcal{P}$ whose transition probability does not depend on the choice of ν.

In other words, the definition requires the existence of a transition matrix $Q = (q(A,B))_{A,B \in \mathcal{P}}$ indexed by $\mathcal{P}$ such that

$$\nu^{(k)}(B) = \sum_{A \in \mathcal{P}} \nu^{(k-1)}(A) q(A,B) \quad k = 1, 2, \ldots, n \quad A \in \mathcal{P}$$

and that Q does not depend on $\nu^{(0)}$ defined by $\nu^{(0)}(A) = \nu(A)$ for all $A \in \mathcal{P}$.

For $x \in X$ and $B \in \mathcal{P}$ set $p(x, B) = \sum_{z \in B} p(x, z)$. Then we have

Proposition 1.10.2 *P is lumpable with respect to $\mathcal{P}$ if and only if for any pair $A, B \in \mathcal{P}$ the function $x \mapsto p(x, B)$ is constant on A. If this is the case, then the transition matrix Q is given by $q(A, B) = p(x, B)$ for $A, B \in \mathcal{P}$ and $x \in A$.*

Proof Suppose that P is lumpable. Take $A, B \in \mathcal{P}$, $x \in A$ and $\nu = \delta_x$ so that $\nu^{(0)}|_{\mathcal{P}} \equiv \delta_A$. Then $\nu^{(1)}(y) = p(x, y)$ and $\nu^{(1)}(B) = \sum_{y \in B} p(x, y) = p(x, B)$ must be equal to $q(A, B)$ and therefore cannot depend on $x \in A$.

Conversely, suppose that $p(x, B)$ does not depend on $x \in A$. Set $q(A, B) = p(x, B)$ and consider an initial distribution ν. Then

$$\nu^{(k)}(B) = \sum_{y \in B} \nu^{(k)}(y) = \sum_{y \in B} \sum_{x \in X} \nu^{(k-1)}(x) p(x, y)$$

$$= \sum_{x \in X} \nu^{(k-1)}(x) p(x, B)$$

$$= \sum_{A \in \mathcal{P}} \sum_{x \in A} \nu^{(k-1)}(x) q(A, B)$$

$$= \sum_{A \in \mathcal{P}} \nu^{(k-1)}(A) q(A, B).$$

$\square$

Corollary 1.10.3 *Suppose that* P *is* $\mathcal{P}$-*lumpable and* Q *is as above. Then*

(i) $p^{(n)}(x, B) = q^{(n)}(A, B)$ *for all* $A, B \in \mathcal{P}$, $x \in A$ *and* $n \in \mathbb{N}$.
(ii) *If* P *is ergodic,* Q *is ergodic as well.*
(iii) *If* P *is reversible and in detailed balance with respect to* π *then* Q *is reversible and in detailed balance with respect to* $\pi|_{\mathcal{P}}$.

Proof (i) and (ii) are trivial. As for (iii) we have

$$\pi(A)q(A, B) = \sum_{x \in A} \pi(x)p(x, B)$$

$$= \sum_{x \in A} \sum_{y \in B} \pi(x)p(x, y)$$

$$= \sum_{y \in B} \sum_{x \in A} \pi(y)p(y, x)$$

$$= \sum_{y \in B} \pi(y)q(y, A)$$

$$= \pi(B)q(B, A).$$

$\square$

As usual, for $f \in L(X)$ and $x \in X$ set $Pf(x) = \sum_{y \in X} p(x, y)f(y)$. Denote by $W_{\mathcal{P}}$ the subspace of $L(X)$ spanned by the functions $\{\mathbf{1}_A : A \in \mathcal{P}\}$, that is $f \in W_{\mathcal{P}}$ if and only if f is constant on each $A \in \mathcal{P}$.

Proposition 1.10.4 P *is* $\mathcal{P}$-*lumpable if and only if* $W_{\mathcal{P}}$ *is* P-*invariant. Moreover* $P(\mathbf{1}_B) = Q(\delta_B)$ *where* $Q(\delta_B)$ *denotes the action of* Q *on the Dirac function in* $L(\mathcal{P})$ *centered at* B. *Therefore there is a bijection between the eigenfunctions of the lumped chain and the eigenfunctions of* P *which are constant on each* $A \in \mathcal{P}$.

Proof

$$(P\mathbf{1}_B)(x) = \sum_{y \in B} p(x, y) = p(x, B)$$

and therefore $P\mathbf{1}_B$ is constant on each $A \in \mathcal{P}$ if and only if $p(x, B)$ does not depend on $x \in A$. $\square$

Example 1.10.5 Let K_n be the complete graph on n vertices and P the transition matrix of the simple random walk on K_n (see Example 1.8.1). Fix a point $x_0 \in K_n$ and consider the partition $K_n = A \coprod B$, where

$A = \{x_0\}$ and $B = K_n \setminus A$. Then P is $\{A, B\}$-lumpable: $p(x, A) = p(x, x_0) = \frac{1}{n-1}$ for every $x \in B$ and $p(x, B) = \frac{n-2}{n-1}$ for all $x \in B$. The lumped chain on $\{A, B\}$ is a two state Markov chain with transition probabilities

$$q(A, B) = 1; \quad q(B, A) = \frac{1}{n-1}; \quad q(A, A) = 0; \quad q(B, B) = \frac{n-2}{n-1}.$$

In virtue of the symmetry of P, that is $p(x, y) = p(y, x)$ for all $x, y \in K_n$, we have that $p^{(k)}(x_0, x) = p^{(k)}(x, x_0) = p^{(k)}(x, A)$ does not depend on $x \in A$ or on $x \in B$ (that is it only depends on the partition class to which x belongs). Therefore if $x \in B$ then

$$p^{(k)}(x_0, x) = \frac{1}{n-1} p^{(k)}(x_0, B)$$

$$= \frac{1}{n-1} q^{(k)}(A, B)$$

$$\text{from (1.21)} = \frac{1}{n} - \frac{1}{n} \left(-\frac{1}{n-1} \right)^k$$

and always in virtue of (1.21) we have

$$p^{(k)}(x_0, x_0) = p^{(k)}(x_0, A) = q^{(k)}(A, A)$$

$$= \frac{1}{n} + \frac{n-1}{n} \left(-\frac{1}{n-1} \right)^k.$$

In the previous example the lumped chain contains all the eigenvalues of the original chain, namely 1 and $-\frac{1}{n-1}$; only the multiplicity of the second one changes. In the following exercise, we show that in general this is not the case.

Exercise 1.10.6 (The star) Let X be a set of cardinality $|X| = n+1$, choose an element $x_0 \in X$ and set $X_1 = X \setminus \{x_0\}$. The *star* T_n is the graph on X with edge set $\{\{x_0, x\} : x \in X_1\}$.

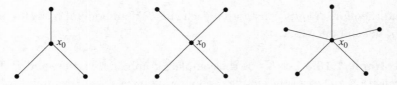

Figure 1.14. The stars T_3, T_4 and T_5

Let P be the transition matrix of the simple random walk on X. Set

- $W_0 = \{f : X \to \mathbb{C} : \text{constant}\}$
- $W_1 = \{f : X \to \mathbb{C} : f(x) = -f(x_0), \ \forall x \in X_1\}$
- $W_2 = \{f : X \to \mathbb{C} : f(x_0) = 0 \text{ and } \sum_{x \in X_1} f(x) = 0\}$.

(1) Show that W_0, W_1 and W_2 are eigenspaces of P and that the corresponding eigenvalues are $\lambda_0 = 1, \lambda_1 = -1$ and $\lambda_2 = 0$.

(2) Show that P is $\{\{x_0\}, X_1\}$-lumpable and that the lumped transition matrix has only the eigenvalues 1 and -1.

Exercise 1.10.7 Let P be the transition matrix for the simple random walk on the discrete circle C_4.

Figure 1.15.

Consider the partition $C_4 = A \coprod B$, where $A = \{0, 3\}$ and $B = \{1, 2\}$. Show that P is $\{A, B\}$-lumpable and that the lumped transition matrix is ergodic, while P is not ergodic.

Exercise 1.10.8 Consider the following transition matrix P on $\{0, 1, 2, 3\} \equiv C_4$: $p(k, k+1) = \frac{1}{2}$ for $k = 0, 1, 2$, $p(k, k) = \frac{1}{2}$ for $k = 0, 1, 2, 3$, $p(3, 0) = \frac{1}{2}$.

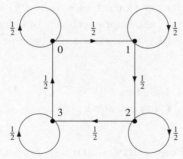

Figure 1.16.

Set $A = \{0, 2\}$ and $B = \{1, 3\}$. Show that P is $\{A, B\}$-lumpable, it is not reversible, but the lumped chain is reversible.

2

Two basic examples on abelian groups

2.1 Harmonic analysis on finite cyclic groups

Let $C_n = \{\overline{0}, \overline{1}, \ldots, \overline{n-1}\}$ be the *cyclic group* of order n. We identify it with the set $\mathbb{Z}/n\mathbb{Z}$ of integers modulo n, so that it is written additively. Therefore $\overline{x} \equiv x + n\mathbb{Z}$ and $\overline{x} + \overline{y} = \overline{x+y}$. For simplicity, in what follows the element $\overline{x} \in C_n$ will be denoted simply by x and we will write $\equiv_n$ for the equality mod n. The symbol $^-$ will be used to denote the conjugate of complex numbers. From the context it will be clear when $x \in \mathbb{Z}$ will denote an element in C_n. Moreover a function $f \in L(C_n)$ may be thought as a function $f : \mathbb{Z} \to \mathbb{C}$ which satisfies the periodicity condition: $f(x+n) = f(x)$ $\forall x \in \mathbb{Z}$. In the following we will use the symbols $\sum_{y \in C_n}$ or $\sum_{y=0}^{n-1}$ to denote the sum over all the elements of C_n.

We introduce the *characters* of C_n by setting $\omega = \exp(\frac{2\pi i}{n}) = \cos(\frac{2\pi}{n}) + i\sin(\frac{2\pi}{n})$ and $\chi_x(y) = \omega^{xy}$, for all $x, y \in C_n$. Clearly $\chi_x(y) = \chi_y(x)$, $\chi_y(-x) = \overline{\chi_y(x)}$ and $\chi_0 \equiv 1$.

The basic identity for the characters is $\chi_z(x+y) = \chi_z(x)\chi_z(y)$ and in the following exercise we ask to prove that it is a characteristic property (we recall that any nth complex root of the unit is of the form ω^k, $k = 0, 1, \ldots, n-1$).

Exercise 2.1.1 *If $\phi : C_n \to \mathbb{T} = \{z \in \mathbb{C} : |z| = 1\}$ satisfies $\phi(x+y) = \phi(x)\phi(y)$ then $\phi(x) = \chi_z(x)$ for some $z \in C_n$.*

Lemma 2.1.2 (Orthogonality relations) *Set $\delta_0(x) = \begin{cases} 1 & \text{if } x \equiv_n 0 \\ 0 & \text{otherwise.} \end{cases}$*

Then

$$\sum_{y=0}^{n-1} \chi_{x_1}(y)\overline{\chi_{x_2}(y)} = n\delta_0(x_1 - x_2).$$

Proof Observe that $\chi_{x_1}(y)\overline{\chi_{x_2}(y)} = \omega^{y(x_1-x_2)} = [\omega^{(x_1-x_2)}]^y$. In the case $x_1 \not\equiv_n x_2$ we have that $z = \omega^{(x_1-x_2)}$ satisfies $z^n - 1 = 0$ but not $z - 1 = 0$ and from the identity

$$z^n - 1 = (z-1)(1 + z + \cdots + z^{n-1}) = (z-1)\sum_{y=0}^{n-1} z^y$$

we deduce that

$$\sum_{y=0}^{n-1} \chi_{x_1}(y)\overline{\chi_{x_2}(y)} = \sum_{y=0}^{n-1} z^y = \frac{z^n - 1}{z - 1} = 0.$$

Clearly if $x_1 \equiv_n x_2$ then $\omega^{(x_1-x_2)} = 1$ and the above sum equals n. □

From the lemma and the fact that $\chi_x(y) = \chi_y(x)$ we immediately deduce the following *dual relation*

$$\sum_{x \in C_n} \chi_x(y_1)\overline{\chi_x(y_2)} = n\delta_0(y_1 - y_2).$$

If we use the standard scalar product (as in the preliminaries) we can express the orthogonality relations as $\langle \chi_{x_1}, \chi_{x_2} \rangle = n\delta_0(x_1 - x_2)$. Note the $\dim(L(C_n)) = n$ and therefore the set $\{\chi_x : x \in C_n\}$ is an orthogonal basis for $L(C_n)$. The *Fourier transform* of a function $f \in L(C_n)$ is the function $\widehat{f}$, still in $L(C_n)$, defined by:

$$\widehat{f}(x) = \langle f, \chi_x \rangle = \sum_{y \in C_n} f(y)\overline{\chi_x(y)}.$$

The following two theorems express in a functional form the fact that the χ_x's are an orthogonal basis of the space $L(C_n)$.

Theorem 2.1.3 (Fourier inversion formula)

$$f = \frac{1}{n}\sum_{x \in C_n} \widehat{f}(x)\chi_x \qquad \forall f \in L(C_n).$$

Proof

$$\frac{1}{n} \sum_{x \in C_n} \widehat{f}(x) \chi_x(y_1) = \frac{1}{n} \sum_{x \in C_n} \sum_{y \in C_n} f(y) \overline{\chi_x(y)} \chi_x(y_1)$$

$$= \frac{1}{n} \sum_{y \in C_n} f(y) \sum_{x \in C_n} \overline{\chi_x(y)} \chi_x(y_1)$$

$$= \frac{1}{n} \sum_{y \in C_n} f(y) \delta_0(y - y_1) n = f(y_1).$$

$\square$

Theorem 2.1.4 (Plancherel's formula) *For $f \in L(C_n)$ we have* $\|\widehat{f}\| = \sqrt{n}\|f\|$.

Proof

$$\|\widehat{f}\|^2 = \langle \widehat{f}, \widehat{f} \rangle = \sum_{x \in C_n} \widehat{f}(x) \overline{\widehat{f}(x)}$$

$$= \sum_{x \in C_n} \left(\sum_{y_1 \in C_n} f(y_1) \overline{\chi_x(y_1)} \right) \left(\sum_{y_2 \in C_n} \overline{f(y_2)} \chi_x(y_2) \right)$$

$$= \sum_{y_1 \in C_n} \sum_{y_2 \in C_n} f(y_1) \overline{f(y_2)} \sum_{x \in C_n} \overline{\chi_x(y_1)} \chi_x(y_2)$$

$$= n \sum_{y \in C_n} f(y) \overline{f(y)} = n\|f\|^2.$$

$\square$

For $f_1, f_2 \in L(C_n)$ define their *convolution* as

$$(f_1 * f_2)(y) = \sum_{x \in C_n} f_1(y - x) f_2(x), \quad y \in C_n.$$

Definition 2.1.5 An *algebra* over a field $\mathbb{F}$ ia a vector space $\mathcal{A}$ over $\mathbb{F}$ endowed with a product such that: $\mathcal{A}$ is a ring with respect to the sum and the product and the following associative laws hold for the product and multiplication by a scalar:

$$\alpha(AB) = (\alpha A)B = A(\alpha B)$$

for all $\alpha \in \mathbb{F}$ and $A, B \in \mathcal{A}$ ([151, 197]).

An algebra $\mathcal{A}$ is *commutative* (or *abelian*) if it is commutative as a ring, namely if $AB = BA$ for all $A, B \in \mathcal{A}$; it is *unital* if it has a *unit*, that is there exists an element $I \in \mathcal{A}$ such that $AI = IA = A$ for all $A \in \mathcal{A}$.

The following proposition gives the main properties of the convolution

Proposition 2.1.6 *For all $f, f_1, f_2, f_3 \in L(C_n)$ one has*

   (i) $f_1 * f_2 = f_2 * f_1$ *(commutativity)*
   (ii) $(f_1 * f_2) * f_3 = f_1 * (f_2 * f_3)$ *(associativity)*
   (iii) $(f_1 + f_2) * f_3 = f_1 * f_3 + f_2 * f_3$ *(distributivity)*
   (iv) $\widehat{f_1 * f_2} = \widehat{f_1} \cdot \widehat{f_2}$
   (v) $\delta_0 * f = f * \delta_0 = f$.

In particular, $L(C_n)$ is a commutative algebra over $\mathbb{C}$ with unit $I = \delta_0$.

Proof We prove only (iv), namely that the Fourier transform of the convolution of two functions equals the pointwise product of their transforms.

$$\widehat{f_1 * f_2}(y) = \sum_{x \in C_n} (f_1 * f_2)(x)\overline{\chi_y(x)}$$

$$= \sum_{x \in C_n} \sum_{t \in C_n} f_1(x - t)f_2(t)\overline{\chi_y(x - t)}\overline{\chi_y(t)}$$

$$= \widehat{f_1}(y)\widehat{f_2}(y).$$

The other identities are left as an exercise. $\qquad\qquad\qquad\square$

Exercise 2.1.7 Show that $\widehat{\delta_x} = \overline{\chi_x}$ for all $x \in C_n$.

The translation operator $T_x : L(C_n) \to L(C_n)$, $x \in C_n$ is defined by:

$$(T_x f)(y) = f(y - x) \qquad \forall x, y \in C_n, \ f \in L(C_n).$$

Clearly $\widehat{T_x f}(y) = \overline{\chi_y(x)}\widehat{f}(y)$ and $T_x f = f * \delta_x$.

Suppose that R is the linear operator on $L(C_n)$ associated with the matrix $\{r(x, y)\}_{x,y \in C_n}$, that is $(Rf)(x) = \sum_{y \in C_n} r(x, y)f(y)$. We say that R is C_n-*invariant* if commutes with all translations, namely

$$RT_x = T_x R \qquad \forall x \in C_n.$$

Also we say that R is a *convolution operator* when there exists an $h \in L(C_n)$ such that $Rf = f * h \ \forall f \in L(C_n)$. The function h is called the *(convolution) kernel* of R.

Lemma 2.1.8 *The linear operator R associated with the matrix $\{r(x,y)\}_{x,y\in C_n}$ is C_n-invariant if and only if*

$$r(x-z, y-z) = r(x,y), \quad \forall x,y,z \in C_n. \tag{2.1}$$

Proof The linear operator R is C_n-invariant if and only if for all $v, z \in C_n$ and $f \in L(C_n)$ one has $[T_z(Rf)](v) = [R(T_z(f))](v)$, that is

$$\sum_{u\in C_n} r(v-z, u)f(u) = \sum_{u\in C_n} r(v, u)f(u-z)$$

which is equivalent to

$$\sum_{u\in C_n} r(v-z, u-z)f(u-z) = \sum_{u\in C_n} r(v, u)f(u-z).$$

$\square$

Theorem 2.1.9 *Let $R : L(C_n) \to L(C_n)$ be a linear operator. The following are equivalent:*

(i) *R is C_n-invariant*
(ii) *R is a convolution operator*
(iii) *Every χ_x is an eigenvector of R.*

Proof (i) $\Rightarrow$ (ii): In this case, by the above lemma, $r(x,y) = r(x-y, 0)$ so that if we put $h(x) = r(x, 0)$ we have

$$(Rf)(x) = \sum_{y\in C_n} h(x-y)f(y) = (h * f)(x)$$

and R is a convolution operator. The converse implication ((ii) $\Rightarrow$ (i)) is trivial since every convolution operator is clearly C_n−invariant. (ii) $\Rightarrow$ (iii): If R is a convolution operator, i.e. there exists $h \in L(C_n)$ such that

$$Rf = f * h \quad \forall f \in L(C_n)$$

then

$$R\chi_x(y) = \sum_{t\in C_n} \chi_x(y-t)h(t) = \chi_x(y) \sum_{t\in C_n} \overline{\chi_x(t)}h(t) = \widehat{h}(x)\chi_x(y)$$

showing that every χ_x is an eigenvector with eigenvalue $\widehat{h}(x)$.

(iii) $\Rightarrow$ (ii): Conversely, if R is a linear operator such that there exists $\lambda \in L(C_n)$ with

$$R\chi_x = \lambda(x)\chi_x, \quad \forall x \in C_n$$

i.e. every χ_x is an eigenvector, then, by the Fourier inversion theorem applied to f and by the linearity of R, we have

$$
\begin{aligned}
(Rf)(z) &= \frac{1}{n} \sum_{x \in C_n} \widehat{f}(x) \lambda(x) \chi_x(z) \\
&= \sum_{y \in C_n} f(y) \frac{1}{n} \sum_{x \in C_n} \lambda(x) \chi_x(z - y) \\
&= (h * f)(z)
\end{aligned}
$$

where $h(y) = \frac{1}{n} \sum_{x \in C_n} \lambda(x) \chi_x(y)$, showing that R is a convolution operator and ending the proof. $\square$

From the proof of the previous theorem we extract the following.

Corollary 2.1.10 Let R be a convolution operator with kernel $h \in L(C_n)$. Then $R\chi_x = \widehat{h}(x)\chi_x$. In particular, the spectrum of R is given by $\sigma(R) = \{\widehat{h}(x) : x \in C_n\}$.

Exercise 2.1.11 A matrix of the form

$$
A = \begin{pmatrix}
a_0 & a_1 & a_2 & \cdots & \cdots & a_{n-1} \\
a_{n-1} & a_0 & a_1 & \cdots & \cdots & a_{n-2} \\
a_{n-2} & a_{n-1} & a_0 & \cdots & \cdots & a_{n-3} \\
\vdots & \vdots & \vdots & \cdots & \cdots & \vdots \\
a_1 & a_2 & a_3 & \cdots & \cdots & a_0
\end{pmatrix}
$$

with $a_0, a_1, \ldots, a_{n-1} \in \mathbb{C}$ is said to be *circulant*. Denote by $\mathcal{C}_n$ the set of all $n \times n$ circulant matrices.

(1) Show that if $A, B \in \mathcal{C}_n$ then $AB = BA$ and $AB \in \mathcal{C}_n$. Since clearly $\alpha A + \beta B \in \mathcal{C}_n$ for all $\alpha, \beta \in \mathbb{C}$, one has that $\mathcal{C}_n$ is a commutative algebra with unit.

(2) For the space $L(C_n)$ take the basis $\mathcal{B} = \{\delta_0, \delta_1, \ldots, \delta_{n-1}\}$. Show that a linear operator $T : L(C_n) \to L(C_n)$ is a convolution operator $Tf = f * h$ if and only if in the basis $\mathcal{B}$ it is represented in the form

$$
\begin{pmatrix}
f(0) \\
f(1) \\
\vdots \\
f(n-1)
\end{pmatrix}
\longrightarrow
\begin{pmatrix}
h(0) & h(n-1) & \cdots & h(1) \\
h(1) & h(0) & \cdots & h(2) \\
\vdots & \vdots & \cdots & \vdots \\
h(n-1) & h(n-2) & \cdots & h(0)
\end{pmatrix}
\begin{pmatrix}
f(0) \\
f(1) \\
\vdots \\
f(n-1)
\end{pmatrix}
$$

that is, if and only if the matrix representing it is circulant. In particular, show that $\mathcal{C}_n$ is isomorphic to $L(C_n)$ as algebras.

(3) Set

$$F = \frac{1}{\sqrt{n}} \begin{pmatrix} 1 & 1 & \cdots & \cdots & 1 \\ 1 & \omega^{-1} & \omega^{-2} & \cdots & \omega^{-(n-1)} \\ 1 & \omega^{-2} & \omega^{-4} & \cdots & \omega^{-2(n-1)} \\ \vdots & \vdots & \vdots & \cdots & \vdots \\ 1 & \omega^{-(n-1)} & \omega^{-2(n-1)} & \cdots & \omega^{-(n-1)^2} \end{pmatrix}.$$

Observe that F is symmetric so that its adjoint F^* is equal to its conjugate $\overline{F}$. Show also that the orthogonality relations in Lemma 2.1.2 are equivalent to say that F is a unitary matrix.

(4) Prove that an $n \times n$ matrix belong to $\mathcal{C}_n$ if and only if FAF^* is diagonal. FAF^* is called the matrix form of the Fourier transform on C_n, or the *discrete Fourier transform* DFT.

A book entirely devoted to circulant matrices is that of Davis [50]. For more on the matrix form of the DFT, see [35]. For applications to number theory see the monographs of Terras [220] and Nathanson [169].

Lemma 2.1.12 *The spectrum of the convolution operator $Rf = f * h$, with h real valued, is real if and only if $h(-y) = h(y)$ $\forall y \in C_n$; if this is the case, we say that h is symmetric.*

Proof By Corollary 2.1.10, the spectrum of R is given by $\{\widehat{h}(x) : x \in C_n\}$ and it is real if and only if, $\forall x \in C_n$,

$$0 = \widehat{h}(x) - \overline{\widehat{h}(x)} = \sum_{y \in C_n} h(y)\overline{\chi_x(y)} - \sum_{y \in C_n} h(y)\chi_x(y)$$

$$= \sum_{y \in C_n} [h(y) - h(-y)]\overline{\chi_x(y)}$$

and this latter vanishes for all $x \in C_n$ if and only if $h(y) = h(-y)$ $\forall y \in C_n$. $\qquad\square$

We say that a Markov chain on C_n is *invariant* when its transition matrix $\{p(x, y)\}_{x,y \in C_n}$ is C_n-invariant: $p(x-z, y-z) = p(x, y)$ $\forall x, y, z \in C_n$. In this case we define a probability measure μ by setting $\mu(x) = p(0, x)$, $x \in C_n$. Then we have $p(x, y) = \mu(y - x)$ and, more generally,

$$p^{(k)}(x, y) = \mu^{*k}(y - x), \tag{2.2}$$

where μ^{*k} is the kth convolution power of μ, that is $\mu^{*k} = \mu * \mu * \cdots * \mu$ (k times). Indeed,

$$p^{(2)}(x,y) = \sum_{z \in C_n} p(x,z)p(z,y)$$

$$= \sum_{z \in C_n} \mu(z-x)\mu(y-z) = \mu^{*2}(y-x)$$

and iterating one obtains (2.2). Moreover,

$$\sum_{y \in X} p(x,y)f(y) = \sum_{y \in X} f(y)\mu(y-x) = \sum_{y \in X} f(y)\widetilde{\mu}(x-y) = (f * \widetilde{\mu})(x)$$

(2.3)

where $\widetilde{\mu}(y) = \mu(-y)$.

From Lemma 2.1.12 we obtain the following.

Lemma 2.1.13 *An invariant transition matrix $p(x,y)$ is reversible if and only if $\mu(x) = p(0,x)$ is symmetric. If this is the case then p is in detailed balance with respect to the uniform measure (or uniform distribution) $\pi(x) = \frac{1}{n}$.*

Proof If $p(x,y)$ is reversible then its eigenvalues are real (Remark 1.5.4) and by Lemma 2.1.12 $\widetilde{\mu}$ in (2.3) (and therefore μ) must be symmetric. Conversely, if μ is symmetric $p(x,y) = p(y,x)$ and therefore p is in detailed balance with respect to the uniform measure. $\square$

If μ is a symmetric probability measure, μ and $\widehat{\mu}$ are real valued (cf. Lemma 2.1.12) and in the formula of the Fourier transform and its inversion applied to μ all the contributes of the imaginary part of χ_x's disappear. Therefore

$$\widehat{\mu}(z) = \sum_{t \in C_n} \mu(t) \cos\left[\frac{2\pi t z}{n}\right]$$

and

$$p^{(k)}(x,y) = \mu^{*k}(x-y) = \frac{1}{n} \sum_{z \in C_n} (\widehat{\mu}(z))^k \cos\left[\frac{2\pi z(x-y)}{n}\right]$$

where in the last equality we have used the Fourier inversion formula and the fact $\widehat{\mu^{*k}}(z) = (\widehat{\mu}(z))^k$ which follows from Proposition 2.1.6 (iv).

Moreover, from the Plancherel formula, if $\nu_x^{(k)}(y) = p^{(k)}(x,y) = \mu^{*k}(x-y)$ and π still denotes the uniform measure, we have

$$
\begin{aligned}
\|\nu_x^{(k)} - \pi\|^2 &= \sum_{y\in C_n} |p^{(k)}(x,y) - \pi(y)|^2 \\
&= \sum_{z\in C_n} |\mu^{*k}(z) - \pi(z)|^2 \\
&= \frac{1}{n}\sum_{z\in C_n} |\widehat{\mu}(z)^k - \widehat{\pi}(z)|^2 \qquad (2.4) \\
&= \frac{1}{n}\sum_{z\in C_n: z\neq 0} \widehat{\mu}(z)^{2k}
\end{aligned}
$$

since $\widehat{\pi}(z) = \frac{1}{n}\langle \chi_0, \chi_z\rangle = \delta_0(z)$ and $\widehat{\mu}(0) = 1$. But $\|f\|_{TV} = \frac{1}{2}\|f\|_{L^1(C_n)} \leq \frac{\sqrt{n}}{2}\|f\|$ (by Cauchy–Schwarz) and therefore the upper bound lemma (Corollary 1.9.2) now becomes:

$$
\|\nu_x^{(k)} - \pi\|_{TV}^2 \equiv \|\mu^{*k} - \pi\|_{TV}^2 \leq \frac{1}{4}\sum_{z\in C_n: z\neq 0} \widehat{\mu}(z)^{2k}. \qquad (2.5)
$$

In the following exercise we ask the reader to analyze some examples of invariant Markov chains on C_n that are not necessarily reversible.

Exercise 2.1.14 (1) Let μ and π denote a probability measure and the uniform measure on C_n, respectively. Show that

$$
\|\mu^{*k} - \pi\|_{TV}^2 \leq \frac{1}{4}\sum_{z\in C_n: z\neq 0} |\widehat{\mu}(z)|^{2k}
$$

and that the random walk with transition matrix $p(x,y) = \mu(y-x)$ is ergodic if and only if 1 is the only eigenvalue of modulus 1 and it has multiplicity 1. Note that if $\widetilde{\mu}$ is as in (2.3), then $\widehat{\widetilde{\mu}}(x) = \overline{\widehat{\mu}(x)}$ for all $x \in C_n$. Moreover, the spectrum of the transition matrix $p(x,y)$ is given by $\{\widehat{\mu}(x) : x \in C_n\}$.

(2) Show that the spectrum of the random walk on C_n associated with the measure $\mu(1) = p$ and $\mu(-1) = 1-p$, with $0 < p < 1$ is given by $\{\cos\frac{2\pi k}{n} + i(2p-1)\sin\frac{2\pi k}{n} : k = 0, 1, \ldots, n-1\}$. Deduce from (1) that the random walk is ergodic if and only if n is odd.

(3) Also prove that the spectrum of the random walk on C_{3n} associated with the measure $\mu(1) = \frac{1}{2}$ and $\mu(-2) = \frac{1}{2}$ contains the three cubic roots of the unity and therefore is not ergodic. Show the nonergodicity of the walk, by observing that, setting $X_l = \{x \in C_{3n} : x \equiv_3 l\}$ for $l = 0, 1, 2$, we have $\mu^{*k}(x) = 0$ if $x \in X_l$ and $k \not\equiv_3 l$.

2.2 Time to reach stationarity for the simple random walk on the discrete circle

We now focus on the Markov chain described by the symmetric measure

$$\mu(-1) = \mu(1) = \frac{1}{2} \quad \text{and} \quad \mu(x) = 0 \quad \text{for } x \not\equiv_n \pm 1.$$

It is clear that μ describes the random walk in Example 1.8.2, the simple random walk on the discrete circle:

$$p(x, y) = \mu(y - x) = \begin{cases} 1/2 & \text{if } y \equiv_n x + 1 \\ 1/2 & \text{if } y \equiv_n x - 1 \\ 0 & \text{otherwise.} \end{cases}$$

The random walk is ergodic if and only if n is odd. From Theorem 2.1.9 we get that its spectrum is equal to the set of numbers

$$\widehat{\mu}(x) = \frac{1}{2}(e^{\frac{2\pi i}{n}x} + e^{\frac{-2\pi i}{n}x}) = \cos\frac{2\pi x}{n} \tag{2.6}$$

with $x = 0, 1, \ldots, n - 1$.

We now give an upper and lower bound for $\|\mu^{*k} - \pi\|_{TV}$. See [55] and [58].

Theorem 2.2.1 *For n odd and $k \geq n^2$ we have*

$$\|\mu^{*k} - \pi\|_{TV} \leq e^{\frac{-\pi^2 k}{2n^2}}.$$

For $n \geq 6$ and any k we have

$$\|\mu^{*k} - \pi\|_{TV} \geq \frac{1}{2}e^{-\pi^2 k/2n^2 - \pi^4 k/2n^4}.$$

Proof First of all note that if we set $h(x) = \log(e^{x^2/2}\cos x)$ then $h'(x) = x - \tan x \leq 0$, $\forall x \in [0, \pi/2)$ and therefore $h(x) \leq h(0) = 0$, $\forall x \in [0, \pi/2)$, that is we have

$$\cos x \leq e^{-x^2/2} \quad \forall x \in [0, \pi/2]. \tag{2.7}$$

From the upper bound lemma (2.5) and from (2.6) we deduce that

$$\|\mu^{*k} - \pi\|_{TV}^2 \leq \frac{1}{4}\sum_{x=1}^{n-1}\left[\cos\frac{2\pi x}{n}\right]^{2k} = \frac{1}{2}\sum_{x=1}^{\frac{n-1}{2}}\left[\cos\frac{2\pi x}{n}\right]^{2k} \tag{2.8}$$

where the last equality follows from the trigonometric equality $\cos\frac{2\pi x}{n} = \cos\frac{2\pi(n-x)}{n}$ (we also recall that n is odd). Moreover,

$$\sum_{x=1}^{\frac{n-1}{2}}\left[\cos\frac{2\pi x}{n}\right]^{2k} = \sum_{x=1}^{\frac{n-1}{2}}\left[\cos\frac{\pi x}{n}\right]^{2k}. \qquad (2.9)$$

Indeed,

$$\sum_{x=1}^{\frac{n-1}{2}}\left[\cos\frac{2\pi x}{n}\right]^{2k} = \sum_{\substack{y=2\\y\text{ even}}}^{n-1}\left[\cos\frac{\pi y}{n}\right]^{2k}$$

$$= \sum_{\substack{y=2\\y\text{ even}}}^{\frac{n-1}{2}}\left[\cos\frac{\pi y}{n}\right]^{2k} + \sum_{\substack{y=\frac{n+1}{2}\\y\text{ even}}}^{n-1}\left[\cos\frac{\pi y}{n}\right]^{2k}$$

and for $\frac{n+1}{2} \leq y \leq n-1$ we can use the identity $\cos\frac{\pi y}{n} = -\cos\frac{\pi(n-y)}{n}$. From (2.8) and (2.9) we get

$$\|\mu^{*k} - \pi\|_{TV}^2 \leq \frac{1}{2}\sum_{x=1}^{\frac{n-1}{2}}\left[\cos\frac{\pi x}{n}\right]^{2k}$$

$$\leq \frac{1}{2}\sum_{x=1}^{\frac{n-1}{2}}e^{-\frac{\pi^2 x^2 k}{n^2}} \quad \text{from}(2.7)$$

$$\leq \frac{1}{2}e^{-\frac{\pi^2 k}{n^2}}\sum_{x=1}^{\infty}e^{-\frac{\pi^2(x^2-1)k}{n^2}}$$

$$\leq \frac{1}{2}e^{-\frac{\pi^2 k}{n^2}}\sum_{x=1}^{\infty}e^{-\frac{3\pi^2(x-1)k}{n^2}}$$

since $x^2 - 1 \geq 3(x-1)$, $x = 1, 2, 3, \dots$

$$= \frac{\frac{1}{2}e^{-\frac{\pi^2 k}{n^2}}}{1 - e^{-\frac{3\pi^2 k}{n^2}}} \quad \text{by the sum of a geometric series.}$$

If $k \geq n^2$ then

$$\frac{1}{2\left(1 - e^{-\frac{3\pi^2 k}{n^2}}\right)} \leq \frac{1}{2(1 - e^{-3\pi^2})} \leq 1$$

and, by taking the square root, this establishes the upper bound.

In order to prove the lower bound we first observe that the sum in (2.8) is dominated by the term for $x = \frac{n-1}{2}$, that is

$$\left| \cos \frac{2\pi x}{n} \right| \leq \left| \cos 2\pi \frac{\frac{n-1}{2}}{n} \right| = \left| -\cos \frac{\pi}{n} \right| = \cos \frac{\pi}{n}$$

for $x = 1, 2, \ldots, n-1$. Setting $\phi(y) = \cos \frac{\pi(n-1)y}{n} = \Re \left[\chi_{\frac{n-1}{2}}(y) \right]$, from the fact that $\widehat{\mu}$ is real valued, we get

$$\sum_{y=0}^{n-1} \mu^{*k}(y)\phi(y) = \left[\widehat{\mu}\left(\frac{n-1}{2} \right) \right]^k = (-1)^k \left(\cos \frac{\pi}{n} \right)^k.$$

Since $|\phi(y)| \leq 1$, $\forall y \in C_n$ and

$$\sum_{y=0}^{n-1} \pi(y)\phi(y) = \sum_{y=0}^{n-1} \frac{1}{n}\phi(y) = \frac{1}{n}\Re\left[\sum_{y=0}^{n-1} \chi_0(y)\overline{\chi_{\frac{n-1}{2}}(y)} \right] = 0.$$

from (1.27) it follows that

$$2\|\mu^{*k} - \pi\|_{TV} \geq \left| \cos \frac{\pi}{n} \right|^k. \tag{2.10}$$

With $h(x) = \log\left[e^{\frac{x^2}{2} + \frac{x^4}{2}} \cos x \right]$, showing that $h(0) = h'(0) = h''(0) = h'''(0) = 0$ and $h^{iv}(x) > 0$ for $0 \leq x \leq \frac{\pi}{6}$, one deduces that $\cos x \geq e^{-\frac{x^2}{2} - \frac{x^4}{2}}$ for $0 \leq x \leq \frac{\pi}{6}$ and therefore (2.10) gives

$$\|\mu^{*k} - \pi\|_{TV} \geq \frac{1}{2} e^{-\frac{\pi^2 k}{2n^2} - \frac{k\pi^4}{2n^4}}$$

for any k and $n \geq 6$. $\qquad\qquad\qquad\qquad\qquad\qquad\qquad\qquad$ $\square$

From Exercise 1.4.3 one could get the estimate

$$\|\mu^{*k} - \pi\|_{TV} \leq 2(2m+1)e^{-\frac{k}{m2^{2m+1}}},$$

which is clearly weaker than the upper estimate in the above theorem (with $n = 2m+1$). This shows the power of the upper bound lemma and of finite harmonic analysis: a complete knowledge of the spectrum leads us to a sharp result for the convergence to the stationary distribution.

Remark 2.2.2 Actually, it is easy to see (exercise) that the upper bound in Theorem 2.2.1 is valid for $k \geq \frac{n^2}{3\pi^2}$. See also Section 2.5.

Remark 2.2.3 We observe that for n even the random walk is not ergodic.

2.3 Harmonic analysis on the hypercube

The n-dimensional hypercube Q_n (see also Example 1.8.3) is the cartesian product of n copies of the cyclic group C_2, that is $Q_n = \{(x_1, x_2, \ldots, x_n) : x_1, x_2, \ldots x_n \in \{0, 1\}\}$, equipped with the commutative product

$$(x_1, x_2, \ldots, x_n) + (y_1, y_2, \ldots, y_n) = (x_1 + y_1, x_2 + y_2, \ldots, x_n + y_n) \mod 2.$$

Note that if $x \in Q_n$ then $x + x = 0$, that is $x = -x$.

We introduce the *characters* of Q_n by setting, for $x, y \in Q_n$,

$$\chi_x(y) = (-1)^{x_1 y_1 + x_2 y_2 + \cdots + x_n y_n}.$$

We also write $\chi_x(y) = (-1)^{x \cdot y}$, where $x \cdot y = x_1 y_1 + x_2 y_2 + \cdots + x_n y_n$. Note that the characters are real valued. Clearly $\chi_z(x + y) = \chi_z(x) \chi_z(y)$. Conversely we have

Exercise 2.3.1 If $\phi : Q_n \to \mathbb{T} = \{z \in \mathbb{C} : |z| = 1\}$ satisfies $\phi(x + y) = \phi(x)\phi(y)$ then $\phi(x) = \chi_z(x)$ for some $z \in Q_n$.

Lemma 2.3.2

$$\sum_{y \in Q_n} \chi_x(y) = \begin{cases} 2^n & \text{if } x = 0 \\ 0 & \text{otherwise.} \end{cases}$$

Proof If $x = 0$ then $\chi_x(y) = 1$ for all $y \in Q_n$ and, since $|Q_n| = 2^n$, the first equality is established.

If $x \neq 0$ there exists $z \in Q_n$ such that $\chi_x(z) = -1$. Therefore

$$-\sum_{y \in Q_n} \chi_x(y) = \chi_x(z) \sum_{y \in Q_n} \chi_x(y) = \sum_{y \in Q_n} \chi_x(y + z) = \sum_{y \in Q_n} \chi_x(y)$$

which gives the second equality. $\qquad\square$

Corollary 2.3.3 (Orthogonality relations)

$$\sum_{x \in Q_n} \chi_y(x) \chi_z(x) = 2^n \delta_y(z) = \sum_{x \in Q_n} \chi_x(y) \chi_x(z).$$

Proof The first equality follows from the previous lemma. Indeed:

$$\sum_{x \in Q_n} \chi_y(x)\chi_z(x) = \sum_{x \in Q_n} \chi_{y+z}(x).$$

The second equality from $\chi_x(y) = \chi_y(x)$. □

Let $L(Q_n) = \{f : Q_n \to \mathbb{C}\}$ be the space of complex valued functions on Q_n. The *Fourier transform* of a function $f \in L(Q_n)$ is defined, for $x \in Q_n$, by:

$$\widehat{f}(x) = \langle f, \chi_x \rangle = \sum_{y \in Q_n} f(y)\chi_x(y).$$

As in Section 2.1, the following theorems express in a functional form the fact that the χ_x's are an orthogonal basis of the space $L(Q_n)$.

Theorem 2.3.4 (Fourier inversion formula) *For $f \in L(Q_n)$ we have*

$$f = \frac{1}{2^n} \sum_{x \in Q_n} \widehat{f}(x)\chi_x.$$

Proof

$$\frac{1}{2^n} \sum_{x \in Q_n} \widehat{f}(x)\chi_x(z) = \frac{1}{2^n} \sum_{y \in Q_n} f(y) \sum_{x \in Q_n} \chi_x(y+z)$$
$$= f(z).$$

□

Theorem 2.3.5 (Plancherel's formula)

$$\|\widehat{f}\| = \sqrt{2^n}\|f\|.$$

Proof

$$\|\widehat{f}\|^2 = \langle \widehat{f}, \widehat{f} \rangle = \sum_{y,z \in Q_n} f(y)\overline{f(z)} \sum_{x \in Q_n} \chi_x(y)\chi_x(z)$$
$$= 2^n\|f\|^2.$$

□

For $f_1, f_2 \in L(Q_n)$ define their *convolution* as

$$(f_1 * f_2)(y) = \sum_{x \in Q_n} f_1(y-x)f_2(x).$$

The convolution is commutative, associative, distributive. Moreover it is easy to check that $\widehat{f_1 * f_2} = \widehat{f_1} \cdot \widehat{f_2}$

The translation operator is defined by:

$$(T_x f)(y) = f(y - x) \quad \forall x, y \in Q_n, f \in L(Q_n).$$

Clearly $\widehat{T_x f}(y) = \chi_y(x) \widehat{f}(y)$ and $T_x f = f * \delta_x$.

We say that a linear operator $R : L(Q_n) \to L(Q_n)$ is Q_n-*invariant* when

$$RT_x = T_x R, \quad \forall x \in Q_n.$$

The proof of the following theorem is similar to that one of Theorem 2.1.9 (for the cyclic group C_n) and it is left as an exercise.

Theorem 2.3.6 *Let* $R : L(Q_n) \to L(Q_n)$ *be a linear operator. The following are equivalent:*

(i) R *is* Q_n-*invariant*

(ii) R *is a convolution operator*

(iii) *Every* χ_x *is an eigenvector of* R.

Moreover if R *is the convolution operator* $Rf = f * h$ *then* $R\chi_x = \widehat{h}(x)\chi_x$.

Now suppose that μ is a probability measure on Q_n. From the Fourier inversion formula (Theorem 2.3.4) we have

$$\widehat{\mu}(z) = \sum_{t \in Q_n} \mu(t)(-1)^{t \cdot z}$$

and

$$\mu^{*k}(y - x) = \frac{1}{2^n} \sum_{z \in Q_n} (\widehat{\mu}(z))^k (-1)^{z \cdot (y-x)}.$$

An invariant Markov chain on Q_n is given by a stochastic matrix $\{p(x, y)\}_{x,y \in Q_n}$ which is Q_n-invariant, namely

$$p(x + z, y + z) = p(x, y) \quad \forall x, y, z \in Q_n.$$

In this case, if we set $\mu(x) = p(0, x)$ for all $x \in Q_n$ then μ is a probability measure on Q_n and we have $p(x, y) = \mu(y - x)$. Therefore, as in formula (2.2),

$$p^{(k)}(x, y) = \mu^{*k}(y - x). \tag{2.11}$$

Note that in this case Lemma 2.1.12 is replaced by the simple remark that any real valued $h : Q_n \to \mathbb{R}$ has a real valued Fourier transform, since every character is real valued (alternatively every $h \in L(Q_n)$ is symmetric : $h(x) = h(-x)$ since $x = -x$) and Lemma 2.1.13 is replaced by the remark that every invariant $p(x, y)$ is symmetric and therefore reversible and in detailed balance with respect to the uniform measure $\pi(x) = \frac{1}{2^n}$.

Moreover, from the Plancherel formula, taking into account that $\widehat{\pi}(z) = \delta_0(z)$ and that $\widehat{\mu}(0) = 1$, and reasoning as in (2.4) we get

$$\|\nu_x^{(k)} - \pi\|^2 = \frac{1}{2^n} \sum_{\substack{z \in Q_n \\ z \neq 0}} \widehat{\mu}(z)^{2k}$$

and therefore the upper bound lemma now becomes

$$\|\nu_x^{(k)} - \pi\|_{TV}^2 = \|\mu^{*k} - \pi\|_{TV}^2 \leq \frac{1}{4} \sum_{\substack{z \in Q_n \\ z \neq 0}} [\widehat{\mu}(z)]^{2k}. \tag{2.12}$$

2.4 Time to reach stationarity in the Ehrenfest diffusion model

Now we consider the measure μ defined by setting

$$\mu(0, 0, \ldots, 0) = \mu(1, 0, \ldots, 0) = \cdots = \mu(0, 0, \ldots, 0, 1) = \frac{1}{n+1} \tag{2.13}$$

and vanishing on others points of Q_n. The corresponding transition matrix coincides with the one in Example 1.8.3

$$p(x, y) = \mu(x - y) = \begin{cases} \frac{1}{n+1} & \text{if } x = y \\ \frac{1}{n+1} & \text{if } x \sim y \\ 0 & \text{otherwise.} \end{cases}$$

For $x = (x_1, x_2, \ldots, x_n) \in Q_n$ we set $w(x) = |\{j : x_j = 1\}| =$ number of ones in x. $w(x)$ is called the *weight* of x.

Proposition 2.4.1 *If μ is as in (2.13) then*

$$\widehat{\mu}(x) = 1 - \frac{2w(x)}{n+1}.$$

Proof

$$\widehat{\mu}(x) = \sum_{y \in Q_n} \mu(y)(-1)^{x \cdot y} = \frac{1}{n+1} + \frac{1}{n+1}\sum_{j=1}^{n}(-1)^{x_j}$$

$$= \frac{1}{n+1} - \frac{w(x)}{n+1} + \frac{n - w(x)}{n+1}$$

$$= 1 - \frac{2w(x)}{n+1}.$$

$\square$

Let V_k be the subspace of $L(Q_n)$ spanned by the χ_x's with $w(x) = k$:

$$V_k = \langle \chi_x : w(x) = k \rangle.$$

Then from Proposition 2.4.1 and Theorem 2.3.6 we get

Corollary 2.4.2 *Each V_k is an eigenspace of the convolution operator $Tf = f * \mu$ and the corresponding eigenvalue is $1 - \frac{2k}{n+1}$. Moreover $\dim V_k = \binom{n}{k}$ and $L(Q_n) = \oplus_{k=0}^{n} V_k$ is the orthogonal decomposition of $L(Q_n)$ into eigenspaces of T. In particular, $1 - \frac{2}{n+1}$ is the second largest eigenvalue and its multiplicity is equal to n.*

We now study the convergence to the stationary measure of the convolution powers μ^{*k}.

The following theorem is taken from [55] and [58]. We have also used a simplification of the proof of the lower bound contained in [185].

Theorem 2.4.3 *Let μ be as in (2.13). For $k = \frac{1}{4}(n+1)(\log n + c)$ we have*

$$\|\mu^{*k} - \pi\|_{TV}^2 \leq \frac{1}{2}(e^{e^{-c}} - 1).$$

Moreover, for $k = \frac{1}{4}(n+1)(\log n - c)$, $0 < c < \log n$ and n large we have

$$\|\mu^{*k} - \pi\|_{TV} \geq 1 - 20e^{-c}.$$

Proof From (2.12), Proposition 2.4.1 and its corollary we get

$$\|\mu^{*k} - \pi\|_{TV}^2 \leq \frac{1}{4} \sum_{\substack{z \in Q_n \\ z \neq 0}} [\hat{\mu}(z)]^{2k}$$

$$= \frac{1}{4} \sum_{j=1}^{n} \dim V_j \left(1 - \frac{2j}{n+1}\right)^{2k}$$

$$= \frac{1}{4} \sum_{j=1}^{n} \binom{n}{j} \left(1 - \frac{2j}{n+1}\right)^{2k}$$

$$\leq \frac{1}{2} \sum_{j=1}^{[\frac{n+1}{2}]} \binom{n}{j} \left(1 - \frac{2j}{n+1}\right)^{2k}$$

$$\text{since } \left(1 - \frac{2j}{n+1}\right)^{2k} = \left(1 - \frac{2j'}{n+1}\right)^{2k} \text{ if } j + j' = n+1$$

$$\leq \frac{1}{2} \sum_{j=1}^{[\frac{n+1}{2}]} \frac{n^j}{j!} e^{-\frac{4kj}{n+1}} \quad (*)$$

since $\binom{n}{j} = \frac{n!}{j!(n-j)!} \leq \frac{n^j}{j!}$ and $1 - x \leq e^{-x}$ yields $\left(1 - \frac{2j}{n+1}\right)^{2k} \leq e^{-\frac{4kj}{n+1}}$, when $j \leq \frac{n+1}{2}$ (if j satisfies this condition, then $1 - \frac{2j}{n+1} > 0$).

Now, if $k = \frac{1}{4}(n+1)(\log n + c)$, $(*)$ becomes

$$\|\mu^{*k} - \pi\|_{TV}^2 \leq \frac{1}{2} \sum_{j=1}^{[\frac{n+1}{2}]} \frac{n^j}{j!} e^{-j \log n - jc}$$

$$= \frac{1}{2} \sum_{j=1}^{[\frac{n+1}{2}]} \frac{1}{j!} e^{-jc}$$

$$\leq \frac{1}{2} \sum_{j=1}^{\infty} \frac{1}{j!} \left(e^{-c}\right)^j$$

$$= \frac{1}{2} \left(e^{e^{-c}} - 1\right).$$

To get the lower bound, consider the function

$$\phi(x) = \sum_{j=1}^{n} (-1)^{x_j} = n - 2w(x) = \sum_{y \in A} \chi_y(x)$$

where $A = \{y \in Q_n : w(y) = 1\}$. The mean of ϕ with respect to π (see Section 1.9) is given by

$$E_\pi(\phi) = \frac{1}{2^n} \sum_{y \in Q_n} \phi(y) = \frac{1}{2^n} \sum_{x \in A} \sum_{y \in Q_n} \chi_x(y) = 0 \qquad (2.14)$$

by the orthogonality relation for the characters. For the variance we have:

$$Var_\pi(\phi) = E_\pi(\phi^2) - E_\pi(\phi)^2 = E_\pi(\phi^2)$$

$$= \frac{1}{2^n} \sum_{x_1, x_2 \in A} \sum_{y \in Q_n} \chi_{x_1}(y) \chi_{x_2}(y) = n \qquad (2.15)$$

again for the orthogonality relations for the characters.

Moreover

$$E_{\mu^{*k}}(\phi) = \sum_{x \in A} \sum_{y \in Q_n} \mu^{*k}(y) \chi_x(y) = \sum_{x \in A} [\widehat{\mu}(x)]^k = n \left(1 - \frac{2}{n+1}\right)^k \qquad (2.16)$$

since $\widehat{\mu}(x) = 1 - \frac{2}{n+1}$ for all $x \in A$.

Setting $B = \{x \in Q_n : w(x) = 2\}$ we find

$$E_{\mu^{*k}}(\phi^2) = \sum_{x_1, x_2 \in A} \sum_{y \in Q_n} \chi_{x_1}(y) \chi_{x_2}(y) \mu^{*k}(y)$$

$$= \sum_{x_1, x_2 \in A} \sum_{y \in Q_n} \chi_{x_1 + x_2}(y) \mu^{*k}(y)$$

$$= n [\widehat{\mu}(0)]^k + 2 \sum_{x \in B} [\widehat{\mu}(x)]^k$$

since $x_1 = x_2 \Rightarrow \chi_{x_1 + x_2} = \chi_0$ and $\sum_{\substack{x_1, x_2 \in A \\ x_1 \neq x_2}} \chi_{x_1 + x_2} = 2 \sum_{x \in B} \chi_x$.

But $\widehat{\mu}(0) = 1$, $\widehat{\mu}(x) = 1 - \frac{4}{n+1}$ for all $x \in B$ and $|B| = \frac{n(n-1)}{2}$. Therefore

$$Var_{\mu^{*k}}(\phi) = E_{\mu^{*k}}(\phi^2) - E_{\mu^{*k}}(\phi)^2$$

$$= n + n(n-1) \left(1 - \frac{4}{n+1}\right)^k - n^2 \left(1 - \frac{2}{n+1}\right)^{2k}$$

$$\leq n. \qquad (2.17)$$

Now define $\mathbf{A}_\beta = \{x \in Q_n : |\phi(x)| < \beta\sqrt{n}\}$, where β is a constant $0 < \beta < \frac{1}{\sqrt{n}} E_{\mu^{*k}}(\phi)$ that will be suitably chosen later. From Markov's inequality (Proposition 1.9.5) it follows that

$$
\begin{aligned}
\pi(\mathbf{A}_\beta) &= 1 - \pi\{x : |\phi(x)| \geq \beta\sqrt{n}\} \\
&= 1 - \pi\{x : \phi(x)^2 \geq \beta^2 n\} \\
&\geq 1 - \frac{1}{\beta^2 n} E_\pi(\phi^2) = 1 - \frac{1}{\beta^2}
\end{aligned}
\tag{2.18}
$$

where the last equality follows from (2.15). In the same way, from Chebyshev's inequality (Corollary 1.9.6) and the fact that $\mathbf{A}_\beta \subseteq \{x \in Q_n : |\phi(x) - E_{\mu^{*k}}(\phi)| \geq E_{\mu^{*k}}(\phi) - \beta\sqrt{n}\}$, we have

$$
\mu^{*k}(\mathbf{A}_\beta) \leq \frac{Var_{\mu^{*k}}(\phi)}{(E_{\mu^{*k}}(\phi) - \beta\sqrt{n})^2}.
\tag{2.19}
$$

Set $k = \frac{1}{4}(n+1)(\log n - c)$, $0 < c < \log n$. From the Taylor expansion of the logarithm it follows that

$$
\log(1 - x) = -x - \frac{x^2}{2}\omega(x)
\tag{2.20}
$$

with $\omega(x) \geq 0$ and $\lim_{x\to 0}\omega(x) = 1$. Then applying (2.20) to the right hand side of (2.16) we get

$$
\begin{aligned}
E_{\mu^{*k}}(\phi) &= n\exp\left[\log\left(1 - \frac{2}{n+1}\right) \cdot \frac{1}{4}(n+1)(\log n - c)\right] \\
&= n\exp\left[\left(-\frac{2}{n+1} - \frac{2}{(n+1)^2}\omega\left(\frac{2}{n+1}\right)\right)\frac{1}{4}(n+1)(\log n - c)\right] \\
&= \sqrt{n}e^{c/2}\exp\left[\frac{c - \log n}{2(n+1)}\,\omega\left(\frac{2}{n+1}\right)\right]
\end{aligned}
$$

and therefore for n large we have

$$
E_{\mu^{*k}}(\phi) \geq \frac{3}{4}\sqrt{n}e^{c/2}
\tag{2.21}
$$

Choosing $\beta = \frac{e^{c/2}}{2}$ and taking into account (2.4) and (2.21), we have that (2.19) becomes

$$
\mu^{*k}(\mathbf{A}_\beta) \leq \frac{n}{(\frac{3}{2}\beta\sqrt{n} - \beta\sqrt{n})^2} = \frac{4}{\beta^2}.
\tag{2.22}
$$

Therefore from (2.18) and (2.22)

$$
\|\mu^{*k} - \pi\|_{TV} \geq \pi(\mathbf{A}_\beta) - \mu^{*k}(\mathbf{A}_\beta) = 1 - \frac{5}{\beta^2} = 1 - \frac{20}{e^c}.
$$

$\square$

Remark 2.4.4 The set $\mathbf{A}_\beta$ in the above proof is a slice of the cube formed by the vertices $x \in Q_n$ with $w(x)$ sufficiently close to $\frac{n}{2}$. Since the random walk starts at zero, if it does not run enough, the probability of being in $\mathbf{A}_\beta$ is close to zero (2.22), but $\mathbf{A}_\beta$ is a consistent part of Q_n; see (2.16).

The analysis of a more general class of random walks on the hypercube is presented in [46].

2.5 The cutoff phenomenon

In the last section, we have proved that the variation distance of μ^{*k} and π for the Ehrenfest model has the following asymptotic behavior.

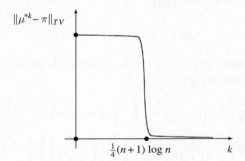

Figure 2.1. The cutoff phenomenon

In other words the variation distance remains close to 1 for a long time, then it drops down to a small value quite suddenly and decreases exponentially fast. This is a quite surprising phenomenon discovered for the first time by Diaconis and Shahshahani [69]. Now we give a precise definition (see [58] and [68]).

Definition 2.5.1 Let (X_n, ν_n, p_n) be a sequence of Markov chains, that is, for all $n \in \mathbb{N}$, X_n is a finite set, ν_n a probability measure on X_n and p_n an ergodic transition probability kernel on X_n. Also denote by π_n the stationary measure of p_n and by $\nu_n^{(k)}$ the distribution of (X_n, ν_n, p_n) after k steps. Suppose that $(a_n)_{n \in \mathbb{N}}$, $(b_n)_{n \in \mathbb{N}}$ are two sequences of positive real numbers such that

$$\lim_{n \to \infty} \frac{b_n}{a_n} = 0.$$

We say that the sequence of Markov chains (X_n, ν_n, p_n) has an (a_n, b_n)-*cutoff* if there exist two functions f_1 and f_2 defined on $[0, +\infty)$ with

- $\lim_{c \to +\infty} f_1(c) = 0$
- $\lim_{c \to +\infty} f_2(c) = 1$

such that, for any fixed $c > 0$, one has

$$\|\nu_n^{(a_n + cb_n)} - \pi_n\|_{TV} \leq f_1(c)$$

and

$$\|\nu_n^{(a_n - cb_n)} - \pi_n\|_{TV} \geq f_2(c),$$

provided that, in both cases, n is sufficiently large.

Clearly, if $a_n + cb_n$ is not an integer, one has to take k as the smallest integer greater than $a_n + cb_n$ and similarly for $a_n - cb_n$.

In other words, when there is a cutoff, the transition of $\|\nu_n^{(k)} - \pi_n\|_{TV}$ from 1 to 0 (that is, from *order* to *chaos*) occurs in a short time: in an interval of length b_n at a scale of length a_n and $\frac{b_n}{a_n} \to 0$ as n tends to $+\infty$.

The cutoff phenomenon occurs in several examples of Markov chains. In general, it can be detected thanks to a careful eigenvalue–eigenvector analysis.

The following lemma is simple but it is important as it clarifies the fact that the function $\|\nu^{(k)} - \pi\|_{TV}$ tends to zero monotonically.

Lemma 2.5.2 *Suppose that p is an ergodic transition kernel on X, with stationary distribution π. Let ν be an initial distribution on X and $\nu^{(k)}(y) = \sum_{x \in X} \nu(x) p^{(k)}(x, y)$ the distribution after n steps. Then*

$$\|\nu^{(k+1)} - \pi\|_{TV} \leq \|\nu^{(k)} - \pi\|_{TV} \quad \forall k \geq 0.$$

Proof From (1.26) we have

$$\|\nu^{(k+1)} - \pi\|_{TV} = \frac{1}{2} \sum_{z \in X} |\nu^{(k+1)}(z) - \pi(z)|$$

$$\text{(using } \pi(z) = \sum_{y \in X} \pi(y) p(y, z)) \quad = \frac{1}{2} \sum_{z \in X} |\sum_{y \in X} \left[\nu^{(k)}(y) - \pi(y) \right] p(y, z)|$$

$$\leq \frac{1}{2} \sum_{z \in X} \sum_{y \in X} |\nu^{(k)}(y) - \pi(y)| p(y, z)$$

$$\text{(using } \sum_{z \in X} p(y, z) = 1) \qquad = \frac{1}{2} \sum_{y \in X} |\nu^{(k)}(y) - \pi(y)|$$

$$= \|\nu^{(k)} - \pi\|_{TV}.$$

□

The following proposition presents a simple necessary condition for the presence of the cutoff phenomenon.

Proposition 2.5.3 *If (X_n, ν_n, p_n) has an (a_n, b_n)-cutoff as in the previous definition, then for every choice of $0 < \epsilon_1 < \epsilon_2 < 1$ there exist $k_2^{(n)} \leq k_1^{(n)}$ such that*

- $k_2^{(n)} \leq a_n \leq k_1^{(n)}$
- *for n large, $k \geq k_1^{(n)} \Rightarrow \|\nu_n^{(k)} - \pi_n\|_{TV} \leq \epsilon_1$*
- *for n large, $k \leq k_2^{(n)} \Rightarrow \|\nu_n^{(k)} - \pi_n\|_{TV} \geq \epsilon_2$*
- $\lim_{n \to \infty} \frac{k_1^{(n)} - k_2^{(n)}}{a_n} = 0$.

Proof It is a straightforward exercise. Indeed there exist c_1 and c_2 such that $f_2(c) \geq \epsilon_2$ for $c \geq c_2$ and $f_1(c) \leq \epsilon_1$ for $c \geq c_1$. Setting $k_2^{(n)} = a_n - c_2 b_n$ and $k_1^{(n)} = a_n + c_1 b_n$ the proposition follows from the definition of cutoff.   □

The simple random walk on the circle C_n does not have any cutoff. Indeed, in the notation of Theorem 2.2.1, we have for $k_2 \geq \frac{n^2}{2}$ (see Remark 2.2.2)

$$\|\mu^{*k_2} - \pi\|_{TV} \leq e^{-\frac{\pi^2}{4}} =: \epsilon_2$$

and for $k_1 \leq \frac{3n^2}{2}$

$$\|\mu^{*k_1} - \pi\|_{TV} \geq \frac{1}{2} e^{-\frac{3\pi^2}{4} - \frac{3\pi^4}{4n^2}} =: \epsilon_1$$

and since $e^{-\frac{\pi^2}{4}} > \frac{1}{2} e^{-\frac{3\pi^2}{4} - \frac{3\pi^4}{4n^2}}$ the transition from ϵ_2 to ϵ_1 is gradual from $\frac{n^2}{2}$ to $\frac{3n^2}{2}$.

The Eherenfest model has a cutoff with $a_n = \frac{1}{4}(n+1) \log n$, $b_n = \frac{1}{4}(n+1)$, $f_1(c) = \frac{1}{2} \left(e^{e^{-c}} - 1 \right)$ and $f_2(c) = 1 - 20 e^{-c}$.

Exercise 2.5.4 ([68]) Show that the simple random walk on the complete graph K_n has a $(1, \epsilon_n)$-cutoff, for any sequence $(\epsilon_n)_{n \in \mathbb{N}}$ such that $\lim_{n \to +\infty} \epsilon_n = 0$.

It is still an open problem to determine what makes a Markov chain cutoff. In general, it depends on the size and the multiplicity of the second eigenvalue of the transition matrix. A clear and comprehensive discussion is in [58]. A recent conjecture is analyzed in [68].

2.6 Radial harmonic analysis on the circle and the hypercube

In this section, we apply the theory of lumpable Markov chains to the random walk on the discrete circle (Example 1.8.2 and Section 2.2) and to the random walk on the hypercube (Example 1.8.3 and Section 2.4). See Sections 5.2 and 5.3 for another approach to the topics presented in this section.

Consider the simple random walk on C_{2n}, the set of integers mod $2n$. Let P be its transition probability. It is easy to check that P is lumpable with respect to the following partition: $C_{2n} = \coprod_{k=0}^{n} \Omega_k$, where $\Omega_0 = \{0\}$, $\Omega_n = \{n\}$ and $\Omega_k = \{n-k, n+k\}$ for $k = 1, 2, \ldots, n-1$. The lumped transition matrix Q is indexed by $0, 1, \ldots, n$ (k corresponds to Ω_k) and is given by:

$$q(0,1) = q(n, n-1) = 1$$

$$q(k, k+1) = q(k, k-1) = \frac{1}{2}, \quad k = 1, 2, \ldots, n-1.$$

For instance, we indeed have $q(0,1) = p(0, \Omega_1) = p(0,1) + p(0, 2n-1) = 1$. This matrix Q is the transition matrix of the simple random walk on the path P_{n+1}, that is the graph with vertex set $\{0, 1, \ldots, n\}$ and edges $\{\{k, k+1\} : k = 0, 1, \ldots, n-1\}$. It is reversible with respect to the measure π defined by

$$\pi(0) = \pi(n) = \frac{1}{2n} \text{ and } \pi(k) = \frac{1}{n} \text{ for } k = 1, 2, \ldots, n-1. \tag{2.23}$$

The sets $\Omega_0, \Omega_1, \ldots, \Omega_n$ are the balls of center 0, namely:

$$\Omega_k = \{x \in C_{2n} : d(x, 0) = k\}$$

where d denotes the distance on the graph C_{2n}. We recall that the distance between two points x, y in a graph X is the length of the shortest path on X joining x and y. We can also introduce the function $\sigma : C_{2n} \to C_{2n}$ defined by $\sigma(n - k) = n + k$ which is a bijective involution whose orbits are the $\Omega_k = \{k, \sigma(k)\}$ for $k = 0, 1, \ldots, n$.

For $h, k \in \mathbb{Z}$, set $c_k(h) = \cos \frac{\pi hk}{n}$ and $s_k(h) = \sin \frac{\pi hk}{n}$. Then the characters of C_{2n} (see Section 2.1) may be expressed as follows:

$$\chi_k = c_k + i s_k \qquad k = 0, 1, 2, \ldots, 2n - 1.$$

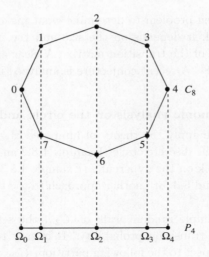

Figure 2.2.

Set

$$\langle f_1, f_2 \rangle_\pi = \sum_{h=0}^{n} \pi(h) f_1(h) \overline{f_2(h)}$$

$$= \frac{1}{2n} f_1(0) \overline{f_2(0)} + \frac{1}{n} \sum_{h=0}^{n-1} f_1(h) \overline{f_2(h)} + \frac{1}{2n} f_1(n) \overline{f_2(n)} \qquad (2.24)$$

where π is as in (2.23).

Theorem 2.6.1 *In the space $L(P_{n+1})$ endowed with the scalar product $\langle \cdot, \cdot \rangle_\pi$, the functions*

$$c_0, \ \sqrt{2} c_1, \ldots, \sqrt{2} c_k, \ldots, \sqrt{2} c_{n-1}, \ c_n$$

form an orthonormal basis of eigenfunctions for the transition matrix Q. The corresponding eigenvalues are

$$1, \ \cos \frac{\pi}{n}, \ldots, \cos \frac{k\pi}{n}, \ldots, \cos \frac{(n-1)\pi}{n}, \ -1.$$

Proof The proof is just elementary and tedious trigonometry. However, there is a fast way to deduce if from the harmonic analysis on C_{2n}. First of all, note that the function c_k, seen as a function on C_{2n}, is the real part of χ_k and therefore it is an eigenfunction of the transition matrix P of the simple random walk on C_{2n}: the eigenvalues of P are real valued. Moreover, $c_k(\sigma(h)) = c_k(h)$ for $h = 0, 1, \ldots, n$ and therefore c_k is constant on each $\Omega_0, \Omega_1, \ldots, \Omega_n$. Then we can apply Proposition 1.10.4. For the orthogonality relations, note that

$$c_k(h) = \frac{\chi_k(h) + \chi_k(-h)}{2},$$

and apply the orthogonality relations for the characters of C_{2n} (Lemma 2.1.2). We leave the details as an exercise.   □

In Appendix 1 we give several other orthonormal systems for $L(P_{n+1})$ that can be expressed in terms of trigonometric functions.

Consider now the simple random walk on the hypercube Q_n. More precisely, we consider the original definition given in Example 1.2.2 rather than its variation in (1.25). Let $P = (p(x,y))_{x,y \in Q_n}$ be its transition probability matrix. Let w be the weight function on Q_n as in Section 2.4 and define $\Omega_k = \{x \in Q_n : w(x) = k\}$, $k = 0, 1, \ldots, n$.

Exercise 2.6.2 Show that $\Omega_k = \{x \in Q_n : d(x, x_0) = k\}$ where $x_0 = (0, 0, \ldots, 0)$.

Proposition 2.6.3 *The transition matrix P is lumpable with respect to the partition $Q_n = \coprod_{k=0}^{n} \Omega_k$. The lumped chain has state space $I_n = \{0, 1, \ldots, n\}$ and transition matrix $Q = (q(k,j))_{k,j \in I}$ defined by*

$$q(k, k+1) = \frac{n-k}{n} \quad \text{for } k = 0, 1, \ldots, n-1$$

$$q(k, k-1) = \frac{k}{n} \quad \text{for } k = 1, 2, \ldots, n$$

$$q(k, j) = 0 \quad \text{otherwise.}$$

Proof First of all, note that if $x, y \in Q_n$ and $x \sim y$ then $|w(x) - w(y)| = 1$. In other words, if we are in Ω_k then we can only jump to Ω_{k-1} or Ω_{k+1}. Moreover, if $x \in \Omega_k$ then

$$p(x, \Omega_{k+1}) = |\{y \in \Omega_{k+1} : x \sim y\}| \frac{1}{n} = \frac{n-k}{n}$$

and similarly one obtains $p(x, \Omega_{k-1}) = \frac{k}{n}$.   □

From Corollary 1.10.3 it follows that Q is reversible and in detailed balance with the measure π defined on $\{0, 1, \ldots, n\}$ by setting

$$\pi(k) = \frac{|\Omega_k|}{|Q_n|} = \frac{\binom{n}{k}}{2^n}.$$

Moreover define for $f_1, f_2 \in L(I_n)$

$$\langle f_1, f_2 \rangle_\pi = \sum_{h=0}^{n} \pi(h) f_1(h) \overline{f_2(h)}. \tag{2.25}$$

We want to describe the eigenfunctions of P that are *radial* on Q_n, namely that are constant on each Ω_k (see Proposition 1.10.4 and Exercise 2.6.5 below). Let $V_h = \langle \chi_x : w(x) = h \rangle$ as in Section 2.4. Set

$$\phi_h(k) = \frac{1}{\binom{n}{h}} \sum_{\ell=\max\{0, h+k-n\}}^{\min\{h,k\}} \binom{k}{\ell} \binom{n-k}{h-\ell} (-1)^\ell$$

and

$$\Phi_h = \sum_{k=0}^{n} \phi_h(k) \mathbf{1}_{\Omega_k}.$$

Then we have:

Proposition 2.6.4

(i) *Each Φ_h is a radial eigenfunction of P.*

(ii) *The set $\{\Phi_h\}_{h \in I}$ is an orthogonal basis for the radial functions on Q_n and*

$$\|\Phi_h\|^2_{L(Q_n)} = \frac{2^n}{\binom{n}{h}}.$$

Proof (i) Clearly, each Φ_h is radial. Moreover we have

$$\Phi_h = \frac{1}{\binom{n}{h}} \sum_{x \in \Omega_h} \chi_x. \tag{2.26}$$

Indeed, if $y \in \Omega_k$ then

$$\sum_{x \in \Omega_h} \chi_x(y) = \sum_{x \in \Omega_h} (-1)^{x \cdot y} = \sum_{\ell=\max\{0, h+k-n\}}^{\min\{h,k\}} \binom{k}{\ell} \binom{n-k}{h-\ell} (-1)^\ell.$$

For, given $y \in \Omega_k$ to get an $x \in \Omega_h$ such that $|\{t : x_t = y_t = 1\}| = \ell$, we have to choose ℓ coordinates in $\{t : y_t = 1\}$ (whose cardinality is k) and $h - \ell$ coordinates in $\{t : y_t = 0\}$ (whose cardinality is $n - k$). But (2.26) implies that $\Phi_h \in V_h$, so that, from Corollary 2.4.2 we deduce that it is an eigenfunction of P and the corresponding eigenvalue is $1 - \frac{2h}{n+1}$.

(ii) From the previous part we have that $\Phi_0, \Phi_1, \ldots, \Phi_n$ are eigenfunctions of the selfadjoint operator P whose corresponding eigenvalues are all distinct. Therefore the Φ_h's are orthogonal. But the dimension of the space of radial functions on Q_n is equal to $n+1$. Therefore the Φ_h's form a basis for this space. Finally, from (2.26) one has

$$\|\phi_n\|^2_{L(Q_n)} = \frac{1}{\binom{n}{h}^2} \sum_{x \in \Omega_h} \langle \chi_x, \chi_x \rangle_{L(Q_n)} = \frac{1}{\binom{n}{h}^2} \cdot 2^n |\Omega_h| = \frac{2^n}{\binom{n}{h}}$$

and the proof is complete. $\qquad\square$

Exercise 2.6.5 (The spherical Fourier transform on Q_n) Show that if $f \in L(Q_n)$ is radial, then its Fourier transform is radial and it is equal to

$$\widehat{f} = \sum_{h=0}^{n} \langle f, \Phi_h \rangle_{L(Q_n)} \mathbf{1}_{\Omega_h}$$

and the inversion formula in Theorem 2.3.5 becomes

$$f = \frac{1}{2^n} \sum_{h=0}^{n} \langle f, \Phi_h \rangle_{L(Q_n)} \binom{n}{h} \Phi_h.$$

Now we may apply Proposition 1.10.4. Set $\phi_h(-1) = \phi_h(n+1) = 0$, for $h = 0, 1, \ldots, n$.

Corollary 2.6.6 *The functions ϕ_h, $h = 0, 1, \ldots, n$ satisfy the finite difference equation*

$$\frac{n-k}{n}\phi_h(k+1) + \frac{k}{n}\phi_h(k-1) = \frac{n-2h}{n}\phi_h(k)$$

for $k = 0, 1, \ldots, n$, and the orthogonality relations

$$\sum_{k=0}^{n} \phi_h(k)\phi_t(k)\frac{\binom{n}{k}}{2^n} = \frac{1}{\binom{n}{h}}\delta_{h,t}.$$

This corollary is just a translation of the fact that the functions ϕ_h constitute an orthogonal basis of eigenfunctions for the lumped chain Q. In particular, the eigenvalues of Q are $\frac{n-2h}{n}$, for $h = 0, 1, \ldots, n$, each with multiplicity one. The coefficients ϕ_h belong to a family of orthogonal polynomials of a discrete variable called Krawtchouk polynomials; see Section 5.3.

Part II
Representation Theory and Gelfand Pairs

3

Basic representation theory of finite groups

This chapter contains the fundamentals of finite group representations. Our approach emphasizes the harmonic analytic point of view (we focus on unitary representations and on Fourier transforms). Our exposition is inspired to Diaconis' book [55] and to Figà-Talamanca lecture notes [88]. We also gained particular benefit from the monographs by: Alperin and Bell [4], Fulton and Harris [95], Isaacs [129], Naimark and Stern [168], Ricci [180], Serre [196], Simon [198] and Sternberg [212].

3.1 Group actions

We start the chapter with a short review on group actions: it is based on class notes by A. Machì [160].

In this section Ω will denote a finite set and G a finite group. 1_G is the identity of G.

Definition 3.1.1 A *(left) action* of G on Ω is a map

$$
\begin{array}{ccc}
G \times \Omega & \longrightarrow & \Omega \\
(g, \omega) & \mapsto & g \cdot \omega
\end{array}
$$

such that

(i) $(gh) \cdot \omega = g \cdot (h \cdot \omega)$ for all $\omega \in \Omega$ and $g, h \in G$;
(ii) $1_G \cdot \omega = \omega$ for all $\omega \in \Omega$.

If we have an action of G on Ω, then we say that G *acts* on Ω.

Remark 3.1.2 A *right* action is defined, *mutatis mutandis*, in the same way. We only consider left actions. Note that, given a left action $(g, \omega) \mapsto g \cdot \omega$ the map $(g, \omega) \mapsto \omega g := g^{-1} \cdot \omega$ defines a right action, and vice versa.

If H is a subgroup of G, then H also acts on Ω: this action is called the *restriction*.

An action is *transitive* if for any $\omega_1, \omega_2 \in \Omega$ there exists $g \in G$ such that $g \cdot \omega_1 = \omega_2$. If G acts transitively on Ω we say that Ω is a (*homogeneous*) *G-space*.

A point $\omega \in \Omega$ is *fixed* by a group element g if $g \cdot \omega = \omega$. If ω is fixed by all elements in G, one says that it is a *fixed point* of the group action.

For $\omega \in \Omega$ we denote by

$$Stab_G(\omega) = \{g \in G : g \cdot \omega = \omega\}$$

the *stabilizer* of ω, and by

$$Orb_G(\omega) = \{g \cdot \omega : g \in G\},$$

also denoted by $G\omega$, the *G-orbit* of ω.

Clearly, if $\omega_1 \in Orb_G(\omega)$ then $Orb_G(\omega) = Orb_G(\omega_1)$. The subsets $Orb_G(\omega), \omega \in \Omega$, are the orbits of G on Ω. If we define the relation $\sim$ on Ω by setting: $\omega_1 \sim \omega_2$ if there exists $g \in G$ such that $\omega_1 = g \cdot \omega_2$, then we have:

Lemma 3.1.3 $\sim$ *is an equivalence relation on Ω and the orbits of G on Ω are the equivalence classes of this relation. In particular, the orbits of Ω form a partition of Ω.*

Proof It is straightforward to see that $\sim$ is an equivalence relation. Moreover, $\omega_1 \sim \omega_2$ if and only if $\omega_1 \in Orb_G(\omega_2)$. $\qquad\square$

An immediate consequence is the following fact: the action of G on Ω is transitive if and only if there is a single orbit, that is $Orb_G(\omega) = \Omega$ (and this holds for any $\omega \in \Omega$). Another consequence is the following equation:

$$\Omega = \coprod_{\omega \in \Gamma} Orb_G(\omega) \qquad (3.1)$$

where Γ is a set of representatives for the orbits.

Example 3.1.4 Let G be a finite group and $K \leq G$ a subgroup. Denote by $X = G/K = \{gK : g \in G\}$ the *homogeneous space* consisting of the *right cosets* of K in G. Then, G acts on X by *left translations*: $h \cdot gK := (hg)K$, for $g, h \in G$. In Lemma 3.1.6 we will prove that every transitive action is indeed of this type.

Observe that the restriction to K of the action of G on X is not transitive (for instance, if $x_0 \in X$ is the point stabilized by K, then clearly $Orb_K(x_0) = \{x_0\}$); the set of K-orbits can be identified with the set of *double cosets* $K\backslash G/K = \{KgK : g \in G\}$.

Denote by S_Ω the group of all bijections $\Omega \to \Omega$. An element $\sigma \in S_\Omega$ is called a *permutation* of Ω. If $\Omega = \{1, 2, \ldots, n\}$ then S_Ω is denoted by S_n and it is called the *symmetric group* on n elements (see Chapter 10).

Proposition 3.1.5 *Suppose that G acts on a set Ω.*

(i) *Any element $g \in G$ induces a permutation $\sigma_g \in S_\Omega$.*

(ii) *The map*

$$G \ni g \mapsto \sigma_g \in S_\Omega \tag{3.2}$$

   is a group homomorphism.

(iii) *There is a natural bijection between the set of all actions of G on Ω and the set of all homomorphisms from G to S_Ω.*

Proof Suppose that G acts on Ω. For all $g \in G$, the map $\sigma_g : \Omega \to \Omega$ defined by $\sigma_g(\omega) = g \cdot \omega$ is a bijection. Indeed $\sigma_{g^{-1}}$ is the inverse of σ_g: for all $\omega \in \Omega$ one has

$$\omega = 1_G \cdot \omega = (g^{-1}g) \cdot \omega = g^{-1} \cdot (g \cdot \omega) = \sigma_{g^{-1}}[\sigma_g(\omega)] = [\sigma_{g^{-1}} \circ \sigma_g](\omega)$$

and similarly, $[\sigma_g \circ \sigma_{g^{-1}}](\omega) = \omega$. This proves (i).

Also, for all $g_1, g_2 \in G$ and $\omega \in \Omega$ one has

$$\sigma_{g_1 g_2}(\omega) = (g_1 g_2) \cdot \omega = g_1 \cdot (g_2 \cdot \omega) = \sigma_{g_1}[\sigma_{g_2}(\omega)] = [\sigma_{g_1} \circ \sigma_{g_2}](\omega)$$

and this shows (ii).

Finally, suppose that $\sigma : G \to S_\Omega$ is a homomorphism. Then the map $G \times \Omega \ni (g, \omega) \mapsto \sigma_g(\omega) \in \Omega$ defines an action of G on Ω as one easily checks. We also leave it to the reader to verify that these two processes are one inverse to the other, thus establishing the desired bijective correspondence. $\square$

Given an action of G on Ω the kernel of the homomorphism (3.2) is the subgroup

$$H = \{g \in G : g \cdot \omega = \omega \text{ for all } \omega \in \Omega\}$$

consisting of all elements in G which fix all elements in Ω. This subgroup is called the *kernel of the action*. If it is trivial, then the action is termed *faithful.* In other words the action is faithful if the homomorphism (3.2) is injective: in this case G is isomorphic to a subgroup of S_Ω.

Two G-spaces Ω and Ω' are *isomorphic* as G-spaces if there exists a bijection $\phi : \Omega \to \Omega'$ which is *G-equivariant*, that is $\phi(g \cdot \omega) = g \cdot [\phi(\omega)]$ for all $\omega \in \Omega$ and $g \in G$.

Lemma 3.1.6 *Let Ω be a G-space, $\omega \in \Omega$ and set $K = Stab_G(\omega)$. Then the map*

$$
\begin{aligned}
\Psi : \quad G/K &\to \quad \Omega \\
gK &\mapsto \quad g \cdot \omega,
\end{aligned}
\tag{3.3}
$$

is a G-equivariant bijection, thus making Ω and $G/Stab_G(\omega)$ isomorphic as G-spaces.

Proof We simultaneously show that Ψ is well defined and injective. For,

$$
\begin{aligned}
g_1 K = g_2 K &\Leftrightarrow g_1^{-1} g_2 \in K \Leftrightarrow g_1^{-1} g_2 \cdot \omega = \omega \\
&\Leftrightarrow g_1^{-1} \cdot (g_2 \cdot \omega) = \omega \Leftrightarrow g_2 \cdot \omega = g_1 \cdot \omega.
\end{aligned}
$$

Moreover Ψ is clearly surjective. Finally,

$$
g_1 \cdot \Psi(g_2 K) = g_1 \cdot (g_2 \cdot \omega) = (g_1 g_2) \cdot \omega = \Psi(g_1 g_2 K) = \Psi(g_1 \cdot g_2 K)
$$

and Ψ is G-equivariant. $\square$

Corollary 3.1.7 *Let G act on a set Ω. Then,*

$$
|G| = |Stab_G(\omega)| \cdot |Orb_G(\omega)|
\tag{3.4}
$$

for all $\omega \in \Omega$.

Proof Applying Lemma 3.1.6 to the action of G on $Orb_G(\omega)$ we get:

$$
|Orb_G(\omega)| = |G/Stab_G(\omega)| = |G|/|Stab_G(\omega)|
$$

and (3.4) follows. $\square$

Exercise 3.1.8 Suppose that $\phi : \Omega \to \Omega'$ is a G-equivariant isomorphism of G-spaces. Show that for all $\omega \in \Omega$ one has

$$
Stab_G(\omega) = Stab_G(\phi(\omega)).
\tag{3.5}
$$

Exercise 3.1.9 Suppose that a finite group G acts transitively on a set X. Also consider the left action of G on itself. Show that the diagonal action of G on $G \times X$ has $|X|$ orbits and that each orbit is isomorphic to G with the left action.

Lemma 3.1.10 *Let Ω be a G-space. Then $Stab_G(g \cdot \omega) = g Stab_G(\omega) g^{-1}$ for all $g \in G$ and $\omega \in \Omega$.*

Proof This follows from the equivalences below.

$$h \in Stab_G(g \cdot \omega) \Leftrightarrow h \cdot (g \cdot \omega) = g \cdot \omega$$
$$\Leftrightarrow g^{-1} \cdot [h \cdot (g \cdot \omega)] \equiv (g^{-1} h g) \cdot \omega = \omega$$
$$\Leftrightarrow g^{-1} h g \in Stab_G(\omega).$$

$\square$

Proposition 3.1.11 *Let $H, K \leq G$ be two subgroups. Then G/H and G/K are isomorphic as G-spaces if and only if H and K are conjugate in G (there exists $g \in G$ such that $K = g^{-1} H g$).*

Proof Suppose first that there exists $g \in G$ such that $K = g^{-1} H g$. Set $\Omega = G/H$ and $\omega = g^{-1} H$. Then $K = Stab_G(\omega)$. By virtue of Lemma 3.1.6, we have that G/K is isomorphic to $\Omega \equiv G/H$ as G-spaces.

Conversely, suppose that $\phi : G/H \to G/K$ is a G-equivariant isomorphism. Let $g \in G$ be such that $\phi(H) = gK$. Then, by (3.5) we have $Stab_G(H) = Stab_G(gK)$. Since, clearly, $Stab_G(gK) = gKg^{-1}$ and $Stab_G(H) = H$ we conclude that $H = gKg^{-1}$. $\square$

Example 3.1.12 (Trivial action) Let G be a group and Ω a set. The action $g \cdot \omega = \omega$ for all $g \in G$ and $\omega \in \Omega$ is the *trivial action*. For all $\omega \in \Omega$ one has $Orb_G(\omega) = \{\omega\}$ and $Stab_G(\omega) = G$.

Example 3.1.13 (Cayley's actions) Let G be a group. We consider two (left) actions of G on itself. The *Cayley action on the left* (resp. *on the right*) is defined by setting $g \cdot h = gh$ (resp. $g \cdot h = hg^{-1}$). These actions are clearly transitive and faithful. In this case G is isomorphic to a subgroup of S_G (Cayley theorem).

Example 3.1.14 (Conjugacy actions) Let G be a group. For $g, h \in G$ we set $g \cdot h = ghg^{-1}$. It is immediate to check that this is an action of G on itself, the *conjugacy action* on G. An orbit is called a *conjugacy class* and the stabilizer of an element h is called the *centralizer* of h. A fixed point $h \in G$ is called a *central element*: it commutes with every element in the group, $gh = hg$ for all $g \in G$.

Clearly, if G is abelian this action is trivial: every element is central.

Let now Ω denote the set of all subgroups $H \leq G$ of G. Define an action of G on Ω by setting $g \cdot H = gHg^{-1} = \{ghg^{-1} : h \in H\}$. This is the *conjugacy action on subgroups*. Clearly, an element $H \in \Omega$ is fixed by this action if and only if it is a *normal* subgroup. The stabilizer of a point H is called the *normalizer* of H in G.

Example 3.1.15 (Product action) Let G_1, G_2 be two finite groups acting on Ω_1 and Ω_2, respectively. Then we define the *product action* of $G_1 \times G_2$ on $\Omega_1 \times \Omega_2$ by setting $(g_1, g_2) \cdot (\omega_1, \omega_2) = (g_1 \cdot \omega_1, g_2 \cdot \omega_2)$.

Example 3.1.16 (Diagonal action) Let G act on a set Ω. We define an action of G on $\Omega \times \Omega$ by setting $g \cdot (\omega_1, \omega_2) = (g \cdot \omega_1, g \cdot \omega_2)$. This is called the *diagonal action* of G on Ω. When the action of G on Ω is transitive, the description of the G-orbits on $\Omega \times \Omega$ is given in Theorem 3.13.1 and Corollary 3.13.2. Note that the diagonal action is the restriction of the product action of $G \times G$ on $\Omega \times \Omega$ to the diagonal subgroup $\widetilde{G} = \{(g, g) : g \in G\}$.

Example 3.1.17 Let G be a group and Ω_1, Ω_2 be two sets. Suppose that G acts on Ω_1. Then G acts on the space $\Omega_2^{\Omega_1} = \{f : \Omega_1 \to \Omega_2\}$ of all functions with domain Ω_1 and range Ω_2 as follows

$$[g \cdot f](\omega_1) = f(g^{-1} \cdot \omega_1)$$

for all $f \in \Omega_2^{\Omega_1}$ and $\omega_1 \in \Omega_1$ and $g \in G$.

When $\Omega_1 = G$ and $\Omega_2 = \mathbb{C}$, then $\Omega_2^{\Omega_1}$ is denoted by $L(G)$. If the action is the Cayley action on the left (resp. on the right) (see Example 3.1.13), then the corresponding action of G on $L(G)$ is called the *left* (resp. *right*) *regular representation* (see Example 3.4.1 and Exercise 3.4.2).

When the action of G on Ω_1 is transitive, so that $\Omega_1 \equiv X = G/K$ (here K denotes the stabilizer of a point, see Lemma 3.1.6) and $\Omega_2 = \mathbb{C}$, then $\Omega_2^{\Omega_1}$ is denoted by $L(X)$ and the corresponding action of G on $L(X)$ is called the *permutation representation*.

Example 3.1.18 A rich source of group actions is geometry. For instance, if we take a regular n-gon, then its isometry group is the dihedral group D_n (see Example 3.8.2) and we may think of D_n as a group that acts on the vertices of the n-gon. For an elementary but thorough introduction to simmetry groups in Eucledean geometry we refer to Artin's book [6]. A complete treatment is in Coxeter's monograph [47].

3.2 Representations, irreducibility and equivalence

Let G be a finite group and V a finite dimensional vector space over $\mathbb{C}$. We denote by $GL(V)$ the *linear group* of V consisting of all invertible linear maps $A : V \to V$.

Definition 3.2.1 A *representation* of G over V is an action $G \times V \ni (g, v) \mapsto \rho(g)v \in V$ such that any $\rho(g)$ is an (invertible) linear map. In other words one has $\rho(g) : V \to V$ is linear and invertible, $\rho(g_1 g_2) = \rho(g_1)\rho(g_2)$ for all $g_1, g_2 \in G$ and $\rho(1_G) = I_V$ where 1_G is the identity element in G and $I_V : V \to V$ is the identity element in $GL(V)$. It may also be seen as a homomorphism $\rho : G \to GL(V)$.

We shall denote the representation by the pair (ρ, V).

Denoting by n the dimension of V, as $GL(V)$ is isomorphic to $GL(n, \mathbb{C})$ the group of all invertible n by n matrices, we can regard a representation of G as a group homomorphism $\rho : G \to GL(n, \mathbb{C})$. Then n is the *dimension* or *degree* of ρ and it will be denoted by d_ρ.

Suppose that (ρ, V) is a representation of G and that $W \leq V$ is a *G-invariant* subspace, namely $\rho(g)w \in W$ for all $g \in G$ and $w \in W$ (sometimes we shall more concisely write $\rho(G)W \leq W$). Then denoting by $\rho_W : G \ni g \mapsto \rho(g)|_W \in GL(W)$ the restriction of ρ to the subspace W, we say that (ρ_W, W) is a *sub-representation* of (ρ, V). We also say that ρ_W is *contained* in ρ. This fact will be also indicated by $\rho_W \preceq \rho$. Clearly every representation is a sub-representation of itself.

An important class of representations consists of the irreducible representations. A representation (ρ, V) of a group G is *irreducible* if the only G-invariant subspaces are trivial: $W \leq V$ and $\rho(G)W \leq W$ implies either $W = \{0\}$ or $W = V$. For instance, any one-dimensional representation is clearly irreducible.

We now introduce the important notion of equivalence of representations. Let (ρ, V) and (σ, W) be two representations of the same group G. Suppose there exists a linear bijection $J : V \to W$ such that, for all $g \in G$, $\sigma(g)J = J\rho(g)$. Then one says that the two representations are *equivalent* and one writes $\rho \sim \sigma$. Note that $\sim$ is an equivalence relation and that two equivalent representations have the same degree.

3.3 Unitary representations

Suppose the vector space V is endowed with a scalar product $\langle \cdot, \cdot \rangle$; then we can define the notion of a *unitary representation*.

Let V and W be two vector spaces endowed with scalar products $\langle \cdot, \cdot \rangle_V$ and $\langle \cdot, \cdot \rangle_W$, respectively. The *adjoint* of a linear operator $T : V \to W$, is the linear operator $T^* : W \to V$ such that $\langle w, Tv \rangle_W = \langle T^*w, v \rangle_V$ for all $v \in V$ and $w \in W$. A linear operator $U : V \to W$ is *unitary* if $U^*U = I = UU^*$, equivalently if $\langle Uv, Uv' \rangle_W = \langle v, v' \rangle_V$ for every $v, v' \in V$. Recall that if U is a *unitary matrix* then $U^* = \overline{U}^t$ is the conjugate transpose of U. Moreover, the spectrum of a unitary operator (or matrix) is contained in the unit circle: $\sigma(U) \subseteq \{z \in \mathbb{C} : |z| = 1\}$.

A representation (ρ, V) is *unitary* if it preserves the inner product, namely $\langle \rho(g)v, \rho(g)v' \rangle = \langle v, v' \rangle$ for all $g \in G$ and $v, v' \in V$. Equivalently, ρ is unitary if $\rho(G) \leq \mathcal{U}(V)$ is a subgroup of the *unitary group* of V.

It is worth to mention that, given an arbitrary representation (ρ, V) it is always possible to endow V with an inner product so that ρ becomes unitary. Given an arbitrary inner product $\langle \cdot, \cdot \rangle$ for V define, for all v and w in V,

$$(v, w) = \sum_{g \in G} \langle \rho(g)v, \rho(g)w \rangle. \tag{3.6}$$

Proposition 3.3.1 *The representation $(\rho, V_{(\cdot, \cdot)})$ is unitary and equivalent to $(\rho, V_{\langle \cdot, \cdot \rangle})$. In particular, every representation of G is equivalent to a unitary representation.*

Proof First of all it is easy to see that $(\cdot, \cdot)$ as defined in (3.6) is an inner product. Moreover, for all $v, w \in V$ and all $h \in G$ one has

$$\begin{aligned}
(\rho(h)v, \rho(h)w) &= \sum_{g \in G} \langle \rho(g)\rho(h)v, \rho(g)\rho(h)w \rangle \\
&= \sum_{g \in G} \langle \rho(gh)v, \rho(gh)w \rangle \\
&= \sum_{k \in G} \langle \rho(k)v, \rho(k)w \rangle \\
&= (v, w).
\end{aligned}$$

The equivalence is trivial: it is given by the identity I_V. $\square$

We are mostly interested in equivalence classes of representations, thus we might confine ourselves to unitary representations. We thus suppose that V is a finite dimensional (complex) vector space with a G-invariant inner product and that ρ is unitary: note that under these assumptions we have $\rho(g^{-1}) = \rho(g)^{-1} = \rho(g)^*$ for all $g \in G$. Finally, two representations (ρ, V) and (σ, W) are *unitarily* equivalent if there exists a unitary linear operator $U : V \to W$ such that $\sigma(g)U = U\rho(g)$ for all $g \in G$.

Lemma 3.3.2 *Suppose that (ρ, V) and (σ, W) are unitary representations of a finite group G. If they are equivalent then they are also unitarily equivalent.*

Proof We want to show that if there exists an invertible linear map $T : V \to W$ such that

$$\rho(g) = T^{-1}\sigma(g)T \tag{3.7}$$

for all $g \in G$, then there exists a unitary $U : V \to W$ such that

$$\rho(g) = U^{-1}\sigma(g)U$$

for all $g \in G$.

Taking adjoints in (3.7), using $\rho(g)^* = \rho(g^{-1})$ and replacing g by g^{-1} one has

$$\rho(g) = T^*\sigma(g)(T^*)^{-1}.$$

As, by (3.7), $\sigma(g) = T\rho(g)T^{-1}$, one has $T^*T\rho(g)(T^*T)^{-1} = T^*\sigma(g)(T^*)^{-1} = \rho(g)$ or, equivalently,

$$\rho(g)^{-1}(T^*T)\rho(g) = T^*T. \tag{3.8}$$

Let $T = U|T|$ be the *polar decomposition* of T (see Exercise 3.3.5): here $|T| : V \to V$ is the square root of the positive linear operator T^*T, while $U : V \to W$ is unitary. We have, by the uniqueness of the square root,

$$\rho(g)^{-1}|T|\rho(g) = |T| \tag{3.9}$$

since the left hand side of (3.9) is the square root of the left hand side of (3.8). Therefore, as $|T|$ is invertible, we have

$$\begin{aligned}
U\rho(g)U^{-1} &= T|T|^{-1}\rho(g)|T|T^{-1} \\
&= T\rho(g)T^{-1} \\
&= \sigma(g)
\end{aligned}$$

where the second equality follows from (3.9) and the third one from (3.7). $\qquad\square$

We now define the *direct sum* of a finite number of representations. If (ρ_j, W_j), $j = 1, 2, \ldots, k$ are representations of a group G and

$V = W_1 \oplus W_2 \oplus \cdots \oplus W_k$ is the direct sum of the corresponding spaces, then $\rho = \rho_1 \oplus \rho_2 \oplus \cdots \oplus \rho_k$ is their direct sum, that is

$$\rho(g)v = \rho_1(g)w_1 + \rho_2(g)w_2 + \cdots + \rho_k(g)w_k$$

for all $v = w_1 + w_2 + \cdots + w_k \in V$, $w_i \in W_i$, and $g \in G$.

Observe that if W is a G-invariant subspace of V under the representation ρ, then $W^\perp = \{v \in V : \langle v, w \rangle = 0, \ \forall w \in W\}$, the *orthogonal complement* of W, is also G-invariant. Indeed, if $v \in W^\perp$ and $g \in G$ one has $\langle \rho(g)v, w \rangle = \langle v, \rho(g^{-1})w \rangle = 0$ for all $w \in W$. Recall that, under the previous assumptions, V can be expressed as the direct sum of the orthogonal subspaces W and $W^\perp$, namely $V = W \oplus W^\perp$ and, denoting by $\rho_1 = \rho_W$ and $\rho_2 = \rho_{W^\perp}$ the corresponding sub-representations, we see that ρ is the direct sum of these: $\rho = \rho_1 \oplus \rho_2$.

Lemma 3.3.3 *Every representation of G is the direct sum of a finite number of irreducible representations.*

Proof Let (ρ, V) be a representation of G. If ρ is irreducible there is nothing to prove. If not, the proof follows an inductive argument on the dimension of V: as before, $V = V_1 \oplus V_2$, with the subspaces V_i G-invariant and $\dim(V_i) < \dim(V)$. $\qquad\square$

Definition 3.3.4 (Dual) Let G be a finite group. We denote by $\widehat{G}$, the *dual* of G, a complete set of irreducible pairwise nonequivalent (unitary) representations of G (in other words, $\widehat{G}$ contains exactly one element belonging to each equivalence class of irreducible representations).

We end this section by sketching a proof of the polar decomposition for an invertible linear operator.

Exercise 3.3.5 Let V be a finite dimensional complex vector space with inner product $\langle \cdot, \cdot \rangle$. We say that a linear operator $P : V \to V$ is *positive* if $\langle Pv, v \rangle > 0$ for all $v \in V$, $v \neq 0$.

(1) Prove that any positive operator is selfadjoint.

(2) Prove that if two selfadjoint operators A and B commute, namely $AB = BA$, then they admit a common basis of eigenvectors, that is there exists an orthogonal decomposition $V = \bigoplus_{i=1}^m W_i$ such that $AW_i = W_i = BW_i$ for all $i = 1, 2, \ldots, m$.

(3) Prove that if P is positive, then there exists a unique positive operator Q such that $Q^2 = P$. This is called the *square root* of P.

(4) Show that any invertible linear operator $A : V \to V$ can be written, in a unique way, as $A = UP$ where U is unitary and P is positive. This is called the polar decomposition of A.

3.4 Examples

Example 3.4.1 Let G be a finite group. Denote by $L(G) = \{f : G \to \mathbb{C}\}$ the space of all complex valued functions on G endowed with the scalar product

$$\langle f_1, f_2 \rangle = \sum_{g \in G} f_1(g)\overline{f_2(g)} \tag{3.10}$$

for all $f_1, f_2 \in L(G)$. The representation $(\lambda_G, L(G))$ defined by

$$[\lambda_G(g)f](h) = f(g^{-1}h) \tag{3.11}$$

for all $g, h \in G$ and $f \in L(G)$, is called the *left regular representation* of G.

Exercise 3.4.2 Show that $\lambda_G(g_1 g_2) = \lambda_G(g_1)\lambda_G(g_2)$ for all g_1, g_2 in G and that λ_G is unitary.

Analogously, the representation $(\rho_G, L(G))$ defined by

$$[\rho_G(g)f](h) = f(hg) \tag{3.12}$$

is again a unitary representation, the *right regular representation*.

The two representations above commute: $\lambda_G(g)\rho_G(h) = \rho_G(h)\lambda_G(g)$ for all $g, h \in G$.

Remark 3.4.3 In many books, the scalar product (3.10) is normalized, that is $\langle f_1, f_2 \rangle = \frac{1}{|G|} \sum_{g \in G} f_1(g)\overline{f_2(g)}$. This changes many formulæ given in the current and following chapters by a factor of $1/|G|$.

Example 3.4.4 Let $C_n = \{1, a, a^2, \ldots, a^{n-1}\}$ denote the cyclic group of order n generated by the element a (here we use multiplicative notation). Set $\omega = e^{2\pi i/n}$ (a primitive nth root of the unity). Then $(\rho_k, \mathbb{C})$ defined by $\rho_k(a^h) = \omega^{kh}$, for each $k = 0, 1, \ldots, n - 1$ is a one-dimensional representation of C_n. As we shall see in Example 3.8.1, every irreducible representation of C_n is one of these, namely $\widehat{C_n} = \{\rho_k : k = 0, 1, \ldots, n - 1\}$. See also Exercise 3.5.3.

Example 3.4.5   Let $G = S_n$ be the symmetric group of degree n, that is the group of permutations on n elements; equivalently, S_n is the group of all bijections $g: \{1, 2, \ldots, n\} \to \{1, 2, \ldots, n\}$. We mention three representations of such a group.

(a) The *trivial representation*. It is one dimensional, that is $V = \mathbb{C}$ and it assigns to each group element the identity: $\rho_1(g) = 1$ for all $g \in G$.

(b) The *sign* (or *alternating*) *representation*. This is also one dimensional ($V = \mathbb{C}$). It assigns to each group element the identity 1 if g is an *even permutation* (that is g is the product of an even number of transpositions, equivalently $g \in A_n$, the *alternating group*), and the opposite of the identity, -1 otherwise, that is if g is an *odd permutation*. In other words, denoting by $sign(g)$ the *sign* of a permutation g, we have $\rho_2(g) = sign(g)I$. As *sign* is a homomorphism $G \to S_n/A_n \equiv C_2$ ($sign(g_1 g_2) = sign(g_1)sign(g_2)$), ρ_2 is indeed a representation. As it is one dimensional, it is also irreducible.

(c) The *permutation representation*. This is n dimensional: fix an orthonormal basis $\{\mathbf{e}_1, \mathbf{e}_2, \ldots, \mathbf{e}_n\}$ of an n-dimensional vector space V and set $\rho_3(g)\mathbf{e}_j = \mathbf{e}_{g(j)}$ for all $g \in S_n$. With respect to the fixed basis for V, $\rho_3(g)$ is a so called *permutation matrix*, that is, a matrix obtained by permuting the rows of the identity matrix. This representation is not irreducible. For instance, an S_n-invariant subspace is the hyperplane $\sum_{i=1}^n x_i = 0$, that is the $n-1$ dimensional subspace $W = \{w = \sum_{i=1}^n x_i \mathbf{e}_i : \sum_{i=1}^n x_i = 0\}$. However it is not difficult to show that $(\rho_3)|_W$, the restriction of ρ_3 to this G-invariant subspace, is irreducible.

Exercise 3.4.6   Prove that $(\rho_3)|_W$ is irreducible.

3.5 Intertwiners and Schur's lemma

Let (ρ, V) and (σ, W) be two representations of a group G. Let $L: V \to W$ be a linear transformation. One says that L *intertwines* ρ and σ if $L\rho(g) = \sigma(g)L$ for all $g \in G$. The following result relates the notion of reducibility of a representation with the existence of intertwiners.

Lemma 3.5.1 (Schur)   *Let (ρ, V) and (σ, W) be two irreducible representations of a group G. If L intertwines ρ and σ, then either L is zero, or it is an isomorphism.*

Proof Let $V_0 = \{v \in V : L(v) = 0\}$ and $W_0 = \{L(v) : v \in V\} \le W$ denote the kernel and the range of L. If L intertwines ρ and σ then V_0 and W_0 are ρ- and σ-invariant, respectively. By irreducibility, either $V_0 = V$ and $W_0 = \{0\}$ or $V_0 = \{0\}$ and, necessarily, $W_0 = W$. In the first case L vanishes, in the second case it is an isomorphism. $\square$

Corollary 3.5.2 *Let (ρ, V) be an irreducible representation of a group G. Let L be a linear transformation of V which intertwines ρ with itself (that is $L\rho(g) = \rho(g)L$ for all $g \in G$). Then L is a multiple of the identity: $L \in \mathbb{C}I$.*

Proof Let λ be an eigenvalue of L. Then $(\lambda I - L)$ also intertwines ρ with itself and, by the previous lemma, it is either invertible or zero. But by definition of eigenvalue it cannot be invertible, therefore necessarily $L = \lambda I$. Note that λ exists because we are dealing with complex vector spaces. $\square$

Exercise 3.5.3 Let G be a (finite) abelian group. A representation (ρ, V) of G is irreducible if and only if it is one dimensional. Use this fact to prove that $\widehat{C_n} = \{\rho_k : k = 0, 1, \ldots, n-1\}$ as in Example 3.4.4 (see also Section 2.1).

Exercise 3.5.4 Show that if $\rho \in \widehat{G}$ and g_0 is in the center $Z(G) = \{g \in G : gh = hg, \ \forall h \in G\}$ of G, then $\rho(g_0) = \lambda I$ for some $\lambda \in \mathbb{C}$.

Exercise 3.5.5 (Converse to Schur's lemma) Suppose that the representation (ρ, V) of G has the property that whenever $T : V \to V$ is a linear mapping and $\rho(g)T = T\rho(g)$ for all $g \in G$ then necessarily $T = \lambda I$ for some $\lambda \in \mathbb{C}$. Show that ρ is irreducible.

3.6 Matrix coefficients and their orthogonality relations

Let (ρ, V) be a unitary representation of a group G. For v and w in V we consider the complex valued function $u_{v,w}(g) = \langle \rho(g)w, v \rangle$, $g \in G$, which is called a *(matrix) coefficient* of the representation ρ. This terminology comes from the fact that if $\{v_1, v_2, \ldots, v_n\}$ is an orthonormal basis for V, then $\rho(g)$, viewed as an n by n matrix coincides with $(u_{v_i,v_j}(g))_{i,j}$, in other words the $u_{v_i,v_j}(g)$'s are the matrix coefficients of $\rho(g)$, for all $g \in G$.

It is interesting to note that any complex valued function $u : G \to \mathbb{C}$ can be realized as a coefficient of a unitary representation. Indeed, denoting by $V = L(G)$, the space of all complex valued functions on G endowed with the scalar product (3.10) we have that $u(g) = \langle \lambda(g) \delta_{1_G}, \overline{u} \rangle$ where λ is the left regular representation of G (see Example 3.4.1) and δ_{1_G} is the Dirac function on the unit element 1_G of G.

We now show that the coefficients of irreducible representations constitute an orthogonal basis for $L(G)$, the space of all complex valued functions on G endowed with the scalar product (3.10).

Lemma 3.6.1 *Let (ρ, V) and (σ, W) be two irreducible, nonequivalent representations of a group G. Then all coefficients of ρ are orthogonal to all coefficients of σ.*

Proof Let $v_1, v_2 \in V$ and $w_1, w_2 \in W$. Our goal is to show that the functions $u(g) = \langle \rho(g)v_1, v_2 \rangle_V$ and $u'(g) = \langle \sigma(g)w_1, w_2 \rangle_W$ are orthogonal in $L(G)$. Consider the linear transformation $L : V \to W$ defined by $Lv = \langle v, v_2 \rangle_V w_2$ for all $v \in V$. Then, the linear transformation $\tilde{L} : V \to W$ defined by

$$\tilde{L} = \sum_{g \in G} \sigma(g^{-1}) L \rho(g)$$

satisfies $\tilde{L}\rho(g) = \sigma(g)\tilde{L}$ for all $g \in G$, that is $\tilde{L}$ intertwines ρ and σ. In virtue of Schur's lemma we have that either $\tilde{L}$ is invertible or $\tilde{L} = 0$. As $\rho \not\sim \sigma$ we necessarily have the second possibility and thus

$$
\begin{aligned}
0 = \left\langle \tilde{L}v_1, w_1 \right\rangle_W &= \sum_{g \in G} \langle L\rho(g)v_1, \sigma(g)w_1 \rangle_W \\
&= \sum_{g \in G} \langle \rho(g)v_1, v_2 \rangle_V \cdot \langle w_2, \sigma(g)w_1 \rangle_W \\
&= \sum_{g \in G} u(g)\overline{u'(g)}.
\end{aligned}
$$

$\square$

Corollary 3.6.2 *Let G be a finite group. Then there exist only finitely many irreducible unitary representations pairwise non equivalent, in other words $|\widehat{G}| < \infty$.*

Proof The space $L(G)$ is finite dimensional and contains only finitely many distinct pairwise orthogonal functions and the statement follows from the previous lemma. $\square$

Let (ρ, V) be an irreducible unitary representation. Let $\{v_1, v_2, \ldots, v_d\}$ be an orthonormal basis of V. Then the coefficients $u_{ij}(g) = \langle \rho(g)v_j, v_i \rangle$ are pairwise orthogonal. More precisely

Lemma 3.6.3 $\langle u_{ij}, u_{kh} \rangle_{L(G)} = \dfrac{|G|}{d} \delta_{ik} \delta_{jh}.$

Proof Fix indices i and k and consider the operator $L_{ik} : V \to V$ defined by $L_{ik}(v) = \langle v, v_i \rangle_V v_k$. Observe that $tr(L_{ik}) = \delta_{ik}$. Now define $\widetilde{L}_{ik} = \frac{1}{|G|} \sum_{g \in G} \rho(g^{-1}) L_{ik} \rho(g)$ and observe that $\widetilde{L}_{ik}\rho(g) = \rho(g)\widetilde{L}_{ik}$. As ρ is irreducible, necessarily $\widetilde{L}_{ik} = \alpha I$ for a suitable $\alpha \in \mathbb{C}$. Actually $\alpha = \delta_{ik}/d$ as $d\alpha = tr(\widetilde{L}_{ik}) = tr(L_{ik}) = \delta_{ik}$ (recall that if $A, B \in GL(V)$ then $tr(AB) = tr(BA)$). It follows that $\widetilde{L}_{ik} = (1/d)\delta_{ik}I$ and therefore $\left\langle \widetilde{L}_{ik}v_j, v_h \right\rangle_V = (1/d)\delta_{jh}\delta_{ik}$. But

$$
\begin{aligned}
\left\langle \widetilde{L}_{ik}v_j, v_h \right\rangle_V &= \frac{1}{|G|} \sum_{g \in G} \langle L_{ik}\rho(g)v_j, \rho(g)v_h \rangle_V \\
&= \frac{1}{|G|} \sum_{g \in G} \langle \rho(g)v_j, v_i \rangle_V \cdot \langle v_k, \rho(g)v_h \rangle_V \\
&= \frac{1}{|G|} \langle u_{ij}, u_{kh} \rangle_{L(G)}
\end{aligned}
$$

which ends the proof. $\qquad \square$

The following lemma presents further properties of the matrix coefficients of a unitary representation.

Lemma 3.6.4 *Under the same hypotheses preceding Lemma 3.6.3 (with no irreducibility assumption), for all $g, g_1, g_2 \in G$ and $1 \le i, j, k \le d$ one has:*

(i) $u_{i,j}(g^{-1}) = \overline{u_{j,i}(g)}$;
(ii) $u_{i,j}(g_1 g_2) = \sum_{h=1}^{d} u_{i,h}(g_1) u_{h,j}(g_2)$;
(iii) $\sum_{j=1}^{d} u_{j,i}(g) u_{j,k}(g) = \delta_{i,k}$ *and* $\sum_{i=1}^{d} u_{j,i}(g)\overline{u_{k,i}(g)} = \delta_{j,k}$ *(dual orthogonality relations).*

Proof (i) follows immediately from $\rho(g)^* = \rho(g^{-1})$ and $\langle x, y \rangle = \overline{\langle y, x \rangle}$ for all $g \in G$ and $x, y \in V$.

(ii) For all $x \in V$ one has $x = \sum_{i=1}^{d} v_i \langle x, v_i \rangle$ so that, for all $g \in G$ and $j = 1, 2, \ldots, d$

$$
\rho(g)v_j = \sum_{h=1}^{d} v_h u_{h,j}(g). \tag{3.13}
$$

Thus $\sum_{h=1}^{d} v_h u_{h,j}(g_1 g_2) = \rho(g_1 g_2) v_j = \sum_{h=1}^{d} \rho(g_1) v_h u_{h,j}(g_2)$ and taking the scalar product with v_i yields the desired equality.

(iii) Recalling that $\rho(g)$ is unitary, we have

$$\sum_{j=1}^{d} u_{j,i}(g)\overline{u_{j,k}(g)} = \langle \sum_{j=1}^{d} v_j u_{j,i}(g), \sum_{j'=1}^{d} v_{j'} u_{j',k}(g) \rangle$$

$$= \langle \rho(g)v_i, \rho(g)v_k \rangle = \langle v_i, v_k \rangle = \delta_{i,k}.$$

□

In the following we shall refer to the matrix $(u_{i,j})$ as a (unitary) *matrix realization* of the (unitary) representation ρ.

3.7 Characters

With each equivalence class of irreducible representations one associates a function, called the *character* of the representation defined by $\chi_\rho(g) = tr(\rho(g))$. We recall that if V is a finite dimensional complex vector space and denoting by $End(V)$ the space of all linear maps $T : V \to V$, the *trace* is the (necessarily unique) linear functional $tr : End(V) \to \mathbb{C}$ such that $tr(TS) = tr(ST)$ for all $T, S \in End(V)$ and $tr(I_V) = \dim V$. In the following exercise we collect the basic properties of the trace.

Exercise 3.7.1 Let V be a finite dimensional vector space. (1) Show the uniqueness of the trace.

(2) Endow V with a scalar product $\langle \cdot, \cdot \rangle$ and fix an orthonormal basis $\{v_1, v_2, \ldots, v_d\}$ for V. Then

$$tr(T) = \sum_{i=1}^{d} \langle Tv_i, v_i \rangle$$

for $T \in End(V)$. This amounts to saying that the trace of a linear operator equals the sum of the diagonal entries of its representing matrix.

(3) Let $A \in GL(V)$ and $T \in End(V)$. Then $tr(T) = tr(A^{-1}TA)$. This is the *central property* of the trace.

(4) Let $\lambda_1, \lambda_2, \ldots, \lambda_d$ denote the eigenvalues of $T \in End(V)$, listed according to their algebraic multiplicities. Then $tr(T) = \sum_{i=1}^{d} \lambda_i$.

(5) With the notation in (2) we have

$$tr(TS) = \sum_{i,j=1}^{d} \langle Tv_j, v_i \rangle \langle Sv_i, v_j \rangle \tag{3.14}$$

for all $T, S \in End(V)$.

From (2) in the previous exercise we then have $tr(\rho(g)) = \sum_{j=1}^{n} u_{jj}(g)$ (where the functions u_{ij} are the matrix coefficients defined in the previous section) and the function χ_ρ does not depend on the particular choice of the orthonormal system $\{v_1, v_2, \ldots, v_d\}$ in V, nor on the particular element ρ in its equivalent class (this in virtue of (3) in Exercise 3.7.1).

We observe that if ρ is a one-dimensional representation, then it coincides with its character: $\rho = \chi_\rho$.

It is also possible to define the character for a representation which is not irreducible. If (ρ, V) is a unitary representation one sets $\chi_\rho(g) = tr(\rho(g))$, for all $g \in G$.

Sometimes we shall write χ_V instead of χ_ρ.

Proposition 3.7.2 *Let χ_ρ be the character of a representation (ρ, V) of a group G. Denote by $d = dim(V)$ its degree. Then, for all $s, t \in G$,*

(i) $\chi_\rho(1_G) = d$;

(ii) $\chi_\rho(s^{-1}) = \overline{\chi_\rho(s)}$;

(iii) $\chi_\rho(t^{-1}st) = \chi_\rho(s)$.

Proof To prove (i) observe that $\rho(1_G) = I$ the identity on V whose trace clearly equals the dimension of V.

We have

$$\chi_\rho(s^{-1}) = tr(\rho(s^{-1})) = tr(\rho(s)^*) = \overline{\chi_\rho(s)}$$

since $\rho(s^{-1}) = \rho(s)^*$ (recall that $\rho(s)$ is unitary) and $tr(A^*) = \overline{tr(A)}$ for all $A \in GL(V)$. This proves (ii).

The last statement follows again from the central property of the trace. Note that $\rho(g)$ is diagonalizable (it is unitary) and its trace coincides with the sum of its eigenvalues. □

Exercise 3.7.3 (1) Let G be a finite group and ρ a G-representation. Show that if $n = |G|$ then the eigenvalues of $\rho(s), s \in G$ are nth roots of the unity.

(2) Deduce that $|\chi_\rho(g)| \leq d$ for all $g \in G$.

From Lemma 3.6.1 it follows that characters relative to inequivalent representations are orthogonal in $L(G)$, while, from Lemma 3.6.3, it follows that the norm of a character equals $|G|$: indeed the square of its norm is the sum of the squares of the norms of the u_{jj}'s, which all have the same norm $|G|/d$. We thus have that the characters of irreducible representations constitute an orthonormal system (not complete!). Therefore they are finite in number and their cardinality equals the number of equivalence classes of irreducible representations. We state this as follows:

Proposition 3.7.4 *Let ρ and σ be irreducible representations of a group G.*

(i) *If ρ and σ are inequivalent then $\langle \chi_\rho, \chi_\sigma \rangle = 0$.*

(ii) $\langle \chi_\rho, \chi_\rho \rangle = |G|$.

Proposition 3.7.5 *Let ρ and σ be two representations of a group G. Suppose that ρ decomposes into irreducibles as $\rho = \rho_1 \oplus \rho_2 \oplus \cdots \oplus \rho_k$ and that σ is irreducible. Then, setting $m_\sigma = |\{j : \rho_j \sim \sigma\}|$, one has*

$$m_\sigma = \frac{1}{|G|} \langle \chi_\rho, \chi_\sigma \rangle . \tag{3.15}$$

In particular, m_σ does not depend on the chosen decomposition of ρ.

Proof If ρ decomposes as the direct sum $\rho = \rho_1 \oplus \rho_2 \oplus \cdots \oplus \rho_k$ of irreducible sub-representations, then $\chi_\rho = \sum_{j=1}^k \chi_{\rho_j}$, that is the character of ρ is the sum of the characters of the components ρ_j's.

Therefore, given an irreducible representation σ, the inner product $\langle \chi_\rho, \chi_\sigma \rangle$ is different from zero if and only if σ is a sub-representation of ρ, equivalently $\sigma = \rho_j$ for some $j = 1, \ldots, k$. □

The number m_σ in (3.15) is called the *multiplicity* of σ as a sub-representation of ρ. If σ is not contained in ρ then clearly $m_\sigma = 0$.

Corollary 3.7.6 *Let ρ be a representation of a group G. Then*

$$\rho = \bigoplus_{\sigma \in \widehat{G}} m_\sigma \sigma$$

where $m_\sigma \sigma = \sigma \oplus \sigma \cdots \oplus \sigma$ is the direct sum of m_σ copies of σ and

$$\chi_\rho = \sum_{\sigma \in \widehat{G}} m_\sigma \chi_\sigma.$$

Corollary 3.7.7 *Let ρ, σ be two representations of G. Suppose that they decompose as $\rho = \oplus_{i \in I} m_i \rho_i$ and $\sigma = \oplus_{j \in J} n_j \rho_j$, respectively, where the*

ρ_i's are irreducible and the m_i's and n_j's are the corresponding multiplicities. Then, denoting by $I \cap J$ the index set of common irreducible representations, we have

$$\frac{1}{|G|} \langle \chi_\rho, \chi_\sigma \rangle = \sum_{i \in I \cap J} m_i n_i.$$

Corollary 3.7.8 *A representation ρ of a group G is irreducible if and only if $\langle \chi_\rho, \chi_\rho \rangle = |G|$.*

Corollary 3.7.9 *Two representations ρ and σ are equivalent if and only if $\chi_\rho = \chi_\sigma$.*

Exercise 3.7.10 Let $n, k \in \mathbb{N}$. Denote by $p(k) = |\{g \in S_k : g(i) \neq i, \forall i = 1, 2, \ldots, k\}|$ the cardinality of the set of elements in the symmetric group of degree k with no fixed points; also, for $g \in S_n$ denote by

$$a(g) = |\{i \in \{1, 2, \ldots, n\} : g(i) = i\}|$$

the number of points fixed by g.

Let now χ_V and χ_W denote the characters of the representations in Example 3.4.5 (c).

(1) Prove that $\sum_{a=0}^n \binom{n}{a} p(n - a) = n!$.

(2) Prove that $\chi_W(g) = \chi_V(g) - 1$ and that $\chi_V(g) = a(g)$ for all $g \in S_n$.

(3) From (1) and (2) deduce that $\sum_{g \in S_n} (\chi_W(g))^2 = n!$. In particular, the representation of S_n on W is irreducible.

Theorem 3.7.11 (Peter–Weyl) *Let G be a finite group and denote by $(\lambda, L(G))$ the left regular representation.*

(i) *Any irreducible representation of $(\rho, V_\rho), \rho \in \widehat{G}$ appears in the decomposition of λ with multiplicity equal to its dimension d_ρ.*

(ii) *Denoting by $u_{i,j}^\rho$ the matrix coefficients of $\rho \in \widehat{G}$ with respect to an orthonormal basis, then the functions*

$$\left\{ \sqrt{\frac{d_\rho}{|G|}} u_{i,j}^\rho : i, j = 1, \ldots, d_\rho, \rho \in \widehat{G} \right\}$$

constitute an orthonormal system in $L(G)$.

(iii) $|G| = \sum_{\rho \in \widehat{G}} d_\rho^2$ *and* $L(G) = \bigoplus_{\rho \in \widehat{G}} d_\rho V_\rho$.

Proof Denote by

$$\lambda = \bigoplus_{\rho \in \widehat{G}} m_\rho \rho$$

the decomposition of λ into irreducibles where the integer m_ρ denotes the multiplicity in λ of the corresponding sub-representation ρ and $m_\rho \rho$ stands for the direct sum of m_ρ copies of ρ.

We observe that using the (complete) orthonormal system $\{\delta_g\}_{g \in G}$ of the Dirac deltas in $L(G)$ we immediately obtain that $\chi_\lambda(1_G) = |G|$ and $\chi_\lambda(g) = 0$ if $g \neq 1_G$: this follows from the fact that if $h, g \in G$, then $\lambda(h)\delta_g = \delta_{hg}$.

On the other hand, if $\rho \in \widehat{G}$, then from Proposition 3.7.5, we get $m_\rho = \frac{1}{|G|} \langle \chi_\lambda, \chi_\rho \rangle = \chi_\rho(1_G)$. This yields $m_\rho = d_\rho$ and $|G| \equiv \dim L(G) = \sum_{\rho \in \widehat{G}} m_\rho d_\rho = \sum_{\rho \in \widehat{G}} d_\rho^2$ which might also be derived from the formula

$$\chi_\lambda(1_G) = |G| = \sum_{\rho \in \widehat{G}} m_\rho \chi_\rho(1_G).$$

From Lemma 3.6.1 and Lemma 3.6.3 we have that the functions $\sqrt{\frac{d_\rho}{|G|}} u_{ij}^\rho$ where $\rho \in \widehat{G}$ and $1 \leq i, j \leq d_\rho$ constitute an orthonormal system in $L(G)$. This system is in fact complete. Indeed as $\sum_{\rho \in \widehat{G}} d_\rho^2 = |G|$ one has that $|\{\sqrt{\frac{d_\rho}{|G|}} u_{ij}^\rho : \rho \in \widehat{G}, 1 \leq i, j \leq d_\rho\}| = |G|$. As the latter equals the dimension of $L(G)$ we are done. $\square$

3.8 More examples

Example 3.8.1 Let $C_n = \langle a \rangle$ be the cyclic group and $\{\rho_k : k = 0, 1, \ldots, n-1\}$ be the representations as in Example 3.4.4. We are now in position to prove that these constitute a complete system, in other words $\{\rho_k : k = 0, 1, \ldots, n-1\} = \widehat{C_n}$. Indeed, observing that the character of a one-dimensional representation coincides with the representation itself, we have that these representations are pairwise inequivalent. Another way to prove this fact is to use the orthogonality relations in Section 2.1 and Proposition 3.7.4:

$$\langle \chi_{\rho_k}, \chi_{\rho_h} \rangle = \sum_{j=0}^{n-1} e^{2\pi i(k-h)j/n} = \begin{cases} n & \text{if } h = k \\ 0 & \text{if } h \neq k. \end{cases}$$

Since the number of the ρ_k's is $n = |C_n|$, by (iii) of the Peter–Weyl theorem (Theorem 3.7.11), these one-dimensional representations exhaust

$\widehat{C_n}$, the irreducible representations of C_n. Alternatively, one can use Exercise 3.5.3.

Example 3.8.2 Let $D_n = \langle a, b : a^2 = b^n = 1, aba = b^{-1} \rangle$ be the *dihedral* group of degree n, i.e. the group of isometries of a regular polygon with n vertices. In the following we outline the determination of all the irreducible representations of D_n.

We consider first the case when n is even. We have four one-dimensional representations (we identify these with the corresponding characters), namely, for all $h = 0, 1$ and $k = 0, 1, \ldots, n - 1$

$$
\begin{aligned}
\chi_1(a^h b^k) &= 1 \\
\chi_2(a^h b^k) &= (-1)^h \\
\chi_3(a^h b^k) &= (-1)^k \\
\chi_4(a^h b^k) &= (-1)^{h+k}.
\end{aligned}
\tag{3.16}
$$

Set $\omega = e^{2\pi i/n}$ and, for $t = 0, 1, \ldots, n$, define the two-dimensional representation ρ_t by setting

$$
\rho_t(b^k) = \begin{pmatrix} \omega^{tk} & 0 \\ 0 & \omega^{-tk} \end{pmatrix} \quad \text{and} \quad \rho_t(ab^k) = \begin{pmatrix} 0 & \omega^{-tk} \\ \omega^{tk} & 0 \end{pmatrix}.
$$

Exercise 3.8.3 (1) Show that ρ_t is indeed a representation.

(2) Show that $\rho_t \sim \rho_{n-t}$.

(3) Show that $\chi_{\rho_0} = \chi_1 + \chi_2$ and $\chi_{\rho_{n/2}} = \chi_3 + \chi_4$.

(4) Show that ρ_t, with $1 \leq t \leq \frac{n}{2} - 1$, are pairwise nonequivalent irreducible representations in two different ways, namely:

    (a) by inspecting for invariant subspaces and intertwining operators;

    (b) by computing the characters and their inner products.

(5) Conclude that $\chi_1, \chi_2, \chi_3, \chi_4, \chi_{\rho_t}$, with $1 \leq t < n/2$ constitute a complete list of irreducible representations of D_n.

Exercise 3.8.4 Determine a complete list of irreducible representations of D_n in the case n odd.

3.9 Convolution and the Fourier transform

Definition 3.9.1 Let P and Q be two complex valued functions defined on the group G. The *convolution* of P with Q is the function defined by

$$[P * Q](g) = \sum_{h \in G} P(gh^{-1})Q(h).$$

Note that the convolution may be also written in the following equivalent forms:

$$[P * Q](g) = \sum_{h \in G} P(h)Q(h^{-1}g) = \sum_{\substack{h,k \in G \\ kh=g}} P(k)Q(h).$$

Remark 3.9.2 The order of the two functions in the convolution is essential, since in general $P*Q \neq Q*P$. Indeed the convolution product is commutative if and only if the group G is abelian. To see this, note that if $a, b \in G$ then

$$\delta_a * \delta_b = \delta_{ab} \qquad \text{and} \qquad \delta_b * \delta_a = \delta_{ba},$$

where $\delta_t(g) = \begin{cases} 1 & \text{if} \quad g = t \\ 0 & \text{if} \quad g \neq t. \end{cases}$ The arbitrarity of the elements a, b infers that the commutativity of the group is a necessary condition for the commutativity of the convolution. On the other hand one immediately checks that if G is abelian then $P * Q = Q * P$ (see also Chapter 2).

Lemma 3.9.3 *The space $L(G)$ of complex valued functions defined on the group G endowed with the convolution product is an algebra (over $\mathbb{C}$: see Definition 2.1.5). Indeed it satisfies the following properties:*

  (i) *$L(G)$ endowed with the (pointwise) sum is a vector space over the field of complex numbers;*
 (ii) *the convolution product is distributive on the right and on the left with respect to the sum: $(P + Q) * R = P * R + Q * R$ and $R * (P + Q) = R * P + R * Q$;*
(iii) *$L(G)$ has δ_{1_G} as a unit element with respect to the convolution product: $P * \delta_{1_G} = P = \delta_{1_G} * P$;*
 (iv) *the convolution product is associative: $(P * Q) * R = P * (Q * R)$.*

Proof (i) and (ii) are obvious.

(iii) One easily checks that δ_{1_G} is the unit element with respect to the convolution product.

(iv) Let P, Q and R be complex valued functions defined on G. Then we have

$$
\begin{aligned}
[P * (Q * R)](g) &= \sum_{h \in G} P(gh^{-1})(Q * R)(h) \\
&= \sum_{h \in G} \sum_{t \in G} P(gh^{-1})Q(ht^{-1})R(t) \\
\text{(setting } h = mt) \quad &= \sum_{t \in G} \sum_{m \in G} P(gt^{-1}m^{-1})Q(m)R(t) \\
&= \sum_{t \in G} (P * Q)(gt^{-1})R(t) = [(P * Q) * R](g).
\end{aligned}
$$

$\square$

In virtue of the above properties, $L(G)$ is called the *group algebra* of the group G.

The *center* of $L(G)$ is the subalgebra made up of all functions $P \in L(G)$ commuting with every element in $L(G)$, that is $P * Q = Q * P$ for all $Q \in L(G)$. Such elements $P \in L(G)$ are termed *central*.

Lemma 3.9.4 *A function $P \in L(G)$ is central if and only if $P(a^{-1}ta) = P(t)$ for all $a, t \in G$, i.e. it is constant on each conjugacy class of G.*

Proof For a function P the condition of belonging to the center is equivalent to

$$
\sum_{h \in G} Q(gh^{-1})P(h) = \sum_{h \in G} P(gh^{-1})Q(h) \quad \forall Q \in L(G), \forall g \in G. \quad (3.17)
$$

If we choose $Q = \delta_a$ and $g = ta$ in (3.17) we obtain: $P(a^{-1}ta) = P(t)$, $\forall a, t \in G$. The converse is trivial. $\square$

Definition 3.9.5 *Let $P \in L(G)$. The Fourier transform of P with respect to the G-representation (ρ, V) is the linear operator $\widehat{P}(\rho) : V \to V$ defined by $\widehat{P}(\rho) = \sum_{g \in G} P(g)\rho(g)$.*

Lemma 3.9.6 *For every couple of functions $P, Q \in L(G)$ and every G-representation (ρ, V) we have:*

$$
\widehat{P * Q}(\rho) = \widehat{P}(\rho)\widehat{Q}(\rho).
$$

Proof

$$\widehat{P * Q}(\rho) = \sum_{g \in G} \left[\sum_{h \in G} P(gh^{-1})Q(h) \right] \rho(g)$$

$$= \sum_{g \in G} \sum_{h \in G} P(gh^{-1})Q(h)\rho(gh^{-1})\rho(h)$$

$$= \sum_{h \in G} \left[\sum_{g \in G} P(gh^{-1})\rho(gh^{-1}) \right] Q(h)\rho(h) = \widehat{P}(\rho)\widehat{Q}(\rho).$$

$\square$

Proposition 3.9.7 *If P is a central function, then its Fourier transform with respect to the irreducible G-representation (ρ, V) is given by:*

$$\widehat{P}(\rho) = \lambda I \quad \text{with} \quad \lambda = \frac{1}{d_\rho} \sum_{g \in G} P(g)\chi_\rho(g) = \frac{1}{d_\rho} \langle P, \overline{\chi_\rho} \rangle.$$

Proof Observe that for any $g \in G$

$$\rho(g)\widehat{P}(\rho)\rho^{-1}(g) = \sum_{h \in G} P(h)\rho(g)\rho(h)\rho(g^{-1}) = \sum_{h \in G} P(h)\rho(ghg^{-1})$$

$$= \sum_{h \in G} P(ghg^{-1})\rho(ghg^{-1}) = \widehat{P}(\rho),$$

so that $\widehat{P}(\rho)$ intertwines ρ with itself. By Corollary 3.5.2 we deduce that $\widehat{P}(\rho) = \lambda I$. Computing the trace of both members we obtain:

$$tr\left(\widehat{P}(\rho)\right) = \sum_{h \in G} P(h)\chi_\rho(h) = \lambda d_\rho.$$

$\square$

Theorem 3.9.8 (Fourier's inversion formula) *For a function $P \in L(G)$ the following formula holds:*

$$P(g) = \frac{1}{|G|} \sum_{\rho \in \widehat{G}} d_\rho tr\left(\rho(g^{-1})\widehat{P}(\rho)\right) \quad \forall g \in G.$$

In particular if $P_1, P_2 \in L(G)$ satisfy the condition $\widehat{P_1}(\rho) = \widehat{P_2}(\rho)$ for every $\rho \in \widehat{G}$, then we have $P_1 = P_2$.

Proof We know, by Theorem 3.7.11, that the coefficients $\sqrt{\frac{d_\rho}{|G|}} u^\rho_{i,j}$, computed with respect to an orthonormal basis $\{v^\rho_1, v^\rho_2, \dots, v^\rho_{d_\rho}\}$ for all

$\rho \in \widehat{G}$, constitute an orthonormal basis in $L(G)$. Then also the conjugates $\sqrt{\frac{d_\rho}{|G|}}\, \overline{u_{i,j}^\rho}$ form an orthonormal basis and therefore we can express every function $P \in L(G)$ as

$$P(g) = \frac{1}{|G|} \sum_{\rho \in \widehat{G}} d_\rho \sum_{i,j=1}^{d_\rho} \left\langle P, \overline{u_{i,j}^\rho} \right\rangle \overline{u_{i,j}^\rho(g)}, \qquad (3.18)$$

where $1 \leq i,\ j \leq d_\rho$. We recall that $\widehat{P}(\rho) = \sum_{g \in G} P(g)\rho(g)$ and $u_{i,j}^\rho(g) = \langle \rho(g)v_j^\rho, v_i^\rho \rangle$, which implies that

$$\langle P, \overline{u_{i,j}^\rho} \rangle = \sum_{g \in G} P(g)u_{i,j}^\rho(g) = \sum_{g \in G} P(g)\langle \rho(g)v_j^\rho, v_i^\rho \rangle = \left\langle \widehat{P}(\rho)v_j^\rho, v_i^\rho \right\rangle$$

$$(3.19)$$

and

$$\sum_{i,j=1}^{d_\rho} \left\langle P, \overline{u_{i,j}^\rho} \right\rangle \overline{u_{i,j}^\rho(g)} = \sum_{i,j=1}^{d_\rho} \left\langle \widehat{P}(\rho)v_j^\rho, v_i^\rho \right\rangle \langle v_i^\rho, \rho(g)v_j^\rho \rangle$$

$$= \sum_{i,j=1}^{d_\rho} \left\langle \widehat{P}(\rho)v_j^\rho, v_i^\rho \right\rangle \langle \rho(g^{-1})v_i^\rho, v_j^\rho \rangle$$

$$= tr\left(\rho(g^{-1})\widehat{P}(\rho) \right),$$

where the last equality follows from (3.14). Therefore, taking into account (3.18), the statement follows. □

The Fourier inversion formula shows that every function in $L(G)$ is uniquely determined by its Fourier transforms with respect to the irreducible representations; moreover it gives an explicit way to express a function P as linear combination of an orthonormal system. Finally from this analysis we can deduce that the algebra $L(G)$ is isomorphic to a direct sum of matrix algebras namely $L(G) \sim \oplus_{\rho \in \widehat{G}} M_{d_\rho}$, where M_{d_ρ} is the algebra of $d_\rho \times d_\rho$ matrices over $\mathbb{C}$ (this fact will be exploited in Section 9.5). In particular, defining $C(G) = \bigoplus_{\rho \in \widehat{G}} \{\widehat{P}(\rho) : P \in L(G)\}$, then the following holds.

Corollary 3.9.9 *The Fourier transform $L(G) \ni P \mapsto \widehat{P} \in C(G)$ and the map $C(G) \ni Q \mapsto \check{Q} \in L(G)$, where $\check{Q}(g) = \frac{1}{|G|} \sum_{\rho \in \widehat{G}} d_\rho tr(\rho(g^{-1})Q(g))$, are bijective and inverses one of each other. Moreover, $C(G) = \bigoplus_{\rho \in \widehat{G}} Hom(V_\rho)$.*

Theorem 3.9.10 *The characters $\{\chi_\rho, \ \rho \in \widehat{G}\}$ constitute an orthogonal basis for the subspace of central functions. In particular $|\widehat{G}|$ equals the number of conjugacy classes in G.*

Proof We have already showed (cf. Proposition 3.7.2 and Proposition 3.7.4) that the characters of irreducible representations are central functions pairwise orthogonal if the representations are inequivalent. If a central function P is orthogonal to these characters, by Proposition 3.9.7 we have $\widehat{P}(\rho) = 0$ for all $\rho \in \widehat{G}$ and therefore by Fourier's inversion formula we have $P = 0$. Therefore the characters are an orthogonal basis of the subspace of central functions which has dimension equal to the number of conjugacy classes (Lemma 3.9.4). $\qquad\qquad\square$

Exercise 3.9.11 (1) Use Theorem 3.9.10 to prove that a finite group G is abelian if and only if its irreducible representations are all one-dimensional.

(2) More generally, prove that if a group G contains a commutative subgroup A, then $d_\rho \leq |G/A|$ for all $\rho \in \widehat{G}$.

Proposition 3.9.12 (Plancherel's formula) *Let P and Q be functions on G. Then we have*

$$\langle P, Q \rangle = \frac{1}{|G|} \sum_{\rho \in \widehat{G}} d_\rho tr\left(\widehat{P}(\rho)\widehat{Q}(\rho)^*\right). \qquad (3.20)$$

Proof It is analogous to that one used for the Fourier inversion formula (Theorem 3.9.8). Indeed, starting from the formula

$$\langle P, Q \rangle = \sum_{\rho \in \widehat{G}} \frac{d_\rho}{|G|} \sum_{i,j=1}^{d_\rho} \left\langle P, \overline{u_{i,j}^\rho} \right\rangle \left\langle \overline{u_{i,j}^\rho}, Q \right\rangle,$$

one can use (3.19) to show that

$$\langle P, Q \rangle = \frac{1}{|G|} \sum_{\rho \in \widehat{G}} d_\rho \sum_{i,j=1}^{d_\rho} \left\langle \widehat{P}(\rho)v_j^\rho, v_i^\rho \right\rangle \cdot \left\langle v_i^\rho, \widehat{Q}(\rho)v_j^\rho \right\rangle$$

$$= \frac{1}{|G|} \sum_{\rho \in \widehat{G}} d_\rho tr\left(\widehat{P}(\rho)\widehat{Q}(\rho)^*\right).$$

$\qquad\qquad\square$

Exercise 3.9.13 From the orthogonality relations for the characters of the irreducible representations:

$$\frac{1}{|G|} \sum_{g \in G} \chi_{\rho_1}(g) \overline{\chi_{\rho_2}(g)} = \delta_{\rho_1, \rho_2}, \quad \rho_1, \rho_2 \in \widehat{G} \tag{3.21}$$

deduce the *dual orthogonality relations*:

$$\sum_{\rho \in \widehat{G}} \chi_\rho(t_1) \overline{\chi_\rho(t_2)} = \frac{|G|}{|C(t_1)|} \delta_{t_1, t_2} \quad t_1, t_2 \in T$$

where $T \subseteq G$ is a set of representatives for the conjugacy classes of G and $C(t) = \{g^{-1} t g : g \in G\}$ denotes the conjugacy class of $t \in T$.

We end this section with a simple property of the convolution of matrix coefficients.

Lemma 3.9.14 Let $\rho, \sigma \in \widehat{G}$ and denote by $u_{i,j}^\rho$ and $u_{h,k}^\sigma$, $1 \leq i, j \leq d_\rho$ and $1 \leq h, k \leq d_\sigma$, the corresponding matrix coefficients (with respect to orthonormal bases on V_ρ and V_σ, respectively). Then

$$u_{i,j}^\rho * u_{h,k}^\sigma = \frac{|G|}{d_\rho} \delta_{j,h} \delta_{\rho,\sigma} u_{i,k}^\rho.$$

Proof Using first (ii) in Lemma 3.6.4 and then the orthogonality relations in Lemma 3.6.3 we have

$$[u_{i,j}^\rho * u_{h,k}^\sigma](g) = \sum_{s \in G} u_{i,j}^\rho(gs) u_{h,k}^\sigma(s^{-1})$$

$$= \sum_{\ell=1}^{d_\rho} u_{i,\ell}^\rho(g) \sum_{s \in G} u_{\ell,j}^\rho(s) \overline{u_{k,h}^\sigma(s)}$$

$$= \sum_{\ell=1}^{d_\rho} u_{i,\ell}^\rho(g) \delta_{\ell,k} \delta_{j,h} \frac{|G|}{d_\rho} \delta_{\rho,\sigma}$$

$$= \frac{|G|}{d_\rho} \delta_{j,h} \delta_{\rho,\sigma} u_{i,k}^\rho(g).$$

$\square$

3.10 Fourier analysis of random walks on finite groups

We recall that a probability measure on a group G is a function $P : G \to \mathbb{R}$ which satisfies the following conditions:

(i) $P(g) \geq 0, \forall g \in G$;

(ii) $\sum_{g \in G} P(g) = 1$.

It is easy to show that if P and Q are probability measures on G, so is their convolution product $Q * P$, defined by

$$Q * P(g) = \sum_{h \in G} Q(h)P(h^{-1}g) = \sum_{h,k \in G : hk=g} Q(h)P(k).$$

Intuitively we can say that, if $Q(g)$ represents the probability that the element g is chosen according with the law defined by Q, then $Q * P(g)$ represents the probability that g equals hk when we first choose k according with the law P and after we independently choose h following the law Q.

The uniform distribution is the probability measure U defined by $U(g) = \frac{1}{|G|} \forall g \in G$. It is characterized by the condition $P * U = U * P = U$ for every probability measure P, as easily follows from $U(gh) = U(g) \forall g, h \in G$.

Let us explain the use of the Fourier analysis for random walks on the group G. Suppose we have a probability measure P on the group G. Then P induces a random walk on G in the sense that $P(h^{-1}g)$ represents the probability to pass from the element h to the element g. Suppose that Q is another probability measure. Interpreting $Q(g)$ as the probability to stay in the element g, we can say that Q represents an initial distribution. If the transition from a state to the following one is determined by P, we have that $Q * P(g) = \sum_{h \in G} Q(h)P(h^{-1}g)$ represents the probability to stay in the element g after one step.

Equivalently, if we define the $|G| \times |G|$ transition matrix $\mathcal{P} = (p(h,g))_{h,g \in G}$ by setting $p(h,g) = P(h^{-1}g)$ and Q represents the initial distribution, then, after one step we are in g with probability $\sum_{h \in G} Q(h) p(h,g) = Q * P(g)$. After n steps we are in g with probability $\sum_{h \in G} Q(h) p^{(n)}(h,g) = Q * P^{*n}(g)$. Here $\left(p^{(n)}(h,g)\right)_{h,g \in G} = \mathcal{P}^n$ is the transition matrix corresponding to the nth convolution powers $P^{*n} = P * \cdots * P$ of P.

In particular, if the initial distribution is the Dirac function δ_{1_G} (that is we start our process from 1_G), then the distribution after n steps is simply P^{*n}.x

If we use the Fourier inversion formula (Theorem 3.9.8) to express P

$$P(g) = |G|^{-1} \sum_{\rho \in \widehat{G}} d_\rho tr\left(\widehat{P}(\rho)\rho(g^{-1})\right)$$

and we take in account Lemma 3.9.6 we obtain

$$P^{*n}(g) = |G|^{-1} \sum_{\rho \in \widehat{G}} d_\rho tr\left(\widehat{P}(\rho)^n \rho(g^{-1})\right)$$

where $\widehat{P}(\rho)^n$ is the nth power of the matrix $\widehat{P}(\rho)$. While the matrix $\mathcal{P} = (p(h,g))_{h,g \in G} = \left(P(h^{-1}g)\right)_{h,g \in G}$ has dimension $|G| \times |G|$, each matrix $\widehat{P}(\rho)$ has smaller dimension $d_\rho \times d_\rho$ and thus, at least in theory, the computation of the nth powers should be easier. Obviously the computation would be easier the smaller the dimensions d_ρ, which in any case should satisfy the constraint $\sum_{\rho \in R} d_\rho^2 = |G|$.

When G is abelian, Schur's lemma (Lemma 3.5.1) implies $d_\rho = 1, \forall \rho \in \widehat{G}$. This is the easiest case (see the examples in Chapter 2) and the setting the most similar to classical Fourier analysis, i.e. Fourier series, that indeed are an important tool in analyzing *classical* random walks (on Euclidean lattices).

We have to say that the computation of the nth powers of the matrices $\widehat{P}(\rho)$ is only in theory easier than the computation of the nth powers of the matrix $P(h^{-1}g)$, since, in order to determine these matrices, a detailed knowledge of the irreducible representations of G is necessary. The computation can be really easier if P satisfies certain symmetry conditions that, in practise, arise very often. It could happen that the matrices $\widehat{P}(\rho)$ reduce to matrices with only one nontrivial entry which is on the diagonal. This radical but frequent simplification can be studied in the framework of the so-called Gelfand pairs, that will be the object of the next chapter.

3.11 Permutation characters and Burnside's lemma

Let G be a group acting on a set Ω. From now on, to denote the action of an element $g \in G$ on a point $\omega \in \Omega$, we shall simply write $g\omega$ instead of $g \cdot \omega$.

Recall that the permutation representation of G on Ω is the G-representation $(\lambda, L(\Omega))$ defined by

$$[\lambda(g)f](\omega) = f(g^{-1}\omega)$$

for all $f \in L(\Omega)$, $g \in G$ and $\omega \in \Omega$. The permutation representation is unitary with respect to the usual scalar product on $L(\Omega)$, namely

$$\langle f_1, f_2 \rangle = \sum_{\omega \in \Omega} f_1(\omega)\overline{f_2(\omega)},$$

for $f_1, f_2 \in L(\Omega)$.

The corresponding character χ_λ, that here we simply denote by χ is called the *permutation character*. Observe that the set $\{\delta_\omega : \omega \in \Omega\}$ is an orthonormal basis in $L(\Omega)$ and therefore the expression

$$\chi(g) \equiv \sum_{\omega \in \Omega} \langle \lambda(g)\delta_\omega, \delta_\omega \rangle = \sum_{\omega \in \Omega} \langle \delta_{g\omega}, \delta_\omega \rangle = |\{\omega \in \Omega : g\omega = \omega\}|, \quad (3.22)$$

which counts the points in Ω fixed by $g \in G$, is called the *fixed point character formula* for a permutation representation.

The following lemma is usually called "the Burnside lemma", but it was known already to Cauchy, see [28, 171, 233].

Lemma 3.11.1 (Burnside's lemma) *Let G act on a set Ω. The following equalities hold true:*

$$\frac{1}{|G|} \sum_{g \in G} \chi(g) = \frac{1}{|G|} \sum_{\omega \in \Omega} |Stab_G(\omega)| = \text{number of } G\text{-orbits in } \Omega. \quad (3.23)$$

Proof Let $\Omega_1, \Omega_2, \ldots, \Omega_h$ denote the orbits of G on Ω. From (3.22) we have

$$\frac{1}{|G|} \sum_{g \in G} \chi(g) = \frac{1}{|G|} \sum_{g \in G} |\{\omega \in \Omega : g\omega = \omega\}|$$

$$= \frac{1}{|G|} |\{(g, \omega) \in G \times \Omega : g\omega = \omega\}|$$

$$= \frac{1}{|G|} \sum_{\omega \in \Omega} |\{g \in G : g\omega = \omega\}|$$

$$= \frac{1}{|G|} \sum_{\omega \in \Omega} |Stab_G(\omega)|$$

$$= \frac{1}{|G|} \sum_{i=1}^{h} \sum_{\omega \in \Omega_i} |Stab_G(\omega)|$$

$$=_{(*)} \frac{1}{|G|} \sum_{i=1}^{h} \sum_{\omega \in \Omega_i} \frac{|G|}{|\Omega_i|}$$

$$= \sum_{i=1}^{h} \frac{1}{|\Omega_i|} \cdot |\Omega_i|$$

$$= h$$

where in $(*)$ we used (3.4). $\qquad \square$

Remark 3.11.2 There is a purely representation theoretical interpretation of the Burnside lemma. We clearly have

$$\frac{1}{|G|} \sum_{g \in G} \chi(g) = \frac{1}{|G|} \langle \chi, \mathbf{1}_G \rangle \qquad (3.24)$$

where $\mathbf{1}_G$ denotes (the character of) the trivial representation. As we have proved in Proposition 3.7.5, the right hand side of (3.24) denotes the multiplicity of the trivial representation as a subrepresentation of the permutation representation. But, by definition, the multiplicity of the trivial representation inside a representation (ρ, V) equals the dimension of the subspace $W = \{w \in V : \rho_g(v) = v, \forall g \in G\}$ of G-invariant vectors in V. In our setting $V = L(\Omega)$ and $W = \oplus_{i=1}^{h} \mathbb{C}\mathbf{1}_{\Omega_i}$ where $\mathbf{1}_{\Omega_i}$ denotes the characteristic function of the orbit Ω_i, $i = 1, 2, \ldots, h$ and this dimension is exactly $dim(W) = h$.

3.12 An application: the enumeration of finite graphs

In this section we present an interesting application of Burnside's lemma, namely the enumeration of isomorphism classes of finite graphs. First of all, we need a preliminary result which is a first step towards the so-called Polya–Redfield theory of enumeration under group actions (see [156]).

Let X and Y be finite sets and denote by $Y^X = \{f : X \to Y\}$ the set of all functions from X to Y.

Suppose that a finite group G acts on X. Then G acts on Y^X by setting

$$[gf](x) = f(g^{-1}x)$$

for all $g \in G$, $f \in Y^X$ and $x \in X$.

Fix $g \in G$ and consider the following equivalence relation on X: $x \sim_g y$ if there exists $t \in \mathbb{Z}$ such that $x = g^t y$. Then denote by

$$X = \coprod_{j=1}^{\ell} X_j(g)$$

the corresponding decomposition into equivalence classes. The subsets $X_j(g)$ are called G-cycles. We also set $\ell(g) :=$ number of cycles of g on X. Denote by $\mathcal{C}_\ell = \{g \in G : \ell(g) = \ell\}$ the subset of elements $g \in G$ which induce a decomposition of X into exactly ℓ g-cycles.

Exercise 3.12.1 Let $g, h \in G$ and let $X = \coprod_{j=1}^{\ell} X_j(g)$ denote the decomposition of X into g-cycles. Show that $X = \coprod_{j=1}^{\ell} hX_j(g)$ is the decomposition of X into hgh^{-1}-cycles and deduce that the number $\ell(g)$ is a conjugacy invariant.

Lemma 3.12.2 *The number of G-orbits on Y^X is equal to*

$$\frac{1}{|G|} \sum_{\ell=1}^{|X|} |\mathcal{C}_\ell| \cdot |Y|^\ell. \tag{3.25}$$

Proof First observe that for $g \in G$ and $f \in Y^X$ one has $gf = f$ if and only if $f(g^{-1}x) = f(x)$ for all $x \in X$, that is if and only if f is constant on each g-cycle on X. Therefore, if g has $\ell(g)$ cycles on X, then the number of G-fixed points in Y^X equals the number $|Y|^{\ell(g)}$. We then deduce

$$\frac{1}{|G|} \sum_{g \in G} |\{f \in Y^X : gf = f\}| = \frac{1}{|G|} \sum_{\ell=1}^{|X|} |\mathcal{C}_\ell| \cdot |Y|^\ell,$$

and the proof follows from Burnside's lemma (Lemma 3.11.1). □

Let K_n be the complete graph on n elements (see Example 1.8.1). Then its edge set is $E_n = \{\{i, j\} : 1 \leq i \neq j \leq n\}$. The set of all simple (and with no loops) graphs on given vertices $1, 2, \dots, n$ may be identified with the set $\{0, 1\}^{E_n}$. Indeed, for $\omega : E_n \to \{0, 1\}$ the corresponding graph has edge set $\{e \in E_n : \omega(e) = 1\}$. The symmetric group S_n acts on E_n and the orbits on $\{0, 1\}^{E_n}$ correspond to the isomorphism classes of finite simple graphs on n vertices.

Example 3.12.3 (Graphs on three vertices) For $n = 3$ the complete graph K_3 is just a triangle.

- The trivial element 1_{S_3} has clearly three cycles, namely

$$E_3 = \{1, 2\} \coprod \{1, 3\} \coprod \{2, 3\}.$$

- A transposition τ has exactly two cycles: for instance for $\tau = (1, 2)$ one has $E_3 = \{1, 2\} \coprod \{\{1, 3\}, \{2, 3\}\}$.
- Finally, a nontrivial even permutation has only one cycle.

The formula (3.25) gives exactly $\frac{1}{6}[2 \times 2 + 3 \times 2^2 + 1 \times 2^3] = 4$ nonisomophic graphs on three vertices, namely

Figure 3.1. The graphs on three vertices

Example 3.12.4 (Graphs on four vertices) For $n = 4$ the complete graph K_4 is as in Example 1.12.

- The trivial element 1_{S_4} has six cycles, namely

$$E_4 = \{1,2\} \coprod \{1,3\} \coprod \{1,4\} \coprod \{2,3\} \coprod \{2,4\} \coprod \{3,4\}.$$

- A transposition τ has exactly four cycles: for instance for $\tau = (1,2)$ one has

$$E_4 = \{1,2\} \coprod \{3,4\} \coprod \{\{1,3\},\{2,3\}\} \coprod \{\{1,4\},\{2,4\}\} .$$

Note that there are six transpositions.

- A nontrivial product of commuting transpositions has four cycles: for instance if $\pi = (12)(34)$ one has

$$E_4 = \{1,2\} \coprod \{3,4\} \coprod \{\{1,3\},\{2,4\}\} \coprod \{\{1,4\},\{2,3\}\} .$$

Note that there are three such permutations.

- A three-cycle has two cycles: for instance for $\pi = (1,2,3)$ one has

$$E_4 = \{\{1,2\},\{2,3\},\{1,3\}\} \coprod \{\{1,4\},\{2,4\},\{3,4\}\} .$$

Note that there are eight three-cycles.

- Finally a four-cycle has again two cycles: for instance if $\pi = (1,2,3,4)$ one has

$$E_4 = \{\{1,2\},\{2,3\},\{3,4\},\{1,4\}\} \coprod \{\{1,3\},\{2,4\}\} .$$

In this case we have six four-cycles.

Grouping together according to the size of the g-cycles we have $\mathcal{C}_2 = 8 + 6 = 14$, $\mathcal{C}_4 = 6 + 3 = 9$, $\mathcal{C}_6 = 1$ and $\mathcal{C}_\ell = 0$ for all other ℓ's. Then the formula (3.25) gives exactly $\frac{1}{24}[14 \times 2^2 + 9 \times 2^4 + 1 \times 2^6] = 11$ nonisomophic graphs on four vertices, namely

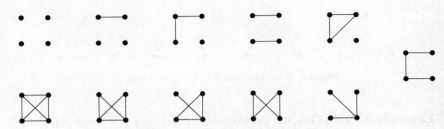

Figure 3.2. The graphs on four vertices

Exercise 3.12.5 (Graphs on five vertices) Work out the analogous computations from previous examples for $n = 5$.

An entire book on applications of group actions to enumeration problems is Kerber's monograph [143] (on the cover of that book there is the picture (above) of the 11 nonisomorphic graphs on four vertices!).

3.13 Wielandt's lemma

Theorem 3.13.3 below was surely known to Schur and possibly even to Frobenius. A standard reference is the book by Wielandt [228]; for convenience we refer to it as to "Wielandt's lemma". Before that, a few observations.

Suppose we are given an action of a group G on a set X and denote by χ its permutation character. Then, the corresponding diagonal action of G on $X \times X$ (see Example 3.1.16) has χ^2 as permutation character. Indeed

$$
\begin{aligned}
\chi(g)^2 &= |\{x \in X : gx = x\}|^2 \\
&= |\{x_1 \in X : gx_1 = x_1\}| \cdot |\{x_2 \in X : gx_2 = x_2\}| \qquad (3.26) \\
&= |\{(x_1, x_2) \in X \times X : g(x_1, x_2) = (x_1, x_2)\}|.
\end{aligned}
$$

Theorem 3.13.1 *Let G act transitively on $X = G/K$. Denote by $x_0 \in X$ the point stabilized by K and by $X = \Omega_0 \coprod \cdots \coprod \Omega_n$ the decomposition into K-orbits of X (with $\Omega_0 = \{x_0\}$). Choose $x_i \in \Omega_i$ for all $i = 0, 1, \ldots, n$. Then the sets*

$$
G(x_i, x_0) = \{(gx_i, gx_0) : g \in G\}
$$

are the orbits of the diagonal action of G on $X \times X$.

Proof First of all, note that if $(x, y) \in X \times X$ then there exist $g \in G$, $k \in K$ and $i \in \{0, 1, \ldots, n\}$ such that

$$(x, y) = (x, gx_0) = (gg^{-1}x, gx_0) = (gkx_i, gkx_0) \in G(x_i, x_0).$$

Indeed G is transitive on X and we denote by $Kx_i = \Omega_i$ the K-orbit containing $g^{-1}x$. This shows that $X \times X = \cup_{i=0}^n G(x_i, x_0)$. It is also easy to show that $G(x_i, x_0) \cap G(x_j, x_0) = \emptyset$ if $i \neq j$, showing that the union above is in fact a disjoint union. $\qquad \square$

In other words, we may also say the following.

Corollary 3.13.2 *Let* Θ *be a* G-*orbit on* $X \times X$. *Then the set* $\Omega = \{x \in X : (x, x_0) \in \Theta\}$ *is an orbit of* K *on* X *and the map* $\Theta \mapsto \Omega$ *is a bijection between the set of orbits of* G *on* $X \times X$ *(with the diagonal action) and those of* K *on* X.

It is also easy to see (exercise) that the map $KgK \mapsto Kgx_0$ is a bijection between the double cosets of G modulo K and the orbits of K on X.

Theorem 3.13.3 (Wielandt's lemma) *Let* G *be a finite group,* $K \leq G$ *a subgroup and denote by* $X = G/K$ *the corresponding homogeneous space. Let* $L(X) = \oplus_{i=0}^N m_i V_i$ *be a decomposition into irreducible* G-*subrepresentations, where* m_i *denotes the multiplicity of* V_i. *Then*

$$\sum_{i=0}^N m_i^2 = \text{number of } G\text{-orbits on } X \times X = \text{number of } K\text{-orbits on } X.$$

$$(3.27)$$

Proof Denote by χ the permutation character associated with the G-action on X. We have

$$\sum_{i=1}^N m_i^2 =_{*)} \frac{1}{|G|} \langle \chi, \chi \rangle$$

$$= \frac{1}{|G|} \sum_{g \in G} \chi(g)^2$$

$$=_{**)} \frac{1}{|G|} \sum_{(x_1, x_2) \in X \times X} |Stab_G(x_1, x_2)|$$

$$=_{***)} \text{number of } G\text{-orbits on } X \times X$$

$$=_{****)} \text{number of } K-\text{orbits on } X$$

where: $=_{*)}$ comes from Corollary 3.7.7; $=_{**)}$ is the combination of (3.26) with the first equality in Lemma 3.11.1; $=_{***)}$ is nothing but the second equality in Lemma 3.11.1 and, finally, $=_{****)}$ follows from previous corollary. $\qquad\square$

Remark 3.13.4 Again, the above formula can be viewed as follows

$$\langle \chi, \chi \rangle = \langle \chi^2, \mathbf{1}_G \rangle$$

so that the right hand side expresses the number of G-invariant subspaces in the representation associated with the action of G on $X \times X$.

The following is a slight but useful generalization of Wielandt's lemma (Theorem 3.13.3).

Exercise 3.13.5 Let G act transitively on two finite sets $X = G/K$ and $Y = G/H$. Define the diagonal action of G on $X \times Y$ by setting, for all $x \in X$, $y \in Y$ and $g \in G$

$$g(x, y) = (gx, gy).$$

(1) Show that the number of G-orbits on $X \times Y$ equals the number of H-orbits on X which in turn equals the number of K-orbits on Y.

(2) Let $L(X) = \oplus_{i \in I} m_i V_i$ and $L(Y) = \oplus_{j \in J} n_j V_j$, $I, J \subseteq \widehat{G}$, be the decomposition of the permutation representations $L(X)$ and $L(Y)$ into irreducible representations. Denoting by $I \cap J$ the set of indices corresponding to common subrepresentations show that the number of G-orbits on $X \times Y$ equals the sum $\sum_{i \in I \cap J} m_i n_i$.

Example 3.13.6 Let now $G = S_n$ be the symmetric group of degree n and $X = \{1, 2, \ldots, n\}$ as in Example 3.4.5 (c). Denote by K the stabilizer of the point 1 which is clearly isomorphic to S_{n-1}. We have that K has exactly two orbits on X, namely $\{1\}$ and $\{2, 3, \ldots, n\}$. From Wielandt's lemma (Theorem 3.13.3) it follows that $L(X)$ decomposes into two irreducible S_n-representations, each with multiplicity one.

On the other hand, we have the decomposition $L(X) = W_0 \oplus W_1$, where $W_0 = \{f : X \to \mathbb{C}, \text{constant}\}$ is the trivial representation and $W_1 = \{f : X \to \mathbb{C}, \sum_{j=1}^n f(j) = 0\}$. Indeed these subspaces are clearly S_n-invariant and orthogonal; moreover, given any $f \in L(X)$ we have

$$f = \frac{1}{|X|} \sum_{x \in X} f(x) + \left[f - \frac{1}{|X|} \sum_{x \in X} f(x) \right]$$

where the first summand belongs to W_0 and the second one to W_1. Therefore W_0 and W_1 are irreducible. Compare with Exercise 3.4.6.

Definition 3.13.7 Suppose that G acts on X. The action is *doubly transitive* if for all (x_1, x_2), $(y_1, y_2) \in (X \times X) \setminus \{(x, x) : x \in X\}$ there exists $g \in G$ such that $gx_i = y_i$ for $i = 1, 2$.

Exercise 3.13.8 Suppose that G acts transitively on X.

(1) Set $W_0 = \{f : X \to \mathbb{C}, \text{constant}\}$ and $W_1 = \{f : X \to \mathbb{C},$ $\sum_{x \in X} f(x) = 0\}$. Prove that $L(X) = W_0 \oplus W_1$ is the decomposition of the permutation representation into irreducibles if and only if G acts doubly transitively on X.
(2) Show that the action of S_n on $\{1, 2, \ldots, n\}$ is doubly transitive.
(3) Prove that if the action of G on $X = G/K$ is doubly transitive, then K is a maximal subgroup ($K \leq H < G$ infers $H = K$).

3.14 Examples and applications to the symmetric group

In this section we apply the results from previous section to some representations of S_n of small dimension. Consider the following homogeneous spaces

$$\Omega_{n-1,1} = S_n/(S_{n-1} \times S_1)$$
$$\equiv \{1, 2, \ldots, n\}$$

$$\Omega_{n-2,2} = S_n/(S_{n-2} \times S_2)$$
$$\equiv \{\{i, j\} : i, j \in \{1, 2, \ldots, n\}, i \neq j\}$$
$$\equiv \{2-\text{subsets in } \{1, 2, \ldots, n\}\}$$
$$\equiv \{\text{unordered pairs of distinct elements in } \{1, 2, \ldots, n\}\}$$

$$\Omega_{n-2,1,1} = S_n/(S_{n-2} \times S_1 \times S_1)$$
$$\equiv \{(i, j) : i, j \in \{1, 2, \ldots, n\}, i \neq j\}$$
$$\equiv \{\text{ordered pairs of distinct elements in } \{1, 2, \ldots, n\}\}$$

where we regard $S_{n-1} \times S_1$ as the stabilizer of $\{1\}$ and $S_{n-2} \times S_2$ and $S_{n-2} \times S_1 \times S_1$ as the stabilizers of $\{1, 2\}$ and $(1, 2)$, respectively (that is, S_{n-2} acts on $\{3, 4, \ldots, n\}$, S_2 acts on $\{1, 2\}$ and $S_1 \times S_1$ reduces to the trivial group acting as the identity on $\{1, 2\}$).

Set

$$M^{n-1,1} = L(\Omega_{n-1,1})$$
$$M^{n-2,2} = L(\Omega_{n-2,2})$$
$$M^{n-2,1,1} = L(\Omega_{n-2,1,1}).$$

1. We have the following decomposition into irreducibles

$$M^{n-1,1} = S^n \oplus S^{n-1,1} \tag{3.28}$$

where $S^n = W_0$ and $S^{n-1,1} = W_1$ as in Examples 3.4.5 (c) and 3.13.6.

2. The orbits of $S_{n-1} \times S_1$ (the stabilizer of $\{1\}$) on $\Omega_{n-2,2}$ are exactly two, namely $\{\{1, j\} : j \in \{2, 3, \dots, n\}\}$ and $\{\{i, j\} : i, j \in \{2, 3, \dots, n\}, i \neq j\}$.

In virtue of Remark 3.11.2 the trivial representation S^n appears with multiplicity one in $M^{n-2,2}$. Also, by (3.28) and Exercise 3.13.5 we have that the number of $(S_{n-1} \times S_1)$-orbits on $\Omega_{n-2,2}$ equals $\sum_{i \in I \cap J} m_i n_i$ (here we use the notation from Exercise 3.13.5) so that, necessarily $|I \cap J| = 2$ (thus $I \subseteq J$) and $m_i = n_i = 1$ for all $i \in I$. This shows that also $S^{n-1,1}$ is contained in $M^{n-2,2}$ with multiplicity one.

But the orbits of $S_{n-2} \times S_2$ (stabilizer of $\{1, 2\}$) on $\Omega_{n-2,2}$ are exactly three, namely $\{\{i, j\} \in \Omega_{n-2,2} : |\{i, j\} \cap \{1, 2\}| = k\}$, $k = 0, 1, 2$. Therefore, by Wielandt's lemma, $M^{n-2,2}$ decomposes into three distinct irreducible representations:

$$M^{n-2,2} = S^n \oplus S^{n-1,1} \oplus S^{n-2,2} \tag{3.29}$$

where $S^{n-2,2}$, the orthogonal complement of $S^n \oplus S^{n-1,1}$ in $M^{n-2,2}$, is a new (irreducible) representation.

3. The orbits of $S_{n-1} \times S_1$ (the stabilizer of $\{1\}$) on $\Omega_{n-2,1,1}$ are exactly three, namely $\{(1, j) : j \neq 1\}$, $\{(j, 1) : j \neq 1\}$ and $\{(i, j) : i, j \neq 1\}$.

Therefore, by Exercise 3.13.5, and recalling that S^n has multiplicity one, one deduces, as before, that

$$S^{n-1,1} \text{ has multiplicity two in } M^{n-2,1,1}. \tag{3.30}$$

Also, $S_{n-2,2}$ has exactly four orbits on $\Omega_{n-2,1,1}$, namely $\{(1, 2), (2, 1)\}$, $\{(1, j), (2, j) : j \neq 1, 2\}$, $\{(j, 1), (j, 2) : j \neq 1, 2\}$ and $\{(i, j) : i, j \neq 1, 2\}$. Therefore, from (3.29) and (3.30) it follows that

$$S^{n-2,2} \text{ has multiplicity one in } M^{n-2,1,1}. \tag{3.31}$$

On the other hand, $S_{n-2} \times S_1 \times S_1$ has exactly seven orbits on $\Omega_{n-2,1,1}$, namely $\{(1, 2)\}$, $\{(2, 1)\}$, $\{(1, j) : j \neq 1\}$, $\{(j, 1) : j \neq 1\}$, $\{(2, j) : j \neq 2\}$,

$\{(j,2) : j \neq 2\}$ and $\{(i,j) : i,j \neq 1,2\}$. Therefore, by Wielandt's lemma, (3.30) and (3.31), as $1 + 2^2 + 1 = 6$ (the first term corresponds to S_n, the second one to $S^{n-1,1}$ and the last one to $S^{n-2,2}$), we have

$$M^{n-2,1,1} = S^n \oplus 2S^{n-1,1} \oplus S^{n-2,2} \oplus S^{n-2,1,1}, \qquad (3.32)$$

where $S^{n-2,1,1}$, the orthogonal complement of $S^n \oplus 2S^{n-1,1} \oplus S^{n-2,2}$ in $M^{n-2,2}$ is a new (irreducible) representation.

Exercise 3.14.1 Let $\chi_n, \chi_{n-1,1}, \chi_{n-2,2}$ and $\chi_{n-2,1,1}$ be the characters of the irreducible representations $S^n, S^{n-1,1}, S^{n-2,2}$ and $S^{n-2,1,1}$, respectively.

For $g \in S_n$ we denote by $a_1(g)$ the number of fixed points of g and by $a_2(g)$ the number of two cycles in g; in other words we have

$$g = (i_1) \cdots (i_{a_1(g)}) \cdot (j_1, j_2) \cdots (j_{2a_2(g)-1}, j_{2a_2(g)})(k_1 k_2 k_3 \cdots) \cdots$$

(1) Show that

- $\chi_{n-1,1}(g) = a_1(g) - 1$ (see also Exercise 3.7.10)
- $\chi_{n-2,2}(g) = \frac{1}{2}(a_1(g) - 1)(a_1(g) - 2) + a_2(g) - 1$
- $\chi_{n-2,1,1}(g) = \frac{1}{2}(a_1(g) - 1)(a_1(g) - 2) - a_2(g)$
- Deduce that, in particular,

$$\dim S^{n-1,1} = n - 1$$
$$\dim S^{n-2,2} = \binom{n}{2} - n$$
$$\dim S^{n-2,1,1} = (n-1)(n-2)/2.$$

(2) Show that

$$(\chi_{n-1,1})^2 = \chi_n + \chi_{n-1,1} + \chi_{n-2,2} + \chi_{n-2,1,1}. \qquad (3.33)$$

Exercise 3.14.2 Set $\Omega_{n-k,k} = S_n/(S_{n-k} \times S_k) \equiv \{k\text{-subsets of } \{1, 2, \ldots, n\}\}$ (we regard $S_{n-k} \times S_k$ as the stabilizer of the k-subset $\{1, 2, \ldots, k\}$) and denote by $M^{n-k,k} = L(\Omega_{n-k,k})$ the corresponding permutation representation.

Let now $k \in \{0, 1, \ldots, [n/2]\}$.

(1) Show that

- $S_{n-k+1} \times S_{k-1}$ has k orbits on $\Omega_{n-k,k}$
- $S_{n-k} \times S_k$ has $k + 1$ orbits on $\Omega_{n-k,k}$.

(2) Use the previous part and the decomposition $M^{n-1,1} = S^n \oplus S^{n-1,1}$ (i.e. (3.28)) to prove, by induction on n, that

- $M^{n-k,k}$ decomposes into $k+1$ distinct irreducible representations;
- $M^{n-k,k}$ contains all the irreducible representations in $M^{n-k+1,k-1}$ together with just one new irreducible representation, say $S^{n-k,k}$;

(3) Show that $\Omega_{n-k,k}$ is isomorphic to $\Omega_{k,n-k}$ as S_n-homogeneous spaces and deduce from (2) that

$$M^{n-h,h} = \oplus_{k=0}^{\min\{n-h,h\}} S^{n-k,k}$$

for all $h = 0, 1, \ldots, n$. Moreover $\dim S^{n-k,k} = \binom{n}{k} - \binom{n}{k-1}$.

The results of this section will be generalized and completed in Chapter 10 (see Example 10.5.15). See also Chapter 6.

4

Finite Gelfand pairs

This chapter contains an exposition on the theory of (finite) Gelfand pairs and their spherical functions. This theory originally was developed in the setting of Lie groups with the seminal paper by I. M. Gelfand [98] (see also [23]); another earlier contribution is the paper by Godement [100].

Expositions of the theory in the setting of locally compact and/or Lie groups are in Dieudonné's treatise on analysis [71] and in the monographs by: Dym and McKean [82], Faraut [86], Figà-Talamanca and Nebbia [91], Helgason [117, 118], Klimyk and Vilenkin [146], Lang [149] and Ricci [180]. See also the papers by Bougerol [30].

Recently, finite and infinite Gelfand pairs have been studied in asymptotic and geometric group theory in connection with the so-called *branch groups* introduced by R.I. Grigorchuk in [107] (see [15, 19]).

Several examples of finite Gelfand pairs, where G is a Weyl group or a Chevalley group over a finite field were studied by Delsarte, Dunkl and Stanton (see the surveys [77, 209, 211] or the book [146] by Klimyk and Vilenkin). Delsarte was motivated by applications to association schemes of coding theory, while Dunkl and Stanton were interested in applications to orthogonal polynomials and special functions. For the point of view of the theory of association schemes see the monographs by Bailey [7], Bannai and Ito [12]; see also the work of Takacs [216], on harmonic analysis on Schur algebras, that contains several applications to probability.

Our exposition is inspired to the book by Diaconis [55] and to Figà-Talamanca's lecture notes [88]. The finite case is also treated in the monographs by: Bump [37], I.G. Macdonald [165] and Terras [220]. See also the papers by Garsia [97] and Letac [152, 153].

Persi Diaconis used Gelfand pairs in order to determine the rate of convergence to the stationary distribution of finite Markov chains. More precisely, given a Markov chain which is invariant under the action of a group G, its transition operator can be expressed as a convolution operator whose kernel can be written, at least in theory, as a "Fourier series" where the classical exponentials $\exp(inx)$ are replaced by the irreducible representations of the group G. This would, in theory, yield an analysis of the convolution powers of the kernel and therefore of the powers of the transition operator, in order to determine their asymptotic behavior and related problems as the rate of convergence to the stationary distribution.

This program, however, is not easy to handle, because on the one hand, the irreducible representations of a finite group, although "known" in theory, most often are in practice not suitable for concrete calculations and, on the other hand, because the representations, being not necessarily just one-dimensional, cannot lead to a complete diagonalization of the transition operator, but just to a block decomposition.

There are however some cases where this Fourier analysis of the action of a group over a finite set can be easily handled and reduced to an essentially commutative analysis, even if the representations involved still remain of dimension higher than one. This is indeed the case when the action of the group on the set corresponds to a Gelfand pair.

4.1 The algebra of bi-K-invariant functions

Let G be a finite group and $K \leq G$ a subgroup of G. We say that a function $f \in L(G)$ is *K-invariant on the right* (resp. *on the left*) if $f(gk) = f(g)$ (resp. $f(kg) = f(g)$) for all $g \in G$ and $k \in K$. Denoting by $X = G/K$ the corresponding homogeneous space, one can identify the space $L(X) = \{f : X \to \mathbb{C}\}$ of all complex valued functions on X with the $L(G)$-subspace of right-K-invariant functions; the isomorphism is induced by the map in (3.3). It is given by the map $f \mapsto \tilde{f}$ which associates with $f \in L(X)$ the right-K-invariant function $\tilde{f} \in L(G)$ defined by

$$\tilde{f}(g) = f(gx_0) \tag{4.1}$$

where $x_0 \in X$ is the point stabilized by K.

We then say that f is *bi-K-invariant* if it is both K-invariant on the right and on the left, that is $f(kgk') = f(g)$ for all $g \in G$ and $k, k' \in K$; in analogy with the previous situation, the space of bi-K-invariant functions can be identified with the space $L(K\backslash G/K) = \{f : K\backslash G/K \to \mathbb{C}\}$ of complex valued functions on $K\backslash G/K$, the set of double cosets $KgK, g \in G$. In view of this identification, we denote by $L(K\backslash G/K)$ the space of bi-K-invariant functions on G. Note also that in virtue of (4.1), $L(K\backslash G/K)$ coincides with the subspace of K-invariant functions on X, that is with $L(X)^K = \{f \in L(X) : f(kx) = f(x), \forall x \in X, k \in K\}$. In other words, the map $KgK \mapsto Kgx_0$ is a bijection between $K\backslash G/K$ and the set of K-orbits on X; compare with Corollary 3.13.2.

It is easy to see that given $f_1, f_2 \in L(G)$, the convolution $f_1 * f_2$ is left invariant if f_1 is left invariant, it is right invariant if f_2 is right invariant. As a consequence of this, the right (resp. left) invariant elements in $L(G)$ form a subalgebra (indeed a right (resp. left) ideal) of $L(G)$. The same is true for bi-invariant elements (in this case we have a two-sided ideal).

Note that if $f_1, f_2 \in L(X)$ and $\widetilde{f}_i$ are the associated right-K-invariant functions in $L(G)$ given by (4.1), $i = 1, 2$, then

$$\langle \widetilde{f}_1, \widetilde{f}_2 \rangle_{L(G)} = |K| \langle f_1, f_2 \rangle_{L(X)}. \tag{4.2}$$

For $f \in L(G)$ set

$$f^K(g) = \frac{1}{|K|} \sum_{k \in K} f(gk).$$

The map $f \mapsto f^K$ is the orthogonal *projection* of $L(G)$ onto $L(X)$.
Similarly we may define the map $f \mapsto {}^K f^K$, where

$$ {}^K f^K(g) = \frac{1}{|K|^2} \sum_{k,k' \in K} f(kgk'). \tag{4.3}$$

This is a *conditional expectation* of $L(G)$ onto $L(K\backslash G/K)$ (that is ${}^K(f_1 * f * f_2)^K = f_1 * {}^K f^K * f_2$ for all $f_1, f_2 \in L(K\backslash G/K)$ and $f \in L(G)$).

Remark 4.1.1 In the sequel we shall make use of the following expression for the total mass of a function in $L(X)$. For $f \in L(X)$ we have

$$\sum_{y \in X} f(y) = \frac{1}{|K|} \sum_{h \in G} f(hx_0).$$

Indeed, if $h_1, h_2, \ldots, h_t$ is a set of representatives for the right cosets of K in G, so that $G = \coprod_{i=1}^{t} h_i K$, then

$$\sum_{h \in G} f(hx_0) = \sum_{i=1}^{t} \sum_{k \in K} f(h_i k x_0)$$

$$= \sum_{i=1}^{t} \sum_{k \in K} f(h_i x_0)$$

$$= |K| \sum_{i=1}^{t} f(h_i x_0)$$

$$= |K| \sum_{y \in X} f(y).$$

4.2 Intertwining operators for permutation representations

Given two representations (ρ_1, V_1) and (ρ_2, V_2) of a group G we denote by

$$Hom_G(V_1, V_2) = \{T : V_1 \to V_2 :$$
$$\rho_2(g)[Tv] = T[\rho_1(g)v] \text{ for all } g \in G, v \in V_1\} \quad (4.4)$$

the space of operators *intertwining* (or, briefly, *intertwiners* of) the representations (ρ_1, V_1) and (ρ_2, V_2). If $T \in Hom_G(V_1, V_2)$, we also say that T is *G-equivariant*. Note also that if $V_1 = V_2$, then $Hom_G(V_1, V_1)$ is an algebra (see Definition 2.1.5): indeed it is closed under composition of maps.

In our context we have the following important fact. First recall that given two algebras $\mathcal{A}$ and $\mathcal{B}$, an *anti-isomorphism* is a linear bijection $\tau : \mathcal{A} \to \mathcal{B}$ such that $\tau(aa') = \tau(a')\tau(a)$ for all $a, a' \in \mathcal{A}$.

Proposition 4.2.1 *The algebra of operators intertwining $L(X)$ with itself is isomorphic to the convolution algebra of bi-K-invariant functions; in formulae: $Hom_G(L(X), L(X)) \cong L(K \backslash G / K)$.*

Proof An operator $T : L(X) \to L(X)$ is expressed by a complex matrix $(r(x, y))_{x, y \in X}$ in the sense that $Tf(x) = \sum_{y \in X} r(x, y)f(y)$. If T is an intertwiner or, what is the same, it is G-equivariant, it satisfies the condition $[\lambda(g)(Tf)] = T[\lambda(g)f]$ which yields $r(g^{-1}x, y) = r(x, gy)$, equivalently, $r(x, y) = r(gx, gy)$ for all $x, y \in X$ and $g \in G$. This means that r is constant on the orbits of G on $X \times X$. Therefore, if x_0 is the point stabilized by K, in virtue of Corollary 3.13.2, the function

$\psi \colon X \to \mathbb{C}$ defined by $\psi(z) = r(z, x_0)$, for all $z \in X$, is constant on the orbits of K on X.

Hence, recalling Remark 4.1.1, if $x = gx_0$, one has

$$
\begin{aligned}
[Tf](x) &= \frac{1}{|K|} \sum_{h \in G} r(x, hx_0) f(hx_0) \\
&= \frac{1}{|K|} \sum_{h \in G} r(h^{-1}gx_0, x_0) f(hx_0) \qquad (4.5) \\
&= \frac{1}{|K|} \sum_{h \in G} f(hx_0)\psi(h^{-1}gx_0).
\end{aligned}
$$

If $\widetilde{\psi} \in L(G)$ is defined as in (4.1), the correspondence $T \equiv (r(x,y))_{x,y \in X} \mapsto \widetilde{\psi}$ is an anti-isomorphism. Indeed if $S \equiv (s(x,y))_{x,y \in X} \mapsto \widetilde{\phi}$, where $\phi \in L(X)$ is defined by $\phi(x) = s(x, x_0)$, then, for $x = hx_0$ and $z = tx_0$, one has

$$
\begin{aligned}
\sum_{y \in X} r(x,y)s(y,z) &= \frac{1}{|K|} \sum_{g \in G} r(hx_0, gx_0)s(gx_0, tx_0) \\
&= \frac{1}{|K|} \sum_{g \in G} r(g^{-1}hx_0, x_0)s(t^{-1}gx_0, x_0) \\
&= \frac{1}{|K|} \sum_{g \in G} \widetilde{\phi}(t^{-1}g)\widetilde{\psi}(g^{-1}h) \\
&= \frac{1}{|K|} \left(\widetilde{\phi} * \widetilde{\psi} \right)(t^{-1}h).
\end{aligned}
$$

In order to have an isomorphism, set, for all $\xi \in L(G)$, $\check{\xi}(g) = \xi(g^{-1})$; then the correspondence $T \mapsto (\widetilde{\psi})^{\vee}$ is the desired isomorphism (note that $\xi \mapsto \check{\xi}$ is an anti-automorphism of the group algebra $L(G)$ with $L(K\backslash G/K)$ an invariant subspace). $\qquad \square$

Exercise 4.2.2 Let $\psi \in L(G)$. Using the notation from Example 3.4.1 and Exercise 3.4.2, show that

(1) $\lambda_G(g)f = \delta_g * f$ for all $f \in L(G)$ and $g \in G$;
(2) $\rho_G(g)f = f * \delta_{g^{-1}}$ for all $f \in L(G)$ and $g \in G$.
    Then deduce the following.
(3) The operator $T : L(G) \to L(G)$, defined by $Tf = f * \psi$, for all $f \in L(G)$, commutes with λ_G.
(4) The operator $T : L(G) \to L(G)$, defined by $Tf = \psi * f$, for all $f \in L(G)$, commutes with ρ_G.

This explains the reason why the map $T \mapsto \widetilde{\psi}$ in the proof of Proposition 4.2.1 is an anti-isomorphism: T acts on the "left", while the convolution is expressed as multiplication on the "right".

Remark 4.2.3 (4.5) may be written in form of convolution

$$\widetilde{Tf} = \frac{1}{|K|}\widetilde{f} * \widetilde{\psi}$$

where we used the notation in (4.1). The function ψ is called the *convolution kernel* of T.

In the following exercise we sketch an alternative and more combinatorial description of $Hom_G\left(L(X), L(X)\right)$.

Exercise 4.2.4 Let G be a group acting transitively on a set X. Let $\Theta_0, \Theta_1, \ldots, \Theta_N$ be the orbits of G on $X \times X$ (with $\Theta_0 = \{(x, x) : x \in X\}$) and set

$$\Delta_i f(x) = \sum_{y \in X : (x,y) \in \Theta_i} f(y)$$

for $f \in L(X)$. Define

$$\Xi_{i,j}(x, y) = \{z \in X : (x, z) \in \Theta_i \text{ and } (z, y) \in \Theta_j\}.$$

Show that

(1) $\Delta_0, \Delta_1, \ldots, \Delta_N$ constitute a basis for $Hom_G\left((L(X), L(X)\right)$.

(2) $|\Xi_{i,j}(x, y)| =: \xi_{i,j}(s)$ only depends on the G-orbit Θ_s containing (x, y).

(3) $\Delta_i \Delta_j = \sum_{s=0}^{N} \xi_{i,j}(s)\Delta_s$ and Δ_0 is the identity.

(4) There is a bijection $i \mapsto i'$ such that $\Delta_i^* = \Delta_{i'}$.

Exercise 4.2.5 Suppose that G acts transitively on $X = G/K$ and $Y = G/H$. Show that the vector space $Hom_G(L(X), L(Y))$ is isomorphic to

$$L(K\backslash G/H) = \{f \in L(G) :$$
$$f(kgh) = f(g) \text{ for all } g \in G, k \in K \text{ and } h \in H\},$$

the space of K-H-invariant functions on G. Compare with Exercise 3.13.5.

4.3 Finite Gelfand pairs: definition and examples

Definition 4.3.1 Let G be a finite group and $K \leq G$ a subgroup. The pair (G, K) is called a *Gelfand pair* if the algebra $L(K \backslash G / K)$ of bi-K-invariant functions is commutative.

More generally, if G acts transitively on a finite space X, we say that the action defines a Gelfand pair if (G, K) is a Gelfand pair, when K is the stabilizer of a point $x_0 \in X$. Note that if $gx_0 = x$ then $K' := gKg^{-1}$ is the stabilizer of x and (G, K) is a Gelfand pair if and only if (G, K') is a Gelfand pair (as it easily follows from the definition of convolution): the algebras $L(K \backslash G / K)$ and $L(K' \backslash G / K')$ are isomorphic (exercise).

Example 4.3.2 (Symmetric Gelfand pairs [153]) Let G and $K \leq G$ be finite groups. Suppose that for any $g \in G$ one has $g^{-1} \in KgK$. Then (G, K) is a Gelfand pair. In this case we say that (G, K) is *symmetric*.

Proof Observe that if $f \in L(K \backslash G / K)$, under our hypothesis we have $f(g^{-1}) = f(g)$. Then for $f_1, f_2 \in L(K \backslash G / K)$ we get

$$[f_1 * f_2](g) = \sum_{h \in G} f_1(gh) f_2(h^{-1})$$

$$= \sum_{h \in G} f_1(gh) f_2(h)$$

setting $t = gh$

$$= \sum_{t \in G} f_1(t) f_2(g^{-1}t)$$

$$= \sum_{t \in G} f_2(g^{-1}t) f_1(t^{-1})$$

$$= [f_2 * f_1](g^{-1}) = [f_2 * f_1](g),$$

showing that $L(K \backslash G / K)$ is commutative. $\qquad \square$

A straightforward generalization is the following:

Exercise 4.3.3 (Gelfand's lemma [55]) Let G and $K \leq G$ be finite groups. Suppose there exists an automorphism τ of G such that $g^{-1} \in K\tau(g)K$ for all $g \in G$. Show that (G, K) is a Gelfand pair. We then say that (G, K) is *weakly symmetric*.

In Example 9.6.1 we present an example of a finite Gelfand pair which is not weakly symmetric.

We now use the following notation: if a group G acts on a set Y, for two elements $x, y \in Y$ we write $x \sim y$ if there exists $g \in G$ such that $gx = y$ (equivalently if x and y belong to the same G-orbit: $Orb_G(x) = Orb_G(y)$).

Lemma 4.3.4 *Let X be a finite set and G a group acting transitively on it. Let $x_0 \in X$ and $K = Stab_G(x_0) = \{k \in G : kx_0 = x_0\}$ be a point of X and its stabilizer. The following are equivalent:*

(i) *For all $x, y \in X$ one has $(x, y) \sim (y, x)$ with respect to the action of G on $X \times X$.*

(ii) *$g^{-1} \in KgK$ for all $g \in G$.*

Proof (i) $\Rightarrow$ (ii): As $(x_0, g^{-1}x_0) = g^{-1}(gx_0, x_0) \sim (gx_0, x_0) \sim_{(i)} (x_0, gx_0)$, there exists $k \in G$ such that $k(x_0, g^{-1}x_0) = (x_0, gx_0)$, that is $kg^{-1}x_0 = gx_0$ and $kx_0 = x_0$, i.e. $k \in K$. The first condition then gives $g^{-1}kg^{-1}x_0 = x_0$ i.e. $g^{-1}kg^{-1} \in K$ and thus $g^{-1} \in KgK$.

(ii) $\Rightarrow$ (i): Let $t, s \in G$ be such that $x = tx_0$ and $y = tsx_0$. Moreover let $k_1, k_2 \in K$ be such that $s^{-1} = k_1sk_2$. Then:

$$
\begin{aligned}
(x, y) &= t(x_0, t^{-1}y) \sim (x_0, t^{-1}y) = (x_0, sx_0) \\
&= s(s^{-1}x_0, x_0) \sim (s^{-1}x_0, x_0) = (k_1sk_2x_0, x_0) \\
&= (k_1sx_0, x_0) = k_1(sx_0, k_1^{-1}x_0) = k_1(sx_0, x_0) \\
&\sim (sx_0, x_0) = (t^{-1}y, x_0) \sim t(t^{-1}y, x_0) = (y, x). \qquad \square
\end{aligned}
$$

Exercise 4.3.5 With the notation of Exercise 4.2.4 show that

(1) **(Orbit criterion)** (G, K) is a Gelfand pair if and only if

$$\xi_{i,j}(s) = \xi_{j,i}(s) \quad \text{for all } i = 0, 1, \ldots, N. \tag{4.6}$$

(2) (G, K) is symmetric if and only if $i' = i$ for all $i = 0, 1, \ldots, N$.

Definition 4.3.6 Let (X, d) be a finite metric space and G a group acting on X by isometries (i.e. $d(gx, gy) = d(x, y)$ for all $x, y \in X$ and $g \in G$). The action is 2-*point homogeneous* (or *distance-transitive*) if for all $(x_1, y_1), (x_2, y_2) \in X \times X$ such that $d(x_1, y_1) = d(x_2, y_2)$ there exists $g \in G$ such that $gx_1 = x_2$ and $gy_1 = y_2$.

From Example 4.3.2 and the previous lemma one deduces immediately the following.

Example 4.3.7 Let (X, d) be a finite metric space on which a finite group G acts isometrically and 2-point homogeneously. Fix $x_0 \in X$ and denote by $K = \{g \in G : gx_0 = x_0\}$ the stabilizer of this point. Then (G, K) is a symmetric Gelfand pair. Indeed $d(x, y) = d(y, x)$ and thus $(x, y) \sim (y, x)$ for all $x, y \in X$.

We observe that the K-orbits of X are exactly the *spheres* $\{x \in X : d(x, x_0) = j\}$, $j = 0, 1, 2, \ldots$ Therefore a function $f \in L(X)$ is K-invariant if and only if it is constant on the spheres (i.e. it is *radial*).

In this setting we can introduce the *Markov operator* $M : L(X) \to L(X)$ defined by

$$[Mf](x) = \frac{1}{|\{y : d(x, y) = 1\}|} \sum_{d(x,y)=1} f(y). \tag{4.7}$$

Exercise 4.3.8 Show that M is the convolution operator (see Remark 4.2.3) with kernel $\frac{1}{|\Omega_1|}\mathbf{1}_{\Omega_1}$, where $\Omega_1 = \{x \in X : d(x, x_0) = 1\}$ is the unit sphere centered at x_0.

4.4 A characterization of Gelfand pairs

Definition 4.4.1 A representation (ρ, V) of a group G is *multiplicity-free* if all the irreducible subrepresentations are pairwise nonequivalent, in symbols $\rho = \oplus_{i=1}^n \rho_i$, ρ_i irreducible, $\rho_i \nsim \rho_j$ if $i \neq j$.

We can now prove the following fundamental characterization of (finite) Gelfand pairs.

Theorem 4.4.2 *Let $K \leq G$ be finite groups. Set $X = G/K$. The following are equivalent:*

    (i) *(G, K) is a Gelfand pair, i.e. $L(K\backslash G/K)$ is commutative;*

    (ii) *$Hom_G(L(X), L(X))$ is commutative;*

    (iii) *the permutation representation $L(X)$ is multiplicity-free.*

Proof The equivalence (i) $\Leftrightarrow$ (ii) follows immediately from Proposition 4.2.1.

(iii) $\Rightarrow$ (ii): Suppose now that $L(X)$ is multiplicity-free, that is $L(X) = \oplus_{i=0}^N V_i$ with $V_0, V_1, \ldots, V_N$ inequivalent irreducible representations. Let $T \in Hom_G(L(X), L(X))$ and denote by $T_i = T|_{V_i}$ the restriction of T to V_i. If T_i is not trivial, then as V_i is irreducible, it has to be injective and therefore $\{Tv : v \in V_i\}$ is a G-invariant subspace isomorphic to V_i. Hence, by multiplicity-freeness, it coincides with V_i and by Schur's lemma (Lemma 3.5.1), there exists $\lambda_i \in \mathbb{C}$ such that $Tv = \lambda_i v$ for all $v \in V_i$.

Since any $f \in L(X)$ decomposes uniquely in the form $f = v_0 + v_1 + \cdots + v_N$, with $v_i \in V_i$, by the above we have that for every $T \in Hom_G(L(X), L(X))$ there exist $\lambda_0, \lambda_1, \ldots, \lambda_N \in \mathbb{C}$ such that

$$Tf = \sum_{i=0}^{N} \lambda_i v_i. \tag{4.8}$$

If $S \in Hom_G(L(X), L(X))$, say $Sf = \sum_{i=0}^{N} \mu_i v_i$, then

$$STf = \sum_{i=0}^{N} \mu_i \lambda_i v_i = TSf$$

for all $f \in L(X)$, showing that $Hom_G(L(X), L(X))$ is commutative.

(ii) $\Rightarrow$ (iii): Suppose that $L(X)$ is not multiplicity-free. Thus there exist two orthogonal irreducible isomorphic subrepresentations V and W in $L(X)$; let $R : V \to W$ be such an isomorphism. Denote by U the orthogonal complement of $V \oplus W$, so that $L(X) = V \oplus W \oplus U$.

We define two linear operators $S, T : L(X) \to L(X)$ by setting

$$T(v + w + u) = Rv$$
$$S(v + w + u) = R^{-1}w$$

for all $v \in V$, $w \in W$ and $u \in U$. We have that $T, S \in Hom_G(L(X), L(X))$; for instance

$$\begin{aligned} T\lambda(g)(v + w + u) &= T(\lambda(g)v + \lambda(g)w + \lambda(g)u) \\ &= R\lambda(g)v \\ &= \lambda(g)Rv \\ &= \lambda(g)T(v + w + u). \end{aligned}$$

But $ST \neq TS$ (for instance, $(ST)|_W = 0$ and $(TS)|_W = I_W$) and therefore $Hom_G(L(X), L(X))$ is not commutative. $\qquad\square$

From Theorem 4.4.2 and its proof, we can deduce further important facts. Regard the $N + 1$ dimensional space $\mathbb{C}^{N+1}$ as an algebra under the coordinatewise multiplication

$$(\alpha_0, \alpha_1, \ldots, \alpha_N) \cdot (\beta_0, \beta_1, \ldots, \beta_N) = (\alpha_0\beta_0, \alpha_1\beta_1, \ldots, \alpha_N\beta_N)$$

for any $(\alpha_0, \alpha_1, \ldots, \alpha_N), (\beta_0, \beta_1, \ldots, \beta_N) \in \mathbb{C}^{N+1}$.

Corollary 4.4.3 *Let (G, K) be a Gelfand pair and $L(X) = \oplus_{i=0}^{N} V_i$ be the decomposition of $L(X)$ into irreducible inequivalent subrepresentations. Then*

(i) *If $T \in Hom_G(L(X), L(X))$ then any V_i is an eigenspace of T.*
(ii) *If $T \in Hom_G(L(X), L(X))$ and λ_i is the eigenvalue of the restriction of T to V_i, then the map*

$$T \mapsto (\lambda_0, \lambda_1, \ldots, \lambda_N)$$

is an isomorphism between $Hom_G(L(X), L(X))$ and $\mathbb{C}^{N+1}$.
(iii) *$N + 1 = \dim(Hom_G(L(X), L(X))) = \dim(L(K\backslash G/K)) = $ number of orbits of K on X.*

Proof It suffices to observe that any $(\lambda_0, \lambda_1, \ldots, \lambda_N) \in \mathbb{C}^{N+1}$ defines, via (4.8), a (unique) $T \in Hom_G(L(X), L(X))$. $\qquad\square$

We end this section with a useful criterion for Gelfand pairs.

Proposition 4.4.4 *Let G be a finite group, $K \leq G$ a subgroup and denote by $X = G/K$ the corresponding homogeneous space. Suppose we have a decomposition $L(X) = \oplus_{t=0}^{h} Z_t$ into pairwise inequivalent G-subrepresentations with $h + 1 = $ the number of K-orbits on X. Then the Z_t's are irreducible and (G, K) is a Gelfand pair.*

Proof Refine if necessary the decomposition with the Z_t's into irreducibles as in the statement of Wielandt's lemma (Theorem 3.13.3). Then $h + 1 \leq \sum_{i=0}^{N} m_i \leq \sum_{i=0}^{N} m_i^2$ (the first inequality comes from the refinement) and the Wielandt lemma forces $h = N$ and $m_i = 1$ for all i's, concluding the proof. $\qquad\square$

4.5 Spherical functions

From now on we suppose that (G, K) is a Gelfand pair. The purpose of this section is to introduce the spherical functions and to characterize the multiplicative linear functionals on the commutative algebra $L(K\backslash G/K)$.

Definition 4.5.1 A bi-K-invariant function ϕ is called *spherical* if it satisfies the following conditions:

(i) for all $f \in L(K\backslash G/K)$ there exists $\lambda_f \in \mathbb{C}$ such that $\phi * f = \lambda_f \phi$;
(ii) $\phi(1_G) = 1$.

Remark 4.5.2 The constant function $\phi(g) \equiv 1$ is clearly spherical. The condition (i) means that ϕ is an eigenfunction for every convolution operator with a bi-K-invariant kernel, equivalently, by Proposition 4.2.1, for every $T \in Hom_G(L(X), L(X))$. From (ii) it follows that the corresponding eigenvalues are the numbers $\lambda_f = [\phi * f](1_G) \equiv T[\phi](1_G)$.

We will show that nonconstant spherical functions exist and we will describe all of them. In particular spherical functions are matrix coefficients of irreducible representations and form an orthogonal basis of the space $L(K\backslash G/K)$ of all bi-K-invariant functions.

Spherical functions may be characterized in terms of a functional equation as follows.

Theorem 4.5.3 *A bi-K-invariant nonidentically zero function ϕ is spherical if and only if*

$$\frac{1}{|K|} \sum_{k \in K} \phi(gkh) = \phi(g)\phi(h) \tag{4.9}$$

for all $g, h \in G$.

Proof Suppose that $\phi \in L(K\backslash G/K)$ satisfies the functional equation (4.9). Taking $h = 1_G$ yields

$$\phi(g) = \frac{1}{|K|} \sum_{k \in K} \phi(gk) = \phi(g)\phi(1_G)$$

so that $\phi(1_G) = 1$ as ϕ is not identically zero.

If $f \in L(K\backslash G/K)$, then, for any $k \in K$

$$[\phi * f](g) = \sum_{h \in G} \phi(gh)f(h^{-1})$$

$$= \sum_{h \in G} \phi(gh)f(h^{-1}k)$$

(setting $t = k^{-1}h$)

$$= \sum_{t \in G} \phi(gkt)f(t^{-1})$$

(averaging over K)

$$= \frac{1}{|K|} \sum_{t \in G} \sum_{k \in K} \phi(gkt)f(t^{-1})$$

$$= \phi(g) \sum_{t \in G} \phi(t)f(t^{-1})$$

$$= \{[\phi * f](1_G)\}\phi(g).$$

Vice versa, if $\phi \in L(K\backslash G/K)$ is a spherical function and $g, h \in G$, set

$$F_g(h) = \sum_{k \in K} \phi(gkh).$$

With $f \in L(K\backslash G/K)$ and $g, g_1 \in G$ we then have

$$
\begin{aligned}
[F_g * f](g_1) &= \sum_{h \in G} \sum_{k \in K} \phi(gkg_1h)f(h^{-1}) \\
&= \sum_{k \in K} [\phi * f](gkg_1) \\
&= [\phi * f](1_G) \sum_{k \in K} \phi(gkg_1) \\
&= [\phi * f](1_G)F_g(g_1).
\end{aligned}
\tag{4.10}
$$

On the other hand the function

$$J_g(h) = \sum_{k \in K} f(hkg)$$

is also bi-K-invariant and

$$
\begin{aligned}
[\phi * J_g](1_G) &= \sum_{h \in G} \phi(h^{-1}) \sum_{k \in K} f(hkg) \\
&= \sum_{k \in K} \sum_{h \in G} \phi(h^{-1})f(hg) \\
&= |K|[\phi * f](g).
\end{aligned}
$$

Therefore,

$$
\begin{aligned}
[F_g * f](g_1) &= \sum_{h \in G} \sum_{k \in K} \phi(gkg_1h)f(h^{-1}) \\
&= \sum_{h \in G} \sum_{k \in K} \phi(gh)f(h^{-1}kg_1) \\
&= [\phi * J_{g_1}](g) \\
&= [\phi * J_{g_1}](1_G)\phi(g) \\
&= |K|[\phi * f](g_1)\phi(g) \\
&= |K|[\phi * f](1_G)\phi(g_1)\phi(g).
\end{aligned}
\tag{4.11}
$$

Since for some $f \in L(K\backslash G/K)$ the value $[\phi * f](1_G) \neq 0$, from (4.10) and (4.11) we conclude that $F_g(g_1) = |K|\phi(g_1)\phi(g)$, i.e. that ϕ satisfies the functional equation (4.9). $\qquad\square$

Lemma 4.5.4 *Let ϕ be a spherical function and let Φ be the linear functional on $L(G)$ defined by*

$$\Phi(f) = \sum_{g \in G} f(g)\phi(g^{-1}). \tag{4.12}$$

Then Φ is multiplicative on $L(K\backslash G/K)$, that is, for any $f_1, f_2 \in L(K\backslash G/K)$,

$$\Phi(f_1 * f_2) = \Phi(f_1)\Phi(f_2).$$

Vice versa every nontrivial multiplicative linear functional on $L(K\backslash G/K)$ is determined by a spherical function as in (4.12).

Proof Let Φ be as in (4.12), with ϕ spherical. Note that $\Phi(f) = [f * \phi](1_G)$. Then, for any $k \in K$,

$$\begin{aligned}
\Phi(f_1 * f_2) &= [(f_1 * f_2) * \phi](1_G) \\
&= [f_1 * (f_2 * \phi)](1_G) \\
&= \{f_1 * [(f_2 * \phi)(1_G)\phi]\}(1_G) \\
&= [f_1 * \phi](1_G)[f_2 * \phi](1_G) \\
&= \Phi(f_1)\Phi(f_2).
\end{aligned}$$

Conversely, suppose that Φ is a nontrivial multiplicative linear functional on $L(K\backslash G/K)$. Observe that Φ may be extended to a linear functional on the whole $L(G)$ by defining $\Phi(f) = \Phi(^K f^K)$ (see (4.3)). Therefore there exists an element $\phi \in L(G)$ such that

$$\Phi(f) = \sum_{g \in G} f(g)\phi(g^{-1}).$$

If f is bi-K-invariant then

$$\Phi(f) = \frac{1}{|K|^2} \sum_{k,k' \in K} \sum_{g \in G} f(kgk')\phi(g^{-1}) = \frac{1}{|K|^2} \sum_{g \in G} \sum_{k,k' \in K} f(g)\phi(k'g^{-1}k).$$

Therefore, the function ϕ may be replaced by the function $^K\phi^K$, that is we may suppose that ϕ is bi-K-invariant. Since Φ is multiplicative on $L(K\backslash G/K)$,

$$\Phi(f_1 * f_2) = \sum_{g \in G} [f_1 * f_2](g^{-1})\phi(g)$$

$$= \sum_{g,h \in G} f_1(g^{-1}h)f_2(h^{-1})\phi(g)$$

$$= \sum_{h \in G} [\phi * f_1](h)f_2(h^{-1})$$

must equal

$$\Phi(f_1)\Phi(f_2) = \sum_{h \in G} \Phi(f_1)\phi(h)f_2(h^{-1})$$

for every bi-K-invariant function f_2. Therefore the equality $[\phi * f_1](h) = \Phi(f_1)\phi(h)$ holds. Taking $h = 1_G$, and observing that, by definition, $[\phi * f_1](1_G) = \Phi(f_1)$, one also obtains $\phi(1_G) = 1$. In conclusion ϕ is a spherical function. $\qquad\square$

Remark 4.5.5 Some authors, for instance Dieudonné [71], define the multiplicative functional in (4.12) with g in place of g^{-1}, in other words $\Phi(f) = \sum_{g \in G} f(g)\phi(g)$. Note that, however, the two definitions are in fact equivalent, up to exchanging the spherical function with its conjugate (cf. Proposition 4.5.7 and Remark 4.8.5).

Corollary 4.5.6 *Let (G, K) be a Gelfand pair. The number of distinct spherical functions equals the number of distinct irreducible subrepresentations in $L(X)$ and therefore equals the number of K-orbits on X.*

Proof Let $N + 1$ denote the number of irreducible subrepresentations in $L(X)$. Then, by Corollary 4.4.3, $L(K \backslash G/K)$ is isomorphic to $\mathbb{C}^{N+1}$ as algebras. Now a linear multiplicative functional Ψ on $\mathbb{C}^{N+1}$ is always of the form

$$\Psi(\alpha_0, \alpha_1, \ldots, \alpha_N) = \alpha_j$$

for some $j = 0, 1, \ldots, N$ (by linearity, $\Psi(\alpha_0, \alpha_1, \ldots, \alpha_N) = \sum_{i=0}^{N} a_i \alpha_i$; then the multiplicativity implies that there exists j such that $a_j = 1$ and $a_i = 0$ if $i \neq j$). Therefore $L(K \backslash G/K) \cong \mathbb{C}^{N+1}$ has exactly $N + 1$ multiplicative linear functionals. Thus, by the previous lemma, the number of spherical functions is exactly $N + 1$. $\qquad\square$

We end this section by proving that distinct spherical functions are orthogonal; indeed we show even more. We state all these important properties as follows.

Proposition 4.5.7 *Let ϕ and ψ be two distinct spherical functions. Then*

    (i) $\phi(g^{-1}) = \overline{\phi(g)}$ *for all $g \in G$;*

    (ii) $\phi * \psi = 0$;

    (iii) $\langle \lambda(g_1)\phi, \lambda(g_2)\psi \rangle = 0$ *for all $g_1, g_2 \in G$;*

    (iv) ϕ *and ψ are orthogonal (i.e. $\langle \phi, \psi \rangle = 0$).*

Proof Set $\phi^*(g) = \overline{\phi(g^{-1})}$ and observe that, by definition of spherical functions, one has $\phi^* * \phi = [\phi^* * \phi](1_G)\phi = \|\phi\|^2\phi$. On the other hand,

$$[\phi^* * \phi](g) = \overline{[\phi^* * \phi](g^{-1})} = \overline{[\phi^* * \phi](1_G) \cdot \phi(g^{-1})} = \|\phi\|^2\overline{\phi(g^{-1})}$$

and we get (i).

Let now ϕ and ψ be spherical functions. By commutativity, $[\phi * \psi](g) = [\phi * \psi](1_G)\phi(g)$ must equal $[\psi * \phi](g) = [\psi * \phi](1_G)\psi(g)$. Therefore, if $\phi \neq \psi$, necessarily $[\phi * \psi](1_G) = [\psi * \phi](1_G)$ must vanish (which, by (i), is nothing but (iv)) and therefore $\phi * \psi = 0$ that is (ii).

(iii) follows immediately from (i) and (ii) as

$$\langle \lambda(g_1)\phi, \lambda(g_2)\psi \rangle = \langle \phi, \lambda(g_1^{-1}g_2)\psi \rangle = [\phi * \psi](g_1^{-1}g_2).$$

$\square$

4.6 The canonical decomposition of $L(X)$ via spherical functions

Suppose again that (G, K) is a Gelfand pair. In what follows, $L(X)$ will be systematically identified with the space of K-invariant functions on G.

Let $N + 1$ denote the number of orbits of K on X. We know that $L(X)$ decomposes into $N + 1$ distinct irreducible subrepresentations and that there exist distinct spherical functions, say $\phi_0 \equiv 1, \phi_1, \ldots, \phi_N$.

Denote by $V_n = \langle \lambda(g)\phi_n : g \in G \rangle$ the subspace of $L(X)$ spanned by the G-translates of ϕ_n, $n = 0, 1, \ldots, N$.

Theorem 4.6.1 $L(X) = \bigoplus_{n=0}^{N} V_n$ *is the decomposition of $L(X)$ into irreducible subrepresentations.*

Proof Each V_n is G-invariant and, by Proposition 4.5.7, if $n \neq m$ then V_n is orthogonal to V_m. By the above comment we have that the V_i's are distinct and that they exhaust the whole $L(X)$: in other words $\oplus_{n=0}^{N} V_n$ is the direct orthogonal sum of $N + 1$ invariant subspaces. $\qquad\Box$

The representation V_n is called the *spherical representation* associated with the spherical function ϕ_n. In particular, V_0 is the trivial representation. Note that we are still dealing with equivalence classes of representations. Therefore V_n is a concrete realization of an irreducible representation as a subspace of $L(X)$. However, any other irreducible G-representation which is isomorphic to some V_n will still be called spherical.

Let (ρ, V) be a representation of G. For a subgroup $K \leq G$ denote by

$$V^K = \{v \in V : \rho(k)v = v, \text{ for all } k \in K\}$$

the space of K-*invariant vectors* in V.

Theorem 4.6.2 (G, K) *is a Gelfand pair if and only if* $\dim V^K \leq 1$ *for all irreducible G-representations V.*

If this is the case, then V is spherical if and only if $\dim V^K = 1$.

Proof Suppose that (G, K) is a Gelfand pair and let (ρ, V) be an irreducible G-representation with $\dim V^K \geq 1$.

Fix a nontrivial invariant vector $u \in V^K$. Then the linear operator $T : V \to L(X)$, defined by setting $Tv(g) = \langle v, \rho(g)u \rangle_V$, is in $Hom_G(V, L(X))$. Indeed Tv is a right K-invariant function and

$$
\begin{aligned}
[T\rho(h)v](g) &= \langle \rho(h)v, \rho(g)u \rangle_V \\
&= \langle v, \rho(h^{-1}g)u \rangle_V \\
&= [\lambda(h)(Tv)](g)
\end{aligned} \tag{4.13}
$$

for all $v \in V$ and $h, g \in G$.

In virtue of Schur's lemma, $V \sim V_n$ for some spherical representation V_n.

But if $L(X) = \oplus_{m=0}^{N} V_m$ is the decomposition of $L(X)$ into irreducible sub-representations, then $N+1$ equals the dimension of the space $L(X)^K$ of K-invariant vectors (Corollary 4.4.3) and must be equal to $\sum_{m=0}^{N} V_m^K$. Since for all m, $\dim V_m^K \geq 1$ (recall that the spherical function $\phi_m \in V_m^K$) we have that $\dim V_m^K = 1$ for all m so that $\dim V^K = \dim V_n^K = 1$.

Conversely, suppose that $dim V^K \leq 1$ for all irreducible representations V. If $L(X) = \sum_{h=0}^{H} m_h W_h$ is the decomposition into irreducible sub-representations, where the m_h's denote multiplicities, and $N+1$ is the number of K-orbits on X, then

$$\sum_{h=0}^{H} m_h^2 = N + 1 = \sum_{h=0}^{H} m_h \dim W_h^K \leq \sum_{h=0}^{H} m_h$$

where the first equality holds in virtue of Wielandt's lemma (see (3.27)) and the inequality comes from the hypothesis that the dimension of W_h^K is at most one. This forces $m_h = 1$ for all $h = 0, 1, \ldots, H$ so that $L(X)$ is multiplicity-free and, by Theorem 4.4.2 we are done. $\qquad\square$

Remark 4.6.3 As $L(X)^K = \oplus_{i=0}^{N} V_i^K$, we deduce that the spherical functions $\phi_0, \phi_1, \ldots, \phi_N$ constitute a basis for the space of bi-K-invariant functions in $L(G)$ or, what is the same, of the K-invariant functions in $L(X)$.

From the above proof, we can extract a few more important results. Often (see for example Sections 6.1 and 6.2) a spherical representation V is not directly realized as a subspace of $L(X)$. Therefore it is important to study the immersion of V into $L(X)$.

Corollary 4.6.4 *Let (ρ, V) be a spherical representation of the Gelfand pair (G, K) and fix a K-invariant vector $u \in V^K$ of norm one: $\|u\|_V = 1$. Then, the following holds.*

(i) *The map $T : V \to L(X)$ defined by*

$$[Tv](g) = \sqrt{\frac{\dim V}{|X|}} \langle v, \rho(g)u \rangle_V$$

is an isometric immersion of V into $L(X)$;

(ii) *$\phi(g) := \langle u, \rho(g)u \rangle_V$ is the spherical function in $T(V) \leq L(X)$ associated with V;*

(iii) *$\|\phi\|_{L(X)}^2 = \frac{|X|}{\dim V}$.*

Proof (i) We already proved that T intertwines the representations V and $L(X)$ in (4.13). Now, set $d = \dim V$ and fix an orthonormal basis $\{u_1, u_2, \ldots, u_d\}$ for V with $u_1 = u$. Then, setting for all $j = 1, 2, \ldots, d$, $\alpha_j = \langle v, u_j \rangle$, we have that for any $v \in V$

$$\|Tv\|_{L(X)}^2 = \frac{1}{|K|}\frac{d}{|X|}\sum_{g\in G}|\langle v,\rho(g)u\rangle_V|^2$$

$$= \frac{d}{|G|}\sum_{g\in G}\sum_{i,j=1}^{d}\alpha_i\overline{\alpha_j}\langle u_i,\rho(g)u_1\rangle_V\langle\rho(g)u_1,u_j\rangle_V$$

$$= \frac{d}{|G|}\|v\|_V^2\frac{|G|}{d}$$

$$= \|v\|_V^2$$

where the last but one equality follows from the orthogonality relations for matrix coefficients in Lemma 3.6.3 and the equality $\|v\|_V^2 = \sum_{j=1}^{d}|\alpha_j|^2$.

(ii) In virtue of the preceding theorem, the spherical function ψ in the spherical representation $T(V)$ may be characterized as the unique left K-invariant function in $T(V)$ such that $\psi(1_G) = 1$. Therefore $\psi \equiv \sqrt{\frac{|X|}{d}}Tu = \phi$.

(iii) is an immediate consequence of (i) and (ii). $\qquad\square$

4.7 The spherical Fourier transform

Let (G,K) be a Gelfand pair and $X = G/K$ the corresponding homogeneous space. In this section we regard $L(X)$ as the space of complex-valued functions on X.

Let $L(X) = \oplus_{i=0}^{N}V_i$ be the decomposition of $L(X)$ into irreducibles. For $i = 0,1,2,\ldots,N$, let ϕ_i be the spherical function in V_i, also viewed as a (K-invariant) function on X, and set $d_i = \dim V_i$. As a consequence of Proposition 4.5.7 and Corollary 4.6.4 (see also Remark 4.6.3) we get:

Proposition 4.7.1 *The spherical functions form a basis for the space of K-invariant functions in $L(X)$ and satisfy the orthogonality relations*

$$\sum_{x\in X}\phi_i(x)\overline{\phi_j(x)} = \frac{|X|}{d_i}\delta_{ij} \quad i,j = 0,1,\ldots,N. \tag{4.14}$$

The *spherical Fourier transform* $\mathcal{F}f$ of a K-invariant function f in $L(X)$ is defined by:

$$(\mathcal{F}f)(i) = \sum_{x\in X}f(x)\overline{\phi_i(x)}, \qquad i = 0,1,\ldots,N$$

i.e. $(\mathcal{F}f)(i) = \langle f,\phi_i\rangle$, where i indicates the ith irreducible G-sub-representation (namely V_i).

The inversion formula

$$f(x) = \frac{1}{|X|} \sum_{i=0}^{N} d_i(\mathcal{F}f)(i)\phi_i(x) \tag{4.15}$$

and the Plancherel formula

$$\langle f_1, f_2 \rangle = \frac{1}{|X|} \sum_{i=0}^{N} d_i(\mathcal{F}f_1)(i)\overline{(\mathcal{F}f_2)(i)}$$

(in particular, $\|f\|^2 = \frac{1}{|X|}\sum_{i=0}^{N} d_i|\mathcal{F}f(i)|^2$) are easy consequences of Proposition 4.7.1.

We now show how the above inversion formula can be deduced from the Fourier inversion formula presented in Theorem 3.9.8.

Proposition 4.7.2 *Let f be a bi-K-invariant function on G and (ρ_i, V_i) an irreducible spherical representation, with $\dim(V_i) = d_i$. Then there exists an orthonormal basis in V_i such that the matrix representing the operator $\widehat{f}(\rho_i)$ takes the form*

$$\begin{pmatrix} * & 0 & \cdots & 0 \\ 0 & 0 & \cdots & 0 \\ \vdots & & \cdots & \vdots \\ 0 & & \cdots & 0 \end{pmatrix}$$

and the nontrivial entry is nothing but $|K|\mathcal{F}f(i)$. Moreover if (π, W) is an irreducible nonspherical representation then $\widehat{f}(\pi) = 0$.

Proof First observe that the operator $\frac{1}{|K|}\sum_{k \in K} \rho_i(k)$ is the orthogonal projector onto the one-dimensional subspace V_i^K. Therefore if $\{v_1, v_2, \cdots, v_{d_i}\}$ is an orthonormal basis of V_i with $v_1 \in V_i^K$ and $(l, m) \neq (1, 1)$ we have, for any bi-K-invariant function f,

$$\begin{aligned}
\langle \widehat{f}(\rho_i)v_l, v_m \rangle &= \sum_{g \in G} \langle f(g)\rho_i(g)v_l, v_m \rangle \\
&= \sum_{g \in G} \frac{1}{|K|} \sum_{k_1 \in K} \frac{1}{|K|} \sum_{k_2 \in K} \langle f(k_1 g k_2)\rho_i(g)v_l, v_m \rangle \\
&= \sum_{g' \in G} f(g')\langle \rho_i(g')\frac{1}{|K|} \sum_{k_2 \in K} \rho_i(k_2^{-1})v_l, \frac{1}{|K|} \sum_{k_1 \in K} \rho_i(k_1)v_m \rangle \\
&= 0.
\end{aligned}$$

Moreover

$$\langle \widehat{f}(\rho_i)v_1, v_1 \rangle = \sum_{g \in G} f(g)\overline{\phi_i(g)} = |K|\mathcal{F}f(i).$$

The fact that $\widehat{f}(\pi)$ vanishes if $\pi \in \widehat{G}$ is not spherical can be proved similarly and this ends the proof. $\quad\square$

Exercise 4.7.3 Prove that the above properties of the Fourier transform of an irreducible representation characterize the bi-K-invariant functions.

Exercise 4.7.4 Deduce the (spherical) Fourier inversion formula for bi-K-invariant functions from Theorem 3.9.8.

Now we give the spectral analysis of a G-invariant operator on $L(X)$ in terms of the spherical Fourier transform of its convolution kernel (see Remark 4.2.3). Recall (Corollary 4.4.3) that each subspace V_i is an eigenspace for every $T \in Hom_G(L(X), L(X))$.

Proposition 4.7.5 *Suppose that $T \in Hom(L(X), L(X))$ and that ψ (which is a K-invariant function in $L(X)$) is its convolution kernel. Then, the eigenvalue of T corresponding to V_i is $[\mathcal{F}\psi](i)$.*

Proof To compute the eigenvalue of $T|_{V_i}$ it clearly suffices to compute $T(\phi_i)$. From Remark 4.2.3 and Definition 4.5.1 it follows that (using the notation from (4.1))

$$[T\phi_i](gx_0) = \frac{1}{|K|}(\widetilde{\phi_i} * \widetilde{\psi})(g)$$

$$= \frac{1}{|K|}(\widetilde{\phi_i} * \widetilde{\psi})(1_G)\widetilde{\phi_i}(g)$$

$$= \frac{1}{|K|}\sum_{h \in G} \widetilde{\psi}(h)\widetilde{\phi_i}(h^{-1})\widetilde{\phi_i}(g)$$

$$= [\mathcal{F}\psi](i)\phi_i(gx_0)$$

where the last equality follows from (i) in Proposition 4.5.7. $\quad\square$

Remark 4.7.6 In particular, if ψ is the characteristic function $\mathbf{1}_\Omega$ of a K-orbit Ω in X, then the eigenvalues of the corresponding convolution operator are given by

$$[\mathcal{F}\mathbf{1}_\Omega](i) = |\Omega|\overline{\phi_i(x)} \tag{4.16}$$

where x is any point in Ω (recall that ϕ_i is constant on Ω).

Exercise 4.7.7 Deduce Remark 4.7.6 from the functional equation (4.9) in Theorem 4.5.3.

Definition 4.7.8 If $f_1, f_2 \in L(X)$ are K-invariant, we can define their convolution as the K-invariant function on X corresponding to $\frac{1}{|K|} \widetilde{f_1} * \widetilde{f_2}$.

With the same argument in the proof of Proposition 4.7.5 one can prove the following.

Proposition 4.7.9 $\mathcal{F}(f_1 * f_2) = \mathcal{F}(f_1)\mathcal{F}(f_2)$.

We end this section by giving a formula for the projection onto the irreducible subspaces V_i's.

Proposition 4.7.10 *The orthogonal projection $E_i : L(X) \to V_i$ is given by the formula:*

$$(E_i f)(gx_0) = \frac{d_i}{|X|} \langle f, \lambda(g)\phi_i \rangle_{L(X)} \tag{4.17}$$

for every $f \in L(X)$ and $g \in G$.

Proof Define E_i by the above formula. From Proposition 4.5.7 it follows that if $j \neq i$ and $h, g \in G$ then

$$\{E_i[\lambda(h)\phi_j]\}(gx_0) = \frac{d_i}{|X|} \langle \lambda(h)\phi_j, \lambda(g)\phi_i \rangle_{L(X)} = 0.$$

Therefore, since V_j is spanned by the functions $\lambda(h)\phi_j$, this proves that $\ker E_i \supseteq \oplus_{j \neq i} V_j$. Moreover

$$\langle \lambda(h)\phi_i, \lambda(g)\phi_i \rangle_{L(X)} = \langle \phi_i, \lambda(h^{-1}g)\phi_i \rangle_{L(X)}$$
$$= \frac{1}{|K|} \sum_{t \in G} \widetilde{\phi}_i(t)\overline{\widetilde{\phi}_i(g^{-1}ht)}$$
$$\text{(by Proposition 4.5.7)}$$
$$= \frac{1}{|K|} \sum_{t \in G} \widetilde{\phi}_i(t)\widetilde{\phi}_i(t^{-1}h^{-1}g)$$
$$= \frac{1}{|K|} [\widetilde{\phi}_i * \widetilde{\phi}_i](h^{-1}g)$$
$$\text{(by Definition 4.5.1)}$$
$$= \frac{1}{|K|} [\widetilde{\phi}_i * \widetilde{\phi}_i](1_G)\widetilde{\phi}_i(h^{-1}g)$$

$$= \frac{\|\widetilde{\phi}_i\|^2}{|K|} \widetilde{\phi}_i(h^{-1}g)$$

$$= \frac{|X|}{d_i} \widetilde{\phi}_i(h^{-1}g)$$

(by Corollary 4.6.4)

$$= \frac{|X|}{d_i} [\lambda(h)\phi_i](gx_0)$$

and this proves that $E_i[\lambda(h)\phi_i] = \lambda(h)\phi_i$ for every $h \in G$, and then $E_i f = f$ for every $f \in V_i$. Hence the above defined E_i is the orthogonal projection from $L(X)$ to V_i. $\square$

Formula (4.17) may also be written in the following convolution form

$$(E_i f)(gx_0) = \frac{d_i}{|G|} \widetilde{f} * \widetilde{\phi}_i(g).$$

The G-invariant matrix $(r_i(x,y))_{x,y \in X}$ representing the operator E_i is given by $r_i(gx_0, y) = \frac{d_i}{|X|} \overline{\phi_i(g^{-1}y)}$, i.e. $(E_i f)(x) = \sum_{y \in X} r_i(x,y) f(y)$. Moreover, since E_i is an orthogonal projection, r_i is selfadjoint: $r_i(x,y) = \overline{r_i(y,x)}$ and idempotent: $\sum_{y \in X} r_i(x,y) r_i(y,z) = r_i(x,z)$. The following formula is an immediate consequence of these properties:

$$\langle E_i f_1, E_i f_2 \rangle = \sum_{x,y \in X} f_1(y) \overline{f_2(x)} r_i(x,y). \tag{4.18}$$

Let now $\Omega_0, \Omega_1, \ldots, \Omega_N$ be the orbits of G on $X \times X$ (with $\Omega_0 = \{(x,x) : x \in X\}$). Define a function $d : X \times X \to \{0, 1, 2, \ldots N\}$ by setting $d(x,y) = j$ if $(x,y) \in \Omega_j$. For instance, if X is a distance-transitive metric space, we can assume that d is the distance function. In any case, we can define a function $\omega_i : \{0, 1, 2, \ldots, N\} \to \mathbb{C}$ by setting, for $(x,y) \in X \times X$, $\omega_i(d(x,y)) = r_i(x,y)$ and we have

$$\omega_i(d(x,x_0)) = \frac{d_i}{|X|} \phi_i(x). \tag{4.19}$$

Under these assumptions, the formulas for $E_i f$ and (4.18) become respectively:

$$E_i f(x) = \sum_{j=0}^{N} \left(\sum_{y:d(x,y)=j} f(y) \right) \omega_i(j) \tag{4.20}$$

$$\langle E_i f_1, E_i f_2 \rangle = \sum_{j=0}^{N} \left(\sum_{x,y:d(x,y)=j} f_1(y) \overline{f_2(x)} \right) \omega_i(j). \tag{4.21}$$

Formula (4.20) is studied from an algorithmic point of view in [67]; see also [188].

In [56] and [67] a spectral analysis that consists in the decomposition $f = f_0 + f_1 + \cdots + f_n$ and the analysis of the most relevant projections was suggested. For instance, it may happen that a few of the projections are much larger than the others, and so their sum is a good approximation of f. From this approximation, it is often possible to derive many properties of the data represented by f. In Section 6.4, we shall treat a specific example.

4.8 Garsia's theorems

In this section we show how the machinery that we have developed can be used to prove two characterizations of Gelfand pairs due to Adriano Garsia [97]. Here we shall regard spherical functions as bi-K-invariant functions on G (rather than functions on X).

A *real form* of $L(K\backslash G/K)$ is a subalgebra $\mathcal{A}$ of $L(K\backslash G/K)$ such that $\dim_{\mathbb{R}}\mathcal{A} = \dim_{\mathbb{C}}L(K\backslash G/K)$ and $L(K\backslash G/K) = \mathcal{A} + i\mathcal{A}$. There are two canonical examples of real forms of $L(K\backslash G/K)$:

(1) $\mathcal{A}_1 \equiv$ the space of real valued bi-K-invariant functions;
(2) $\mathcal{A}_2 \equiv$ the real linear span of the spherical functions (this is a real form because the spherical functions satisfy the identity $\phi_i * \phi_j = \|\phi_i\|^2 \phi_i \delta_{i,j}$).

In this section an $\mathbb{R}$-*linear anti-automorphism* of the group algebra $L(G)$ will be an $\mathbb{R}$-linear map $\tau : L(G) \to L(G)$ such that $\tau(f_1 * f_2) = \tau(f_2) * \tau(f_1)$ for all $f_1, f_2 \in L(G)$.

With each real form defined above we associate an $\mathbb{R}$-linear anti-automorphism. For $f \in L(G)$ and $g \in G$ we set, with the notation of Theorem 3.9.8,

$$f^\sharp(g) = \frac{1}{|G|} \sum_{\rho \in \widehat{G}} d_\rho \sum_{i,j=1}^{d_\rho} \overline{\langle f, \overline{u^\rho_{i,j}} \rangle} u^\rho_{i,j}(g^{-1})$$

$$\equiv \frac{1}{|G|} \sum_{\rho \in \widehat{G}} d_\rho \sum_{i,j=1}^{d_\rho} \overline{\langle f, \overline{u^\rho_{i,j}} \rangle} u^\rho_{j,i}(g);$$

and

$$f^\flat(g) = \overline{f(g^{-1})}.$$

It is obvious that $f \mapsto f^\flat$ is an $\mathbb{R}$-linear anti-automorphism, while for $f \mapsto f^\sharp$ this follows from the Fourier inversion and the properties of the matrix coefficients. Indeed for $f_s \in L(G)$, $s = 1, 2$, formula (3.18) becomes

$$f_s(g) = \frac{1}{|G|} \sum_{\rho \in \widehat{G}} d_\rho \sum_{i,j=1}^{d_\rho} \left\langle f_s, \overline{u_{i,j}^\rho} \right\rangle \overline{u_{i,j}^\rho(g)}$$

and therefore, taking into account Lemma 3.9.14 we have

$$(f_1 * f_2)^\sharp = \left(\frac{1}{|G|} \sum_{\rho \in \widehat{G}} d_\rho \sum_{i,j,m=1}^{d_\rho} \left\langle f_1, \overline{u_{i,j}^\rho} \right\rangle \left\langle f_2, \overline{u_{j,m}^\rho} \right\rangle \overline{u_{i,m}^\rho(g)} \right)^\sharp$$

$$= \frac{1}{|G|} \sum_{\rho \in \widehat{G}} d_\rho \sum_{i,j,m=1}^{d_\rho} \overline{\left\langle f_1, \overline{u_{i,j}^\rho} \right\rangle} \, \overline{\left\langle f_2, \overline{u_{j,m}^\rho} \right\rangle} u_{m,i}^\rho(g)$$

$$= f_2^\sharp * f_1^\sharp.$$

Theorem 4.8.1 (G, K) *is a Gelfand pair if and only if there exists an $\mathbb{R}$-linear anti-automorphism $\tau : L(G) \to L(G)$ of the group algebra $L(G)$ and a real form $\mathcal{A}$ of $L(K \backslash G / K)$ such that $\tau(f) = f$ for all $f \in \mathcal{A}$. Moreover, we can always choose $\mathcal{A}$ both as $\mathcal{A}_1$ and $\mathcal{A}_2$.*

Proof If $\mathcal{A}$ is a real form of $L(K \backslash G / K)$, then every f and $h \in L(K \backslash G / K)$ can be expressed as $f = f_1 + if_2$ and $h = h_1 + ih_2$ with $f_i, h_i \in \mathcal{A}$, $i = 1, 2$. Therefore if τ is an anti-automorphism fixing any element $f \in \mathcal{A}$ we have

$$f * h = (f_1 + if_2) * (h_1 + ih_2)$$
$$= (f_1 * h_1 - f_2 * h_2) + i(f_1 * h_2 + f_2 * h_1)$$
$$= (\tau(f_1) * \tau(h_1) - \tau(f_2) * \tau(h_2)) + i(\tau(f_1) * \tau(h_2) + \tau(f_2) * \tau(h_1))$$
$$= (\tau(h_1 * f_1) - \tau(h_2 * f_2)) + i(\tau(h_2 * f_1) + \tau(h_1 * f_2))$$
$$= (h_1 * f_1 - h_2 * f_2) + i(h_2 * f_1 + h_1 * f_2) = h * f,$$

showing the commutativity of the subalgebra $L(K \backslash G / K)$.

Vice versa suppose that (G, K) is a Gelfand pair. Then, recalling that for a spherical representation $\rho_i \in \widehat{G}$ one has $\phi_i \equiv \overline{u_{1,1}^{\rho_i}}$ and $\phi_i(g^{-1}) = \overline{\phi_i(g)}$, for all $g \in G$, it is easy to check that for $f \in L(K \backslash G / K)$ and $g \in G$, one has

$$f^\sharp(g) = \frac{1}{|X|} \sum_{i=0}^{N} d_i \overline{\mathcal{F}f(i)} \phi_i(g^{-1}) = \frac{1}{|X|} \sum_{i=0}^{N} d_i \overline{\mathcal{F}f(i)} \phi_i(g)$$

and from the Fourier inversion formula (4.15) it follows that the map
$f \mapsto f^\sharp$ fixes any element of $\mathcal{A}_1$.

Again, the identity $\overline{\phi_i(g^{-1})} = \phi_i(g)$ immediately implies that all
elements in $\mathcal{A}_2$ are fixed under the map $f \mapsto f^\flat$.   □

Theorem 4.8.2 *A Gelfand pair is symmetric if and only if the spherical
functions are real valued (i.e., if and only if $\mathcal{A}_1 = \mathcal{A}_2$).*

Proof Let $\mathbf{1}_{KgK}$ be the characteristic function of the double coset KgK.
Its spherical Fourier transform is given by:

$$\mathcal{F}(\mathbf{1}_{KgK})(i) = \sum_{x \in KgK} \overline{\phi_i(x)} = |KgK|\overline{\phi_i(g)},$$

while

$$\mathcal{F}(\mathbf{1}_{Kg^{-1}K})(i) = |Kg^{-1}K|\overline{\phi_i(g^{-1})} = |KgK|\phi_i(g),$$

where the last equality follows from Proposition 4.5.7. By the inversion
formula it follows that $KgK = Kg^{-1}K$ for every $g \in G$ if and only if
the spherical functions are real valued.   □

Example 4.8.3 (Non-symmetric Gelfand pairs) The simplest
Gelfand pair, namely $(C_n, \{e\})$, where C_n denotes the cyclic group of
order n and e is the unit element, is nonsymmetric for $n \geq 3$; note that
the spherical functions are the characters $\phi_j(x) = \exp(2\pi ijx/n)$.

Another example of a nonsymmetric Gelfand pair is provided by
(A_4, K) where A_4 is the alternating group on $\{1, 2, 3, 4\}$ and $K =
\{e, (1, 2)(3, 4)\}$. Indeed, letting A_4 act on the set X of all 2-subsets
of $\{1, 2, 3, 4\}$ we have that this action is transitive and K is the stabilizer of the point $\{1, 2\}$. By simple calculations one shows that there are
exactly four K-orbits on X, namely

$$\Lambda_0 = \{1, 2\}$$
$$\Lambda_1 = \{\{2, 3\}, \{1, 4\}\}$$
$$\Lambda_2 = \{\{1, 3\}, \{2, 4\}\}$$
$$\Lambda_3 = \{3, 4\}.$$

From Wielandt's lemma (Theorem 3.13.3) one deduces that (G, K) is a
Gelfand pair (in fact $4 = 1 + \sum_{i=1}^3 m_i^2 \Rightarrow m_1 = m_2 = m_3 = 1$). However
it is not symmetric as $(\{1, 2\}, \{1, 3\})$ and $(\{1, 3\}, \{1, 2\})$ do not belong
to the same G-orbit in $X \times X$.

Exercise 4.8.4 Show that the Gelfand pair (A_4, K) of the previous exercise is weakly symmetric.

Remark 4.8.5 Suppose that (G, K) is a Gelfand pair and let $N + 1$ denote the number of spherical functions. Observe that if ϕ_i is a spherical function, then also $\overline{\phi_i}$ is a spherical function: this follows immediately from the definition. Therefore there exists a bijection $\{0, 1, \ldots, N\} \ni i \mapsto i' \in \{0, 1, \ldots, N\}$ such that $\overline{\phi_i} = \phi_{i'}$. Clearly, by Theorem 4.8.2, this map is the identity if and only if (G, K) is symmetric.

A spherical representation V_i is said to be *real* if the corresponding spherical function is real, equivalently if $i' = i$. In general, we call the representation $V_{i'}$ the *conjugate* (or the *adjoint*) of V_i. This notion will be explored in Chapter 9.

We say that a K-invariant function $\psi \in L(X)$ is a *symmetric* when $\overline{\widetilde{\psi}(g)} = \widetilde{\psi}(g^{-1})$ for all $g \in G$.

Exercise 4.8.6 Let $\psi \in L(X)$.

(1) Show that ψ is symmetric if and only if the corresponding convolution operator $T : L(X) \to L(X)$ is selfadjoint.

(2) Show that if ψ is real valued, then ψ is symmetric if and only if its spherical Fourier transform $\mathcal{F}\psi$ is real valued.

4.9 Fourier analysis of an invariant random walk on X

Let G be a finite group, $K \leq G$ a subgroup and denote by $X = G/K$ the corresponding homogeneous space.

We say that a matrix $P = (p(x, y))_{x,y \in X}$ is *G-invariant stochastic* if it satisfies the following conditions

$$\begin{cases} p(x, y) \geq 0 & \forall x, y \in X \\ \sum_{y \in X} p(x, y) = 1 \\ p(gx, gy) = p(x, y) & \forall x, y \in X, g \in G. \end{cases}$$

For instance, if G acts isometrically on a metric space (X, d) and $P = (p(x, y))_{x,y \in X}$ is a stochastic matrix, then P is G-invariant if and only if the transition probability $p(x, y)$ only depends on the distance $d(x, y)$ for all $x, y \in X$.

Let $x_0 \in X$ denote the point stabilized by K and set $\nu(x) = p(x_0, x)$. Then ν is a K-invariant probability measure on X. Moreover, recalling Definition 4.7.8, it is easy to check that

$$p^{(k)}(x_0, x) = \nu^{*k}(x)$$

for all $x \in X$ and $k = 0, 1, 2, \ldots$, that is, the distribution probability after k steps of the Markov chain with initial point x_0 and transition probability matrix P coincides with the kth convolution power of ν.

We denote by π the uniform distribution on X, that is $\pi(x) = \frac{1}{|X|}$ for all $x \in X$.

From now on we suppose that (G, K) is a Gelfand pair. Let $\phi_0 \equiv 1, \phi_1, \ldots, \phi_N$ be the associated spherical functions and $d_0, d_1, \ldots, d_N$ the dimensions of the corresponding spherical representations.

Proposition 4.9.1 (k-step iterate)

(i) $\nu^{*k} = \frac{1}{|X|} \sum_{i=0}^{N} d_i \left[(\mathcal{F}\nu)(i) \right]^k \phi_i$;

(ii) $\|\nu^{*k}(x) - \pi\|_{L(X)}^2 = \frac{1}{|X|} \sum_{i=1}^{N} d_i |(\mathcal{F}\nu)(i)|^{2k}$.

Proof (i) This is just an immediate consequence of the inversion formula (4.15) and of Proposition 4.7.9.

(ii) As $\pi(x) = \frac{\phi_0}{|X|}$, from (i) and from the fact that $d_0 = 1$, one deduces that

$$\nu^{*k}(x) - \pi(x) = \frac{1}{|X|} \sum_{i=1}^{N} d_i \left[(\mathcal{F}\nu)(i) \right]^k \phi_i(x)$$

and therefore (ii) follows from the orthogonality relations for the spherical functions (4.14). $\qquad \square$

We can now give the upper bound lemma for Gelfand pairs (compare with Corollary 1.9.2).

Corollary 4.9.2 (Upper bound lemma)

$$\|\nu^{*k} - \pi\|_{TV}^2 \leq \frac{1}{4} \sum_{i=1}^{N} d_i |(\mathcal{F}\nu)(i)|^{2k}. \tag{4.22}$$

Proof

$$\|\nu^{*k} - \pi\|_{TV}^2 = \frac{1}{4}\|\nu^{*k} - \pi\|_{L^1(X)}^2$$

$$\text{by Cauchy–Schwarz} \ \leq \frac{1}{4}\|\nu^{*k} - \pi\|^2 \cdot |X|$$

$$= \frac{1}{4}\sum_{i=1}^{N} d_i|(\mathcal{F}\nu)(i)|^{2k}.$$

$\square$

Exercise 4.9.3 Show that a G-invariant stochastic matrix P is reversible if and only if it is symmetric: $p(x,y) = p(y,x)$ for all $x, y \in X$. Compare with Lemma 2.1.13.

Remark 4.9.4 Suppose that (G, K) is not symmetric and that the G-invariant stochastic matrix P is not reversible. Set $Pf(x) = \sum_{y \in X} p(x,y) f(y)$ and $\nu(x) = p(x_0, x)$ for all $x \in X$, as before. Then, following the notation introduced at the end of Section 4.8, $(\mathcal{F}\nu)(i)$ is the eigenvalue of P corresponding to the conjugate spherical representation $V_{i'}$. Indeed

$$P\phi_{i'}(x_0) = \sum_{x \in X} \nu(x)\phi_{i'}(x) = \sum_{x \in X} \nu(x)\overline{\phi_i(x)} = (\mathcal{F}\nu)(i)\phi_{i'}(x_0).$$

Note that ν is the convolution kernel (cf. Remark 4.2.3) of the transposed P^t of P and $(\mathcal{F}\nu)(i)$ is the eigenvalue of P^t corresponding to V_i (compare with Proposition 4.7.5). Moreover, P is normal (P and its adjoint P^t commute: we are in a Gelfand pair) and the spectrum of P is the conjugate of the spectrum of P^t.

Exercise 4.9.5 Suppose that P is reversible.

(1) Show that if $f \in V_i$ is orthogonal to ϕ_i then $f(x_0) = 0$.
(2) Use the above result to derive (i) in Proposition 4.9.1 from Proposition 1.5.5.

Let $\Omega_0 \equiv \{x_0\}, \Omega_1, \ldots, \Omega_N$ denote the orbits of K on X. Denote by $L([N])$ the space $L(\{0, 1, \ldots, N\})$ and set $\langle f_1, f_2 \rangle_{[N]} = \sum_{k=0}^{N} |\Omega_k| f_1(k) \overline{f_2(k)}$ for all $f_1, f_2 \in L([N])$. Also set $\widetilde{\phi_i}(j) = \phi_i(x)$ whenever $x \in \Omega_j$ and $\widetilde{\pi}(i) = \frac{|\Omega_i|}{|X|}$, $i, j = 0, 1, \ldots, N$.

Suppose that $P = (p(x,y))_{x,y \in X}$ is a G-invariant stochastic matrix. Assume also (to avoid technicalities) that P is reversible. We show that P is lumpable (see Section 1.10) and even more.

Set $Pf(x) = \sum_{y \in X} p(x,y)f(y)$ and denote by λ_i the eigenvalue of P corresponding to the spherical representation V_i.

Proposition 4.9.6

(i) P is $\{\Omega_0, \Omega_1, \ldots, \Omega_N\}$-*lumpable.*

(ii) *Let* $\widetilde{p}(i,j) = p(x, \Omega_j)$, $x \in \Omega_i$ *(recall Definition 1.10.1) be the lumped chain and set* $\widetilde{P}f(i) = \sum_{j=0}^{N} \widetilde{p}(i,j)f(j)$ *for all* $f \in L([N])$. *Then,* $\widetilde{\phi}_i$ *is an eigenvector of* $\widetilde{P}$ *with eigenvalue* λ_i.

(iii) $\widetilde{\phi}_0, \widetilde{\phi}_1, \ldots, \widetilde{\phi}_N$ *constitute an orthogonal basis for* $L([N])$. *Moreover,* $\|\widetilde{\phi}_i\|^2_{L([N])} = \frac{|X|}{d_i}$ *for all* i's.

(iv) *If* $\widetilde{\nu}^{(k)}(j) = \widetilde{p}^{(k)}(0, j)$ *(the distribution probability after k steps, given that we started from 0), then*

$$\|\nu^{*k} - \pi\|^2_{L(X)} = \|\widetilde{\nu}^{*k} - \widetilde{\pi}\|^2_{L([N])}.$$

The proof is just an easy exercise. The main point in the above statement is that the spectral analysis of the lumped matrix $\widetilde{P}$ is equivalent to that of the original chain P (note that this is in fact much more than just lumpability).

5

Distance regular graphs and the Hamming scheme

In this chapter we explore a theory which gives an alternative approach to some of the diffusion processes presented in the introduction (namely the random walk on the discrete circle, the Ehrenfest and the Bernoulli–Laplace models). In some sense, this can be regarded as a theory of (finite) Gelfand pairs without group theory. Thus, Sections 5.1, 5.2, and 5.3 (as well as Sections 6.1 and 6.3 in the next chapter) do not rely on group representation theory and can be read independently of Chapters 3 and 4. The connection with group theory will be presented in the final part of Section 5.4 and in Section 6.2.

5.1 Harmonic analysis on distance-regular graphs

In this section we focus our attention on a remarkable class of finite graphs for which it is possible to develop a nice harmonic analysis. Our exposition is inspired to the monographs by Bailey [7] and by Bannai and Ito [12]. We would like to mention that during our preparation of this book we attended a minicourse by Rosemary Bailey on association schemes [8] which undoubtedly turned out to be very useful and stimulating for us.

We shall denote by X a finite, connected (undirected) graph without self-loops. Recall that given two vertices $x, y \in X$, their distance $d(x, y)$ is the length of the shortest path joining x and y. This way, (X, d) becomes a metric space. Its *diameter* is the integer number

$$\text{diam}\,(X) = \max\{d(x, y) : x, y \in X\}.$$

Definition 5.1.1 A graph X is *distance-regular* if there exist two sequences of constants $b_0, b_1, \ldots, b_N$ and $c_0, c_1, \ldots, c_N$, where $N = \text{diam}(X)$,

147

such that, for any pair of vertices $x, y \in X$ with $d(x, y) = i$ one has

$$|\{z \in X : d(x, z) = 1, d(y, z) = i + 1\}| = b_i$$
$$|\{z \in X : d(x, z) = 1, d(y, z) = i - 1\}| = c_i$$

for all $i = 0, 1, \ldots, N$. In other words, if $d(x, y) = i$, then x has b_i neighbors at distance $i + 1$ from y and c_i neighbors at distance $i - 1$ from y. In particular, taking $x = y$ we get $b_0 = |\{z \in X : d(x, z) = 1\}|$, for all $x \in X$, that is X is regular of degree b_0.

Exercise 5.1.2 Let X be a distance-regular graph. Then

(1) $b_N = 0 = c_0$;
(2) $c_1 = 1$;
(3) for $x, y \in X$ with $d(x, y) = i$ one has

$$|\{z \in X : d(x, z) = 1, d(y, z) = i\}| = b_0 - b_i - c_i.$$

More generally one has:

Lemma 5.1.3 *For any $x \in X$, the cardinality of the sphere of radius i centered at x is given by $|\{y \in X : d(x, y) = i\}| = k_i$, where $k_i = b_0 b_1 \cdots b_{i-1}/c_2 c_3 \cdots c_i$, $i = 2, 3, \ldots, N$.*

Proof Let $k_i = |\{y \in X : d(x, y) = i\}|$. For any $x \in X$,

$$|\{(y, z) \in X \times X : d(x, y) = i, d(x, z) = i + 1, d(y, z) = 1\}|$$

equals both $k_i b_i$ (choose first y and then z) and $k_{i+1} c_{i+1}$ (choose first z and then y). Therefore $k_i b_i = k_{i+1} c_{i+1}$ and since $k_1 = b_0$, the lemma follows. $\square$

In the following we shall use the convention that $b_{-1} = c_{N+1} = k_{-1} = k_{N+1} = 0$.

On the vector space $L(X)$ we define the linear operators $\Delta_0, \Delta_1, \ldots,$ $\Delta_N : L(X) \to L(X)$ by setting

$$(\Delta_j f)(x) = \sum_{y \in X : d(x, y) = j} f(y) \tag{5.1}$$

for all $f \in L(X)$ and $x \in X$.

Note that, for $j = 0$, the operator Δ_0 is nothing but the identity operator $I : L(X) \to L(X)$. On the other hand, for $j = 1$ the operator Δ_1, often simply denoted by Δ, is called the *adjacency operator* or the *Laplace operator* of the graph.

Proposition 5.1.4

  (i) *For $j = 0, 1, \ldots, N$ one has*

$$\Delta_j \Delta_1 = b_{j-1}\Delta_{j-1} + (b_0 - b_j - c_j)\Delta_j + c_{j+1}\Delta_{j+1} \qquad (5.2)$$

  where $\Delta_{N+1} = 0$.

  (ii) *For $j = 0, 1, \ldots, N$ there exists a real polynomial p_j of degree j such that*

$$\Delta_j = p_j(\Delta_1). \qquad (5.3)$$

Proof (i) For $f \in L(X)$ and $y \in X$ one clearly has

$$(\Delta_j \Delta_1 f)(y) = \sum_{\substack{z \in X: \\ d(z,y)=j}} (\Delta_1 f)(z)$$

$$= \sum_{\substack{z \in X: \\ d(z,y)=j}} \sum_{\substack{x \in X: \\ d(x,z)=1}} f(x)$$

$$= \sum_{\substack{z \in X: \\ d(z,y)=j}} \left(\sum_{\substack{x \in X: \\ d(x,z)=1 \\ d(x,y)=j-1}} f(x) + \sum_{\substack{x \in X: \\ d(x,z)=1 \\ d(x,y)=j}} f(x) + \sum_{\substack{x \in X: \\ d(x,z)=1 \\ d(x,y)=j+1}} f(x) \right)$$

$$= b_{j-1} \sum_{\substack{x \in X: \\ d(x,y)=j-1}} f(x)$$

$$+ (b_0 - b_j - c_j) \sum_{\substack{x \in X: \\ d(x,y)=j}} f(x)$$

$$+ c_{j+1} \sum_{\substack{x \in X: \\ d(x,y)=j+1}} f(x)$$

$$= b_{j-1}(\Delta_{j-1} f)(y) + (b_0 - b_j - c_j)(\Delta_j f)(y)$$
$$+ c_{j+1}(\Delta_{j+1} f)(y)$$

because for any x with $d(x,y) = j - 1$ there exist b_{j-1} elements $z \in X$ such that $d(x,z) = 1$ and $d(z,y) = j$, and therefore $f(x)$ appears b_{j-1} times in the above sums. A similar argument holds for $d(x,y) = j$ or $j + 1$ (also recall (3) in Exercise 5.1.2).

  (ii) From (i) we get

$$\Delta_1^2 = b_0 \Delta_0 + (b_0 - b_1 - c_1)\Delta_1 + c_2 \Delta_2 \qquad (5.4)$$

that is

$$\Delta_2 = \frac{1}{c_2}\Delta_1^2 - \frac{b_0 - b_1 - c_1}{c_2}\Delta_1 - \frac{b_0}{c_2}I$$

and the general case follows by induction (note that as X is connected, one always has $c_2, c_3, \ldots, c_N > 0$). $\qquad\square$

Definition 5.1.5 The algebra $\mathcal{A}$ generated by the operators $\Delta_0 = I$, $\Delta_1, \ldots, \Delta_N$ is called the *Bose-Mesner algebra* associated with X.

Theorem 5.1.6

  (i) $\mathcal{A} = \{p(\Delta_1) : p \text{ polynomial over } \mathbb{C}\}$.
 (ii) *$\mathcal{A}$ is commutative and its dimension is $N + 1$.*
(iii) *There is an orthogonal decomposition $L(X) = \oplus_{i=0}^N V_i$ such that $V_0, V_1, \ldots, V_N$ are the distinct eigenspaces of Δ_1. The V_i's are also invariant for all operators $A \in \mathcal{A}$. Moreover, if V_0 is the subspace of constant functions, the eigenvalue λ_0 of Δ_1 corresponding to V_0 is equal to the degree of X: $\lambda_0 = b_0$.*
 (iv) *Denote by E_i the orthogonal projection onto V_i and let λ_i denote the eigenvalue of Δ_1 corresponding to V_i. Then,*

$$\Delta_j = \sum_{i=0}^N p_j(\lambda_i)E_i \tag{5.5}$$

  where p_j is the polynomial in (5.3). Similarly, the projection $E_i = q_i(\Delta_1)$ for some polynomial q_i.

Proof (i) This is an immediate consequence of Proposition 5.1.4.

(ii) From (i) it follows that $\mathcal{A}$ is commutative. Moreover $\{\Delta_0 = I, \Delta_1, \ldots, \Delta_N\}$ is a linear basis for $\mathcal{A}$. Indeed, for any polynomial p one has that $p(\Delta_1)$ is a linear combination of $\Delta_0 = I, \Delta_1, \ldots, \Delta_N$ (this is a converse to (ii) in the previous proposition: as in (5.4) it follows from a repeated application of (5.2)). Moreover, if $\alpha_0, \alpha_1, \ldots, \alpha_N \in \mathbb{C}$ and $x, y \in X$, one has

$$\left(\sum_{j=0}^N \alpha_j \Delta_j \delta_y\right)(x) = \alpha_{d(x,y)}$$

thus showing that $\Delta_0, \Delta_1, \ldots, \Delta_N$ are also independent.

(iii) First note that Δ_1 is selfadjoint with respect to the scalar product $\langle f_1, f_2 \rangle = \sum_{x \in X} f_1(x)\overline{f_2(x)}$, for all $f_1, f_2 \in L(X)$. Let then $L(X) = \oplus_{i=0}^M V_i$ be the decomposition of $L(X)$ into eigenspaces of Δ_1 and denote by $\lambda_0 > \lambda_1 > \ldots > \lambda_M$ the corresponding distinct (real) eigenvalues. Since $\Delta_j = p_j(\Delta_1)$ we have that V_i is also an eigenspace of the selfadjoint operator Δ_j with corresponding eigenvalue $p_j(\lambda_i)$ (below we shall prove that $M = N$).

The last statement, that is that the eigenvalue λ_0 corresponding to the eigenspace V_0 consisting of constant functions equals the degree of X, is nothing but a reformulation of the fact that a graph X is connected (if and) only if 1 is an eigenvalue of multiplicity 1 of the Markov operator $M = \frac{1}{b_0}\Delta_1$ (see Proposition 1.7.3).

(iv) Denote by E_i the orthogonal projection onto V_i. From the preceding facts we deduce that $\Delta_j = \sum_{i=0}^{M} p_j(\lambda_i) E_i$ for all $j = 0, 1, \ldots, N$. As the spaces V_i's are orthogonal, the corresponding projections E_i's are independent. Moreover they belong to $\mathcal{A}$ as they are expressed as polynomials in Δ_1:

$$E_i = \frac{\prod_{j \neq i}(\Delta_1 - \lambda_j I)}{\prod_{j \neq i}(\lambda_i - \lambda_j)}. \tag{5.6}$$

Indeed, if E_i is defined as in (5.6) then, for $v \in V_i$ we have $\Delta_1 v = \lambda_i v$ so that $E_i v = v$ whereas, if $v \in V_j$, with $j \neq i$, $\Delta_1 v = \lambda_j v$ so that $E_i v = 0$.

As a consequence of this, the operators $E_0, E_1, \ldots, E_M$ constitute another linear basis for $\mathcal{A}$ and therefore $M = N$. $\qquad\square$

Set $d_i = \dim V_i$. From the above theorem it follows that there exist *real* coefficients $\phi_i(j)$, $i, j = 0, 1, \ldots, N$ such that

$$E_i = \frac{d_i}{|X|} \sum_{j=0}^{N} \phi_i(j) \Delta_j \tag{5.7}$$

for $i = 0, 1, \ldots, N$. The function ϕ_i is called the *spherical function* of X associated with V_i. The factor $\frac{d_i}{|X|}$ is just a normalization constant.

The matrices

$$P = \left(p_j(\lambda_i)\right)_{j, i = 0, 1, \ldots, N}$$

and

$$Q = \left(\frac{d_i}{|X|}\phi_i(j)\right)_{i, j = 0, 1, \ldots, N}$$

are called the *first* and the *second eigenvalue matrix* of X, respectively.

Remark 5.1.7 In the book by Bannai and Ito [12], $Q = (d_i\phi_i(j))_{i,j=0}^N$, while P is as above. In Bayley's book [7], P and Q are denoted by C and D, respectively.

Lemma 5.1.8

(i) $P^{-1} = Q$ (that is $\frac{d_i}{|X|}\sum_{i=0}^N \phi_i(j)p_j(\lambda_h) = \delta_{i,h}$).

(ii) $\phi_i(j) = \frac{1}{k_j}p_j(\lambda_i)$, where k_j is as in Lemma 5.1.3, for all $i, j = 0, 1, \ldots, N$.

(iii) $\phi_0(j) = 1$ for all $j = 0, 1, \ldots, N$.

(iv) $\phi_i(0) = 1$ for all $i = 0, 1, \ldots, N$.

(v) $\lambda_i = b_0\phi_i(1)$, for all $i = 0, 1, \ldots, N$.

(vi) $(E_i\delta_x)(y) = \frac{d_i}{|X|}\phi_i(d(x,y))$, for all $x, y \in X$ and $i = 0, 1, \ldots, N$.

Proof (i) This is an immediate consequence of (5.7) and (5.5).

(ii) We shall compute the trace of $E_i\Delta_j$ in two different ways. First, from (5.5) we get

$$
\begin{aligned}
tr(E_i\Delta_j) &= tr(\sum_{h=0}^N p_j(\lambda_h)E_iE_h)\\
&= tr(p_j(\lambda_i)E_i)\\
&= p_j(\lambda_i)d_i
\end{aligned}
\tag{5.8}
$$

where we used the facts that $E_iE_h = \delta_{h,i}E_i$ and $tr(E_i) = d_i$. On the other hand, using (5.7) we get

$$tr(E_i\Delta_j) = \sum_{t=0}^N \frac{d_i}{|X|}\phi_i(t)tr(\Delta_t\Delta_j). \tag{5.9}$$

But using the orthonormal basis $\{\delta_x : x \in X\}$ we get

$$
\begin{aligned}
tr(\Delta_t\Delta_j) &= \sum_{x\in X}\langle\Delta_t\Delta_j\delta_x, \delta_x\rangle\\
&= \sum_{x\in X}\langle\Delta_j\delta_x, \Delta_t\delta_x\rangle\\
&= \sum_{x\in X}\delta_{j,t}k_j\\
&= |X|k_j\delta_{j,t}
\end{aligned}
$$

where we used the fact that Δ_j is selfadjoint and $\Delta_j\delta_x$ is the characteristic function of $\{z \in X : d(x,z) = j\}$, the sphere of radius j centered at x whose cardinality is equal to k_j (cf. Lemma 5.1.3). It follows that the right hand side in (5.9) equals $d_i k_j \phi_i(j)$ and, comparing with (5.8), one obtains $\phi_i(j) = \frac{1}{k_j} p_j(\lambda_i)$.

(iii) From (5.7) we have $E_0 = \sum_{j=0}^{N} \frac{1}{|X|} \phi_0(j) \Delta_j$. Setting $E = \frac{1}{|X|} \sum_{j=0}^{N} \Delta_j$ we have

$$(Ef)(x) = \frac{1}{|X|} \sum_{j=0}^{N} (\Delta_j f)(x) = \frac{1}{|X|} \sum_{y \in X} f(y)$$

so that $E = E_0$ as both are the orthogonal projection onto the space V_0 of constant functions. Then, necessarily $\phi_0(j) = 1$ for all $j = 0, 1, \ldots, N$.

(iv) As

$$tr(\Delta_j) = \sum_{x \in X} \langle \Delta_j \delta_x, \delta_x \rangle = |X| \delta_{0,j},$$

from (5.7) we obtain

$$d_i = tr(E_i) = \frac{d_i}{|X|} \sum_{j=0}^{N} \phi_i(j) tr(\Delta_j) = d_i \phi_i(0).$$

This gives $\phi_i(0) = 1$ for all $i = 0, 1, \ldots, N$.

(v) From (ii) and the fact that $p_1(\lambda_i) = \lambda_i$ (cf. (5.5)) we get $\lambda_i = k_1 \phi_i(1) = b_0 \phi_i(1)$.

(vi) From (5.7) one has

$$(E_i \delta_x)(y) = \frac{d_i}{|X|} \sum_{j=0}^{N} \phi_i(j)(\Delta_j \delta_x)(y) = \frac{d_i}{|X|} \phi_i(d(y,x)).$$

$\square$

Corollary 5.1.9

(i) *The spherical functions satisfy the following orthogonality relations:*

$$\sum_{j=0}^{N} k_j \phi_i(j) \phi_h(j) = \frac{|X|}{d_i} \delta_{i,h}$$

for all $i, h = 0, 1, \ldots, N$.

(ii) *We have the following finite difference equations:*

$$c_j \phi_i(j-1) + (b_0 - b_j - c_j)\phi_i(j) + b_j \phi_i(j+1) = \lambda_i \phi_i(j) \quad (5.10)$$

for all $i, j = 0, 1, \ldots, N$ (we use the convention that $\phi_i(-1) = \phi_i(N+1) = 0$).

(iii) *A function* $\phi : \{0, 1, \ldots, N\} \to \mathbb{C}$ *is the spherical function associated with* V_i *if and only if for any fixed* $x_0 \in X$ *the function*
$\widetilde{\phi}(x) = \phi(d(x_0, x))$ *satisfies*

    (a) $\widetilde{\phi}(x_0) = 1$;
    (b) $\Delta_1 \widetilde{\phi} = \lambda_i \widetilde{\phi}$.

Proof (i) This is easily established by explicitly writing the coefficients
in $QP = I$ and then using Lemma 5.1.8 to express $p_j(\lambda_i) = k_j \phi_i(j)$.

(ii) From (5.5) and Lemma 5.1.8 one obtains

$$\Delta_j = \sum_{i=0}^{N} k_j \phi_i(j) E_i. \tag{5.11}$$

From Proposition 5.1.4 and (5.11) we deduce

$$\Delta_1 \Delta_j = b_{j-1} \Delta_{j-1} + (b_0 - b_j - c_j) \Delta_j + c_{j+1} \Delta_{j+1}$$

$$= \sum_{i=0}^{N} \left[b_{j-1} k_{j-1} \phi_i(j-1) + (b_0 - b_j - c_j) k_j \phi_i(j) + \right. \tag{5.12}$$

$$\left. + c_{j+1} k_{j+1} \phi_i(j+1) \right] E_i.$$

On the other hand, as $\Delta_1 E_i = E_i \Delta_1 = \lambda_i E_i$ (recall that $\Delta_1 = \sum_{i=0}^{N} \lambda_i E_i$), multiplying both sides of (5.11) by Δ_1 we obtain

$$\Delta_1 \Delta_j = \sum_{i=0}^{N} k_j \phi_i(j) \lambda_i E_i. \tag{5.13}$$

Equating the two expressions of $\Delta_1 \Delta_j$ in (5.12) and (5.13) we obtain

$$b_{j-1} \frac{k_{j-1}}{k_j} \phi_i(j-1) + (b_0 - b_j - c_j) \phi_i(j) + c_{j+1} \frac{k_{j+1}}{k_j} \phi_i(j+1) = \lambda_i \phi_i(j).$$

Then (5.10) follows from Lemma 5.1.3. Note that for $j = 0$ (5.10)
corresponds to (v) in Lemma 5.1.8.

(iii) If we set $\widetilde{\phi}_i(x) = \phi_i(d(x_0, x))$ then $\widetilde{\phi}_i(x_0) = 1$ ((iv) in Lemma 5.1.8)
and $(E_i \delta_{x_0})(x) = \frac{d_i}{|X|} \widetilde{\phi}_i(x)$ ((vi) in Lemma 5.1.8) so that $\widetilde{\phi}_i \in V_i$ and
$\Delta_1 \widetilde{\phi}_i = \lambda_i \widetilde{\phi}_i$.

On the other hand, the functions $\widetilde{\phi}_0, \widetilde{\phi}_1, \ldots, \widetilde{\phi}_N$ clearly form an orthogonal basis for the functions in $L(X)$ that are radial with respect to
x_0, that is whose values at $x \in X$ only depend on the distance $d(x, x_0)$
(recall that $N = \text{diam}(X)$). Therefore, all radial functions in V_i are
multiples of $\widetilde{\phi}_i$. $\qquad\qquad\square$

Remark 5.1.10 By abuse of language, the function $\widetilde{\phi}$ is called the spherical function in V_i with respect to the fixed point $x_0 \in X$. Indeed, it is often useful to think of a spherical function as a function defined on X.

Exercise 5.1.11 (1) Prove the *dual orthogonality relations*:

$$\sum_{i=0}^{N} d_i \phi_i(j) \phi_i(t) = \frac{|X|}{k_j} \delta_{j,t}.$$

(2) Prove that the polynomials p_j's in (5.3) satisfy the following recurrence relation:

$$b_{j-1} p_{j-1}(x) + (b_0 - b_j - c_j) p_j(x) + c_{j+1} p_{j+1}(x) = x p_j(x)$$

with $p_0(x) \equiv 1$ and $p_1(x) \equiv x$.

Using the orthogonality relations in Corollary 5.1.9 we can define a *Fourier transform* for functions $f \in L(\{0, 1, \ldots, N\})$ by setting

$$(\mathcal{F}f)(i) = \sum_{j=0}^{N} \phi_i(j) f(j) k_j$$

for all $i = 0, 1, \ldots, N$, with *inversion formula*

$$f(j) = \frac{1}{|X|} \sum_{i=0}^{N} d_i (\mathcal{F}f)(i) \phi_i(j).$$

We now apply this theory to a class of random walks on X.

Let $p = (p(x, y))_{x,y \in X}$ be a stochastic matrix on X. We say that p is *invariant* if $p(x, y)$ depends only on the distance $d(x, y)$. This means that there exist coefficients $\overline{\nu}(0), \overline{\nu}(1), \ldots, \overline{\nu}(N)$ such that

$$p(x, y) = \overline{\nu}(d(x, y)) \tag{5.14}$$

for all $x, y \in X$. Note that $\sum_{x \in X} p(x_0, x) = 1$ yields $\sum_{j=0}^{N} k_j \overline{\nu}(j) = 1$.

For $x_0, x \in X$ denote by $\nu_{x_0}^{(k)}(x) = p^{(k)}(x_0, x)$ the probability of going from x_0 to x in k steps, also denote by $P : L(X) \to L(X)$, where $(Pf)(x) = \sum_{y \in X} p(x, y) f(y)$, the corresponding operator and by $\pi(x) \equiv \frac{1}{|X|}$, for all $x \in X$, the uniform measure.

Clearly, p is reversible: indeed it is symmetric ($p(x, y) = p(y, x)$). Thus it is in detailed balance with π.

Theorem 5.1.12 (k-step iterate) *If $\overline{\nu}$ is as in (5.14) one has*

(i) *For $i = 0, 1, \ldots, N$, V_i is an eigenspace of P with corresponding eigenvalue $(\mathcal{F}\overline{\nu})(i)$ and*

$$\nu_{x_0}^{(k)}(x) = \frac{1}{|X|} \sum_{i=0}^{N} d_i \left[(\mathcal{F}\overline{\nu})(i)\right]^k \phi_i(d(x_0, x));$$

(ii) $\|\nu_{x_0}^{(k)} - \pi\|^2 = \frac{1}{|X|} \sum_{i=1}^{N} d_i \left[(\mathcal{F}\overline{\nu})(i)\right]^{2k}$.

Proof It is easy to check that $P = \sum_{j=0}^{N} \overline{\nu}(j)\Delta_j$. Therefore, in virtue of Theorem 5.1.6, V_i is an eigenspace of P and the corresponding eigenvalue is

$$\sum_{j=0}^{N} \overline{\nu}(j) p_j(\lambda_i) = \sum_{j=0}^{N} \overline{\nu}(j) k_j \phi_i(j) = (\mathcal{F}\overline{\nu})(i),$$

where the first equality follows from Lemma 5.1.8.

This means that $P = \sum_{i=0}^{N} (\mathcal{F}\overline{\nu})(i) E_i$ and $P^k = \sum_{i=0}^{N} [(\mathcal{F}\overline{\nu})(i)]^k E_i$. But then

$$\nu_{x_0}^{(k)}(x) = p^{(k)}(x_0, x) = (P^k \delta_x)(x_0) = \sum_{i=0}^{N} [(\mathcal{F}\overline{\nu})(i)]^k (E_i \delta_x)(x_0)$$

$$= \frac{1}{|X|} \sum_{i=0}^{N} [(\mathcal{F}\overline{\nu})(i)]^k d_i \phi_i(d(x_0, x))$$

where the last equality follows from (vi) in Lemma 5.1.8. This proves (i).

Observe that π may be written

$$\pi(x) = \frac{d_0}{|X|} \phi_0(d(x_0, x))$$

so that

$$\nu_{x_0}^{(k)}(x) - \pi(x) = \frac{1}{|X|} \sum_{i=1}^{N} [(\mathcal{F}\overline{\nu})(i)]^k d_i \phi_i(d(x_0, x))$$

and one can then apply the orthogonality relations for the ϕ_i's. $\quad\square$

Corollary 5.1.13 (Upper bound lemma)

$$\|\nu_{x_0}^{(k)} - \pi\|_{TV} \leq \frac{1}{4} \sum_{i=1}^{N} d_i [(\mathcal{F}\overline{\nu})(i)]^{2k}.$$

Proof We have

$$\|\nu_{x_0}^{(k)} - \pi\|_{TV} = \frac{1}{4}\|\nu_{x_0}^{(k)} - \pi\|_{L^1(X)}^2 \quad \text{see (1.26)}$$

$$\leq \frac{1}{4}\|\nu_{x_0}^{(k)} - \pi\|^2 |X| \quad \text{(by Cauchy–Schwarz)}$$

$$\leq \frac{1}{4}\sum_{i=1}^{N} d_i [(\mathcal{F}\bar{\nu})(i)]^{2k}.$$

$\square$

Suppose again that $p(x,y)$ is invariant. We now examine the lumpability of p (cf. Section 1.10). Fix $x_0 \in X$ and denote by $\Omega_h = \{x \in X : d(x_0, x) = h\}$ the sphere of radius h centered at x_0, for $h = 0, 1, 2, \ldots, N$. Set $[N] = \{0, 1, \ldots, N\}$ and on $L([N])$ introduce the scalar product $\langle f_1, f_2 \rangle = \sum_{i=0}^{N} f_1(i)\overline{f_2(i)}|\Omega_i|$ (recall that $|\Omega_i| = k_i$; see Lemma 5.1.3). Also set $\pi(i) = \frac{|\Omega_i|}{|X|}$ for $i = 0, 1, \ldots, N$.

Theorem 5.1.14 *With the above notation we have the following.*

(i) *p is $\{\Omega_0, \Omega_1, \ldots, \Omega_N\}$-lumpable.*

(ii) *Let $\widetilde{p}(i,j) = p(x, \Omega_j)$, for $x \in \Omega_i$, be the lumped chain and set $\left(\widetilde{P}f\right)(i) = \sum_{j=0}^{N} \widetilde{p}(i,j)f(j)$ for all $f \in L([N])$. Then the spherical function ϕ_i is an eigenvector of $\widetilde{P}$ with eigenvalue λ_i.*

(iii) *$\phi_0, \phi_1, \ldots, \phi_N$ constitute an orthogonal basis for $L([N])$ and $\|\phi_i\|^2 = \frac{|X|}{d_i}$, for $i = 0, 1, \ldots, N$.*

(iv) *If $\widetilde{\nu}^{(k)}(j) = \widetilde{p}^{(k)}(0,j)$ denotes the distribution probability after k steps given that we start from 0, then $\|\nu^{*k} - \pi\|_{L(X)}^2 = \|\widetilde{\nu}^{(k)} - \widetilde{\pi}\|_{L([N])}^2$.*

The proof is just an easy exercise. The main fact is that the spectral analysis of the lumped matrix $\widetilde{p}$ is equivalent to that of the original chain p (this is even more than lumpability). In other words, the eigenvalues are the same, only the corresponding multiplicities do change.

Exercise 5.1.15 Let K_n denote the complete graph on n vertices (see Example 1.8.1).

(1) Show that K_n is distance-regular and that $L(K_n) = V_0 \oplus V_1$ – with V_0 the space of constant functions and $V_1 = \{f : K_n \to \mathbb{C} : \sum_{x \in K_n} f(x) = 0\}$ the space of mean zero functions – is the decomposition into eigenspaces of Δ_1.

(2) Show that $\phi_0 \equiv 1$ (constant function) and

$$\phi_1(x) = \begin{cases} 1 & \text{if } x = x_0 \\ -\frac{1}{n-1} & \text{if } x \neq x_0 \end{cases}$$

are the spherical functions.

(3) Deduce the results in Example 1.10.5 from the previous theorem.

In the final part of this section we analyze the connections with group theory and finite Gelfand pairs. We will also briefly give the more general notion of an association scheme.

Let X be a finite, simple, connected graph without loops. Denote by E its edge set. An *automorphism* of X is a bijection $\alpha : X \rightarrow X$ such that $\{\alpha(x), \alpha(y)\} \in E$ if and only if $\{x, y\} \in E$, for all $x, y \in X$.

Example 5.1.16 Consider the graph with vertices $1, 2, 3, 4$ and edges $\{1, 2\}$, $\{2, 4\}$, $\{2, 3\}$ and $\{3, 4\}$ (see Figure 5.1). Then the map $\alpha : X \rightarrow X$ given by $\alpha(1) = 1$, $\alpha(2) = 2$, $\alpha(3) = 4$ and $\alpha(4) = 3$ is an automorphism.

Figure 5.1.

The set of all automorphisms of a graph X is clearly a group and it is denoted by $Aut(X)$.

An *isometry* of the graph X is a bijection $\beta : X \rightarrow X$ such that $d(\beta(x), \beta(y)) = d(x, y)$ for all $x, y \in X$. Since we have excluded the presence of loops in X, it is clear that β is an isometry if and only if it is an automorphism; in other words, the notions of automorphism and isometry coincide.

Exercise 5.1.17 Show that for X as in Figure 5.2 one has $Aut(X) = \{1\}$ and that X is, in fact, the smallest (i.e. $|X|$ minimal) graph with no nontrivial automorphisms.

Figure 5.2.

Exercise 5.1.18 ([25]) Show that for $n \geq 7$ the graph in Figure 5.3 has trivial automorphism group.

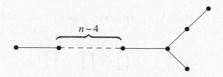

Figure 5.3.

Definition 5.1.19 A graph X is *distance-transitive* if for any (x, y) and $(u, z) \in X \times X$ with $d(x, y) = d(u, z)$ there exists $\alpha \in Aut(X)$ such that $(\alpha(x), \alpha(y)) = (u, z)$.

If G is a group of isometries of X (i.e. $G \leq Aut(X)$) and the isometry α above can be found inside G, we say that the action of G on X is *distance-transitive*.

We then know that if we fix $x_0 \in X$ and set $K = \{k \in G : kx_0 = x_0\}$, then (G, K) is a symmetric Gelfand pair (see Examples 4.3.2 and 4.3.7).

Exercise 5.1.20 Show that a distance-transitive graph is also distance-regular. Check that the harmonic analysis on distance-regular graphs developed in this section coincides with that developed in Chapter 4 when the graph is distance-transitive. Show, in particular, that the subspaces $V_0, V_1, \ldots, V_N$ in Theorem 5.1.6 are the spherical representations of the pair (G, K) (in particular they are irreducible G-representations). Moreover, $\phi_0, \phi_1, \ldots, \phi_N$ are the spherical functions on the Gelfand pair (G, K).

The "new" aspect in the harmonic analysis on distance-regular graphs is that the algebra $Hom_G(L(X), L(X)) \cong L(K \backslash G / K)$ is singly generated (it is the polynomial algebra in Δ_1).

Exercise 5.1.21 ([1, 25]) Let X be the graph with vertices $x_1, x_2, \ldots, x_{13}, y_1, y_2, \ldots, y_{13}$ and edge set

$$E = \{\{x_i, x_j\} : |i - j| = 1, 3, 4 \mod 13\}$$
$$\cup \{\{y_i, y_j\} : |i - j| = 2, 5, 6 \mod 13\}$$
$$\cup \{\{x_i, y_j\} : i - j = 0, 1, 3, 9 \mod 13\}.$$

Show that X is distance-regular with diameter $N = 2$ and $b_0 = 10$, $b_1 = 6, c_1 = 1, c_2 = 4$ but it is *not* distance transitive (there is no automorphism carrying x_i to y_j).

Let X be a finite set. A *symmetric association scheme* on X is a partition of $X \times X$

$$X \times X = \mathcal{C}_0 \coprod \mathcal{C}_1 \coprod \cdots \coprod \mathcal{C}_N$$

where the sets $\mathcal{C}_i$ (called the *associate classes*) satisfy the following properties:

  (i) $\mathcal{C}_0 = \{(x, x) : x \in X\}$.
  (ii) For $i = 1, 2, \ldots, N$, $\mathcal{C}_i$ is *symmetric*, that is $(x, y) \in \mathcal{C}_i$ if and only if $(y, x) \in \mathcal{C}_i$.
  (iii) There exist nonnegative integers $p_{i,j}^k$, $i, j, k = 0, 1, \ldots, N$, such that, for all $(x, y) \in \mathcal{C}_k$ we have

$$|\{z \in X : (x, z) \in \mathcal{C}_i, (z, y) \in \mathcal{C}_j\}| = p_{i,j}^k.$$

Exercise 5.1.22

  (1) (1) Let X be a distance-regular graph with diameter N. Show that, taking $\mathcal{C}_i = \{(x, y) : d(x, y) = i\}$, $i = 0, 1, \ldots, N$, then $\mathcal{C}_0, \mathcal{C}_1, \ldots, \mathcal{C}_N$ form a symmetric association scheme over X.
  (2) Suppose that (G, K) is a symmetric Gelfand pair. Set $X = G/K$ and let $\mathcal{C}_0, \mathcal{C}_1, \ldots, \mathcal{C}_N$ (with $\mathcal{C}_0 = \{(x, x) : x \in X\}$) be the orbits of G on $X \times X$. Show that $\mathcal{C}_0, \mathcal{C}_1, \ldots, \mathcal{C}_N$ form a symmetric association scheme over X (cf. Exercise 4.2.4).

The peculiarity of a distance-regular graph is that its Bose–Mesner algebra is singly generated, namely by Δ_1. This is no longer true for general symmetric association schemes: see Chapter 7.

The fundamental reference on association scheme is Delsarte's epochal thesis [51]. A recent survey by the same author is [54]. Another useful survey is Bannai's [11]. There are many books devoted to this subject, or with some chapters devoted to it: we mention, among others those by Bannai and Ito [12], Bayley [7], Godsil [102], van Lint and Wilson [156], Cameron [39] Cameron and van Lint [40], MacWilliams and Sloane [159] and by P.-H. Zieschang [234]. Also, the book by Browuer, Cohen and Neumaier [36] is an encyclopedic treatment on distance-regular graphs. A very interesting paper on random walks on distance-regular graphs is [22].

5.2 The discrete circle

As a first example of distance-regular (but in fact of a distance-transitive) graph we examine the discrete circle (see Example 1.2.1 and Sections 2.1 and 2.2), also called the *cycle graph* (cf. Biggs [25]) or the *regular n-gon* (cf. Stanton [208]).

As we have already studied this example, one should regard this section just as an exercise of translation of the concepts into the new setting.

Observe that C_n has diameter $N = \left[\frac{n}{2}\right]$. Moreover it is distance-regular: it is easy to check that $b_0 = 2, b_1 = b_2 = \ldots = b_{N-1} = 1$, $c_1 = c_2 = \ldots = c_{N-1} = 1$ and, finally, $c_N = 1$ if n is odd and $c_N = 2$ if n is even.

In this case, the operator Δ_1 coincides with $2P$ where P is the operator associated with the simple random walk on C_n. Therefore if we set (according to the notation of Sections 2.1 and 2.2) V_j equal to the space spanned by χ_j and χ_{-j}, then $V_0, V_1, \ldots, V_N$ are the eigenspaces of Δ_1 and the corresponding eigenvalues $\lambda_j = 2\cos(\frac{2\pi j}{n})$ for $j = 0, 1, \ldots, N$. Moreover, the spherical functions are

$$\phi_j(h) = \frac{1}{2}\left(\chi_j(h) + \chi_j(-h)\right) = \cos\left(\frac{\pi h j}{n}\right).$$

Indeed, $\phi_i \in V_i$ and it is radial with respect to the base point 0 ($\phi_j(h) = \phi_j(-h)$) and $\phi_j(0) = 0$. Note also that ϕ_j coincides with the function c_j in Section 2.6.

The graph C_n is also distance-transitive with automorphism group given by the dihedral group $D_n = \langle a, b : a^2 = b^n = 1, aba = b^{-1}\rangle$. Note that a is the reflection with respect to the diameter passing through 0, that is $a(h) = -h \mod n$ for $h = 0, 1, \ldots, n-1$, and b is the rotation $b(h) = h + 1 \mod n$ for $h = 0, 1, \ldots, n-1$.

The stabilizer of 0 is the cyclic group of order two $C_2 = \langle a\rangle$ and clearly (D_n, C_2) is a Gelfand pair. Moreover, $V_0, V_1, \ldots, V_N$ are all irreducible representations of D_n.

Exercise 5.2.1 Show that, with the notation in Example 3.8.2 and in Exercise 3.8.3 and Exercise 3.8.4, that the representation on V_0 coincides with χ_0, the representation on V_i, $i = 1, 2, \ldots, N-1$, coincides with ρ_i and the representation on V_N coincides with χ_3 if n is even and with ρ_N if n is odd.

5.3 The Hamming scheme

Set $X_{n,m+1} = \{(x_1, x_2, \ldots, x_n) : x_1, \ldots, x_n \in \{0, 1, \ldots, m\}\}$. In other words, $X_{n,m+1}$ is the cartesian product of n copies of $\{0, 1, 2, \ldots, m\} = X_{1,m+1}$.

Define the *Hamming distance* d on $X_{n,m+1}$ by setting

$$d((x_1, x_2, \ldots, x_n), (y_1, y_2, \ldots, y_n)) = |\{k : x_k \neq y_k\}|$$

that is, the distance between two points equals the number of coordinates at which they differ.

We can define on $X_{n,m+1}$ a graph structure by taking $\{(x, y) : d(x, y) = 1\}$ as the set of edges, that is, two points $x, y \in X_{n,m+1}$ are connected if and only if they differ in exactly one coordinate. It is easy to check (exercise) that such a structure induces exactly the metric d.

Proposition 5.3.1 $X_{n,m+1}$ *is a distance regular graph with diameter n and coefficients*

$$
\begin{aligned}
c_i &= i, & i &= 1, 2, \ldots, n \\
b_i &= (n - i)m, & i &= 0, 1, \ldots, n - 1.
\end{aligned}
$$

In particular, its degree is $b_0 = nm$.

Proof Let $x, y \in X_{n,m+1}$ with $d(x, y) = i$. Set $A = \{j : x_j = y_j\}$. Then $|A| = n - i$ and any $z \in X_{n,m+1}$ with $d(x, z) = 1$ and $d(y, z) = i + 1$ is obtained by changing a coordinate x_j with $j \in A$; there are $|A| = (n-i)$ such coordinates and m different ways of performing such a change. This gives $b_i = (n - i)m$. The other cases are similar and left to the reader as an easy exercise. $\quad\square$

Note that $X_{1,m+1}$ coincides with the complete graph K_{m+1} on $m + 1$ vertices. Moreover, in this case we always have $b_i + c_i = b_0$ (cf. Exercise 5.1.2).

Let Δ denote the adjacency operator, that is, $(\Delta f)(x) = \sum_{y:d(x,y)=1} f(y)$ for all $x \in X_{n,m+1}$ and $f \in L(X_{n,m+1})$. We have already seen that the harmonic analysis on $X_{n,m+1}$ is equivalent to the spectral analysis of Δ.

Therefore our task is to obtain the decomposition of $L(X_{n,m+1})$ into eigenspaces of Δ. For $f_1, f_2, \ldots, f_n \in L(X_{1,m+1})$, define their coordinatewise product $f_1 \otimes f_2 \otimes \cdots \otimes f_n \in L(X_{n,m+1})$ by setting: $(f_1 \otimes f_2 \otimes \cdots \otimes f_n)(x_1, x_2, \ldots, x_n) = f_1(x_1)f_2(x_2) \cdots f_n(x_n)$. We recall the following orthogonal decomposition: $L(X_{1,m+1}) = V_0 \oplus V_1$ where V_0 is the space of constant functions and $V_1 = \{f : \sum_{i=0}^m f(i) = 0\}$, see Example 1.8.1 ($V_0$ and V_1, in the representation theory of the symmetric group S_{m+1}, are usually denoted by S^{m+1} and $S^{m,1}$, respectively; see, for instance, Section 3.14).

The *weight* of $(i_1, i_2, \ldots, i_n) \in \{0,1\}^n$ is $w(i_1, i_2, \ldots, i_n) = |\{k : i_k = 1\}|$. Define W_j as the subspace of $L(X_{n,m+1})$ spanned by the products $f_1 \otimes f_2 \otimes \cdots \otimes f_n$ with each $f_i \in V_0$ or V_1 and $|\{i : f_i \in V_0\}| = n - j$ and $|\{i : f_i \in V_1\}| = j$ and set $\overline{x} = (0, 0, \ldots, 0)$.

Taking into account Remark 5.1.10 (with the fixed point x_0 replaced by $\overline{x}$) we have the following:

Theorem 5.3.2

(i) $L(X_{n,m+1}) = \oplus_{j=0}^n W_j$ is the decomposition of $L(X_{n,m+1})$ into distinct Δ-eigenspaces. The eigenvalue corresponding to W_j is equal to $nm - j(m+1)$ and $\dim W_j = m^j \binom{n}{j}$.

(ii) The spherical function $\Phi_j \in W_j$ is given by

$$\Phi_j(x) = \frac{1}{\binom{n}{j}} \sum_{t=\max\{0,j-n+\ell\}}^{\min\{\ell,j\}} \binom{\ell}{t} \binom{n-\ell}{j-t} \left(-\frac{1}{m}\right)^t \text{ if } d(x, \overline{x}) = \ell.$$

Proof First of all note that for $f_1, f_2, \ldots, f_n, f'_1, f'_2, \ldots, f'_n \in L(X_{1,m+1})$ we have

$$\langle f_1 \otimes f_2 \otimes \cdots \otimes f_n, f'_1 \otimes f'_2 \otimes \cdots \otimes f'_n \rangle_{L(X_{n,m+1})}$$

$$= \sum_{x_1, \ldots, x_n \in X_{1,m+1}} f_1(x_1) f_2(x_2) \cdots f_n(x_n) \overline{f'_1(x_1)} \overline{f'_2(x_2)} \cdots \overline{f'_n(x_n)}$$

$$= \prod_{i=1}^n \left(\sum_{x_i \in X_{1,m+1}} f_i(x_i) \overline{f'_i(x_i)} \right)$$

$$= \prod_{i=1}^n \langle f_i, f'_i \rangle_{L(X_{1,m+1})}. \tag{5.15}$$

It follows that if $\psi_0, \psi_1, \ldots, \psi_m$ is an orthonormal basis in $L(X_{1,m+1})$ with $\psi_0 \in V_0$ and $\psi_1, \psi_2, \ldots, \psi_m \in V_1$, then

$$\{\psi_{i_1} \otimes \psi_{i_2} \otimes \cdots \otimes \psi_{i_n} : i_1, i_2, \ldots, i_n \in \{0, 1, \ldots, m\}\}$$

is an orthonormal basis in $L(X_{n,m+1})$.

It is also clear that the set $\{\psi_{i_1} \otimes \psi_{i_2} \otimes \cdots \otimes \psi_{i_n} : |\{k : i_k = 0\}| = n-j\}$ is an orthonormal basis for W_j. This shows that $L(X_{n,m+1}) = \oplus_{j=0}^n W_j$ with orthogonal direct sum and that moreover $\dim(W_j) = m^j \binom{n}{j}$.

Now observe that if $f \in V_0$ then $\sum_{y \neq x} f(y) = m f(x)$ while if $f \in V_1$ then $\sum_{y \neq x} f(y) = -f(x)$. Therefore if $w(i_1, \ldots, i_n) = j$ and $f_k \in V_{i_k}$, $k = 1, 2, \ldots, n$, then

$$[\Delta(f_1 \otimes \cdots \otimes f_n)](x_1, \ldots, x_n) = \left(\sum_{y_1 \neq x_1} f_1(y_1)\right) f_2(x_2) \cdots f_n(x_n)$$
$$+ f_1(x_1) \left(\sum_{y_2 \neq x_2} f_2(y_2)\right) f_3(x_3) \cdots f_n(x_n)$$
$$+ \cdots +$$
$$+ f_1(x_1) \cdots f_{n-1}(x_{n-1}) \left(\sum_{y_n \neq x_n} f_n(y_n)\right)$$
$$= [nm - j(m+1)] \cdot [(f_1 \otimes \cdots \otimes f_n)(x_1, \ldots, x_n)]$$

and (i) is proved.

Now define $\phi_0, \phi_1 \in L(X_{1,m+1})$ by setting $\phi_0(x) = 1$ for $x = 0, 1, \ldots, m$, $\phi_1(0) = 1$ and $\phi_1(x) = -\frac{1}{m}$ for $x = 1, 2, \ldots, m$. Clearly $\phi_0 \in V_0$, $\phi_1 \in V_1$ and they are the spherical functions on $X_{1,m+1}$ (see Exercise 5.1.15). Set

$$\Phi_j = \frac{1}{\binom{n}{j}} \sum_{w(i_1,\ldots,i_n)=j} \phi_{i_1} \otimes \phi_{i_2} \otimes \cdots \otimes \phi_{i_n}.$$

Then $\Phi_j \in W_j$; moreover, if $d(x, \overline{x}) = \ell$ then

$$\Phi_j(x) = \frac{1}{\binom{n}{j}} \sum_{t=\max\{0,j-n+\ell\}}^{\min\{\ell,j\}} \binom{\ell}{t} \binom{n-\ell}{j-t} \left(-\frac{1}{m}\right)^t.$$

Indeed, the number of $(i_1, \ldots, i_n) \in \{0,1\}^n$ such that $w(i_1, \ldots, i_n) = j$ and $|\{k : \phi_{i_k}(x_k) = -\frac{1}{m}\}| = t$ is equal to $\binom{\ell}{t}\binom{n-\ell}{j-t}$: this corresponds to choosing t indices $\{k : x_k \neq 0$ and $i_k = 1\}$ inside the set (of cardinality ℓ) $\{k : x_k \neq 0\}$ simultaneously with other $j-t$ indices $\{k : x_k = 0$ and $i_k = 1\}$ inside the set (of cardinality $n - \ell$) $\{k : x_k = 0\}$.

In particular, Φ_j is constant on the spheres $\{x \in X_{n,m+1} : d(x, \overline{x}) = \ell\}$, $\ell = 0, 1, 2, \ldots, m$; it is obvious that $\Phi_j(\overline{x}) = 1$ and therefore (Corollary 5.1.9) it is the spherical function in W_j. $\square$

Remark 5.3.3 Setting

$$K_j(\ell; m/(m+1), n) = \frac{1}{\binom{n}{j}} \sum_{t=\max\{0,j-n+\ell\}}^{\min\{\ell,j\}} \binom{\ell}{t} \binom{n-\ell}{j-t} \left(-\frac{1}{m}\right)^t$$

then $\{K_j(\,\cdot\,; m/(m+1), n) : j = 0, 1, \ldots, n\}$ are orthogonal polynomials on $\{0, 1, \ldots, n\}$ with respect to the mass $\binom{n}{\ell}\frac{m^\ell}{(m+1)^n}$ at $\ell \in \{0, 1, \ldots, n\}$, normalized by $K_j(0; m/(m+1), n) = 1$; they are called *Krawtchouk polynomials* (see also Section 2.6).

Indeed, from Corollary 5.1.9 (i), since $K_j(\ell; m/(m+1), n)$ is the value of Φ_j at a point x with $d(x, \overline{x}) = \ell$, we have

$$\sum_{\ell=0}^{n} K_i(\ell, m/(m+1), n) K_j(\ell, m/(m+1), n) m^\ell \binom{n}{\ell} = \frac{(m+1)^n}{m^j \binom{n}{j}} \delta_{i,j}$$

as $|\{x : d(x, \overline{x}) = \ell\}| = m^\ell \binom{n}{\ell}$.

Moreover, again from Corollary 5.1.9, it follows that $\Delta \Phi_j = [nm - j(m+1)]\Phi_j$ may be translated into the second order finite difference equation for the Krawtchouk polynomials:

$$\begin{aligned}
\ell K_j(\ell - 1, m/(m+1), n) \quad + \quad & \ell(m-1)K_j(\ell, m/(m+1), n) \\
+ \quad & (n-\ell)m K_j(\ell+1, m/(m+1), n) \\
= \quad & [nm - j(m+1)]K_j(\ell, m/(m+1), n).
\end{aligned}$$

5.4 The group-theoretical approach to the Hammming scheme

In this section, we reformulate the results of the previous section in the setting of Gelfand pairs. We first need a suitable group action on $X_{n,m+1}$.

Let S_{m+1} and S_n be the symmetric groups respectively on $\{0, 1, 2, \ldots, m\}$ and $\{1, 2, \ldots, n\}$. The *wreath product* of S_{m+1} by S_n is the set $S_{m+1} \wr S_n = \{(\sigma_1, \sigma_2, \ldots, \sigma_n; \theta) : \sigma_1, \ldots, \sigma_n \in S_{m+1}, \theta \in S_n\}$ together with the composition law

$$\begin{aligned}
(\sigma_1, \sigma_2, \ldots, \sigma_n; \theta) \cdot (\tau_1, \tau_2, \ldots, \tau_n; \xi) \\
= (\sigma_1 \tau_{\theta^{-1}(1)}, \sigma_2 \tau_{\theta^{-1}(2)}, \ldots, \sigma_n \tau_{\theta^{-1}(n)}; \theta \xi).
\end{aligned}$$

Lemma 5.4.1 $S_{m+1} \wr S_n$ *is a group with respect to the above defined composition law.*

Proof The identity element is clearly $(1_{S_{m+1}}, 1_{S_{m+1}}, \ldots, 1_{S_{m+1}}; 1_{S_n})$. The inverse of $(\sigma_1, \sigma_2, \ldots, \sigma_n; \theta)$ is $(\sigma_{\theta(1)}^{-1}, \sigma_{\theta(2)}^{-1}, \ldots, \sigma_{\theta(n)}^{-1}; \theta^{-1})$. For the associativity check that

$$[(\sigma_1, \sigma_2, \ldots, \sigma_n; \theta) \cdot (\tau_1, \tau_2, \ldots, \tau_n; \xi)] \cdot (\eta_1, \eta_2, \ldots, \eta_n; \mu)$$

$$= (\sigma_1 \tau_{\theta^{-1}(1)} \eta_{\xi^{-1}(\theta^{-1}(1))}, \sigma_2 \tau_{\theta^{-1}(2)} \eta_{\xi^{-1}(\theta^{-1}(2))},$$

$$\ldots, \sigma_n \tau_{\theta^{-1}(n)} \eta_{\xi^{-1}(\theta^{-1}(n))}; \theta\xi\mu)$$

$$= (\sigma_1, \sigma_2, \ldots, \sigma_n; \theta) \cdot [(\tau_1, \tau_2, \ldots, \tau_n; \xi) \cdot (\eta_1, \eta_2, \ldots, \eta_n; \mu)].$$

$\square$

Lemma 5.4.2 *The set* $X_{n,m+1} = \{(x_1, x_2, \ldots, x_n) : x_1, \ldots, x_n \in \{0, 1, \ldots, m\}\}$ *is a homogenous space for* $S_{m+1} \wr S_n$ *under the action*

$$(\sigma_1, \sigma_2, \ldots, \sigma_n; \theta)(x_1, x_2, \ldots, x_n) = (\sigma_1 x_{\theta^{-1}(1)}, \sigma_2 x_{\theta^{-1}(2)}, \ldots, \sigma_n x_{\theta^{-1}(n)})$$

(i.e. x_i *is moved by* θ *to the position* $\theta(i)$ *and then it is changed by the action of* $\sigma_{\theta(i)}$*).*

Proof This is indeed an action:

$$(\sigma_1, \sigma_2, \ldots, \sigma_n; \theta) \cdot [(\tau_1, \tau_2, \ldots, \tau_n; \xi)(x_1, x_2, \ldots, x_n)]$$

$$= (\sigma_1, \sigma_2, \ldots, \sigma_n; \theta) \cdot (\tau_1 x_{\xi^{-1}(1)}, \tau_2 x_{\xi^{-1}(2)}, \ldots, \tau_n x_{\xi^{-1}(n)})$$

$$= (\sigma_1 \tau_{\theta^{-1}(1)} x_{\xi^{-1}(\theta^{-1}(1))}, \sigma_2 \tau_{\theta^{-1}(2)} x_{\xi^{-1}(\theta^{-1}(2))},$$

$$\ldots, \sigma_n \tau_{\theta^{-1}(n)} x_{\xi^{-1}(\theta^{-1}(n))}),$$

while

$$[(\sigma_1, \sigma_2, \ldots, \sigma_n; \theta) \cdot (\tau_1, \tau_2, \ldots, \tau_n; \xi)](x_1, x_2, \ldots, x_n)$$

$$= [(\sigma_1 \tau_{\theta^{-1}(1)}, \sigma_2 \tau_{\theta^{-1}(2)}, \ldots, \sigma_n \tau_{\theta^{-1}(n)}; \theta\xi)](x_1, x_2, \ldots, x_n)$$

$$= (\sigma_1 \tau_{\theta^{-1}(1)} x_{\xi^{-1}(\theta^{-1}(1))}, \sigma_2 \tau_{\theta^{-1}(2)} x_{\xi^{-1}(\theta^{-1}(2))},$$

$$\ldots, \sigma_n \tau_{\theta^{-1}(n)} x_{\xi^{-1}(\theta^{-1}(n))}).$$

It is easy to check that this action is transitive. $\square$

Note that the stabilizer of $\bar{x} = (0, 0, \ldots, 0) \in X_{n,m+1}$ coincides with the wreath product $S_m \wr S_n$, where $S_m \leq S_{m+1}$ is the stabilizer of 0. Therefore we can write $X_{n,m+1} = (S_{m+1} \wr S_n)/(S_m \wr S_n)$.

The group $S_{m+1} \wr S_n$ acts isometrically on $X_{n,m+1}$ and the action is distance transitive. Indeed, suppose that $x, y, u, v \in X_{n,m+1}$ and $d(x, y) = d(u, v)$. If we set $A = \{k : x_k = y_k\}$ then there exists $\theta \in S_n$ such that $\{k : u_{\theta^{-1}(k)} = v_{\theta^{-1}(k)}\} = A$. Therefore it is possible to find

$\sigma_1, \ldots, \sigma_n \in S_{m+1}$ such that $x_k = \sigma_k u_{\theta^{-1}(k)}$ and $y_k = \sigma_k v_{\theta^{-1}(k)}$, $k = 1, 2, \ldots, n$. Then $(\sigma_1, \sigma_2, \ldots, \sigma_n; \theta)u = x$ and $(\sigma_1, \sigma_2, \ldots, \sigma_n; \theta)v = y$. It follows that $(S_{m+1} \wr S_n, S_m \wr S_n)$ is a (symmetric) Gelfand pair.

Combining the above discussion with Theorem 5.3.2(i), we conclude that

Proposition 5.4.3 $L(X_{n,m+1}) = \oplus_{j=0}^n W_j$ *is the decomposition of* $L(X_{n,m+1})$ *into its* $(S_{m+1} \wr S_n)$-*irreducible pairwise inequivalent representations.*

Remark 5.4.4 Let S_{m+1} act on $L(X_{1,m+1})$ by $(\sigma f)(x) = f(\sigma^{-1}x)$ for any $\sigma \in S_{m+1}$, $x \in X_{1,m+1}$ and $f \in L(X_{1,m+1})$. Then, it is easy to check that the action of an element $(\sigma_1, \ldots, \sigma_n; \theta) \in S_{m+1} \wr S_n$ on a tensor product $f_1 \otimes \cdots \otimes f_n$ is given by: $(\sigma_1, \ldots, \sigma_n; \theta)(f_1 \otimes \cdots \otimes f_n) = (\sigma_1 f_{\theta^{-1}(1)}) \otimes \cdots \otimes (\sigma_n f_{\theta^{-1}(n)})$. This directly shows that each W_j is $S_{m+1} \wr S_n$-invariant.

Note that the hypercube Q_n coincides with $X_{n,1}$ (in particular it is a group). Moreover, $S_2 \wr S_n$ is the group of all isometries of Q_n and the simple random walk on Q_n (the Ehrenfest diffusion model) is $(S_2 \wr S_n)$-invariant. Compare also with Section 2.6. For a similar situation for classical Fourier analysis we refer to Section 4.15 in the book by Dym and McKean [82].

6

The Johnson scheme and the
Bernoulli–Laplace diffusion model

As mentioned in the preceding chapter, Sections 6.1 and 6.3 below do
not use group representation theory and only rely on harmonic analysis
on distance-regular graphs (Section 5.1).

6.1 The Johnson scheme

In this section we present the Johnson scheme. The main sources are
Delsarte's thesis [51] (see also [53]) and the papers of Dunkl [73, 75, 76]
and Stanton [209, 210, 211].

Fix once and for all a positive integer n and, for $0 \leq m \leq n$, denote by
Ω_m the family of all m-subsets of $\{1, 2, \ldots, n\}$. As usual, $L(\Omega_m)$ denotes
the space of all complex valued functions defined on Ω_m. If $A \in \Omega_m$ then
δ_A denotes the Dirac function centered at A and therefore any function
$f \in L(\Omega_m)$ may be expressed in the form $f = \sum_{A \in \Omega_m} f(A)\delta_A$.

A standard notation ([182]; see also Chapter 10) for the space $L(\Omega_m)$
is $M^{n-m,m}$; clearly $M^{m,n-m}$ is isomorphic to $M^{n-m,m}$ (via $\delta_A \leftrightarrow \delta_{A^c}$,
where $A^c \in \Omega_{n-m}$ is the complement $\{1, 2, \ldots, n\} \setminus A$ of A).

On Ω_m we introduce the *Johnson distance* δ by setting

$$\delta(A, B) = m - |A \cap B|, \quad A, B \in \Omega_m$$

that is $\delta(A, B) = |A \setminus B| \equiv |B \setminus A|$. It is easy to check that (Ω_m, δ) is a
metric space.

Moreover, Ω_m may be endowed with the graph structure whose edges
are the (A, B) with $\delta(A, B) = 1$; this way δ coincides with the geodesic
distance in this graph. Note that Ω_1 is nothing but the complete graph
K_n (see Example 1.8.1).

Lemma 6.1.1 Ω_m *with the above defined graph structure is a distance-regular graph with diameter* $\min\{m, n-m\}$ *and with parameters* $c_i = i^2$ *and* $b_i = (n-m-i)(m-i)$ *for* $i = 0, 1, \ldots, \min\{m, n-m\}$*. Moreover, the map*

$$\Omega_m \ni A \mapsto A^c \in \Omega_{n-m} \tag{6.1}$$

is a graph isomorphism.

Proof Suppose that $A, B \in \Omega_m$ are at distance $\delta(A, B) \equiv m - |A \cap B| = i$. Then the number of $C \in \Omega_m$ such that $\delta(A, C) = i+1$ and $\delta(B, C) = 1$ is equal to $(n-m-i)(m-i)$. Indeed any such C is of the form

$$C = (B \setminus \{b\}) \cup \{a\}$$

with $b \in B \cap A$ and $a \in \{1, 2, \ldots, n\} \setminus (A \cap B) \equiv (A \cup B)^c$. As $|B \cap A| = m - i$ and $|(A \cup B)^c| = n - m - i$ we have exactly $m - i$ ways of choosing b and $n - m - i$ ways for choosing a. We leave to the reader as an easy exercise to check that $c_i = i^2$.

Finally, if $A, B \in \Omega_m$ then

$$\delta(A, B) = |A \setminus B| = |B^c \setminus A^c| = \delta(A^c, B^c)$$

and therefore the map (6.1) is an isometry. $\qquad\square$

We now introduce some important linear operators between the spaces $M^{n-m,m}$. The operator $d : M^{n-m,m} \to M^{n-m+1,m-1}$ is defined by setting

$$(df)(A) = \sum_{\substack{B \in \Omega_m: \\ A \subseteq B}} f(B) \tag{6.2}$$

for all $A \in \Omega_{m-1}$ and $f \in M^{n-m,m}$. In other words, the value of df at A is equal to the sum of all the values of f at the points $B \in \Omega_m$ that contain A.

The adjoint operator $d^* : M^{n-m+1,m-1} \to M^{n-m,m}$ is then given by

$$(d^* f)(B) = \sum_{\substack{A \in \Omega_{m-1}: \\ A \subseteq B}} f(A) \tag{6.3}$$

for all $B \in \Omega_m$ and $f \in M^{n-m+1,m-1}$. Indeed, it is easy to check that

$$\langle df_1, f_2 \rangle_{M^{n-m+1,m-1}} = \sum_{\substack{(A,B) \in \Omega_{m-1} \times \Omega_m: \\ A \subseteq B}} f_1(B)\overline{f_2(A)} = \langle f_1, d^* f_2 \rangle_{M^{n-m,m}}$$

for all $f_1 \in M^{n-m,m}$ and $f_2 \in M^{n-m+1,m-1}$.

Using Dirac functions as bases for the spaces $M^{n-m,m}$ the operators d and d^* have the following alternative (equivalent) description.

$$d\delta_B = \sum_{\substack{A\in\Omega_{m-1}: \\ A\subseteq B}} \delta_A = \sum_{j\in B} \delta_{B\setminus\{j\}}$$

and

$$d^*\delta_A = \sum_{\substack{B\in\Omega_m: \\ A\subseteq B}} \delta_B = \sum_{j\notin A} \delta_{A\cup\{j\}}$$

for all $B \in \Omega_m$ and $A \in \Omega_{m-1}$, that is $d\delta_B$ is the characteristic function of the set $\{A \in \Omega_{m-1} : A \subset B\}$ and $d^*\delta_A$ is the characteristic function of the set $\{B \in \Omega_m : A \subset B\}$.

Remark 6.1.2 We denote by d the linear operator

$$d : \bigoplus_{m=0}^{n} M^{n-m,m} \to \bigoplus_{m=0}^{n-1} M^{n-m,m}$$

defined by setting, for $f = \sum_{m=1}^{n} f_m$, $f_m \in M^{n-m,m}$,

$$df = \sum_{m=1}^{n} df_m$$

where df_m is as in (6.2). In a similar way one has the adjoint

$$d^* : \bigoplus_{m=0}^{n-1} M^{n-m,m} \to \bigoplus_{m=0}^{n} M^{n-m,m}.$$

Moreover, for $f \in M^{n,0}$ we set $df = 0$; similarly, for $f \in M^{0,n}$ we set $d^*f = 0$.

The adjacency operator Δ on Ω_m is given by

$$\Delta f(A) = \sum_{B\in\Omega_m:\delta(A,B)=1} f(B), \quad \text{for every } A \in \Omega_m \text{ and } f \in M^{n-m,m},$$

that is, if $A \in \Omega_m$, the Δ-image of δ_A is given by the characteristic function of the set consisting of all $B \in \Omega_m$ such that $\delta(A, B) = 1$.

Lemma 6.1.3 *Let* $0 \leq k \leq n$. *For* $f \in M^{n-k,k}$ *we have*

(i) $dd^*f = \Delta f + (n-k)f$,
(ii) $d^*df = \Delta f + kf$.

Proof If $A \in \Omega_k$, then

$$dd^* \delta_A = d \sum_{j \notin A} \delta_{A \cup \{j\}}$$

$$= \sum_{j \notin A} \delta_A + \sum_{j \notin A} \sum_{i \in A} \delta_{(A \cup \{j\}) \setminus \{i\}}$$

$$= (n - |A|)\delta_A + \Delta\delta_A.$$

This proves (i); (ii) is proved similarly. $\qquad\square$

In the sequel we shall use the following notation (*Pochhammer symbol*) for $a \in \mathbb{C}$ and $i \in \mathbb{N}$: $(a)_i = a(a+1)(a+2)\cdots(a+i-1)$. In particular, $(1)_n = n!$ and $\binom{n}{k} = \frac{(n-k+1)_k}{k!}$. Also note that $(a+1)_{i-1}a = (a)_i$.

Lemma 6.1.4 *Let $f \in M^{n-k,k}$ and $1 \le p \le q \le n-k$. Then*

(i) $d(d^*)^q f = (d^*)^q df + q(n - 2k - q + 1)(d^*)^{q-1}f;$

(ii) *if $df = 0$, then $d^p(d^*)^q f = (q - p + 1)_p(n - 2k - q + 1)_p(d^*)^{q-p}f$.*

Proof In the case $q = 1$, the first identity immediately follows by subtracting (ii) from (i) in the statement of previous lemma. The general case of (i) then follows by induction on q:

$$d(d^*)^q f = d^* d(d^*)^{q-1}f + (n - 2k - 2q + 2)(d^*)^{q-1}f$$
$$= (d^*)^q df + q(n - 2k - q + 1)(d^*)^{q-1}f$$

where in the first equality we have applied the case $q = 1$ to $(d^*)^{q-1}f$ and in the second equality we have applied the inductive step (namely, case $q - 1$) directly to f.

Regarding (ii), the case $p = 1$ reduces simply to (i) and, again, the general case follows easily by induction on p. $\qquad\square$

We are now in position to decompose the space $M^{n-m,m}$ into the Δ-eigenspaces.

Definition 6.1.5 For $0 < k \le n/2$ define $S^{n-k,k} = M^{n-k,k} \cap \operatorname{Ker} d \equiv \{f \in M^{n-k,k} : df = 0\}$. Also set $S^{n,0} = M^{n,0}$.

Clearly, since $\dim M^{n-k,k} = \binom{n}{k} > \binom{n}{k-1} = \dim M^{n-k+1,k-1}$ (so that d cannot be injective!), $S^{n-k,k}$ is a nontrivial subspace of $M^{n-k,k}$.

Theorem 6.1.6 *We have:*

(i) dim $S^{n-k,k} = \binom{n}{k} - \binom{n}{k-1}$, $0 < k \leq n/2$.

(ii) *If* $0 \leq m \leq n$, $0 \leq k \leq \min\{n - m, m\}$ *and* $f \in S^{n-k,k}$, *then*

$$\|(d^*)^{m-k}f\|^2_{M^{n-m,m}} = (m - k)!(n - k - m + 1)_{m-k}\|f\|^2_{M^{n-k,k}};$$

in particular $(d^*)^{m-k}$ *is injective from* $S^{n-k,k}$ *to* $M^{n-m,m}$.

(iii) *For* $0 \leq m \leq n$,

$$M^{n-m,m} = \bigoplus_{k=0}^{\min\{m,n-m\}} (d^*)^{m-k}S^{n-k,k}$$

is the decomposition of $M^{n-m,m}$ *into the* Δ-*eigenspaces.*

(iv) *The eigenvalue of* Δ *corresponding to* $(d^*)^{m-k}S^{n-k,k}$ *is* $m(n - m) - k(n - k + 1)$.

(v) *If* $0 \leq m \leq n$ *and* $\max\{1, 1 - n + 2m\} \leq p$ *then*

$$Ker\ (d^p) \cap M^{n-m,m} = \bigoplus_{k=\max\{m-p+1,0\}}^{\min\{m,n-m\}} (d^*)^{m-k}S^{n-k,k}.$$

Proof First of all, note that from Lemma 6.1.4 (ii) it follows that, for $f_1 \in S^{n-h,h}$ and $f_2 \in S^{n-k,k}$, $0 < h \leq k \leq m \leq n$, we have

$$\langle(d^*)^{m-h}f_1, (d^*)^{m-k}f_2\rangle_{M^{n-m,m}} = \langle d^{m-k}(d^*)^{m-h}f_1, f_2\rangle_{M^{n-k,k}}$$
$$= (k - h + 1)_{m-k}(n - h - m + 1)_{m-k}\langle(d^*)^{k-h}f_1, f_2\rangle_{M^{n-k,k}}. \tag{6.4}$$

Then, identity (ii) is an easy consequence of (6.4) for $h = k$.

Moreover (iv) is a consequence of Lemma 6.1.4 (i) and of Lemma 6.1.3 (i): if $f \in M^{n-k,k}$ and $df = 0$ then

$$\Delta(d^*)^{m-k}f = dd^*(d^*)^{m-k}f - (n - m)(d^*)^{m-k}f$$
$$= [m(n - m) - k(n - k + 1)](d^*)^{m-k}f. \tag{6.5}$$

Suppose now that $0 \leq m \leq n/2$. To get the decomposition in (iii), first observe that

$$M^{n-1,1} = S^{n-1,1} \oplus d^*S^{n,0} \tag{6.6}$$

coincides with the decomposition into constant and mean zero functions: $f \in M^{n-1,1}$ and $df = 0$ implies that $(df)(\emptyset) \equiv \sum_{j=1}^{n} f(\{j\}) = 0$; compare with the spectral decomposition for the simple random walk on the complete graph (Example 1.8.1).

More generally, we have the orthogonal decomposition

$$M^{n-m,m} = S^{n-m,m} \oplus d^* M^{n-m+1,m-1}. \tag{6.7}$$

Indeed, a function $f_1 \in M^{n-m,m}$ belongs to $S^{n-m,m}$ if and only if $0 = \langle df_1, f_2 \rangle_{M^{n-m+1,m-1}} = \langle f_1, d^* f_2 \rangle_{M^{n-m,m}}$ for all $f_2 \in M^{n-m+1,m-1}$, that is if and only if f_1 is orthogonal to $d^* M^{n-m+1,m-1}$.

Then the decomposition $M^{n-m,m} = \oplus_{k=0}^{m} (d^*)^{m-k} S^{n-k,k}$ may be obtained by induction on m, starting from (6.6) and using (6.7). The orthogonality of the decomposition also follows from (6.7) and (6.4).

Note also that the function $k \mapsto [m(n-m) - k(n-k+1)]$ is decreasing for $0 \le k \le n/2$ and so (iv) gives $m+1$ distinct eigenvalues. As $M^{n-m,m}$ decomposes into $m+1 = \text{diam}(\Omega_m)$ Δ-eigenspaces (cf. Theorem 5.1.6), this ends the proof of (iii). Note that by (6.7) and the injectivity of $d^* : M^{n-k+1,k-1} \to M^{n-k,k}$, for $0 \le k \le n/2$, we have

$$\dim S^{n-k,k} = \dim(M^{n-k,k} \cap \ker d) = \dim M^{n-k,k} - \dim d^* M^{n-k+1,k-1}$$

$$= \dim M^{n-k,k} - \dim M^{n-k+1,k-1} = \binom{n}{k} - \binom{n}{k-1}.$$

This proves (i).

Finally, (v) is a consequence of (ii), (iii) and of Lemma 6.1.4 (ii). The case $n/2 \le m$ follows from the isomorphism between Ω_{n-m} and Ω_m and applying again (6.4), (6.5), (ii) and (iii). □

If $A \subseteq \{1, 2, \ldots, n\}$ and ℓ is a nonnegative integer not greater than the cardinality of A, we denote by $\sigma_\ell(A)$ the characteristic function of the set $\{C \in \Omega_\ell : C \subseteq A\}$, that is $\sigma_\ell(A) = \sum_{C \in \Omega_\ell, C \subseteq A} \delta_C$. We have

$$d\sigma_\ell(A) = (|A| - \ell + 1)\sigma_{\ell-1}(A). \tag{6.8}$$

Indeed, for any $C' \subseteq A$ with $|C'| = \ell - 1$, there exist $|A| - \ell + 1 \equiv |A \setminus C'|$ many C's with $C' \subset C \subset A$ and $|C| = \ell$. This follows from the fact that any such C is of the form $C = C' \cup \{j\}$ with $j \in A \setminus C'$.

Once and for all in this section we fix an element $A \in \Omega_m$ and denote by $B = A^c$ its complement. Let $0 \le h \le n$. For $\max\{0, h - m\} \le \ell \le \min\{n - m, h\}$, we use the product $\sigma_\ell(B)\sigma_{h-\ell}(A)$ to denote the characteristic function of the set $\{C \in \Omega_h : |B \cap C| = \ell, |A \cap C| = h - \ell\}$. We also set $\sigma_{-1}(B)\sigma_k(A) = \sigma_k(B)\sigma_{-1}(A) \equiv 0$. In other words,

$$\sigma_\ell(B)\sigma_{h-\ell}(A) = \sum_{\substack{C \in \Omega_h: \\ |B \cap C| = \ell, \\ |A \cap C| = h - \ell}} \delta_C.$$

If C satisfies the conditions expressed in the above sum, we say that δ_C appears in $\sigma_\ell(B)\sigma_{h-\ell}(A)$. Note also that for $h = m$ we have that $\{\sigma_\ell(B)\sigma_{m-\ell}(A) : \ell = 0, 1, \ldots, m\}$ are the characteristic functions of the *spheres* in Ω_m centered in A.

Lemma 6.1.7 $d[\sigma_i(B)\sigma_{k-i}(A)] = (n-m-i+1)\sigma_{i-1}(B)\sigma_{k-i}(A)+(m-k+i+1)\sigma_i(B)\sigma_{k-i-1}(A)$.

Proof If $C' \in \Omega_{k-1}$, $|C' \cap B| = i - 1$ and $|C' \cap A| = k - i$, then there are $|B \setminus (C' \cap B)| = n - m - i + 1$ many C's in Ω_k such that $C' \subset C$, $|C \cap B| = i$ and $|C \cap A| = k - i$.

Analogously, the coefficient $m-k+i+1$ is equal to $|A \setminus C'|$ if $|C' \cap A| = k - i - 1$ (and $|C' \cap B| = i$). □

Corollary 6.1.8 *If* $0 \leq k \leq \min\{n-m, m\}$, *the solutions of the equation*

$$d \sum_{i=0}^{k} \phi(i)\sigma_i(B)\sigma_{k-i}(A) = 0$$

with $\phi(0), \phi(1), \ldots, \phi(k) \in \mathbb{C}$, *are given by the multiples of the vector*

$$\sum_{i=0}^{k} \frac{(n-m-k+1)_{k-i}}{(-m)_{k-i}}\sigma_i(B)\sigma_{k-i}(A). \tag{6.9}$$

Proof From the previous lemma it follows that

$$d\sum_{i=0}^{k}\phi(i)\sigma_i(B)\sigma_{k-i}(A) = \sum_{i=1}^{k}\phi(i)(n-m-i+1)\sigma_{i-1}(B)\sigma_{k-i}(A)$$

$$+ \sum_{i=0}^{k-1}\phi(i)(m-k+i+1)\sigma_i(B)\sigma_{k-i-1}(A)$$

$$= \sum_{i=0}^{k-1}[(n-m-i)\phi(i+1) + (m-k+i+1)\phi(i)]\sigma_i(B)\sigma_{k-i-1}(A)$$

and therefore $d \sum_{i=0}^{k} \phi(i)\sigma_i(B)\sigma_{k-i}(A) = 0$ if and only if

$$(n - m - i)\phi(i+1) + (m - k + i + 1)\phi(i) = 0, \quad \text{for } i = 0, 1, 2, \ldots, k - 1$$

(the functions $\sigma_i(B)\sigma_{k-i}(A)$ are clearly linearly independent). The solutions of this recurrence relations are given by the multiples of the function ϕ defined by setting $\phi(k) = 1$ and

$$
\phi(i) = \frac{(n-m-k+1)(n-m-k+2)\cdots(n-m-i)}{(-m)(-m+1)\cdots(-m+k-i-1)}
$$

$$
= \prod_{j=i}^{k-1} \frac{(n-m-j)}{(-n+k-j-1)},
$$

for $i = 0, 1, 2, \ldots, k-1$, as one easily checks. $\qquad\square$

Lemma 6.1.9

(i) *If $C \in \Omega_k$ and $k \leq h$, then $\frac{1}{(h-k)!}(d^*)^{h-k}\delta_C$ equals the characteristic function of the set of all $D \in \Omega_h$ such that $C \subset D$.*

(ii) *For $\max\{0, k-m\} \leq i \leq \min\{k, n-m\}$,*

$$
\frac{1}{(h-k)!}(d^*)^{h-k}\sigma_i(B)\sigma_{k-i}(A)
$$

$$
= \sum_{\ell=\max\{i, h-m\}}^{\min\{h-k+i, n-m\}} \binom{\ell}{i}\binom{h-\ell}{k-i}\sigma_\ell(B)\sigma_{h-\ell}(A). \qquad (6.10)
$$

Proof (i) The value of $(d^*)^{h-k}\delta_C$ on D (with $|D| = h$) is equal to the number of ordered sequences (of distinct elements) $(i_1, i_2, \ldots, i_{h-k})$ such that one can express D in the form $D = (\cdots((C \cup \{i_1\}) \cup \{i_2\}) \cup \cdots) \cup \{i_{h-k}\}$ (that is, D is obtained by sequentially adding the elements $i_1, i_2, \ldots, i_{h-k}$ to C) and therefore it equals $|D \setminus C|! \equiv (h-k)!$.

(ii) If δ_D appears in $\sigma_\ell(B)\sigma_{h-\ell}(A)$ then $|D \cap B| = \ell$ and $|D \cap A| = h-\ell$. Thus the number of δ_C's appearing in $\sigma_i(B)\sigma_{k-i}(A)$ such that $C \subset D$ is $\binom{\ell}{i}\binom{h-\ell}{k-i}$, since any such a C is the union of an i-subset of $D \cap B$ and of a $(k-i)$-subset of $D \cap A$. Then we can apply (i). $\qquad\square$

We can now calculate the spherical functions. Taking into account Remark 5.1.10 (with the fixed point x_0 replaced by A) we have the following:

Theorem 6.1.10 *Fix $A \in \Omega_m$ and set $B = A^c$. For $0 \leq k \leq \min\{m, n-m\}$ the spherical function in $(d^*)^{m-k}S^{n-k,k}$ is given by*

$$
\Phi(n, m, k) = \sum_{\ell=0}^{\min\{m, n-m\}} \phi(n, m, k; \ell)\sigma_\ell(B)\sigma_{m-\ell}(A) \qquad (6.11)
$$

where

$$\phi(n,m,k;\ell) = \frac{(-1)^k}{\binom{n-m}{k}} \sum_{i=\max\{0,\ell-m+k\}}^{\min\{\ell,k\}} \binom{m-\ell}{k-i}\binom{\ell}{i}\frac{(n-m-k+1)_{k-i}}{(-m)_{k-i}}.$$

Proof Apply the operator $\frac{1}{(m-k)!}(d^*)^{m-k}$ to the function (6.9) obtained in Corollary 6.1.8; by (ii) in Lemma 6.1.9 we have

$$\frac{1}{(m-k)!}(d^*)^{m-k}\sum_{i=0}^{k}\frac{(n-m-k+1)_{k-i}}{(-m)_{k-i}}\sigma_i(B)\sigma_{k-i}(A)$$

$$= \sum_{\ell=0}^{\min\{m,n-m\}}\left[\sum_{i=\max\{0,\ell-m+k\}}^{\min\{\ell,k\}}\binom{m-\ell}{k-i}\binom{\ell}{i}\frac{(n-m-k+1)_{k-i}}{(-m)_{k-i}}\right]$$

$$\cdot \sigma_\ell(B)\sigma_{m-\ell}(A)$$

$$(6.12)$$

since

$$\sum_{i=0}^{k}\sum_{\ell=i}^{\min\{m-k+i,n-m\}} = \sum_{\ell=0}^{\min\{n-m,m\}}\sum_{i=\max\{0,\ell-m+k\}}^{\min\{\ell,k\}}.$$

Corollary 6.1.8 ensures that $\Phi(n,m,k) \in (d^*)^{m-k}S^{n-k,k} \subseteq M^{n-m,m}$. But $\Phi(n,m,k)$ is constant on the spheres of Ω_m centered at A (recall that $\sigma_\ell(B)\sigma_{m-\ell}(A)$ are the characteristic functions of these spheres) and therefore, in order to apply Corollary 5.1.9, it suffices to prove that the value at A is equal to 1, that is the coefficient of $\sigma_0(B)\sigma_m(A)$ is 1. Since the coefficient of $\sigma_0(B)\sigma_m(A)$ in (6.12) is equal to $\binom{m}{k}\frac{(n-m-k+1)_k}{(-m)_k} = (-1)^k\binom{n-m}{k}$, the proof follows. $\square$

6.2 The Gelfand pair $(S_n, S_{n-m} \times S_m)$ and the associated intertwining functions

In this section we give the group theoretic approach to the Johnson scheme. Let S_n be the symmetric group. Then Ω_m is an S_n-homogeneous space. The action of S_n on Ω_m is given as follows: for $g \in S_n$ and $\{i_1, i_2, \ldots, i_m\} \in \Omega_m$ one has $g\{i_1, i_2, \ldots, i_m\} = \{g(i_1), g(i_2), \ldots, g(i_m)\}$. Therefore, if $A \in \Omega_m$ then the stabilizer of A is $S_{n-m} \times S_m$ where S_m is the group of all permutations of the elements of A and S_{n-m} the group of all permutations of the elements of A^c (cf. Section 3.14).

The action of S_n on the metric space (Ω_m, δ) is 2-point homogeneous: if $A, B, A', B' \in \Omega_m$ are such that $\delta(A, B) = \delta(A', B')$, then there exists

$g \in S_n$ such that $gA = A'$ and $gB = B'$. Indeed such a g may be constructed by taking a bijection between $A \setminus B$ and $A' \setminus B'$, a bijection between $B \setminus A$ and $B' \setminus A'$, a bijection between $B \cap A$ and $B' \cap A'$ and finally a bijection between $(A \cup B)^c$ and $(A' \cup B')^c$. Therefore $(S_n, S_{n-m} \times S_m)$ is a Gelfand pair. Viewing $S_{n-m} \times S_m$ as the *stabilizer* of a point, say A_0 in Ω_m, the orbits are the spheres $\{A \in \Omega_m : \delta(A_0, A) = r\}$ for $r = 0, 1, 2 \ldots, \min\{m, n - m\}$. It follows from the general theory (see Section 4.6) that $M^{n-m,m}$ decomposes into $\min\{m, n - m\} + 1$ irreducible and inequivalent representations of S_n. It is easy to check that the operators d and d^* intertwine the permutation representations of S_n on Ω_m and Ω_{m-1}.

Now we give the decomposition of $M^{n-m,m}$ into its irreducible components. The following is the group theoretic analog of Theorem 6.1.6 (compare also with Exercise 3.14.2).

Theorem 6.2.1 *We have:*

(i) $\{S^{n-k,k} : 0 \le k \le n/2\}$ *is a family of distinct irreducible representations of the symmetric group* S_n.

(ii) *For* $0 \le m \le n$,

$$M^{n-m,m} = \bigoplus_{k=0}^{\min\{m,n-m\}} (d^*)^{m-k} S^{n-k,k} \cong \bigoplus_{k=0}^{\min\{m,n-m\}} S^{n-k,k}$$

is the decomposition of $M^{n-m,m}$ *into its irreducible components.*

Now we fix an h-subset A of $\{1, 2, \ldots, n\}$, that is an element of Ω_h, and suppose that $S_{n-h} \times S_h$ is its stabilizer; if $B = A^c$ is the complement of A, then B is also stabilized by $S_{n-h} \times S_h$. In the notation of the previous section, the characteristic functions of the orbits of $S_{n-m} \times S_m$ on Ω_h are then given by the products $\sigma_\ell(A)\sigma_{m-\ell}(B)$, for $\max\{0, m-n+h\} \le \ell \le \min\{m, h\}$. Then Corollary 6.1.8 may be reformulated in the following form:

Lemma 6.2.2 *If* $0 \le k \le \min\{n - h, h\}$, $A \in \Omega_h$ *and* $B = A^c$, *then the space of* $S_{n-h} \times S_h$-*invariant vectors in* $S^{n-k,k}$ *consists of the multiples of the vector*

$$\sum_{i=0}^{k} \frac{(n - h - k + 1)_{k-i}}{(-h)_{k-i}} \sigma_i(B)\sigma_{k-i}(A).$$

Proof A function $\psi \in M^{n-k,k}$ is $S_{n-h} \times S_h$-invariant if and only if there exist coefficients $\phi(0), \phi(1), \ldots, \phi(k)$ such that $\psi = \sum_{i=0}^{k} \phi(k)\sigma_i(B)\sigma_{k-i}(A)$. If we impose $d\psi = 0$ then we can conclude as in Corollary 6.1.8 (with m replaced by h). $\qquad\square$

Theorem 6.2.3 *If* $0 \leq k \leq \min\{h, n-h, m, n-m\}$, $A \in \Omega_h$ *and* $B = A^c$ *then the space of all* $S_{n-h} \times S_h$*-invariant functions in the subspace of* $M^{n-m,m}$ *isomorphic to* $S^{n-k,k}$ *consists of the multiples of the function*

$$\Phi(n, h, m, k) = \sum_{\ell=\max\{0,m-h\}}^{\min\{m,n-h\}} \phi(n, h, m, k; \ell)\sigma_\ell(B)\sigma_{m-\ell}(A)$$

where

$$\phi(n, h, m, k; \ell) = \frac{(-1)^k}{\binom{n-h}{k}} \sum_{i=\max\{0,\ell-m+k\}}^{\min\{\ell,k\}} \binom{m-\ell}{k-i}\binom{\ell}{i}\frac{(n-h-k+1)_{k-i}}{(-h)_{k-i}}.$$

In particular, $\Phi(n, h, h, k)$ *is the spherical function of* $(S_n, S_{n-h} \times S_h)$ *corresponding to the representation* $S^{n-k,k}$*; in other words, it coincides with* $\Phi(n, h, k)$ *(see (6.11)).*

Proof Apply the operator $\frac{1}{(m-k)!}(d^*)^{m-k}$ to the function obtained in Lemma 6.2.2; by (ii) in Lemma 6.1.9 we have

$$\frac{1}{(m-k)!}(d^*)^{m-k} \sum_{i=0}^{k} \frac{(n-h-k+1)_{k-i}}{(-h)_{k-i}}\sigma_i(B)\sigma_{k-i}(A)$$

$$= \sum_{\ell=\max\{0,m-h\}}^{\min\{m,n-h\}} \left[\sum_{i=\max\{0,\ell-m+k\}}^{\min\{\ell,k\}} \binom{m-\ell}{k-i}\binom{\ell}{i}\frac{(n-h-k+1)_{k-i}}{(-h)_{k-i}} \right]$$

$$\cdot \sigma_\ell(B)\sigma_{m-\ell}(A).$$

$$(6.13)$$

Theorem 6.1.6 and Lemma 6.2.2 ensure that the $S_{n-h} \times S_h$-invariant functions in the subspace of $M^{n-m,m}$ isomorphic to $S^{n-k,k}$ are all multiples of the function (6.13). Observe that the factor $(-1)^k/\binom{n-h}{k}$ is just a normalization constant (see Theorem 6.1.10). $\qquad\square$

Clearly, if k does not satisfy the conditions in the statement of Theorem 6.2.3, then $S^{n-k,k}$ is not contained in $M^{n-m,m}$ and/or $S^{n-k,k}$ does not contain nontrivial $S_h \times S_{n-h}$-invariant vectors (cf. Theorem 4.6.2) and so (6.13) is identically zero.

Note that from (6.13) and the definition of $\Phi(n, h, m, k)$ in Theorem 6.2.3 it follows that

$$\Phi(n, h, m, k) = \frac{1}{(m-k)!}(d^*)^{m-k}\Phi(n, h, k, k). \tag{6.14}$$

Remark 6.2.4 The coefficients $\phi(n, h, m, k)$ can be expressed by means of the *Hahn polynomials*. These are defined, for integers m, a, b with $a \geq m$, $b \geq m \geq 0$ by setting

$$Q_k(x; -a-1, -b-1, m) = \frac{1}{\binom{m}{k}}\sum_{j=0}^{k}(-1)^j\frac{\binom{b-k+j}{j}}{\binom{a}{j}}\binom{m-x}{k-j}\cdot\binom{x}{j}$$

for $k = 0, 1, \ldots, m$ and $x = 0, 1, \ldots, m$; see [135] and [75]. These, as functions of x, form a set of orthogonal polynomials of a discrete variable. We limit ourselves to observe that

$$\phi(n, h, m, k; \ell) = \frac{\binom{m}{k}}{\binom{h}{k}}Q_k(\ell; -n+h-1, -h-1, m)$$

and that most of the properties of the Hahn polynomials, i.e. the coefficients $\phi(n, h, m, k; \ell)$, may be easily deduced from their interpretation as intertwining functions. See also [78].

For instance, it is easy to check that the second order finite difference equation

$$[(n-h-\ell)\ell + (h-m+\ell)(m-\ell)]\phi(n, h, m, k; \ell)$$
$$+ (n-h-\ell)(m-\ell)\phi(n, h, m, k; \ell+1)$$
$$+ (h-m+\ell)\ell\phi(n, h, m, k; \ell-1)$$
$$= [m(n-m) - k(n-k+1)]\phi(n, h, m, k; \ell)$$

corresponds to

$$\Delta\Phi(n, h, m, k) = [m(n-m) - k(n-k+1)]\Phi(n, h, m, k)$$

and this follows from Theorem 6.2.1 and Theorem 6.2.3. The orthogonality relations

$$\sum_{\ell=\max\{0, m-h\}}^{\min\{m, n-h\}}\binom{n-h}{\ell}\binom{h}{m-\ell}\phi(n, h, m, k; \ell)\phi(n, h, m, t; \ell)$$

$$= \delta_{k,t}\frac{(h-k)!(n-h-k)!}{(n-k-m)!(m-k)!}\cdot\frac{\binom{n}{h}}{\binom{n}{k} - \binom{n}{k-1}}$$

which coincide with those for the Hahn polynomials Q_k, follow from

$$\langle \Phi(n, h, m, k), \Phi(n, h, m, t) \rangle_{M^{n-m,m}}$$

$$= \delta_{k,t} \frac{(h-k)!(n-h-k)!}{(n-k-m)!(m-k)!} \cdot \frac{\binom{n}{h}}{\binom{n}{k} - \binom{n}{k-1}}.$$

To obtain the norm of Φ, first note that by Corollary 4.6.4

$$\|\Phi(n, h, h, k)\|^2 = \binom{n}{h} \frac{1}{\binom{n}{k} - \binom{n}{k-1}}.$$

Then, using (6.14) and Theorem 6.1.6 (ii) we can first compute the norm of $\Phi(n, h, k, k)$ and then the norm of $\Phi(n, h, m, k)$.

6.3 Time to reach stationarity for the Bernoulli–Laplace diffusion model

In this section Ω_n will denote the family of all n-subsets of $\{1, 2, \ldots, 2n\}$. It is clear that the Bernoulli–Laplace diffusion model coincides with the simple random walk on Ω_n equipped with the graph structure of the Johnson scheme (see Exercise 1.8.4). Therefore the corresponding linear operator coincides with Δ/n^2. The corresponding eigenvalues are

$$\lambda_i = 1 - \frac{i(2n - i + 1)}{n^2}$$

with multiplicities $d_i = \binom{2n}{i} - \binom{2n}{i-1}$ for all $i = 0, 1, \ldots, n$. Denote by π the uniform distribution on Ω_n (that is $\pi(A) = \frac{1}{\binom{2n}{n}}$ for all $A \in \Omega_n$) and by $\nu_{A_0}^{(k)}$ the distribution probability of the Bernoulli–Laplace diffusion model after k steps (A_0 is the starting point corresponding to the initial configuration).

Now we show that the Bernoulli–Laplace diffusion model has a cutoff after $k = \frac{1}{4}log(2n)$ steps. First we give an upper bound.

Theorem 6.3.1 (Diaconis and Shahshahani [70]) *There exists a universal positive constant a such that if $k = \frac{n}{4}(\log(2n) + c)$ with $c \geq 0$, then we have*

$$\|\nu_{A_0}^{(k)} - \pi\|_{TV} \leq ae^{-c/2}.$$

Proof In virtue of the upper bound Lemma (Corollary 4.9.2 or, equivalently, Corollary 5.1.13) we must estimate the quantity

$$\sum_{i=1}^{n} d_i \lambda_i^{2k} \equiv \sum_{i=1}^{n} \left[\binom{2n}{i} - \binom{2n}{i-1} \right] \left[1 - \frac{i(2n-i+1)}{n^2} \right]^{2k}. \qquad (6.15)$$

To bound (6.15) we first observe that

$$\text{if } x \leq 1 + e^{-2} \text{ then } |1 - x| \leq e^{-x}. \qquad (6.16)$$

Indeed, if $x \leq 1$ this is just elementary calculus: $1 - x \leq e^{-x}$. If $1 \leq x \leq 1 + e^{-2}$ then $\frac{1}{x-1} \geq e^2 > e^x$ and therefore we still have $|1 - x| = x - 1 \leq e^{-x}$.

The quantity $\frac{i(2n-i+1)}{n^2}$ is ≤ 1 for $i \leq \frac{2n+1-\sqrt{1+4n}}{2}$; but in any case $\frac{i(2n-i+1)}{n^2}$ is increasing for $i \leq n + \frac{1}{2}$ and for $i = n$ takes the value $\frac{n+1}{n}$ (in other terms, the negative eigenvalues are bounded from below by $-\frac{1}{n}$). It follows that for $n \geq 8 > e^2$ we can apply (6.16) to $|\lambda_i|$, with $x = \frac{i(2n-i+1)}{n^2}$, obtaining for $i = 0, 1, \ldots, n$

$$\lambda_i^{2k} \leq \exp\left[-\frac{2ki(2n-i+1)}{n^2} \right].$$

This estimate together with

$$d_i \equiv \left[\binom{2n}{i} - \binom{2n}{i-1} \right] \leq \frac{(2n)^i}{i!}$$

for all $i = 0, 1, \ldots, n$, ensures that (6.15) is bounded from above by

$$\sum_{i=1}^{n} \exp\left[-\frac{2ki(2n-i+1)}{n^2} + i \log(2n) - \log(i!) \right]. \qquad (6.17)$$

The first term in (6.17) is

$$\exp\left[-\frac{4k}{n} + \log(2n) \right]$$

which becomes e^{-c} if we set $k = \frac{n}{4}(\log(2n) + c)$. If k is equal to this critical value, then (6.17) may be written in the form

$$\sum_{i=1}^{n} \exp(a(i) + b(i))$$

where

$$a(i) = ci\left[\frac{i-1}{2n} - 1 \right] \quad \text{and} \quad b(i) = \frac{i(i-1)\log(2n)}{2n} - \log(i!).$$

Since $a(i)$ is decreasing for $i \leq \frac{2n+1}{2}$ we have that $a(i) \leq a(1) = -c$ for all $i = 1, 2, \ldots, n$. We are left to show that $\sum_{i=1}^{n} \exp(b(i)) \leq B$ with B independent of n. Observe that

$$b(i) \leq i^2 \frac{\log(2n)}{2n} - i \log i + i$$

as follows from the inequality

$$\log(i!) = \log(i) + \log(i-1) + \cdots + \log(1) \geq \int_1^i \log(x)dx = i \log i - i + 1.$$

On the other hand if $22 \leq i \leq n$ we have

$$i^2 \frac{\log(2n)}{2n} - i \log i + i < -i. \tag{6.18}$$

Indeed, the derivative of $f(x) = \frac{\log x - 2}{x}$ is negative for $x > e^3$ so that if $n \geq i > 21 > e^3$ we have

$$\frac{\log i - 2}{i} > \frac{\log n - 2}{n}.$$

Since the right hand side of the last inequality is bigger than $\log(2n)/2n$ if $n \geq 110 > 2e^4$ we get (6.18). Thus

$$\sum_{i=1}^{n} \exp(b(i)) \leq \sum_{i=1}^{110} \exp(b(i)) + \sum_{i=111}^{\infty} \exp(-i) \leq B,$$

and this ends the proof. $\qquad \square$

We now show that if we make less than $\frac{1}{4}n[\log(2n)]$ switches we are not close to the uniform distribution.

Theorem 6.3.2 *For n large, if $k = \frac{1}{4}n[\log(2n) - c]$ with $0 < c < \log(2n)$, then we have*

$$\|\nu_{A_0}^{(k)} - \pi\|_{TV} \geq 1 - 32e^{-c}.$$

Proof It suffices to find a subset $\mathcal{A} \subset \Omega_n$ such that $\nu_{A_0}^{(k)}(\mathcal{A})$ is close to 1 and $\pi(\mathcal{A})$ is close to 0.

For $f \in M^{n,n}$ denote by $E_k(f) = \sum_{B \in \Omega_n} f(B)\nu_{A_0}^{(k)}(B)$ and by $Var_k(f) = E_k(f^2) - E_k(f)^2$ the mean value and the variance of f with respect to $\nu_{A_0}^{(k)}$, respectively. Observe that if ϕ_i is a spherical function then from the orthogonality relations and Proposition 4.9.1 (or, equivalently, from Theorem 5.1.12) we get

$$E_k(\phi_i) = \lambda_i^k. \tag{6.19}$$

We recall the explicit expression of the first three spherical functions (cf. Theorem 6.1.10 with n in place of m, $2n$ in place of n and $k = 0, 1, 2$):

$$\phi_0(\ell) = 1$$

$$\phi_1(\ell) = 1 - \frac{2\ell}{n}$$

$$\phi_2(\ell) = \frac{(n - \ell)(n - \ell - 1)}{n(n - 1)} - \frac{2(n - \ell)\ell}{n^2} + \frac{\ell(\ell - 1)}{n(n - 1)}$$

from which we deduce

$$\phi_1^2 = \frac{\phi_0}{2n - 1} + \frac{2n - 2}{2n - 1}\phi_2. \tag{6.20}$$

Let $f = (2n - 1)^{1/2}\phi_1$. Then, by (6.19), we obtain

$$E_k(f) = (2n - 1)^{1/2}\left(1 - \frac{2}{n}\right)^k. \tag{6.21}$$

Moreover in virtue of (6.19), (6.20) and the fact that E_k is linear we have that

$$E_k(f^2) = (2n - 1)E_k\left(\frac{\phi_0}{2n - 1} + \frac{2n - 2}{2n - 1}\phi_2\right)$$

$$= 1 + 2(n - 1)\left(1 - \frac{2(2n - 1)}{n^2}\right)^k$$

and therefore

$$Var_k(f) = E_k(f^2) - E_k(f)^2$$

$$= 1 + 2(n - 1)\left(1 - \frac{2(2n - 1)}{n^2}\right)^k - (2n - 1)\left(1 - \frac{2}{n}\right)^{2k} \leq 1. \tag{6.22}$$

Now we can proceed as in the last part of Theorem 2.4.3. Set $\mathcal{A}_\alpha = \{B \in \Omega_n : |f(B)| < \alpha\}$ where $0 < \alpha < E_k(f)$. From Markov's inequality (Proposition 1.9.5) and the orthogonality relations for the spherical functions, it follows that

$$\pi(\mathcal{A}_\alpha) = 1 - \pi\{B \in \Omega_n : f(B)^2 \geq \alpha^2\}$$

$$\geq 1 - \frac{1}{\alpha^2}E_\pi(f^2) = 1 - \frac{1}{\alpha^2}. \tag{6.23}$$

From Chebyshev's inequality (Corollary 1.9.6), the inclusion $\mathcal{A}_\alpha \subseteq \{B \in \Omega_n : |f(B) - E_k(f)| \geq E_k(f) - \alpha\}$ and (6.22) we get

$$\nu_{A_0}^{(k)}(\mathcal{A}_\alpha) \leq \frac{Var_k(f)}{(E_k(f) - \alpha)^2} \leq \frac{1}{(E_k(f) - \alpha)^2}. \tag{6.24}$$

Now set $k = \frac{1}{4}n(\log(2n) - c)$, with $0 < c < \log(2n)$. From the Taylor expansion of the logarithm it follows that

$$\log(1 - x) = -x - \frac{x^2}{2}\omega(x) \tag{6.25}$$

with $\omega(x) \geq 0$ and $\lim_{x \to 0} \omega(x) = 1$. Then applying (6.25) to the right hand side of (6.21) we get

$$E_k(f) = (2n - 1)^{1/2} \exp\left[\log\left(1 - \frac{2}{n}\right) \cdot \frac{1}{4}n(\log(2n) - c)\right]$$

$$= (2n - 1)^{1/2} \exp\left[\left(-\frac{2}{n} - \frac{2}{n^2}\omega(2/n)\right) \cdot \frac{1}{4}n(\log(2n) - c)\right]$$

$$= \sqrt{\frac{2n - 1}{2n}}e^{c/2} \exp\left[\frac{c - \log(2n)}{2n}\omega(2/n)\right]$$

and therefore for n large we obtain

$$E_k(f) \geq \sqrt{\frac{2n - 1}{2n}} \frac{3}{4}e^{c/2} \geq \frac{1}{2}e^{c/2}.$$

Choosing $\alpha = \frac{e^{c/2}}{4}$ we have $E_k(f) \geq 2\alpha$ and therefore (6.24) yields

$$\nu_{A_0}^{(k)}(\mathcal{A}_\alpha) \leq \frac{1}{\alpha^2};$$

this and (6.23) imply that

$$\|\nu_{A_0}^{(k)} - \pi\|_{TV} \geq \pi(\mathcal{A}_\alpha) - \nu_{A_0}^{(k)}(\mathcal{A}_\alpha) \geq 1 - \frac{2}{\alpha^2} = 1 - 32e^{-c}.$$

$\square$

Remark 6.3.3 Observe that $\mathcal{A}_\alpha$ in the preceding proof is the set of all $B \in \Omega_n$ such that $|\delta(B, A_0) - \frac{n}{2}| \leq \frac{n\alpha}{2\sqrt{2n-1}}$.

6.4 Statistical applications

In [55] and [56] Diaconis developed a spectral analysis for data on groups and permutation spaces, inspired by the classical analysis of time series and the analysis of variance (ANOVA). We will describe a particular case which is an application of the Gelfand pair $(S_n, S_{n-k} \times S_k)$.

Consider an election where there are n candidates. Each voter must indicate k candidates, without ranking within (thus each voter chooses a k-subset between the n candidates, or he/she indicates k preferences). Therefore the data is a function $f : \Omega_k \to \mathbb{N}$ where Ω_k, as in Section 6.1, is the homogeneous space of all k-subsets of $\{1, 2, \ldots, n\}$ (the set of candidates). If $A \in \Omega_k$, then $f(A)$ is the number of voters that have chosen A. The winner (or the final ranking of the candidates) is usually determined by the number of preferences taken by each candidate.

For $t = 1, 2, \ldots, n$, define v_t as the characteristic function of all $A \in \Omega_k$ containing t. Then $\{v_1, v_2, \ldots, v_n\}$ is clearly a basis for the subspace of $M^{n-k,k}$ isomorphic to $S^n \oplus S^{n-1,1} = M^{n-1,1}$. It is not an orthogonal basis:

$$\langle v_t, v_s \rangle = \begin{cases} \binom{n-1}{k-1} & \text{if } t = s \\ \binom{n-2}{k-2} & \text{otherwise.} \end{cases} \tag{6.26}$$

The number of preferences taken by the candidate t is clearly equal to

$$\beta_t := \sum_{\substack{A \in \Omega_k: \\ t \in A}} f(A) = \langle f, v_t \rangle.$$

Moreover, if P is the projection from $M^{n-k,k}$ to $S^n \oplus S^{n-1,1}$ ($P = E_0 + E_1$ in the notation of the end of Section 4.7), then $\beta_t = \langle Pf, v_t \rangle$, i.e. *the value of β_t is determined only by the projection of f on $S^n \oplus S^{n-1,1}$.* Since the outcome of the election is determined by the values of the coefficients β_t, a lot of (possibly relevant) information (that contained in the projection on the subspace orthogonal to $S^n \oplus S^{n-1,1}$) is lost. Now we give an elementary formula that expresses Pf in terms of the values β_t. A generalization will be given at the end of Section 6.5.

Theorem 6.4.1 $Pf(A) = \frac{1}{\binom{n-1}{k-1}} \sum_{t \in A} \beta_t - \frac{k-1}{(n-k)\binom{n-1}{k-1}} \sum_{w \notin A} \beta_w$

First proof If $Pf = \sum_{t=1}^{n} \alpha_t v_t$ then from (6.26) it follows that $\beta_t = \langle f, v_t \rangle = \langle Pf, v_t \rangle = \binom{n-1}{k-1} \alpha_t + \binom{n-2}{k-2} \sum_{s \neq t} \alpha_s$. Then an elementary but tedious calculation shows that

$$\alpha_t = \frac{nk - 2k + 1}{k(n-k)\binom{n-1}{k-1}} \beta_t - \frac{k-1}{k(n-k)\binom{n-1}{k-1}} \sum_{u \neq t} \beta_u$$

and therefore

$$Pf(A) = \sum_{t \in A} \alpha_t = \sum_{t \in A} \left[\frac{nk - 2k + 1}{k(n-k)\binom{n-1}{k-1}} \beta_t - \frac{k-1}{k(n-k)\binom{n-1}{k-1}} \sum_{u \neq t} \beta_u \right]$$

$$= \frac{1}{\binom{n-1}{k-1}} \sum_{t \in A} \beta_t - \frac{k-1}{(n-k)\binom{n-1}{k-1}} \sum_{w \notin A} \beta_w.$$

Second proof In this case, the function ω_1 of formula (4.19) is given by: $\omega_1(\ell) = \frac{n-1}{\binom{n}{k}} - \frac{\ell}{\binom{n-2}{k-1}}$; see Theorem 6.1.10. Therefore, applying formula (4.20) we obtain that

$$Pf(A) = E_0 f(A) + E_1 f(A)$$

$$= E_0 f(A) + \sum_{\ell=0}^{k} \left(\sum_{B \in \Omega_k : \delta(A,B)=\ell} f(B) \right) \left[\frac{n-1}{\binom{n}{k}} - \frac{\ell}{\binom{n-2}{k-1}} \right]$$

$$= E_0 f(A) + \frac{n-1}{\binom{n}{k}} \sum_{B \in \Omega_k} f(B)$$

$$- \frac{1}{\binom{n-2}{k-1}} \sum_{\ell=1}^{k} \ell \left(\sum_{B \in \Omega_k : \delta(B,A)=\ell} f(B) \right).$$

Since

$$k \binom{n}{k} E_0 f = k \sum_{B \in \Omega_k} f(B) = \sum_{t=1}^{n} \beta_t = \sum_{t \in A} \beta_t + \sum_{w \notin A} \beta_w$$

and $\sum_{\ell=1}^{k} \ell \left(\sum_{B : \delta(B,A)=l} f(B) \right) = \sum_{w \notin A} \beta_w$, the formula of the theorem follows from an easy calculation. □

Now we give an example that shows that if we analyze the results of an election only counting the votes taken by each candidate we may not understand how the voters are grouped or other structures of the data.

Example 6.4.2 Suppose that in an election there are N voters and eight candidates (we indicate these latter by $\{1,2,3,4,5,6,7,8\}$). Each voter must indicate two candidates, without ranking within (therefore $n = 8$ and $k = 2$). Moreover, suppose that the voters are divided into three groups: the first group (the smallest) has two leaders, 1 and 2, the second group has one leader, 3, and two subordinates, 4 and 5 and the third group (approximately equals to the second) has one leader, 6 and two subordinates, 7 and 8. To fix things up and simplify calculations, we suppose that the first group consists of M members and they all vote $\{1,2\}$, the second group consists of $2M$ members: M of them vote $\{3,4\}$ and the others vote $\{3,5\}$, the third group is of $2M$ members: M of them vote $\{6,7\}$ and the others vote $\{6,8\}$; clearly $N = 5M$.

Therefore, the function representing the outcome of the election is $f = M\delta_{\{1,2\}} + M\delta_{\{3,4\}} + M\delta_{\{3,5\}} + M\delta_{\{6,7\}} + M\delta_{\{6,8\}}$. Clearly, $\|f\| = \sqrt{5}M$,

$f_0 = \frac{5M}{28}$, $\|f_0\|^2 = \frac{25}{784}M^2$. Moreover, $\beta_1 = \beta_2 = \beta_4 = \beta_5 = \beta_7 = \beta_8 = M$ and $\beta_3 = \beta_6 = 2M$. From Theorem 6.4.1 it follows that

$$Pf(\{i,j\}) = \frac{1}{7}(\beta_i + \beta_j) - \frac{1}{42}\sum_{w\neq i,j}\beta_w$$

and therefore

$$Pf(\{3,6\}) = \frac{3}{7}M$$

$$Pf(\{i,j\}) = \frac{11}{42}M \qquad \text{if} \qquad |\{i,j\}\cap\{3,6\}| = 1$$

$$Pf(\{i,j\}) = \frac{2}{21}M \qquad \text{if} \qquad \{i,j\}\cap\{3,6\} = \emptyset.$$

It follows that $\|Pf\| = \frac{2\sqrt{2}}{\sqrt{7}}M$ and thus $\|f_1\| = \sqrt{\|Pf\|^2 - \|f_0\|^2} = \frac{1}{2}M$ and $\|f_2\| = \sqrt{\|f\|^2 - \|Pf\|^2}M = \frac{3\sqrt{3}}{\sqrt{7}}M$. This means that the most relevant component of f (the largest) is f_2: *counting the preferences, most of the information contained in f, that is the structure of the set of voters, is lost.* By counting the preferences, one can only say that 3 and 6 have taken $2M$ votes, and all the other candidates have taken M votes.

6.5 The use of Radon transforms

In the general case $k \geq 2$, the spectral decomposition $f = f_0 + f_1 + \cdots + f_k$ has a simple interpretation: f_0 is the mean value, f_1 is the effect of the popularity of individual members, f_2 is the effect of the popularity of pairs, f_3 is the effect of the popularity of triples, and so on. Note that such effects are "pure": $f_0 + f_1 + f_2$ is the projection onto the subspace isomorphic to $M^{n-2,2}$, and thus it represents the most natural dat on pairs; therefore, f_2 is depurated by the effect of the popularity of singles. We end this section by giving a generalization of Theorem 6.4.1. Suppose that $0 \leq j < k \leq n/2$ and define the intertwining operator $D : M^{n-k,k} \to M^{n-j,j}$ by setting

$$Df(A) = \sum_{B\in\Omega_k:A\subset B} f(B) \tag{6.27}$$

for every $f \in M^{n-k,k}$ and $A \in \Omega_j$. In the notation of Section 6.1, $D = \frac{1}{(k-j)!}(d)^{k-j}$.

We also define the intertwining operator $D^- : M^{n-j,j} \to M^{n-k,k}$ by setting

$$D^- f(B) = \sum_{\ell=k-j}^{k} \frac{(-1)^{k-j}(k-j)}{(-1)^{\ell}\ell\binom{n-j}{\ell}} \sum_{A \in \Omega_j : |B \setminus A| = \ell} f(A)$$

for every $f \in M^{n-j,j}$ and $B \in \Omega_k$.

We show that D^- is a right-inverse of D and that $D^- D$ projects $M^{n-k,k}$ onto its subspace isomorphic to $M^{n-j,j}$. We first give the following elementary combinatorial lemma (see identity (5.24) in [104]). Recall that, by definition, $\binom{n}{k} = 0$ if $k < 0$.

Lemma 6.5.1 *For integers m, r, s, t with $t \geq 0$ we have*

$$\sum_{\ell=\max\{s,m+r\}}^{s+t} (-1)^{\ell} \binom{\ell-r}{m} \binom{t}{\ell-s} = (-1)^{s+t} \binom{s-r}{m-t}. \tag{6.28}$$

In particular, if $m < t$ then the left hand side is equal to zero.

Proof We prove this identity by induction on t. For $t = 0$ it is trivial. The inductive step from $t-1$ to t is an easy consequence of the elementary identity

$$\binom{n}{k} = \binom{n-1}{k} + \binom{n-1}{k-1}. \tag{6.29}$$

Indeed,

$$\sum_{\ell=\max\{s,m+r\}}^{s+t} (-1)^{\ell} \binom{\ell-r}{m} \binom{t}{\ell-s}$$

$$= \sum_{\ell=\max\{s,m+r\}}^{s+t-1} (-1)^{\ell} \binom{\ell-r}{m} \binom{t-1}{\ell-s}$$

$$+ \sum_{\ell=\max\{s+1,m+r\}}^{s+t} (-1)^{\ell} \binom{\ell-r}{m} \binom{t-1}{\ell-s-1}$$

by induction

$$= (-1)^{t+s-1} \binom{s-r}{m-t+1} + (-1)^{t+s} \binom{s+1-r}{m-t+1}$$

by (6.29)

$$= (-1)^{s+t} \binom{s-r}{m-t}.$$

where the first equality follows from (6.29).

If $m < t$ then $\binom{s-r}{m-t} = 0$. □

Theorem 6.5.2 (i) D^- is a right-inverse of D, i.e. DD^- is the identity of $M^{n-j,j}$;

(ii) $Q_j = D^-D$ is the orthogonal projection from $M^{n-k,k}$ onto the subspace isomorphic to $M^{n-j,j}$.

Proof (i) If $f \in M^{n-j,j}$ and $A \in \Omega_j$ then

$$(DD^-f)(A) = \sum_{\substack{B \in \Omega_k: \\ A \subset B}} D^-f(B)$$

$$= \sum_{\substack{B \in \Omega_k: \\ A \subset B}} \sum_{\ell=k-j}^{k} \frac{(-1)^{k-j}(k-j)}{(-1)^\ell \ell \binom{n-j}{\ell}} \sum_{\substack{C \in \Omega_j: \\ |B \setminus C| = \ell}} f(C)$$

$$= (-1)^{k-j}(k-j) \sum_{t=0}^{j} \sum_{\ell=\max\{t,k-j\}}^{k+t-j} (-1)^\ell \frac{\binom{t}{k-\ell-j+t}\binom{n-j-t}{\ell-t}}{\ell\binom{n-j}{\ell}}$$

$$\cdot \sum_{\substack{C \in \Omega_j: \\ \delta(C,A)=t}} f(C)$$

because, once we fixe $A, C \in \Omega_j$ such that $\delta(A,C) = t$, then the number of $B \in \Omega_k$ such that $A \subseteq B$ and $|B \setminus C| = \ell$ is equal to $\binom{t}{k-\ell-j+t}\binom{n-j-t}{\ell-t}$. But the expression

$$(-1)^{k-j}(k-j) \sum_{\ell=\max\{t,k-j\}}^{k+t-j} (-1)^\ell \frac{\binom{t}{k-\ell-j+t}\binom{n-j-t}{\ell-t}}{\ell\binom{n-j}{\ell}} \qquad (6.30)$$

is equal to 1 for $t = 0$; for $t \geq 1$, by mean of the elementary transformation

$$\frac{\binom{t}{k-\ell-j+t}\binom{n-j-t}{\ell-t}}{\ell\binom{n-j}{\ell}} = \frac{\binom{\ell-1}{t-1}\binom{t}{\ell+j-k}}{(n-j-t+1)\binom{n-j}{t-1}}$$

is reduced to

$$\frac{(-1)^{k-j}(k-j)}{(n-j-t+1)\binom{n-j}{t-1}} \sum_{\ell=\max\{t,k-j\}}^{k+t-j} (-1)^\ell \binom{\ell-1}{t-1}\binom{t}{\ell+j-k}$$

which is equal to zero in virtue of the preceding lemma. It follows that $(DD^-f)(A) = f(A)$.

(ii) From (i) it follows that D^-D is idempotent (i.e. $D^-DD^-D = D^-D$), D^- is injective and D is surjective (the last assertion has been also proved in Theorem 6.1.6). If $f \in M^{n-k,k}$ and $f = f_1 + f_2$ with $f_1 \in S^0 \oplus S^{n-1,1} \oplus \cdots \oplus S^{n-j,j} = \operatorname{Im} D^-$ and $f_2 \in S^{n-j-1,j+1} \oplus \cdots S^{n-k,k} = \ker D = (\operatorname{Im} D^-)^\perp$ then there exists $f_3 \in M^{n-j,j}$ such that $D^- f_3 = f_1$. Therefore

$$D^- Df = D^- Df_1 + D^- Df_2 = D^- DD^- f_3 = D^- f_3 = f_1$$

and $D^- D$ is the orthogonal projection onto the subspace of $M^{n-k,k}$ isomorphic to $M^{n-j,j}$.   □

The last lemma was proved in [105] and [67], but working with the adjoint of D. See also [188]. The formula

$$Q_j f = D^- Df(B) = \sum_{\ell=k-j}^{k} \frac{(-1)^{k-j}(k-j)}{(-1)^\ell \ell \binom{n-j}{\ell}} \sum_{A \in \Omega_j : |B \setminus A| = \ell} (Df)(A)$$

(6.31)

is clearly a generalization of Theorem 6.4.1: $Df(A) = \sum_{B \in \Omega_k : A \subset B} f(B)$ are the votes taken by the subset A; they correspond to the β_t in the case $j = 1$.

Therefore we have the following practical way to analyze the data f of an election. First of all, one can compute the values (6.27) for every $A \in \cup_{j=0}^{k} \Omega_j$: (6.27) represents the votes taken by A. It is a very simple and natural operation: (6.27) is the number of ballot cards that contain all the candidates in A. Then one can use (6.31) to find the projection onto the subspace isomorphic to $M^{n-j,j}$. Finally, $Q_j f - Q_{j-1} f$ is exactly the projection onto $S^{n-j,j}$.

Remark 6.5.3 The operator D (together with its adjoint action $\frac{(d^*)^{k-j}}{(k-j)!}$: $M^{n-j} \to M^{n-k}$) is usually called the *Radon transform* [29], a name that comes from classical integral geometry; see [212] and the recent monograph [161].

7

The ultrametric space

This chapter is based on the papers by Letac [153] and Figà-Talamanca [90]. An earlier appearance of the spherical functions related to the ultrametric space is also in Mauceri [162]. See also the papers by Nebbia [170], Figà-Talamanca [89] and Stanton [210].

7.1 The rooted tree $\mathbb{T}_{q,n}$

Let $\Sigma = \{0, 1, \ldots, q - 1\}$, where $q \in \mathbb{N}$. We call the set Σ the *alphabet*. A *word* over Σ is a finite sequence $w = \sigma_1 \sigma_2 \cdots \sigma_k$ where $\sigma_i \in \Sigma$ for $i = 1, 2, \ldots, k$. The quantity $\ell(w) = k$ is called the *length* of w. We denote by $\emptyset$ the *empty word* (which clearly has length $\ell(\emptyset) = 0$), by Σ^k the set of words of length $k \geq 0$ (in particular, $\Sigma^0 = \{\emptyset\}$), and by $\Sigma^* = \coprod_{k=0}^{\infty} \Sigma^k$ the set of all (finite) words over Σ.

If X is a graph, a *circuit* in X is a closed path $p = (x_0, x_1, \ldots, x_\ell)$ with $x_0 = x_\ell$, $\ell \geq 3$ and such that $x_0, x_1, \ldots, x_{\ell-1}$ are all distinct. A *tree* is a (finite) simple connected graph without loops and circuits.

The *q-homogeneous rooted tree of depth n* (where $n \in \mathbb{N}$) is the graph $\mathbb{T}_{q,n}$ whose vertices are the finite (possibly empty) words $w \in \Sigma^*$ of length not exceeding n. Moreover two vertices $u = \sigma_1 \sigma_2 \cdots \sigma_k$ and $v = \rho_1 \rho_2 \cdots \rho_h$ are neighbors, i.e. they form an edge and we write $u \sim v$, if $|k - h| = 1$ and $\sigma_i = \rho_i$ for all $i = 1, 2, \ldots, \min\{h, k\}$. If $u \sim v$ and $\ell(u) < \ell(v)$ (thus, necessarily, $\ell(v) = \ell(u) + 1$), we say that u is the *father* (or *predecessor*) of v and, conversely v is a *son* (or *successor*) of u; also, if v and u have the same predecessor (so that, in particular, $\ell(u) = \ell(v)$), we call them *brothers*. This way, $\emptyset$, the vertex corresponding to the empty word, the common *ancestor*, is called the *root* of the tree and the vertices at distance k from $\emptyset$, which correspond to the kth generation, constitute the kth *level* of the tree $\mathbb{T}_{q,n}$. These are clearly identified with Σ^k. The vertices at distance n (the maximum distance from $\emptyset$) are called the *leaves* of the tree. Recall that the degree $\deg(v)$ of a vertex v in a graph $G = (V, E)$ is the number of neighboring vertices, namely $\deg(v) = |\{u \in V : u \sim v\}|$. In our setting we have $\deg(\emptyset) = q$, $\deg(v) = q + 1$, if $1 \le \ell(v) \le n - 1$ and $\deg(v) = 1$ if v is a leaf.

If $v = \sigma_1 \sigma_2 \cdots \sigma_k \in \Sigma^k$, we denote by $\mathbb{T}_v = \{w : w = \rho_1 \rho_2 \cdots \rho_h, h \ge k, \rho_i = \sigma_i \text{ for all } 1 \le i \le k\}$ the *subtree rooted at v*.

We finally denote by d' the geodesic distance on $\mathbb{T}_{q,n}$.

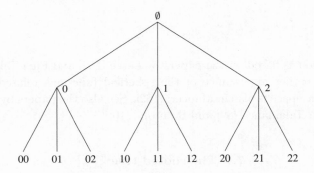

Figure 7.1. The rooted tree $\mathbb{T}_{3,2}$

7.2 The group $Aut(\mathbb{T}_{q,n})$ of automorphisms

Let $Aut(\mathbb{T}_{q,n})$ denote the group of all automorphisms of $\mathbb{T}_{q,n}$.

Note that an automorphism necessarily preserves the degrees of the vertices: $\deg(g(v)) = \deg(v)$ and preserves the distances, namely it is an *isometry*: $d'(g(v), g(u)) = d'(u, v)$ for all $u, v \in \mathbb{T}_{q,n}$. Therefore, if $g \in Aut(\mathbb{T}_{q,n})$ one necessarily has $g(\emptyset) = \emptyset$ and, more generally, $g(\Sigma^k) = \Sigma^k$ for all k's, that is, g globally fixes the levels of the tree.

Denote by S_q the symmetric group on q elements. Observe that if $g \in Aut(\mathbb{T}_{q,n})$ then g is uniquely determined by a *labeling*, that we continue to denote by g, namely by a map $g : \mathbb{T}_{q,n-1} \ni v \mapsto g_v \in S_q$ such that

$$g(x_1 x_2 \cdots x_k) = g_\emptyset(x_1) g_{x_1}(x_2) \cdots g_{x_1 x_2 \cdots x_{k-1}}(x_k) \qquad (7.1)$$

$k = 1, 2, \ldots, n - 1$, $x_1 x_2 \cdots x_k \in \mathbb{T}_{q,n}$. Indeed, if $g \in Aut(\mathbb{T}_{q,n})$ and $g(x_1 x_2 \cdots x_{k-1}) = y_1 y_2 \cdots y_{k-1}$, then, for any $x_k \in \Sigma$ there exists $y_k \in \Sigma$ such that $g(x_1 x_2 \cdots x_{k-1} x_k) = y_1 y_2 \cdots y_{k-1} y_k$ and the correspondence $x_k \mapsto y_k$ is a bijection of Σ, that is an element of S_q, and we denote it by $g_{x_1 x_2 \cdots x_{k-1}}$.

For example, if $q = n = 3$, so that $\mathbb{T}_{q,n}$ is the binary tree of depth 3, an automorphism $g \in Aut(\mathbb{T}_{3,3})$ is determined by the data $(g_\emptyset, g_0, g_1, g_2, g_{00}, \dots, g_{22}) \in S_3^{13}$ where $g_\emptyset$ is the permutation induced by g on the first level $\Sigma^1 = \{0, 1, 2\}$. Then g_i is the permutation induced by g on the first level of the subtree rooted at $g_\emptyset(i)$, $i = 0, 1, 2$, and so on.

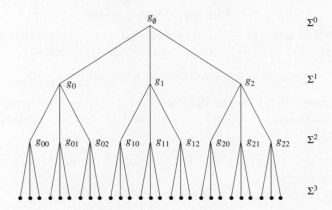

Figure 7.2. The labeling of an automorphism $g \in Aut(\mathbb{T}_{3,3})$

Conversely, it is clear that any labeling gives rise to an automorphism of $\mathbb{T}_{q,n}$ defined by (7.1). In particular, if $h, g : \mathbb{T}_{q,n-1} \to S_q$ are automorphisms of $\mathbb{T}_{q,n}$ represented as labelings, then the label of hg is given by $(hg)_\emptyset = h_\emptyset \cdot g_\emptyset$ and, for $k \geq 1$,

$$(hg)_{x_1 x_2 \cdots x_k} = h_{g_\emptyset(x_1) g_{x_1}(x_2) \cdots g_{x_1 x_2 \cdots x_{k-1}}(x_k)} g_{x_1 x_2 \cdots x_k}$$
$$\equiv h_{g(x_1 x_2 \dots x_k)} \cdot g_{x_1 x_2 \dots x_k}. \tag{7.2}$$

This formula follows by a direct application of (7.1) to $hg(x_1 x_2 \cdots x_k)$: for instance, if $g(x_1 x_2) = y_1 y_2$ and $h(y_1 y_2) = z_1 z_2$, then $g_\emptyset(x_1) = y_1$, $g_{x_1}(x_2) = y_2$, $h_{y_1}(y_2) = z_2$ and therefore $(hg)_{x_1}(x_2) = z_2 = h_{g_\emptyset(x_1)} \cdot g_{x_1}(x_2)$.

In other words, $Aut(\mathbb{T}_{q,n})$ coincides with the set of all labelings endowed with the composition law (7.2).

Finally, note that if $v \in \Sigma^k$ we have $\mathbb{T}_v \cong \mathbb{T}_{q,n-k}$ and $Aut(\mathbb{T}_v) \cong Aut(\mathbb{T}_{q,n-k})$.

Exercise 7.2.1 (1) Directly verify that the set of all labelings with composition law (7.2) indeed defines a group and that (7.1) is an action.

(2) Show that $Aut(\mathbb{T}_{q,2})$ is isomorphic to the wreath product $S_q \wr S_q$ (see Section 5.4).

Remark 7.2.2 $Aut(\mathbb{T}_{q,n})$ is isomorphic to the n-iterated wreath product $S_q \wr S_q \wr \cdots \wr S_q$ (see the monograph by Bass et al. [16]). The labeling representation of an element in $Aut(\mathbb{T}_{q,n})$ was introduced by Grigorchuk [107].

7.3 The ultrametric space

The set Σ^n of leaves of $\mathbb{T}_{q,n}$ can be endowed with a metric structure as follows. For $x = x_1 x_2 \cdots x_n$ and $y = y_1 y_2 \cdots y_n \in \Sigma^n$ define

$$d(x, y) = n - \max\{k : x_i = y_i \text{ for all } i \leq k\}. \tag{7.3}$$

In other words $d(x, y)$ is the distance from x and y of the most recent common ancestor. It is easy to see that the distance d satisfies the conditions

(a) $d(x, y) \geq 0$ and $d(x, y) = 0$ if and only if $x = y$,
(b) $d(x, y) = d(y, x)$,
(c) $d(x, z) \leq \max\{d(x, y), d(y, z)\}$, for all $x, y, z \in \Sigma^n$.

Condition (c) which is called the *ultrametric inequality* clearly implies the triangular inequality

(c′) $d(x, z) \leq d(x, y) + d(y, z)$ for all $x, y, z \in \Sigma^n$;

thus (Σ^n, d) is a metric space; in particular, d is called an *ultrametric distance* and (Σ^n, d) is called an *ultrametric space*.

We observe that in Σ^n we have two metric distances: the one which is induced by $\mathbb{T}_{q,n}$, we continue to denote it by d' and the ultrametric distance d defined above. We clearly have $d = d'/2$: this shows, in particular, that the action of $Aut(\mathbb{T}_{q,n})$ on (Σ^n, d) is isometric. In addition we observe that the diameter of (Σ^n, d) is n and indeed the range of the function d is precisely

$$d(\Sigma^n, \Sigma^n) = \{0, 1, \ldots, n\}. \tag{7.4}$$

In Figure 7.3 we denote by $\Omega_j = \{x \in \Sigma^3 : d(x, x_0) = j\}$, $0 \leq j \leq 3$, the *sphere* of radius j centered at the point $x_0 = 000$. Note that $|\Omega_j| = (q - 1)q^{j-1}$.

We now let $Aut(\mathbb{T}_{q,n})$ act on Σ^n: the action is clearly transitive. Let $x_0 = 00 \cdots 0 \in \Sigma^n$ be the leftmost leaf and denote by $K(q, n) = \{g \in Aut(\mathbb{T}_{q,n}) : g(x_0) = x_0\}$ its *stabilizer*. We now prove that $(Aut(\mathbb{T}_{q,n}), K(q, n))$ is a Gelfand pair by showing that the above action is indeed 2-point homogeneous (see Definition 4.3.6 and Example 4.3.7).

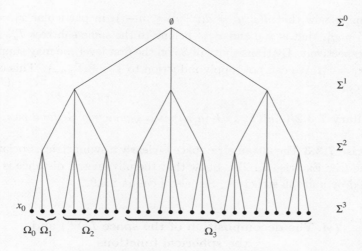

Figure 7.3. The orbit decomposition of Σ^3

Lemma 7.3.1 *The action of $Aut(\mathbb{T}_{q,n})$ on (Σ^n, d) is 2-point homogeneous.*

Proof We have to show that given $x, y, x', y' \in \Sigma^n$ such that $d(x,y) = d(x',y')$, then there exists $g \in Aut(\mathbb{T}_{q,n})$ such that $g(x) = x'$ and $g(y) = y'$. We proceed with an induction argument on the depth n of the tree $\mathbb{T}_{q,n}$.

For $n = 1$ we have $\Sigma^1 = \{0, 1, \ldots, q-1\}$ and $Aut(\mathbb{T}_{q,1}) = S_q$ and the statement is obvious: if $d(x,y) = d(x',y') = 0$ then $x = y$ and $x' = y'$ and it suffices to take any $g \in S_q$ such that $g(x) = x'$; otherwise we have $x \neq y$ and $x' \neq y'$ and therefore there exists a $g = g(x,y,x',y')$ (depending on x, y, x' and y') in S_q such that $g(x) = x'$ and $g(y) = y'$ (see Definition 3.13.7).

Suppose, by induction that the statement holds for $Aut(\mathbb{T}_{q,k})$ with $1 \leq k \leq n-1$. Set $x = x_1 x_2 \ldots x_n$, $x' = x'_1 x'_2 \cdots x'_n$ and similarly for y and y'.

If $d(x,y) = d(x',y') = n$, i.e. $x_1 \neq y_1$ and $x'_1 \neq y'_1$, namely the distance is maximal, let $g \in Aut(\mathbb{T}_{q,n})$ be an automorphism with label $g_\emptyset = g(x_1, y_1, x'_1, y'_1)$ (where we used the notation from the above paragraph) and with trivial label elsewhere. We now have that $g(x)$ and x' belong to the same subtree rooted at x'_1 while $g(y)$ and y' belong to the subtree rooted at y'_1 which is different from the previous one (as $x'_1 \neq y'_1$) and the statement follows by the transitivity of the action of $Aut(\mathbb{T}_{q,n-1}) \cong Aut(T_{x'_1}) \cong Aut(T_{y'_1})$ on the leaves of $\mathbb{T}_{q,n-1}$.

Suppose now that $d(x,y) = d(x',y') \leq n - 1$, in particular $x_1 = y_1$ and $x_1' = y_1'$, that is x, y and x', y' belong to the same subtrees T_{x_1} and $T_{x_1'}$, respectively. By transitivity of S_q on the first level, we may suppose that $x_1 = x_1'$. We can now apply induction to $\mathbb{T}_{x_1} \cong \mathbb{T}_{q,n-1}$. This ends the proof. $\qquad\square$

Corollary 7.3.2 $(Aut(\mathbb{T}_{q,n}), K(q,n))$ *is a symmetric Gelfand pair.*

Exercise 7.3.3 The ultrametric space is clearly a symmetric association scheme (see Exercise 5.1.22). Show that the ultrametric distance is not induced by a graph structure. See also Lemma 8.3.9.

7.4 The decomposition of the space $L(\Sigma^n)$ and the spherical functions

In order to describe the decomposition into irreducible subrepresentations of the permutation representation $L(\Sigma^n)$ we introduce some subspaces. Note that a function $f \in L(\Sigma^n)$ may be regarded as a function $f = f(x_1, x_2, \ldots, x_n)$ of the Σ-valued variables $x_1, x_2, \ldots, x_n$.

Set $W_0 = L(\emptyset) = \mathbb{C}$ and, for $j = 1, \ldots, n$, set

$$W_j = \{f \in L(\Sigma^n) : f = f(x_1, x_2, \ldots, x_j) \text{ and}$$

$$\sum_{x=0}^{q-1} f(x_1, x_2, \ldots, x_{j-1}, x) \equiv 0\}. \qquad (7.5)$$

Note that, for $j \geq 1$ one has $\dim(W_j) = q^{j-1}(q-1)$.

In other words, W_j is the set of all functions $f \in L(\Sigma^n)$ that only depend on $x_1 x_2 \cdots x_j$ and whose mean on the sets $\{x_1 x_2 \cdots x_{j-1} x : x \in \Sigma\}$, is equal to zero for any $x_1 x_2 \cdots x_{j-1} \in \Sigma^{j-1}$.

In a more geometrical language, we can say that if $x_1 x_2 \cdots x_{j-1} \in \Sigma^{j-1}$ and $f \in W_j$, then, for each $x \in \Sigma$, f is constant on the set

$$A_x := \{x_1 x_2 \cdots x_{j-1} x x_{j+1} x_{j+2} \cdots x_n : x_{j+1}, x_{j+2}, \ldots, x_n \in \Sigma\}$$

of descendants of $x_1 x_2 \cdots x_{j-1} x$ in Σ^n and, if f_x is the value of f on A_x, then $\sum_{x=0}^{q-1} f_x = 0$.

Theorem 7.4.1 *We have that*

$$L(\Sigma^n) = \oplus_{j=0}^{n} W_j, \qquad (7.6)$$

is the (multiplicity free) decomposition into $Aut(\mathbb{T}_{q,n})$-irreducibles of $L(\Sigma^n)$.

Proof We first show that these subspaces are $Aut(\mathbb{T}_{q,n})$-invariant. The first condition, namely the dependence of $f \in W_j$ only on the first j variables is clearly invariant as $Aut(\mathbb{T}_{q,n})$ preserves the levels.

The induced action of $Aut(\mathbb{T}_{q,n})$ on $L(\Sigma^n)$ (cf. with (7.1)) is given by

$$[g^{-1}f](x_1, x_2, \ldots, x_n) = f(g_\emptyset(x_1), g_{x_1}(x_2), g_{x_1 x_2}(x_3), \ldots, g_{x_1 x_2 \cdots x_{n-1}}(x_n))$$

for $f \in L(\Sigma^n)$ and $g \in Aut(\mathbb{T}_{q,n})$; thus if $f \in W_j$ we have

$$\sum_{x=0}^{q-1} [g^{-1}f](x_1, x_2, \ldots, x_{j-1}, x)$$

$$= \sum_{x=0}^{q-1} f(g_\emptyset(x_1), g_{x_1}(x_2), g_{x_1 x_2 \cdots x_{j-2}}(x_{j-1}), g_{x_1 x_2 \cdots x_{j-1}}(x))$$

$$= \sum_{x'=0}^{q-1} f(g_\emptyset(x_1), g_{x_1}(x_2), g_{x_1 x_2 \cdots x_{j-2}}(x_{j-1}), x') \equiv 0.$$

This shows that also the second defining condition for an f to be in W_j is invariant. This is also clear from the geometrical interpretation in Figure 7.4.

We now show that these spaces are pairwise orthogonal: if $f \in W_j$, $f' \in W_{j'}$, then $\langle f, f' \rangle = 0$ if $j \neq j'$. To fix the ideas suppose that $j < j'$. We then have

$$\langle f, f' \rangle = \sum_{x_1=0}^{q-1} \sum_{x_2=0}^{q-1} \cdots \sum_{x_n=0}^{q-1} f(x_1, x_2, \ldots, x_n) \overline{f'(x_1, x_2, \ldots, x_n)}$$

$$= q^{n-j'} \sum_{x_1=0}^{q-1} \sum_{x_2=0}^{q-1} \cdots \sum_{x_{j'-1}=0}^{q-1} f(x_1, x_2, \ldots, x_j)$$

$$\cdot \sum_{k=0}^{q-1} \overline{f'(x_1, x_2, \ldots, x_{j'-1}, k)} = 0.$$

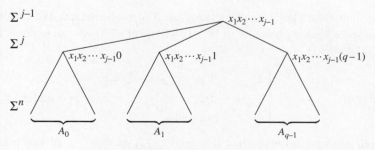

Figure 7.4. $f_0 + f_1 + \cdots + f_{q-1} = 0$

We now show that the W_j's fill up the whole space $L(\Sigma^n)$. We use induction on n. For $n = 1$ it is a standard fact that any function $f(x_1)$ can be expressed as $f(x_1) = c + g(x_1)$ where $c \in \mathbb{C}$ is a constant (indeed $c = \frac{1}{q} \sum_{x=0}^{q-1} f(x)$) and g is a function of mean zero: $\sum_{x=0}^{q-1} g(x) = 0$ (this is again the decomposition in Example 1.8.1). Suppose we have the assertion true for $n - 1$; analogously, a function $f \in \Sigma^n$ can be expressed as $f(x_1, x_2, \ldots, x_n) = c(x_1, x_2, \ldots, x_{n-1}) + g(x_1, x_2, \ldots, x_{n-1}, x_n)$ where c does not depend on the last variable x_n and g has mean zero with respect to x_n. Applying the inductive step to c we are done.

Denote by $\Omega_j = \{x \in \Sigma^n : d(x, x_0) = j\}$ the sphere of radius j centered at $x_0 = 00 \cdots 0$; these clearly are the $K_{q,n}$-orbits (recall that $K(q, n)$ is the stabilizer of the point x_0). In virtue of (7.4) we have that the number of the $K(q, n)$-orbits is exactly $n + 1$.

In virtue of Wielandt's lemma (Theorem 3.13.3) and Proposition 4.4.4 we have that the W_j's are irreducible subspaces and this ends the proof.
□

We remark that, incidentally, the previous arguments offer an alternative proof that $(Aut(\mathbb{T}_{q,n}), K(q, n))$ is a Gelfand pair (compare with Corollary 7.3.2).

Our next step is the determination of the spherical functions $\phi_0, \phi_1, \ldots,$ ϕ_n relative to $(Aut(\mathbb{T}_{q,n}), K(q, n))$; we combine the defining conditions of the W_j's with the $K(q, n)$-invariance (ϕ_j is constant on the spheres Ω_k's centered at $x_0 = 00 \cdots 0$).

Proposition 7.4.2 *The spherical function* $\phi_j \in W_j$ *is given by*

$$\phi_j(x) = \begin{cases} 1 & if \ d(x, x_0) < n - j + 1 \\ -\frac{1}{q-1} & if \ d(x, x_0) = n - j + 1 \\ 0 & if \ d(x, x_0) > n - j + 1. \end{cases} \tag{7.7}$$

Proof It is clear that these functions are $K(q, n)$-invariant. We are only left to show that ϕ_j belongs to W_j. In virtue of (7.3) we can express the ϕ_j in (7.7) by

$$\phi_j(x_1, x_2, \ldots, x_n) = \begin{cases} 1 & if \ x_1 = x_2 = \cdots = x_j = 0 \\ -\frac{1}{q-1} & if \ x_1 = x_2 = \cdots = x_{j-1} = 0 \ and \ x_j \neq 0 \\ 0 & otherwise. \end{cases}$$

Indeed, we first observe that if $x_1 x_2 \cdots x_{j-1} \neq 00 \cdots 0$ all the points of the form $x_1 x_2 \cdots x_n$ have the same distance ($> n - j + 1$) from the base point $x_0 = 00 \cdots 0$ (in fact all the points in the spheres Ω_h with $h > n - j + 1$ are of this type).

The spherical function ϕ_j in W_j is constant on the Ω_h's and thus if $x_1 x_2 \ldots x_{j-1} \neq 00 \cdots 0$, then $\phi_j(x_1, x_2, \cdots, x_{j-1}, x)$ does not depend on x because all the $(x_1, x_2, \ldots, x_{j-1}, x)$'s, with $x = 0, 1, \ldots, q-1$, belong to the same Ω_h. This, coupled with the condition $\sum_{x=0}^{q-1} \phi_j(x_1, x_2, \ldots, x_{j-1}, x) = 0$, infers that ϕ_j vanishes on all points at distance $> n-j+1$ from x_0.

Similarly, all the points of the form $\underbrace{00 \cdots 0}_{j-1} x y_{j+1} \cdots y_n$ with $x = 1, 2, \ldots, q - 1$ constitute the ball of radius $n - j + 1$. Moreover, since, by definition, $\phi_j(0, 0, \ldots, 0, 0) = 1$ and ϕ_j depends only on the first j variables (it belongs to W_j), the condition $\phi_j(0, 0, \ldots, 0, 0) + \sum_{x=1}^{q-1} \phi_j(0, 0, \ldots, 0, x) = 0$, coupled with the condition that ϕ_j is constant on Ω_{n-j+1}, uniquely determines the value of ϕ_j on points at distance $n - j + 1$, namely $-\frac{1}{q-1}$.

Finally, if $d(x, x_0) < n - j + 1$, then $x = \underbrace{00 \cdots 0}_{h} y_{h+1} \cdots y_n$ with $h > j - 1$ and therefore $\phi_j(x) = \phi_j(00 \cdots 0) = 1$. $\qquad \square$

7.5 Recurrence in finite graphs

In the next section we analyze a Markov chain on Σ^n. For this purpose we need some more tools that we now independently develop in this section.

Let X be a finite set and $(p(x, y))_{x,y \in X}$ be a stochastic matrix on X. We say that P is *irreducible* if for any pair $x, y \in X$ there exists $k = k(x, y) \in \mathbb{N}$ such that $p^{(k)}(x, y) > 0$. Observe that this condition is weaker than ergodicity (cf. Section 1.4) where the k above can be chosen independently on $x, y \in X$. On the other hand, introducing a suitable notion of *period*, then one has that P is ergodic if and only if it is irreducible and aperiodic; in the setting of reversible Markov chains this corresponds to Theorem 1.7.6. We shall not use the general notion of period but we refer to the monographs by Behrends [18], Bremaud [32] and Norris [172] for a more general treatment.

A *path* of length n in X is a sequence $x_0 x_1 \cdots x_n$ of elements (states) in X such that $p(x_i, x_{i+1}) > 0$ for all $i = 0, 1, \ldots, n-1$. Now fix $x_0 \in X$. We shall consider the Markov chain with transition matrix P and with starting point x_0.

Let $\mathcal{P}_n$ denote the set of all paths of length n starting at x_0. On $\mathcal{P}_n$ the following is a natural probability measure $\mathbb{P}_n$ given by the construction in Example 1.3.8

$$\mathbb{P}_n(x_0 x_1 \cdots x_n) = p(x_0, x_1) p(x_1, x_2) \cdots p(x_{n-1}, x_n)$$

for all $x_0 x_1 \cdots x_n \in \mathcal{P}_n$.

Suppose now that Y is a proper subset of X. For all $y \in Y$ consider the following subset of $\mathcal{P}_n$

$$\mathcal{P}_n^y = \{x_0 x_1 \cdots x_n : \text{ there exists } 1 \leq k \leq n \text{ s.t. } x_k = y \text{ and}$$
$$x_1, \ldots, x_{k-1} \notin Y\}.$$

In other words, $\mathcal{P}_n^y$ is the set of all paths of length n in X such that y is the first element of Y *visited* by the path. Note that x_0 may belong to Y; the visit is considered only at time $t \geq 1$.

Theorem 7.5.1 *The limit*

$$\mu(y) := \lim_{n \to \infty} \mathbb{P}_n(\mathcal{P}_n^y) \tag{7.8}$$

exists for any $y \in Y$ and μ is a probability measure Y.

Proof First of all note that if $x_0 x_1 \cdots x_n \in \mathcal{P}_n^y$, then all paths of the form $x_0 x_1 \cdots x_n z$ belong to $\mathcal{P}_{n+1}^y$. Therefore we have

$$\mathbb{P}_{n+1}(\mathcal{P}_{n+1}^y) \geq \sum_{x_0 x_1 \cdots x_n \in \mathcal{P}_n^y} \sum_{\substack{z \in X: \\ p(x_n, z) > 0}} \mathbb{P}_n(x_0 x_1 \cdots x_n) p(x_n, z)$$

$$= \sum_{x_0 x_1 \cdots x_n \in \mathcal{P}_n^y} \mathbb{P}_n(x_0 x_1 \cdots x_n)$$

$$= \mathbb{P}_n(\mathcal{P}_n^y).$$

Thus the sequence $(\mathbb{P}_n(\mathcal{P}_n^y))_n$ is increasing and bounded above (by 1). This ensures that the limit in (7.8) exists. It remains to prove that $\sum_{y \in Y} \mu(y) = 1$.

For $x \in X$ denote by $d(x, Y) = \min\{d(x, y) : y \in Y\}$ the length of the shortest path connecting x to Y and set

$$m = \max\{d(x, Y) : x \in X\} \tag{7.9}$$

and

$$\varepsilon = \min\{p(z_0, z_1) \cdots p(z_{m-1}, z_m) : z_0 \cdots z_m \text{ is a path in } X \text{ of length } m\}. \tag{7.10}$$

Also set

$$\mathcal{C}_n = \{x_0 x_1 \cdots x_n \in \mathcal{P}_n : x_1, x_2 \ldots, x_n \notin Y\}$$

so that

$$\mathcal{P}_n = \mathcal{C}_n \coprod \left(\coprod_{y \in Y} \mathcal{P}_n^y \right).$$

To end the proof we show that $\lim_{n \to \infty} \mathbb{P}_n(\mathcal{C}_n) = 0$. Denote by

$$\mathcal{C}_n^k = \{x_0 x_1 \cdots x_n x_{n+1} \cdots x_{n+k} \in \mathcal{P}_n : x_0 x_1 \cdots x_n \in \mathcal{C}_n\}$$

the set of all paths in $\mathcal{P}_{n+k}$ that do not visit Y in the first n steps. Clearly

$$
\begin{aligned}
\mathbb{P}_{n+k}(\mathcal{C}_n^k) &= \sum_{\substack{x_0 x_1 \cdots x_n \in \mathcal{C}_n \\ x_{n+1}, \ldots, x_{n+k} \in X}} \mathbb{P}_n(x_0 x_1 \cdots x_n) p(x_n, x_{n+1}) \cdots p(x_{n+k-1}, x_{n+k}) \\
&= \mathbb{P}_n(\mathcal{C}_n). \tag{7.11}
\end{aligned}
$$

But if m is as in (7.9), then, for any $x \in X$ there exists a path $z_m(x) \equiv z_1 \cdots z_m$ with $p(x, z_1) > 0$ which visits Y (i.e. there exists $i \geq 1$ such that $z_i \in Y$). We thus have

$$\mathcal{C}_{n+m} \subseteq \mathcal{C}_n^m \setminus \{x_0 x_1 \cdots x_n z_m(x_n) : x_0 x_1 \cdots x_n \in \mathcal{C}_n\}$$

and therefore

$$
\begin{aligned}
\mathbb{P}_{n+m}&(\mathcal{C}_{n+m}) \\
&\leq \mathbb{P}_{n+m}(\mathcal{C}_n^m) - \mathbb{P}_{n+m}\left(\{x_0 x_1 \cdots x_n z_m(x_n) : x_0 x_1 \cdots x_n \in \mathcal{C}_n\}\right) \\
&\quad \text{(by (7.10) and (7.11))} \\
&\leq \mathbb{P}_n(\mathcal{C}_n) - \sum_{x_0 x_1 \cdots x_n \in \mathcal{C}_n} \mathbb{P}_n(x_0 x_1 \cdots x_n) \cdot \varepsilon \\
&= (1 - \varepsilon) \mathbb{P}_n(\mathcal{C}_n).
\end{aligned}
$$

Thus, for any $0 \leq r < m$ and $k \geq 1$,

$$\mathbb{P}_{r+km}(\mathcal{C}_{r+km}) \leq (1 - \varepsilon)^k \mathbb{P}_r(\mathcal{C}_r)$$

and this shows that $\mathbb{P}_n(\mathcal{C}_n) \to 0$. $\qquad\square$

There is a simple way to express the statement of the theorem: *the Markov chain will visit Y with probability 1 and $\mu(y)$ is the probability that y is the first state in Y to be visited.*

7.6 A Markov chain on Σ^n

The ultrametric space (Σ^n, d) is not a graph, in the sense that there is no graph structure that induces the ultrametric distance (see Exercise 7.3.3). Moreover, given an invariant Markov chain on Σ^n, then necessarily its transition probabilities are given by

$$p(x, y) = \frac{1}{(q-1)q^{d(x,y)-1}} \mu(d(x, y))$$

with μ a probability measure on $\{0, 1, \ldots, n\}$. Set $k = \max\{j : \mu(j) > 0\}$. If $k < n$, then we have that starting from x one never reaches a point y with $d(x, y) > k$. Indeed, given a path $p = (x_0, x_1, \ldots, x_m)$ with $d(x_i, x_{i+1}) \leq k$, for $i = 0, 1, \ldots, m-1$, then also $d(x_0, x_i) \leq k$ for all i's (ultrametric inequality). Therefore we consider a Markov chain on Σ^n with $k = n$: it is taken from [90].

Consider the simple random walk on $\mathbb{T}_{q,n}$. It is clearly irreducible (as $\mathbb{T}_{q,n}$ is connected). If we start at a point $x \in \Sigma^n$ and, with the notation from the preceding section, take $Y \subseteq \mathbb{T}_{q,n}$ equal to Σ^n, then there is a probability distribution μ_x on Σ^n such that, for $y \in \Sigma^n$, $\mu_x(y)$ is the probability that y is the first point in Σ^n visited by the random walk. Setting $P(x, y) = \mu_x(y)$ we clearly have that $(P(x, y))_{x,y \in \Sigma^n}$ is stochastic and $Aut(\mathbb{T}_{q,n})$-invariant: the simple random walk on $\mathbb{T}_{q,n}$ is invariant.

Therefore, there is no loss of generality in considering $x_0 = 00 \cdots 0$ as the starting point of the simple random walk on $\mathbb{T}_{q,n}$.

In order to avoid confusion between the random walk in $\mathbb{T}_{q,n}$ and the Markov chain in Σ^n we concentrate only on a single step in the last one, equivalently we think of Σ^n as an *absorbing* subspace for the first one: the walk stops if it reaches a point in Σ^n.

Also, we shall not develop all the theoretical background underlying the analysis of this Markov chain. The interested reader can easily fill in all the gaps of a theoretical nature using the tools developed Theorem 7.5.1; see also Exercise 7.6.2 at the end of the section.

Set $\xi_n = \emptyset$ and, for $0 \leq i < n$, $\xi_i = 00 \cdots 0$ ($n - i$ times); this way $(\xi_0 = x_0, \xi_1, \cdots, \xi_n = \emptyset)$ is the unique geodesical path connecting x_0 with the root of $\mathbb{T}_{q,n}$.

For $j \geq 0$ call α_j the probability of *ever reaching ξ_{j+1} given that ξ_j is reached at least once* (recall that we assume that the random walk stops whenever it reaches Σ^n). This way, $\alpha_0 = 1$ and $\alpha_1 = (q+1)^{-1}$: indeed with probability one ξ_1 is reached at the first step while, starting from ξ_1, with probability $(1+q)^{-1}$ one reaches ξ_2 and with probability $q(1+q)^{-1}$ one returns to Σ^n and the walk stops. Moreover $\alpha_{n+1} = 0$.

For $1 < j < n$ we have the recurrence relation

$$\alpha_j = (1+q)^{-1} + \alpha_{j-1}\alpha_j q(1+q)^{-1}. \tag{7.12}$$

Indeed starting from ξ_j, with probability $(1+q)^{-1}$ one reaches in one step ξ_{j+1}, otherwise with probability $q(1+q)^{-1}$ one reaches either ξ_{j-2} or one of its brothers (with equal probability); but then with probability α_{j-1} one reaches again ξ_{j-1} and one starts the recursive argument. The solution of (7.12) is given by

$$\alpha_j = \frac{q^j - 1}{q^{j+1} - 1}, \qquad 1 \le j \le n-1. \tag{7.13}$$

Our next step is the computation of $P(x_0, x_0)$, namely the probability that the first point of Σ^n visited by the random walk is x_0. We have

$$P(x_0, x_0) = q^{-1}(1-\alpha_1) + q^{-2}\alpha_1(1-\alpha_2) + \cdots +$$
$$+ q^{-n+1}\alpha_1\alpha_2\cdots\alpha_{n-2}(1-\alpha_{n-1}) + q^{-n}\alpha_1\alpha_2\cdots\alpha_{n-1}.$$

Indeed the jth summand represents the probability of returning back to x_0 if the corresponding random walk in the tree reaches ξ_j but not ξ_{j+1}; observe that in this case, once ξ_j is reached all the q^j leaves of the subtree $\mathbb{T}_{\xi_j}$ (namely all points $x \in \Sigma^n$ at distance $d(x, x_0) \le j$) are equiprobable.

We now compute $P(x_0, x)$ where x is a point at distance $d(x_0, x) = j$. For $j = 1$ we clearly have $P(x_0, x) = P(x_0, x_0)$.

Suppose that $j > 1$. We observe that, in order to reach x one is forced to first reach ξ_j. Therefore,

$$P(x_0, x) = q^{-j}\alpha_1\alpha_2\cdots\alpha_{j-1}(1-\alpha_j) + \cdots +$$
$$+ q^{-n+1}\alpha_1\alpha_2\cdots\alpha_{n-2}(1-\alpha_{n-1})$$
$$+ q^{-n}\alpha_1\alpha_2\cdots\alpha_{n-1}.$$

By the $Aut(\mathbb{T}_{q,n})$-invariance of the random walk(s) we have

$$P(x, y) = q^{-j}\alpha_1\alpha_2\cdots\alpha_{j-1}(1-\alpha_j) + \cdots +$$
$$+ q^{-n+1}\alpha_1\alpha_2\cdots\alpha_{n-2}(1-\alpha_{n-1})$$
$$+ q^{-n}\alpha_1\alpha_2\cdots\alpha_{n-1} \tag{7.14}$$

for all $x, y \in \Sigma^n$ with $0 < d(x, y) = j$.

We now determine the eigenvalues of the transition matrix $P(\cdot, \cdot)$. In virtue of the equivalence between G-invariant operators and bi-K-invariant functions, these eigenvalues are given by the spherical Fourier transform (see Section 4.7) of the convolver that represents P, namely

$$\lambda_j = \sum_{x \in \Sigma^n} P(x_0, x) \phi_j(x),$$

for all $j = 0, 1, \ldots, n$.

For $j = 0$ we have $\phi_0 \equiv 1$ and $\lambda_0 \equiv \sum_{x \in \Sigma^n} P(x_0, x) = 1$, as $P(x, y)$ is stochastic.

For $j = n$ from (7.7) we have that $\phi_n(x_0) = 1$, $\phi_n(x) = -1/(q-1)$ if $d(x, x_0) = 1$ and $\phi_n(x) = 0$ otherwise. Since $P(x_0, x) = P(x_0, x_0)$ if $d(x, x_0) = 1$, we conclude that $\lambda_n = 0$.

For $1 \leq j < n$ and choosing $x_i \in \Omega_i$ we have

$$\begin{aligned}
\lambda_j &= qP(x_0, x_1) + (q^2 - q)P(x_0, x_2) + \cdots + \\
&\quad + (q^{n-j} - q^{n-j-1})P(x_0, x_{n-j}) \\
&\quad + (1-q)^{-1}(q^{n-j+1} - q^{n-j})P(x_0, x_{n-j+1}) \\
&= q(P(x_0, x_1) - P(x_0, x_2)) + q^2(P(x_0, x_2) - P(x_0, x_3)) + \cdots + \\
&\quad + q^{n-j-1}(P(x_0, x_{n-j-1}) - P(x_0, x_{n-j})) \\
&\quad + q^{n-j}P(x_0, x_{n-j}) \\
&\quad + (1-q)^{-1}(q^{n-j+1} - q^{n-j})P(x_0, x_{n-j+1}) \\
&= \sum_{h=1}^{n-j} q^h \left(P(x_0, x_h) - P(x_0, x_{h+1})\right) =_{*)} \\
&=_{*)} (1 - \alpha_1) + \alpha_1(1 - \alpha_2) + \cdots + \alpha_1\alpha_2 \cdots \alpha_{n-j-1}(1 - \alpha_{n-j}) \\
&= 1 - \alpha_1\alpha_2 \cdots \alpha_{n-j} =_{**)} \\
&=_{**)} 1 - (q-1)(q^{n-j+1} - 1)^{-1},
\end{aligned}$$

where $=_{*)}$ comes from (7.14) and $=_{**)}$ from the calculation for the α_j's given by (7.13).

Exercise 7.6.1 Let $\mathbf{m} = (m_0, m_1, \ldots, m_{n-1})$ be an n-tuple of integers ≥ 1. Denote by $\mathbb{T_m}$ the *spherically homogeneous rooted tree of type* $\mathbf{m}$. In other words, denoting by $\emptyset$ the root of $\mathbb{T_m}$, then $\emptyset$ has exactly m_0 descendants, which constitute the first level of the tree, call it Σ^1. Then every element $x \in \Sigma^1$ has exactly m_1 descendants at level two, and so on. Note that the set of leaves, that is the last level Σ^n, consists of exactly $m_0 m_1 \cdots m_{n-1}$ elements. When $m_0 = m_1 = \cdots m_{n-1} = q$ this reduces to the q-homogeneous rooted tree $\mathbb{T}_{q,n}$.

Denote by $Aut(\mathbb{T_m})$ the full automorphism group of $\mathbb{T_m}$. It is well known that $Aut(\mathbb{T_m})$ is isomorphic to the wreath product $S_{m_{n-1}} \wr S_{m_{n-2}} \wr \cdots \wr S_{m_1} \wr S_{m_0}$ (see, e.g. [16, 114]). The definition of labeling for automorphisms from the beginning of this chapter naturally extends to this more general setting.

Work out the corresponding computations for a spherically homogeneous tree. See also [90, 153, 19].

Exercise 7.6.2 Justify the definition of the α_j's and the recurrence equation (7.12) with arguments similar to those developed for Theorem 7.5.1.

Part III
Advanced Theory

8

Posets and the q-analogs

All the examples of Gelfand pairs previously studied (the Hamming scheme, the Johnson scheme and the ultrametric space) have an underlying geometric/combinatorial structure that comes from an order relation. In this chapter, we investigate closely these kinds of structures and give further examples. Our main sources are the papers by Delsarte [52], Dunkl [73, 74] and Stanton [206, 207, 210]. For general references on posets and their combinatorial applications, see the books by Stanley [204] and Aigner [3].

8.1 Generalities on posets

A *partially ordered set* (briefly, a *poset*) is a set X with a binary relation $\preceq$ (the (partial) *order*) satisfying the usual axioms:

$$x \preceq x \text{ (reflexivity)}$$

$$x \preceq y, y \preceq x \Rightarrow x = y \text{ (anti-symmetry)}$$

$$x \preceq y, y \preceq z \Rightarrow x \preceq z \text{ (transitivity)}$$

for all x, y and $z \in X$.

Let X be a poset. We assume that X is finite.

If $x \preceq y$ or $y \preceq x$ we say that x and y are *comparable*; otherwise we say that they are *uncomparable*.

If $x \preceq y$ and $x \neq y$ we write $x \prec y$. Moreover, if $x \prec y$ and $x \prec z \preceq y$ implies that $z = y$ (in other words there exists no intermediate element between x and y) we say that y *covers* x and x is *covered* by y and we write $x \lessdot y$.

An element $x_0 \in X$ such that $x_0 \preceq x$ for all $x \in X$ is clearly unique (if it exists) and it is called the 0-*element* (it is usually denoted by 0).

A *chain* of *length* n in X is a sequence $x_1, x_2, \ldots, x_{n+1}$ of elements in X such that $x_1 \prec x_2 \prec \cdots \prec x_{n+1}$. The chain is *saturated* if $x_i \lessdot x_{i+1}$ for all $i = 1, 2, \ldots, n$. The chain is called *maximal* if it is not contained in any longer chain, that is it is saturated and there are no $y, z \in X$ such that $y \prec x_1$ or $x_{n+1} \prec z$.

The *Hasse diagram* of the poset X is the graph with vertex set X and edge set $E = \{\{x, y\} : x \lessdot y\}$. In its graphical representation, the Hasse diagram is drawn in such a way that if $x \prec y$ then the vertex x is represented below the vertex y.

Example 8.1.1 The hypercube Q_2, that is the set of all subsets of $\{1, 2\}$, is a poset with respect to the inclusion and its Hasse diagram is shown in Figure 8.1.

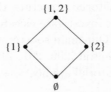

Figure 8.1. The Hasse diagram of Q_2

Let $x, y \in X$. An element $z \in X$ such that

$$z \preceq x \quad \text{and} \quad z \preceq y \tag{8.1}$$

and such that, for every other $z' \in X$ satisfying $z' \preceq x$ and $z' \preceq y$ one has $z' \preceq z$, is called the *meet* (or the *greatest lower bound*) of x and y and it is denoted by $x \wedge y$.

If the meet (which is clearly unique) $x \wedge y$ exists for all $x, y \in X$ we say that X is a *meet semi-lattice*.

Symmetrically, an element $u \in X$ such that

$$x \preceq u \quad \text{and} \quad y \preceq u \tag{8.2}$$

and such that, for every other $u' \in X$ satisfying $x \preceq u'$ and $y \preceq u'$ one has $u \preceq u'$, is called the *join* (or the *least upper bound*) of x and y and it is denoted by $x \vee y$.

If the join (which is clearly unique) $x \vee y$ exists for all $x, y \in X$ we say that X is a *join semi-lattice*.

Finally, if X is both a meet and a join semi-lattice we say that it is a *lattice*.

Exercise 8.1.2 Let X be a lattice. Show that the binary operations $\wedge$ and $\vee$ satisfy the properties:

(1) $(x \wedge y) \wedge z = x \wedge (y \wedge z)$ and $(x \vee y) \vee z = x \vee (y \vee z)$ (associativity);

(2) $x \wedge y = y \wedge x$ and $x \vee y = y \vee x$ (commutativity);

(3) $x \wedge x = x$ and $x \vee x = x$ (idempotence);

(4) $x \wedge (x \vee y) = x \vee (x \wedge y) = x$ (absorbtion);

(5) $x \wedge y = x$ if and only if $x \vee y = y$ if and only if $x \preceq y$,

for all x, y and $z \in X$.

Also show, by providing a counterexample, that the distributive laws

$$x \wedge (y \vee z) = (x \wedge y) \vee (x \wedge z)$$
$$x \vee (y \wedge z) = (x \vee y) \wedge (x \vee z) \tag{8.3}$$

do not hold in general (hint: use the example in Exercise 8.1.12).

Exercise 8.1.3 Let X be a set with two binary operations $\wedge$ and $\vee$ satisfying properties (1)–(4) from the previous exercise. Show that X is a lattice with respect to the binary relation $\preceq$ defined by setting $x \preceq y$ if $x \wedge y = x$. Note that, *a posteriori* this is equivalent, by (5) in the previous exercise, to defining $x \preceq y$ if $x \vee y = y$.

Let X again be a finite poset.

Suppose that for all distinct $x, y \in X$ the following hold: if $z \in X$ covers x and y, then there exists $u \in X$ covered by both x and y. We then say that X is *lower semi-modular*. Symmetrically, if whenever $u \in X$ is covered by x and y there exists $z \in X$ which covers both x and y, we then say that X is *upper semi-modular*.

A poset that is both lower and upper semi-modular is termed *modular*.

Also, X is *graded* (or *ranked*) of *rank* N if there exists a positive integer N and a surjective function (called the *rank*) $r : X \to \{0, 1, \ldots, N\}$ such that if $x \lessdot y$ then $r(y) = r(x) + 1$, for all $x, y \in X$. If this is the case, setting $X_n = \{x \in X : r(x) = n\}$ we have $X = \coprod_{n=0}^{N} X_n$, the X_n's are called the *fibers* and X_N is the *top level* of X.

Clearly, if $x \in X_n$ and $y \in X_k$ with $x \prec y$ then $n < k$.

Exercise 8.1.4 Suppose that X admits a 0-element. Show that X is ranked if and only if it satisfies the *Jordan–Dedekind condition*: for all x and y in X, all satured chains between x and y have the same length.

Proposition 8.1.5 *A finite ranked lattice X is lower semi-modular if and only if the rank function satisfies the inequality*

$$r(x) + r(y) \leq r(x \wedge y) + r(x \vee y) \qquad (8.4)$$

for all $x, y \in X$.

Proof Let $x \neq y$ belong to X and suppose that $z \in X$ covers both x and y. Then, $z = x \vee y$ and $r(x) = r(y) = r(z) - 1$.

Suppose that (8.4) holds. Then,

$$r(x \wedge y) \geq r(x) + r(y) - r(z) = r(x) - 1$$

and x, y cover $x \wedge y$. This shows that X is lower semi-modular.

Conversely let X be lower semi-modular. By contradiction suppose that there exist $x \neq y \in X$ such that

$$r(x) + r(y) > r(x \wedge y) + r(x \vee y). \qquad (8.5)$$

We can choose x and y in such a way that firstly $r(x \vee y) - r(x \wedge y)$ is minimal and then, in addition, $r(x) + r(y)$ is maximal.

Clearly x and y cannot be both covered by $x \vee y$ (otherwise (8.5) would be violated by the lower semi-modularity of X) and therefore we may suppose that there exists $x' \in X$ such that (up to exchanging the role of x and y, if necessary) $x \prec x' \prec x \vee y$. Then, $x' \vee y = x \vee y$ and $x \wedge y \preceq x' \wedge y$; in particular, $r(x' \vee y) - r(x' \wedge y) \leq r(x \vee y) - r(x \wedge y)$

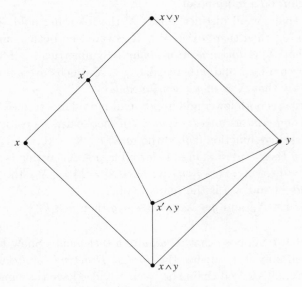

Figure 8.2.

and $r(x') + r(y) > r(x) + r(y)$ and therefore, by the above minmax assumptions, we have

$$r(x' \wedge y) + r(x' \vee y) \geq r(x') + r(y). \tag{8.6}$$

Adding (8.5) and (8.6) together, after simplifications one gets

$$r(x) + r(x' \wedge y) > r(x') + r(x \wedge y). \tag{8.7}$$

Hence, setting $u = x$ and $v = x' \wedge y$ we have $u \wedge v = x \wedge y$ and $u \vee v = x \vee (x' \wedge y) \preceq x'$ and therefore

$$r(u) + r(v) = r(x) + r(x' \wedge y) > r(x') + r(x \wedge y) \geq r(u \vee v) + r(u \wedge v)$$

where the strict inequality follows from (8.7) and this contradicts the minmax assumptions (because $r(u \vee v) - r(u \wedge v) < r(x \vee y) - r(x \wedge y)$). $\square$

Corollary 8.1.6 *Let X be a ranked lattice. Then it is modular if and only if $r(x) + r(y) = r(x \wedge y) + r(x \vee y)$ for all $x, y \in X$.*

Proof State and prove the upper semimodular version of Proposition 8.1.5. $\square$

Exercise 8.1.7 Show that the ranked lattice in Figure 8.3 is lower semi-modular but not modular.

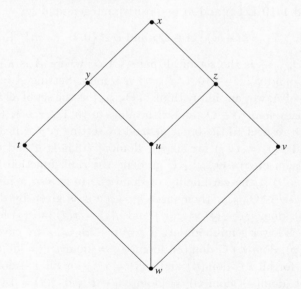

Figure 8.3.

Example 8.1.8 (The Boolean lattice) Let Q_n denote, as usual, the family of all subsets of the set $\{1, 2, \ldots, n\}$ (see Example 1.8.3). Then Q_n is a poset by setting $A \preceq B$ if A is contained in B, for all $A, B \in Q_n$. It is easy to check that it is ranked (for $A \in Q_n$, $r(A) = |A|$, the cardinality of A), it is a lattice ($A \wedge B = A \cap B$ and $A \vee B = A \cup B$ for all $A, B \in Q_n$) and it is modular.

Example 8.1.9 (The homogegenous rooted tree) Let $\mathbb{T}_{q,n}$ denote the homogeneous rooted tree of degree q and depth n (see Chapter 7). The set $X_{q,n}$ of vertices of the tree is a poset with the order $x \succ y$ if x is a descendent of y, for all $x, y \in X$. The corresponding Hasse diagram coincides with the tree $\mathbb{T}_{q,n}$ itself but it is drawn with the root at the bottom level and the leaves are at the top. It is easy to check that $X_{q,n}$ is a ranked, lower semi-modular meet semi-lattice (but it is neither upper semi-modular nor join).

Figure 8.4. The rooted tree $\mathbb{T}_{2,3}$

Example 8.1.10 Let n and m be positive integers and set

$$\Theta_{n,m+1} = \{(A, \phi) : A \in Q_n \text{ and } \phi \in \{0, 1, \ldots, m\}^A\}.$$

In words, $\Theta_{n,m+1}$ is the set of all pairs (A, ϕ) where A is a subset of $\{1, 2, \ldots, n\}$ and $\phi : A \to \{0, 1, \ldots, m\}$ is a map. Setting $A = \text{dom}(\phi)$, the *domain* of A, we can simply think of $\Theta_{n,m+1}$ as the set of all functions ϕ with domain $\text{dom}(\phi) \in Q_n$ and with values in $\{0, 1, \ldots, m\}$. $\Theta_{n,m+1}$ is a poset with respect to the order defined by setting $\phi \preceq \psi$ if $\text{dom}(\phi) \subseteq \text{dom}(\psi)$ and $\phi(j) = \psi(j)$ for all $j \in \text{dom}(\phi)$ (that is if and only if ψ is an extension of ϕ). $\Theta_{n,m+1}$ is graded: the rank function is simply $r(\phi) = |\text{dom}(\phi)|$, the cardinality of $\text{dom}(\phi)$. It is also a meet semi-lattice: if $\phi, \psi \in \Theta_{n,m+1}$ their meet $\eta = \phi \wedge \psi$ is given by $\text{dom}(\eta) = \{j \in \text{dom}(\phi) \cap \text{dom}(\psi) : \phi(j) = \psi(j)\}$ and $\eta(j) = \phi(j) = \psi(j)$ for all $j \in \text{dom}(\eta)$. It is lower semi-modular: if $\phi \neq \psi \in \Theta_{n,m+1}$ are covered by ϵ then, $\text{dom}(\phi), \text{dom}(\psi) \subseteq \text{dom}(\epsilon)$, $|\text{dom}(\phi)| = |\text{dom}(\psi)| = |\text{dom}(\epsilon)| - 1$, $\phi(i) = \epsilon(i)$ for all $i \in \text{dom}(\phi)$ and $\psi(j) = \epsilon(j)$ for all $j \in \text{dom}(\psi)$. It follows that $|\text{dom}(\phi) \cap \text{dom}(\psi)| = |\text{dom}(\phi)| - 1$ and $\phi(j) = \psi(j)$ for all

$j \in \mathrm{dom}(\phi) \cap \mathrm{dom}(\psi)$. Therefore $\phi \wedge \psi = \eta$ is given by the restriction of ϕ to $\mathrm{dom}(\phi) \cap \mathrm{dom}(\psi)$ and η is covered by both ϕ and ψ. We leave it as an exercise to prove that $\Theta_{n,m+1}$ is not upper semi-modular nor a join semi-lattice.

Example 8.1.11 Let $\mathbb{T}(n, m+1)$ denote the finite rooted tree of depth two where the root has degree n and all vertices at level one have $m+1$ children. Any element $\phi \in \Theta_{n,m+1}$ may be identified with a (rooted) subtree $\mathbb{T}_\phi$ of $\mathbb{T}(n, m+1)$ as explained below (see Figure 8.5):

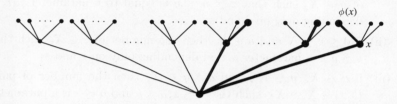

Figure 8.5. An element $\phi \in \Theta_{n,m+1}$ coincides with a subtree $\mathbb{T}_\phi$ of $\mathbb{T}(n, m+1)$

The root of $\mathbb{T}_\phi$ coincides with the root of $\mathbb{T}(n, m+1)$ and the first level of $\mathbb{T}_\phi$ is the subset $\mathrm{dom}(\phi)$ inside the first level of $\mathbb{T}(n, m+1)$; finally the leaves of $\mathbb{T}_\phi$ are the leaves y in $\mathbb{T}(n, m+1)$ whose fathers x belong to $\mathrm{dom}(\phi)$ and $y = \phi(x)$. Note that this way the order relation $\preceq$ in $\Theta_{n,m+1}$ corresponds to the inclusion of subtrees.

Exercise 8.1.12 Let $\prod_n$ be the set of all *partitions* of the set $\{1, 2, \ldots, n\}$. That is, an element $\pi \in \prod_n$ is a family $\pi = \{A_1, A_2, \ldots, A_k\}$ of nonempty subsets of $\{1, 2, \ldots, n\}$ (the *blocks* of the partition) such that $\{1, 2, \ldots, n\} = A_1 \coprod A_2 \coprod \cdots \coprod A_k$ (disjoint union). We introduce an order relation in $\prod_n$ by defining $\sigma \preceq \pi$ if any block of σ is contained in a block of π (that is if σ is a *refinement* of π). Show that $\prod_n$ is a lattice, that it is upper semi-modular, but it is not lower semi-modular.

Exercise 8.1.13 Let $\mathbb{Y}$ be the set of all *partitions* of nonnegative integers, that is an element $\mu \in \mathbb{Y}$ is a finite sequence $\mu = (\mu_1, \mu_2, \ldots, \mu_k)$ of positive integers such that $\mu_1 \geq \mu_2 \geq \cdots \geq \mu_k$ or $\mu = (0)$. We introduce an order in $\mathbb{Y}$ by setting $\mu = (\mu_1, \mu_2, \ldots, \mu_k) \preceq \nu = (\nu_1, \nu_2, \ldots, \nu_h)$ when $h \geq k$ and $\mu_j \leq \nu_j$ for $j = 1, 2, \ldots, k$. Clearly $\mathbb{Y}$ is infinite but it is locally finite: for any $\sigma, \pi \in \mathbb{Y}$ the set $\{\nu \in \mathbb{Y} : \sigma \preceq \nu \preceq \pi\}$ is finite. Show that $\mathbb{Y}$ is a modular lattice with zero element. It is called the *Young lattice*. More on this lattice can be found in Section 10.8.

8.2 Spherical posets and regular semi-lattices

In this section we present two important classes of posets.

Definition 8.2.1 (Delsarte [52]) Let X be a meet semi-lattice. We say that X is a *regular semi-lattice* if it has a zero element, it is ranked and the rank function $r : X \to \{0, 1, \ldots, N\}$ satisfies the following conditions

- (i) if $0 \leq r \leq N$, $y \in X_N, z \in X_r$ and $z \preceq y$ then the number of $u \in X_s$ such that $z \preceq u \preceq y$ is equal to a parameter $\mu(r, s)$ depending only on r and s;

- (ii) if $0 \leq r \leq N$ and $u \in X_s$ then the number of $z \in X_r$ such that $z \preceq u$ is a parameter $\nu(r, s)$ depending only on r and s;

- (iii) if $a \in X_r, y \in X_N$ and $a \wedge y \in X_j$ then the number of pairs $(b, z) \in X_s \times X_N$ such that $b \preceq z, b \preceq y$ and $a \preceq z$ is a parameter $\pi(j, r, s)$ depending only on j, r, s.

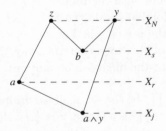

Figure 8.6.

The numbers μ, ν, π are called the *parameters* of the regular semi-lattice.

Clearly, $\mu(r, s)$ and $\nu(r, s)$ are defined and positive whenever $0 \leq r \leq s \leq N$, while π is only defined when there exists a pair (a, y) with $a \in X_r$, $y \in X_N$ and $a \wedge y \in X_j$. Also note that $\mu(r, r) = \nu(r, r) = 1$. We set $\nu(r, s) = \mu(r, s) = 0$ if $r > s$.

Example 8.2.2 The tree $\mathbb{T}_{q,n}$ (see Example 8.1.9) is a regular semi-lattice with $N = n$,

$$\mu(r, s) = \nu(r, s) = \begin{cases} 1 & \text{if } r \leq s \\ 0 & \text{otherwise} \end{cases}$$

and

$$\pi(j, r, s) = \begin{cases} q^{n-r} & \text{if } s \leq j \leq r \\ q^{n-s} & \text{if } r = j < s \\ 0 & \text{otherwise.} \end{cases}$$

Lemma 8.2.3 *Suppose that any $z \in X$ may be written in the form $z = x \wedge y$ with $x, y \in X_N$. Then in Definition 8.2.1 (iii) implies (ii) and we have $\nu(s, j) = \pi(j, N, s)$.*

Proof Take $u \in X_j$. By the assumptions, there exist $a, y \in X_N$ such that $u = a \wedge y$. Observe that in (iii) of Definition 8.2.1 if $a \in X_N$ we necessarily have $z = a$ and therefore $\pi(j, N, s) = |\{b \in X_s : b \preceq a \text{ and } b \preceq y\}|$ which coincides with $|\{b \in X_s : b \preceq a \wedge y\}| = \nu(s, j)$. $\square$

Lemma 8.2.4 *Let X be a regular semi-lattice. Then*

(i) *If $a \in X_r$ then $|\{z \in X_N : a \preceq z\}|$ is a constant equal to $\theta(r) = \pi(r, r, 0)$.*

(ii) *If $y \in X_N, u \in X_s$ and $u \preceq y$ then $|\{z \in X_r : z \preceq y \text{ and } u \wedge z \in X_j\}|$ is a constant $\psi(j, r, s)$ depending only on j, r and s.*

Proof (i) This is a particular case of (iii) in Definition 8.2.1, namely when $a \preceq y$ and $s = 0$ so that $a \wedge y = a \in X_r$.

(ii) For a fixed $0 \leq k \leq \min\{r, s\}$, the number of pairs $(x, z) \in X_k \times X_r$ such that $x \preceq u$ and $x \preceq z \preceq y$ may be counted in two ways.

First of all, the number of $x \in X_k$ with $x \preceq u$ is $\nu(k, s)$. For any such an x, the number of $z \in X_r$ with $x \preceq z \preceq y$ is equal to $\mu(k, r)$. Therefore the number of pairs (x, z) with the above properties equals $\nu(k, s)\mu(k, r)$. On the other hand, we may first determine z: in particular the number of $z \in X_r$ with $z \wedge u \in X_j$ and $z \preceq y$ is equal to $\psi(j, r, s)$; then for any such a z there exist $\nu(k, j)$ x's with $x \preceq u \wedge z$. Therefore, if $h = \min\{r, s\}$, the number of pairs (x, z) as above is also equal to $\sum_{j=k}^{h} \nu(k, j)\psi(j, r, s)$. From the identity

$$\sum_{j=k}^{h} \nu(k, j)\psi(j, r, s) = \nu(k, s)\mu(k, r) \tag{8.8}$$

one can get an expression of $\psi(j, r, s)$ in terms of the coefficients $\nu(k, j)$ and $\mu(k, r)$: the matrix $(\nu(k, j))_{k,j=0,1,\ldots,h}$ is not singular since $\nu(k, k) = 1$ and $\nu(k, j) = 0$ if $k > j$. This shows that $\psi(j, r, s)$ only depends on j, r and s. $\square$

Let X be a poset. An *automorphism* of X is a bijection $g : X \to X$ such that $x \preceq y$ if and only if $g(x) \preceq g(y)$. The set of all automorphisms of X is clearly a group. Moreover, any automorphism also preserves $\wedge$ and $\vee$, when they exist.

Definition 8.2.5 (Stanton [210]) Let X be a ranked poset with $X = \coprod_{n=0}^{N} X_n$ the decomposition into fibers. Let G be the group of automorphisms of X. Fix $x^{(0)} \in X_N$ and set $H = \{g \in G : gx^{(0)} = x^{(0)}\}$. We say that X is *spherical* when

  (i) X is a meet semi-lattice;

  (ii) X is lower semi-modular;

  (iii) G is transitive on X_N;

  (iv) the sets $\Omega_{n,i} := \{\alpha \in X_n : \alpha \wedge x^{(0)} \in X_{n-i}\}$ for $i = 0, 1, \ldots, n$ are nonempty and they are the orbits of H on X_n.

The transitivity assumption ensures that the property of being spherical does not depend on the particular choice of $x^{(0)}$.

Example 8.2.6 Let Q_n be the Boolean lattice. Set $Q_n^m = \{A \in Q_n : |A| \leq m\}$. Then for $i = 0, 1, \ldots, [\frac{n}{2}]$, Q_n^i is a spherical poset with automorphisms group S_n (the action is as in the Johnson scheme). We leave the simple check as an exercise. Moreover, for $m > \frac{n}{2}$, Q_n^m is not spherical. Indeed if we fix $A \in Q_n$ with $|A| = m$ we have that $\Omega_{m,i} = \{B \in Q_n : |B| = m \text{ and } |A \cap B| = m-i\}$ is empty for $i > n-m$.

Example 8.2.7 The tree $\mathbb{T}_{q,n}$ (see Chapter 7 and Example 8.1.9) with group of automorphisms $G = Aut(\mathbb{T}_{q,n})$ is a spherical poset.

Example 8.2.8 Let $\Theta_{n,m+1}$ be as in Example 8.1.10. We have seen that $\Theta_{n,m+1}$ is a lower modular meet semi-lattice. In order to identify its group of automorphisms we first define an action of $G = S_{m+1} \wr S_n$ (cf. Section 5.4) on it. If $(\sigma_1, \sigma_2, \ldots, \sigma_n; \theta) \in S_{m+1} \wr S_n$ and $\phi \in \Theta_{n,m+1}$ we may define $(\sigma_1, \sigma_2, \ldots, \sigma_n; \theta)\phi$ by setting $\mathrm{dom}[(\sigma_1, \sigma_2, \ldots, \sigma_n; \theta)\phi] = \theta \mathrm{dom}(\phi)$ (the action of S_n on Q_n is as in the Boolean lattice) and $[(\sigma_1, \sigma_2, \ldots, \sigma_n; \theta)\phi](j) = \sigma_j[\phi(\theta^{-1}(j))]$, where $\sigma_j \in S_{m+1}$ acts on $\phi(\theta^{-1}(j)) \in \{0, 1, \ldots, m\}$ in the obvious way.

Exercise 8.2.9 Show that this is indeed an action of $S_{m+1} \wr S_n$ on $\Theta_{n,m+1}$ and that $S_{m+1} \wr S_n$ is the group of automorphisms of $\Theta_{n,m+1}$.

In this example, the fibers of the poset are

$$X_s = \{\phi \in \Theta_{n,m+1} : |\text{dom}(\phi)| = s\}.$$

In particular the top level is $X_n = \{\phi : \{1, 2, \ldots, n\} \to \{0, 1, \ldots, m\}\}$ and coincides with the Hamming scheme $X_{n,m+1}$: a function $\phi : \{1, 2, \ldots, n\} \to \{0, 1, \ldots, m\}$ is just an n-tuple $(\phi(1), \phi(2), \ldots, \phi(n)) \in X_{n,m+1}$. Therefore the group H that stabilizes a fixed $\phi_0 \in X_n$ is just $S_m \wr S_n$ (see Section 5.4). Using this fact it is easy to prove that $\Theta_{n,m+1}$ is a spherical poset.

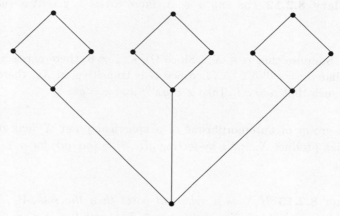

Figure 8.7.

Exercise 8.2.10 Prove that the poset in Figure 8.7 satisfies axioms (i), (ii), (iii) in Definition 8.2.5, but not (iv).

Lemma 8.2.11 *Let X be a spherical poset. Then*

(i) *for any $\alpha \in X$ there exists $x \in X_N$ such that $\alpha \preceq x$;*
(ii) *G is transitive on X_n, $n = 0, 1, \ldots, N$;*
(iii) *X has a zero element, that is $|X_0| = 1$.*

Proof (i) Suppose that $\alpha \in \Omega_{n,i} \subseteq X_n$, that is $\alpha \wedge x^{(0)} \in X_{n-i}$. As $\Omega_{N,N-n+i}$ is not empty we can take an element $x \in X_N$ such that $x \wedge x^{(0)} \in X_{n-i}$. Also consider a chain $\beta_{n-i} = x \wedge x^{(0)} \prec \beta_{n-i-1} \prec \cdots \prec \beta_1 \prec \beta_0 = x$. Then $\beta_{n-i} \wedge x^{(0)} = x \wedge x^{(0)} \in X_{n-i}$ that is $\beta_{n-i} \in \Omega_{n,i}$ and $\beta_{n-i} \preceq x$. As H is transitive on $\Omega_{n,i}$, we can take $h \in H$ such that $h\beta_{n-i} = \alpha$ and we get $\alpha = h\beta_{n-i} \preceq hx \in X_N$.

(ii) By definition 8.2.5, H is transitive on $\Omega_{n,0}$ and therefore it suffices to show that for any $\alpha \in X_n$ there exists $g \in G$ such that $g\alpha \in \Omega_{n,0}$.

But $\alpha \in \Omega_{n,i}$ for some i and, by (i), there exists $x \in X_N$ such that $\alpha \preceq x$. Then if $gx = x^{(0)}$ (such an element x exists since G is transitive on X_N) we also have $g\alpha \preceq x^{(0)}$ and therefore $g\alpha \in \Omega_{n,0}$.

(iii) Observe that $x \succeq \alpha$ for any $\alpha \in X_0$ and $x \in X_N$, since X is a meet semi-lattice and $x \wedge \alpha$ is necessarily equal to α. But $\Omega_{N,N} \neq \emptyset$ and if $x \in \Omega_{N,N}$ then $x \wedge x^{(0)} \in X_0$ and the uniqueness of $x \wedge x^{(0)}$ force $|X_0| = 1$. $\qquad\square$

Corollary 8.2.12 *For any $\alpha \in X$ there exists $x, y \in X_N$ such that $\alpha = x \wedge y$.*

Proof Suppose that $\alpha \in X_s$. Since $\Omega_{N,N-s} \neq \emptyset$, there exists $z \in X_N$ such that $\beta := x^{(0)} \wedge z \in X_s$. Since G is transitive on X_s, there exists $g \in G$ such that $\alpha = g\beta$. Take $x = gx^{(0)}$ and $y = gz$. $\qquad\square$

The group of automorphisms of a spherical poset X acts on each cartesian product $X_n \times X_k$ by setting $g(\alpha, \beta) = (g\alpha, g\beta)$ for $\alpha \in X_n$ and $\beta \in X_k$.

Lemma 8.2.13 *If X is a spherical poset then the sets $A(r, N, s) = \{(\alpha, x) : \alpha \in X_r, x \in X_N$ and $\alpha \wedge x \in X_s\}$ with $0 \leq s \leq r \leq N$ are nonempty and they constitute the orbits of G on $X_r \times X_N$.*

Proof The set $A(r, N, s)$ is not empty since it contains the (nonempty) set $\{(\alpha, x^{(0)}) : \alpha \in \Omega_{r,r-s}\}$. Moreover, if $(\alpha, x) \in A(r, N, s)$ and $g \in G$ then clearly $g(\alpha, x)$ is still in $A(r, N, s)$. It remains to show that the action of G on $A(r, N, s)$ is transitive. Suppose that $(\alpha, x), (\alpha', x') \in A(r, N, s)$. Then we can first take $g, g' \in G$ such that $gx = x^{(0)} = g'x'$. Then $g\alpha, g'\alpha \in \Omega_{r,r-s}$ and the transitivity of H on $\Omega_{r,r-s}$ yields an $h \in H$ such that $hg'\alpha' = g\alpha$. Therefore

$$g^{-1}hg'(\alpha', x') = g^{-1}h(g'\alpha', x^{(0)}) = g^{-1}(g\alpha, x^{(0)}) = (\alpha, x)$$

ending the proof. $\qquad\square$

Corollary 8.2.14 *If X is a spherical poset, then (G, H) is a symmetric Gelfand pair and $L(G/H)$ decomposes in $N + 1$ distinct irreducible representations.*

Proof The G-orbits on $X_N \times X_N$ are $A(N, N, s) = \{(x, y) \in X_N \times X_N : x \wedge y \in X_s\}$ for $s = 0, 1, \ldots, N$. In particular, $(x, y) \in A(N, N, s) \Leftrightarrow (y, x) \in A(N, N, s)$ and we can apply Lemma 4.3.4. $\qquad\square$

Exercise 8.2.15 Set $A(r, k, s) = \{(\alpha, \beta) \in X_r \times X_k : \alpha \wedge \beta \in X_s\}$ for $s = 0, 1, \ldots \min\{k, r\}$. Prove that $A(r, k, s)$ is nonempty but in general the group G is not transitive on $A(r, k, s)$. Use Example 8.1.10 to give a counterexample.

Now we show that any spherical poset also satisfies the axioms of a regular semi-lattice.

Theorem 8.2.16 *A spherical poset X is also a regular semi-lattice.*

Proof We have to show that (i), (ii) and (iii) of Definition 8.2.1 are satisfied. For $y \in X_N$, $z \in X_r$ with $z \preceq y$ (that is $(z, y) \in A(r, N, r)$) set $B_s(z, y) = \{u \in X_s : z \preceq u \preceq y\}$. Then $gB_s(z, y) = B_s(gz, gy)$ for any $g \in G$ and Lemma 8.2.13 applied to $A(r, N, r)$ ensures that $\mu(r, s) = |B_s(z, y)|$ does not depend on z and y. This proves (i). Now suppose that $(a, y) \in A(r, N, j)$ and set $C_s(y, a) = \{(b, z) \in X_s \times X_N : b \preceq z, b \preceq y \text{ and } a \preceq z\}$. Then again G is transitive on $A(r, N, j)$ (by Lemma 8.2.13) and therefore $|C_s(y, a)| = |gC_s(y, a)| = |C_s(gy, ga)|$ depends only on j, r, s and coincides with $\pi(j, r, s)$ in (iii).

Finally, (ii) is a consequence of (iii): we can apply Lemma 8.2.3 and Corollary 8.2.12. $\qquad\square$

Till now we have not used the axiom of lower semi-modularity in Definition 8.2.5. This will now be used to endow X_N with a metric space structure.

Lemma 8.2.17 *Let X be a ranked lower modular meet semi-lattice. If $w, w' \in X$ and w covers w', then, for any $x \in X$, one has $r(x \wedge w') \geq r(x \wedge w) - 1$.*

Proof Set $k = r(w) - r(x \wedge w)$ and let $w = w_0 > w_1 > \cdots > w_k = x \wedge w$ be a saturated chain between $x \wedge w$ and w. We have two possibilities for w_1. If $w_1 = w'$ then necessarily $x \wedge w' = x \wedge w$. If $w_1 \neq w'$ then w covers both w' and w_1 and, by the lower semi-modularity, there exists $v_2 \in X$ covered by both w_1 and w'. If $v_2 = w_2$ then, as before, $w' \succeq x \wedge w$ and therefore $x \wedge w' = x \wedge w$. Iterating this argument, if we always find $v_j \neq w_j$, we can construct a saturated chain $w' > v_2 > \cdots > v_{k+1}$ with v_{i+1} covered by w_i, $i = 1, 2, \ldots, k$.

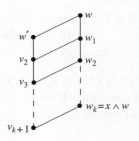

Figure 8.8.

But in this case $x \wedge w' \succeq v_{k+1}$ and $r(x \wedge w') \geq r(v_{k+1}) = r(x \wedge w) - 1$.

□

Proposition 8.2.18 *Suppose that X is a ranked lower semi-modular meet semi-lattice. Then $\delta(x, y) = N - r(x \wedge y)$ is a metric distance on X_N.*

Proof It is clear that $\delta(x, y) = 0$ if and only if $x = y$ and $\delta(x, y) = \delta(y, x)$, for all $x, y \in X_N$. Thus we only need to prove the triangle inequality.

For $x, y, z \in X_N$ set $k = \delta(x, y)$, $j = \delta(y, z)$ and let $y > y_1 > \cdots > y_j = y \wedge z$ be a saturated chain between y and $y \wedge z$.

An iterated application of Lemma 8.2.17 yields

$$r(y_j \wedge x) \geq r(y_{j-1} \wedge x) - 1 \geq \cdots \geq r(y \wedge x) - j = N - k - j.$$

But $x \succeq y_j \wedge x$ and $z \succeq y_j \wedge x$ and therefore $r(x \wedge z) \geq N - k - j$ so that

$$\delta(x, z) = N - r(x \wedge z) \leq k + j = \delta(x, y) + \delta(y, z).$$

□

Remark 8.2.19 δ is indeed a distance on X_n for all $n = 0, 1, \ldots, N$. But, in general, it can be used to parameterize the G-orbits on $X_n \times X_n$ only when $n = N$; see Lemma 8.2.13 and Exercise 8.2.15.

8.3 Spherical representations and spherical functions

Let X be a ranked poset with $X = \coprod_{n=0}^{N} X_n$ the decomposition into fibers. We define a linear map $d : L(X) \to L(X)$ by setting

$$(df)(x) = \sum_{\substack{y \in X: \\ x \lessdot y}} f(y) \tag{8.9}$$

for any $f \in L(X)$ and $x \in X$. In other words, $(df)(x)$ is the sum of the values of f on all points y covering x. In particular, if $x \in X_N$, we set $(df)(x) = 0$.

Using the basis of $L(X)$ consisting of the Dirac functions $\{\delta_x : x \in X\}$, then (8.9) may be expressed in the equivalent form

$$(d\delta_y) = \sum_{\substack{x \in X: \\ x \lessdot y}} \delta_x. \tag{8.10}$$

Indeed

$$(d\delta_y)(x) = \sum_{\substack{z \in X: \\ x \lessdot z}} \delta_y(z) = \begin{cases} 1 & \text{if } y \gtrdot x \\ 0 & \text{otherwise.} \end{cases}$$

In other words, the matrix $(J(x,y))_{x,y \in X}$ representing the linear operator d with respect to the basis $\{\delta_x : x \in X\}$ is given by $J(x,y) = 1$ if $x \lessdot y$ and $J(x,y) = 0$ otherwise.

The adjoint of d is the operator $d^* : L(X) \to L(X)$

$$(d^*f)(y) = \sum_{\substack{x \in X: \\ x \lessdot y}} f(x)$$

for any $f \in L(X)$ and $y \in X$. Equivalently

$$(d^*\delta_x) = \sum_{\substack{y \in X: \\ x \lessdot y}} \delta_y.$$

Clearly, in the decomposition $L(X) = \oplus_{n=0}^{N} L(X_n)$, we have that $d(L(X_n)) \subseteq L(X_{n-1})$ and $d^*(L(X_{n-1})) \subseteq L(X_n)$ for all $n = 1, 2, \ldots, N$.

Suppose now that X is a spherical poset and denote by $F_{n,i}$, for $i = 0, 1, \ldots, n$, the characteristic function of the subset $\Omega_{n,i} \subseteq X_n$ (cf. Definition 8.2.5). Since the $\Omega_{n,i}$'s are the nonempty orbits of H on X_n, it follows that the set $\{F_{n,i} : i = 0, 1, \ldots, n\}$ is a basis for the space of H-invariant functions in $L(X_n)$.

We now study the action of d on the $F_{n,i}$'s. Fix $\beta \in \Omega_{n,i}$ and set

$$\begin{aligned} a_{n,i} &= |\{\gamma \in \Omega_{n+1,i} : \gamma \succ \beta\}|, \\ b_{n,i} &= |\{\gamma \in \Omega_{n+1,i+1} : \gamma \succ \beta\}|, \end{aligned} \tag{8.11}$$

for $n = 0, 1, \ldots, N-1$ and $i = 0, 1, \ldots, n$. Clearly, as the group H is transitive on each $\Omega_{n,i}$, the numbers $a_{n,i}$ and $b_{n,i}$ do not depend on the particular choice of β.

Lemma 8.3.1 *For $n \geq 1$ we have*

$$dF_{n,i} = a_{n-1,i}F_{n-1,i} + b_{n-1,i-1}F_{n-1,i-1}.$$

In particular, $dF_{n,n} = b_{n-1,n-1}F_{n-1,n-1}$ and $dF_{n,0} = a_{n-1,0}F_{n-1,0}$.

Proof We have

$$dF_{n,i} = d \sum_{\alpha \in \Omega_{n,i}} \delta_\alpha = \sum_{\alpha \in \Omega_{n,i}} \sum_{\substack{\beta \in X_{n-1}: \\ \alpha > \beta}} \delta_\beta$$

$$= \sum_{\beta \in X_{n-1}} |\{\alpha \in \Omega_{n,i} : \alpha > \beta\}| \delta_\beta.$$

But from Lemma 8.2.17 we know that if $\alpha \in \Omega_{n,i}$ (that is $r(\alpha \wedge x^{(0)}) = n - i$) and $\alpha > \beta$, then $r(\beta \wedge x^{(0)}) \geq n - i - 1$. On the other hand, $\beta \wedge x^{(0)} \preceq \alpha \wedge x^{(0)}$ and therefore either $\beta \in \Omega_{n-1,i}$ (and $|\{\alpha \in \Omega_{n,i} : \alpha > \beta\}| = a_{n-1,i}$) or $\beta \in \Omega_{n-1,i-1}$ (and $|\{\alpha \in \Omega_{n,i} : \alpha > \beta\}| = b_{n-1,i-1}$). $\square$

Set $W_n = L(X_n) \cap \ker d$ for $n = 1, 2, \ldots, N$.

Lemma 8.3.2 *For $n = 1, 2, \ldots, N$ and $i = 0, 1, \ldots, n - 1$ one has $b_{n-1,i} \neq 0$; moreover W_n contains a unique (up to a scalar multiple) H-invariant function which is given by*

$$\phi_n = \sum_{i=0}^{n} c_i F_{n,i}$$

where $c_0 = 1$ and

$$c_i = (-1)^i \prod_{j=0}^{i-1} \frac{a_{n-1,j}}{b_{n-1,j}}, \qquad i = 1, 2, \ldots, n. \tag{8.12}$$

Proof Suppose $\alpha \in \Omega_{n,i+1}$. Then $\alpha \wedge x^{(0)} \in X_{n-i-1}$ and there exists $\beta \in \Omega_{n-1,i}$ covered by α. This shows that $b_{n-1,i} \neq 0$.

Any H-invariant function in $L(X_n)$ is clearly of the form $\phi = \sum_{i=0}^{n} c_i F_{n,i}$ with $c_i \in \mathbb{C}$. Applying the preceding lemma we have that

$$d \sum_{i=0}^{n} c_i F_{n,i} = \sum_{i=0}^{n-1} [c_i a_{n-1,i} + c_{i+1} b_{n-1,i}] F_{n-1,i}$$

and therefore $d\phi = 0$ if and only if

$$a_{n-1,i} c_i + b_{n-1,i} c_{i+1} = 0 \qquad i = 0, 1, \ldots, n - 1.$$

The solution of these recurrence relation with $c_0 = 1$ is given by (8.12).

□

Before stating the main theorem of this section we need some more definitions. For $0 \leq n \leq N$, $0 \leq j \leq N$ and $0 \leq i \leq n$ we set

$$C_{i,j}^n = |\{\beta \in \Omega_{n,i} : \beta \preceq x\}| \tag{8.13}$$

where $x \in \Omega_{N,j}$. Clearly, $C_{i,j}^n$ does not depend on x. Moreover, since $\beta \preceq x$ implies that $x^{(0)} \wedge \beta \preceq x^{(0)} \wedge x$ we conclude that if $n-i > N-j$ then $C_{i,j}^n = 0$. In particular, for $j = N$ we have $C_{i,N}^n = \delta_{i,n} \nu(n, N)$, where $\nu(n, N) = |\{\beta \in X_n : \beta \preceq x^{(0)}\}|$ as in Definition 8.2.1. Analogously, for $j = 0$ we have $C_{i,0}^n = \delta_{i,0} \nu(n, N)$. Note also that $\nu(n, N)$ is strictly positive.

For $0 \leq n \leq N$, $\alpha \in X_n$, $x \in X_N$ with $\alpha \preceq x$ denote by

$$m_n = \text{ the number of saturated chains from } \alpha \text{ to } x. \tag{8.14}$$

The number m_n does not depend on the particular choice of α and x: this is a consequence of Lemma 8.2.13 (note that $\{(\alpha, x) \in X_n \times X_N : \alpha \preceq x\}$ is a G-orbit). Moreover it is easy to check that if $\alpha \in X_n$ then

$$(d^*)^{N-n} \delta_\alpha = m_n \sum_{\substack{x \in X_N: \\ \alpha \preceq x}} \delta_x. \tag{8.15}$$

Indeed,

$$(d^*)^{N-n} \delta_\alpha = (d^*)^{N-n-1} \sum_{\substack{\alpha_1 \in X_{n+1}: \\ \alpha_1 \succ \alpha}} \delta_{\alpha_1} = \cdots =$$

$$= \sum_{\substack{\alpha_1 \in X_{n+1}: \\ \alpha_1 \succ \alpha}} \sum_{\substack{\alpha_2 \in X_{n+2}: \\ \alpha_2 \succ \alpha_1}} \cdots \sum_{\substack{\alpha_{N-n} \in X_N: \\ \alpha_{N-n} \succ \alpha_{N-n-1}}} \delta_{\alpha_{N-n}}$$

and therefore the number of times that δ_x, $x \in X_N$, appears in this sum is equal to the number of saturated chains from α to x.

The following lemma is an easy consequence of the above combinatorial definitions.

Lemma 8.3.3 *If $x \in \Omega_{N,j}$ then*

$$\left[(d^*)^{N-m} F_{n,i}\right](x) = m_n C_{i,j}^n.$$

We have already seen that (G, H) is a Gelfand pair and that $L(X_N)$ decomposes into $N + 1$ distinct irreducible representations. We are now in a position to describe the spherical representations and the spherical functions.

Theorem 8.3.4 *Set $W_n = L(X_n) \cap \ker d$ and $V_n = (d^*)^{N-n} W_n$ for $n = 1, 2, \ldots, N$ and denote by V_0 the space of constant functions on X_N. Then*

(i) *$L(X_N) = \oplus_{n=0}^N V_n$ is the decomposition of $L(X_N)$ into spherical representations;*

(ii) *taking $c_0, c_1, \ldots, c_n$ as in Lemma 8.3.2, then the spherical function corresponding to V_n has the expression:*

$$\Phi_n(x) = \sum_{i=0}^n c_i \frac{C_{i,k}^n}{C_{0,0}^n}$$

for $x \in \Omega_{N,k}$, and $k = 0, 1, \ldots, N$.

Proof Since d and d^* commute with the action of G, then any V_n is G-invariant. By Lemma 8.3.2, the unique (up to a constant) H-invariant function in V_n is $(d^*)^{N-n} \phi_n$. By Lemma 8.3.3, $(d^*)^{N-n} \phi_n = m_n C_{0,0}^n \Phi_n$, where Φ_n is as in the statement of the theorem. In particular, V_n is nontrivial (as $\Phi_n \neq 0$ and $C_{0,0}^n = \nu(n, N) > 0$) and contains a unique (up to a constant) H-invariant function. Therefore each V_n is irreducible and Φ_n (which is suitably normalized: $\Phi_n(x^{(0)}) = 1$) is its spherical function.

It remains to show that $V_0, V_1, \ldots, V_N$ are distinct or, equivalently, that their sum is all of $L(X_N)$. First note that

$$L(X_n) = W_n \oplus d^* L(X_{n-1}). \tag{8.16}$$

Indeed $W_n = L(X_n) \cap \ker d$ and $d^* : L(X_{n-1}) \to L(X_n)$ is the adjoint of $d : L(X_n) \to L(X_{n-1})$ and therefore $f_1 \in W_n$ if and only if $0 = \langle df_1, f_2 \rangle_{L(X_{n-1})} = \langle f_1, d^* f_2 \rangle_{L(X_n)}$ for all $f_2 \in L(X_n)$, that is if and only if f_1 is orthogonal to $d^* L(X_{n-1})$.

Iterating (8.16) one gets that

$$L(X_N) = V_N \oplus d^* L(X_{N-1}) = \cdots = V_N \oplus V_{N-1} \oplus \cdots \oplus V_0.$$

We know that this decomposition is an orthogonal direct sum as the V_n are irreducible (and $L(X_N)$ is multiplicity free). Since their sum is all of $L(X)$ they must be also distinct. $\qquad\square$

Remark 8.3.5 Let $X = \coprod_{n=0}^{N} X_n$ be again a spherical poset. For $x \in X_n$ set $\alpha_n = |\{y \in X_{n+1} : y > x\}|$, $\beta_n = |\{u \in X_{n-1} : x > u\}|$ and for $n \leq N - 1$,

$$D_n \delta_x = \sum \delta_z,$$

where the sum is over all $z \in X_n$, $z \neq x$, such that there exists $y \in X_{n+1}$ that covers both x and z, and finally, for $n \geq 1$,

$$\Delta_n \delta_x = \sum_{\substack{z \in X_n: \\ x \wedge z \in X_{n-1}}} \delta_z.$$

In general we have

$$dd^* \delta_x = \alpha_n \delta_x + D_n \delta_x$$

and

$$d^* d \delta_x = \beta_n \delta_x + \Delta_n \delta_x$$

but D_n is not necessarily equal to Δ_n, as it is shown in the exercise below.

Therefore the commutation relations proved in the case of the Johnson scheme (see Lemma 6.1.4) in general do not hold and only the weaker theory of this section can be developed.

In the following exercise this point is investigated further.

Exercise 8.3.6 (1) Show that on the tree $\mathbb{T}_{q,n}$ we have $\beta_m = 1, \alpha_m = q$ and $D_m \equiv 0$.

(2) Show that, denoting by $X_0, X_1, \ldots, X_n$ the levels of the tree, we have

(a) for $1 \leq p \leq r \leq n$ and $f \in L(X_k)$

$$d^p (d^*)^r f = q^p (d^*)^{r-p} f;$$

(b) for $0 \leq k \leq m \leq n$ and $f \in L(X_k)$

$$\|(d^*)^{m-k} f\|_{L(X_m)}^2 = q^{m-k} \|f\|_{L(X_k)}^2;$$

(c) for $0 \leq k \leq t \leq n$ and $f \in L(X_k) \cap \ker d$

$$\Delta_t (d^*)^{t-k} f = \begin{cases} -f & \text{if } t = k \\ (q-1)f & \text{if } t > k. \end{cases}$$

In particular, the spherical representation W_k in Theorem 7.4.1 coincides with $(d^*)^{n-k} (L(X_k) \cap \ker d)$.

(3) Show that the key ingredient in the proof of the commutation relations of Lemma 6.1.4 is the fact that the cube Q_n is both lower and upper semi-modular (and therefore $D_m = \Delta_m$).

In the case of the tree we have the commutation relations ((a) of the previous exercise) because $D_m \equiv 0$. In the case of the cube the commutations relations come from the fact that $\Delta_m = D_m$.

In the following exercise we show that both $0 \neq D_m$ and $D_m \neq \Delta_m$ may hold true.

Exercise 8.3.7 Let $\Theta_{n,m+1}$ be the poset in Example 8.1.10. Denote by $X_0, X_1, \ldots, X_n$ its levels; in particular X_n is the Hamming scheme. Show that $D_k \neq 0$ and $D_k \neq \Delta_k$. Also show that the spherical representations V_k in Theorem 8.3.4 coincides with the representation W_k in the Hamming scheme, but $L(X_k) \cap \ker d$ is, in general, not irreducible and $d^* : L(X_k) \cap \ker d \to V_k$ is not injective.

Exercise 8.3.8 Use Theorem 8.3.4 to compute the spherical functions of the ultrametric Space (cf. Section 7.4).

Let (X, δ) be a finite metric space with the distance δ taking only integer values (that is $\delta(x, y) \in \mathbb{N}$ for all $x, y \in X$). Then we can define a graph structure on X by taking as edge set $E = \{\{x, y\} : \delta(x, y) = 1\}$. We can ask when δ coincides with the distance induced by the graph structure. The following lemma answers this question.

Lemma 8.3.9 δ *is the distance induced by the graph structure if and only if the following condition is satisfied: for any pair* $x, y \in X$ *with* $\delta(x, y) \geq 2$ *there exists* $z \in X$ *such that* $\delta(x, z) = 1$ *and* $\delta(z, y) = \delta(x, y) - 1$.

Proof The "only if" part is obvious. For the "if" part, it suffices to show that when the above condition is satisfied then $\delta(x, y)$ equals the length of the shortest path from x to y. A repeated application of the condition yields the existence of a path $x_0 = x, x_1, \ldots, x_k = y$ with $k = \delta(x, y)$. On the other hand, if $y_0 = x, y_1, \ldots, y_h = y$ is a path from x to y then, by the triangle inequality, we have

$$\delta(x, y) \leq \delta(y_0, y_1) + \delta(y_1, y_2) + \cdots + \delta(y_{h-1}, y_h) = h.$$

$\square$

For instance, if δ is an ultrametric (see Section 7.3) then it cannot be induced by a graph structure.

Now let X be a spherical poset. On the top level we can introduce a graph structure by taking as edge set $E = \{\{x, y\} : \delta(x, y) = 1\}$ where δ is the distance given by $\delta(x, y) = N - r(x \wedge y)$. It is natural to ask if δ is induced by the graph structure; if this is the case, then by Lemma 8.2.13 (X_N, E) is distance-transitive. To answer this question, consider the graph structure on the whole X with edge set $\{\{x, y\} : x$ covers $y\}$. Then we have

Proposition 8.3.10 *The distance δ on X_N is induced by the graph structure (X_N, E) if and only if $X_{N-1} \coprod X_N$ is connected as a (bipartite) subgraph of X.*

Proof Suppose that δ is induced by the graph structure (X_N, E). From the preceding lemma, we have that if $x, y \in X_N$ and $r(x \wedge y) = N - k$ then there exists $z \in X_N$ such that $r(x \wedge z) = N - 1$ and $r(z \wedge y) = N - k + 1$. Iterating this fact we can produce a path in $X_{N-1} \coprod X_N$ connecting x and y:

$$x > \alpha_0 < x_1 > \alpha_1 < x_2 > \cdots < x_{k-1} > \alpha_{k-1} < y$$

$x, x_1, \ldots, x_{k-1}, y \in X_N$ and $\alpha_0, \alpha_1, \ldots, \alpha_{k-1} \in X_{N-1}$. Thus $X_{N-1} \coprod X_N$ is connected.

The proof of the reverse implication is left to the reader (hint: by Lemma 8.2.13 it suffices to show that for any $k = 0, 1, \ldots, N$ there exist $x, y, z \in X_N$ with $\delta(x, y) = k$, $\delta(x, z) = 1$ and $\delta(z, y) = k - 1$). $\qquad\square$

8.4 Spherical functions via Moebius inversion

We now present an alternative approach, due to Delsarte [52], to the computation of the spherical functions associated with a regular poset.

Let $X = \coprod_{n=0}^{N} X_n$ be a spherical poset and let m_n be as in (8.14). We set

$$C_n = \frac{1}{(m_n)^2} (d^*)^{N-n} d^{N-n} \colon L(X_N) \to L(X_N).$$

In other words, if $x \in X_N$ then

$$C_n \delta_x = \sum_{\substack{\alpha \in X_n: \\ \alpha \prec x}} \sum_{\substack{y \in X_N: \\ y \succ \alpha}} \delta_y \tag{8.17}$$

(compare with (8.15)).

Define the linear operators $\Delta_j : L(X_N) \to L(X_N)$, $j = 0, 1, \ldots, N$ by setting, as in (5.1),

$$(\Delta_j f)(x) = \sum_{\substack{y \in X_N: \\ \delta(x,y)=j}} f(y), \qquad \text{for all } f \in L(X_N), x \in X_N.$$

Let $\mathcal{A}$ be the algebra generated by $\Delta_0, \Delta_1, \ldots, \Delta_N$ (the Bose–Mesner algebra of X_N). With $(\nu(r,s))_{r,s=0}^N$ as in Definition 8.2.1 we have:

Lemma 8.4.1 *The operators* $C_0, C_1, \ldots, C_N$ *generate* $\mathcal{A}$ *and we have:*

$$C_t = \sum_{k=t}^N \nu(t,k) \Delta_{N-k}$$

for all $t = 0, 1, \ldots, N$.

Proof From (8.17) we get that if $x, y \in X_N$ then $(C_t \delta_x)(y)$ is equal to the number of $\alpha \in X_t$ such that $\alpha \preceq x$ and $\alpha \preceq y$. If $\delta(x,y) = N - k$, that is $x \wedge y \in X_k$, then this number is equal to $\nu(t,k)$. ☐

With π, as in Definition 8.2.1 and denoting by $(\nu'(i,k))_{i,k=0}^N$ the inverse of the matrix $(\nu(r,s))_{r,s=0}^N$ (see the proof of Lemma 8.2.4) we have:

Lemma 8.4.2

$$C_r C_s = \sum_{t=0}^r \mu(t,r) \left[\sum_{j=0}^t \nu'(j,t)\pi(j,r,s) \right] C_t$$

for all $r, s = 0, 1, \ldots, N$.

Proof First of all we prove that

$$C_r C_s = \sum_{k=0}^N \left[\sum_{j=0}^{\min\{r,k\}} \psi(j,r,k)\pi(j,r,s) \right] \Delta_{N-k} \qquad (8.18)$$

where ψ is as in Lemma 8.2.4. Indeed, if $x, y \in X_N$ and $k = r(x \wedge y)$, then the coefficient of Δ_{N-k} in $C_r C_s$ is equal to $[C_r C_s \delta_y](x)$, which, in turn, equals the number of triples $(\alpha, \beta, z) \in X_r \times X_s \times X_N$ such that $\alpha \preceq x \wedge z$ and $\beta \preceq y \wedge z$.

For any $j = 0, 1, \ldots, \min\{r, k\}$ we count the number of triples above that satisfy $j = r(\alpha \wedge y)$: we can choose α in $\psi(j,r,k)$ different ways

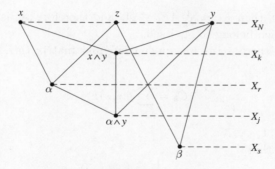

Figure 8.9.

(by Lemma 8.2.4, with y, u and z replaced by $x, x \wedge y$ and α, respectively) and then we can choose the pair (β, z) in $\pi(j, r, s)$ different ways. Summing over j, we get (8.18).

From (8.8) it follows that

$$\psi(j, r, k) = \sum_{t=j}^{\min\{r,k\}} \nu(t, k)\nu'(j, t)\mu(t, r) \qquad (8.19)$$

(recall that the matrices $(\nu(t, k))_{t,k=0}^{N}$ and $(\mu(t, r))_{t,r=0}^{N}$ are upper-triangular with unit diagonal and therefore so is $(\nu'(j, t))_{j,t=0}^{N}$).

From (8.18) and (8.19) and Lemma 8.4.1 the statement follows. $\qquad \square$

Remark 8.4.3 In the statement of previous lemma, the first sum, namely $\sum_{t=0}^{r}$, may be replaced by the sum $\sum_{t=0}^{\min\{r,s\}}$ because $\mathcal{A}$ is commutative so that $C_r C_s = C_s C_r$.

For $r = 0, 1, \dots, N$ let $\mathcal{A}_r$ be the subspace of $\mathcal{A}$ spanned by the operators $C_0, C_1, \dots, C_r$. Also set $\mathcal{A}_{-1} = \{0\}$.

Lemma 8.4.4 *The vector space $\mathcal{A}_r$ is an ideal of $\mathcal{A}$, that is, for $A \in \mathcal{A}$ and $B \in \mathcal{A}_r$ one has $AB \in \mathcal{A}_r$.*

Proof This is an immediate consequence of Lemma 8.4.2: $C_r C_s$ is a linear combination of $C_0, C_1, \dots, C_r$. $\qquad \square$

Now let $E_0, E_1, \dots, E_N$ be the projections onto the spherical representations in $L(X_N)$ (as in Theorem 5.1.6). We can suppose that these operators are numbered in such a way that $\mathcal{A}_r$ coincides with the subspace spanned by $E_0, E_1, \dots, E_r$ (indeed, if $A \in \mathcal{A}_r$ and $A = \sum_{k=0}^{N} \lambda_k E_k$

with $\lambda_k \neq 0$, then $E_k = \frac{1}{\lambda_k} A E_k \in \mathcal{A}_r$ and therefore $\mathcal{A}_r$ is spanned by the E_k's which belong to $\mathcal{A}_r$ itself).

Therefore there exists an upper triangular matrix $(\rho(i,s))_{i,s=0}^N$ such that

$$C_s = \sum_{i=0}^{s} \rho(i,s) E_i \tag{8.20}$$

for all $s = 0, 1, \ldots, N$.

Note also that

$$C_s = \frac{1}{(m_s)^2} (d^*)^{N-s} d^{N-s}$$

is a positive operator, that is

$$\langle C_s f, f \rangle_{L(X_N)} = \frac{1}{(m_s)^2} \langle d^{N-s} f, d^{N-s} f \rangle_{L(X_s)} \geq 0$$

for all $f \in L(X_N)$.

Thus, the eigenvalues of the C_s's, namely the $\rho(i,s)$, $0 \leq i \leq s \leq N$ are all nonnegative. Moreover, since $\mathcal{A}_s$ is spanned by both $\{C_0, C_1, \ldots, C_s\}$ and $\{E_0, E_1, \ldots, E_s\}$, necessarily $\rho(s,s) > 0$.

Lemma 8.4.5 *For $0 \leq r \leq s \leq N$, the coefficients $\rho(r,s)$ are given by the formula*

$$\rho(r,s) = \sum_{j=0}^{r} \nu'(j,r) \pi(j,r,s).$$

Proof From Lemma 8.4.2 we get (for $r \leq s$)

$$C_r C_s = \left[\sum_{j=0}^{r} \nu'(j,r) \pi(j,r,s) \right] C_r + A \tag{8.21}$$

where $A \in \mathcal{A}_{r-1}$. On the other hand, from (8.20) we get

$$C_r C_s = \left(\sum_{i=0}^{r} \rho(i,r) E_i \right) \left(\sum_{h=0}^{s} \rho(h,s) E_h \right)$$

(as $E_i E_h = E_i \delta_{i,h}$) $$= \sum_{i=0}^{r} \rho(i,r) \rho(i,s) E_i$$

$$= \rho(r,s) C_r + \sum_{i=0}^{r-1} \rho(i,r) \left[\rho(i,s) - \rho(r,s) \right] E_i$$

that is

$$C_r C_s = \rho(r,s) C_r + B \tag{8.22}$$

where, again, $B \in \mathcal{A}_{r-1}$. Comparing (8.21) and (8.22) the lemma follows (in particular, $A = B$).   □

Now let $(p_j(\lambda_i))_{i,j=0}^{N}$ be as in Theorem 5.1.6, that is

$$\Delta_j = \sum_{i=0}^{N} p_j(\lambda_i) E_i. \tag{8.23}$$

In other words, the $p_j(\lambda_i)$'s are the eigenvalues of Δ_j. We recall (see Lemma 5.1.8) that the spherical function ϕ_i is equal to

$$\phi_i(x) = \frac{1}{k_j} p_j(\lambda_i)$$

for $\delta(x, x^{(0)}) = j$, where $k_j = |\{x \in X : \delta(x, x^{(0)}) = j\}|$ (see also Lemma 5.1.3).

We are now in a position to give Delsarte's formula for the eigenvalues $p_j(\lambda_i)$ (and therefore for the spherical functions).

Theorem 8.4.6 *The eigenvalues $p_j(\lambda_r)$'s are given by the formula*

$$p_j(\lambda_r) = \sum_{s=\max\{N-j,r\}}^{N} \nu'(N-j,s)\rho(r,s) \tag{8.24}$$

for $r, j = 0, 1, \ldots, N$.

Proof From Lemma 8.4.1 and (8.23) we get

$$C_s = \sum_{r=0}^{N} \left[\sum_{k=s}^{N} \nu(s,k) p_{N-k}(\lambda_r) \right] E_r.$$

Then, from (8.20) we get the systems of linear equations

$$\sum_{k=s}^{N} \nu(s,k) p_{N-k}(\lambda_r) = \rho(r,s), \qquad r, s = 0, 1, \ldots, N.$$

Solving these systems we get (8.24).   □

Corollary 8.4.7 *The eigenvalues $p_j(\lambda_i)$ are integer numbers.*

Proof We first recall that if an eigenvalue λ_0 of an $n \times n$ matrix A with integer entries is a rational number, then it is necessarily an integer number. Indeed, λ_0 is a root of the characteristic polynomial of A which is of the form $p_A(\lambda) = \lambda^n + a_1\lambda^{n-1} + \cdots + a_{n-1}\lambda + a_n$ were the a_i's are integers. If $\lambda_0 = h/k$ with $h, k \in \mathbb{Z}$ and relatively prime, then one has $h^n + a_1 h^{n-1}k + \cdots a_{n-1}hk^{n-1} + a_n k^n = 0$. Therefore k divides h and the claim follows.

Recall that $p_j(\lambda_i)$ is a eigenvalue of Δ_j which is represented by an integer matrix (namely by the matrix $(\delta_j(x,y))_{x,y \in X_N}$ with $\delta_j(x,y) = 1$ if and only if $\delta(x, y) = j$, and 0 otherwise). By the above considerations, we are thus only left to show that $p_j(\lambda_i) \in \mathbb{Q}$. But this follows from the fact that all the coefficients in (8.24) are in fact rational numbers (for $\rho(r, s)$ use Lemma 8.4.5). □

We observe that Lemma 8.4.5 and Theorem 8.4.6 reduce the computation of the eigenvalues $p_j(\lambda_i)$'s to the computation of the inverse of the matrix $(\nu(r, s))_{r,s,=0}^{N}$.

We illustrate a general technique to perform this last computation which is called the *Moebius inversion*. First we give some definitions.

For the moment, let X be a finite poset. For $x, y \in X$ we set

$$\zeta(x, y) = \begin{cases} 1 & \text{if } x \preceq y \\ 0 & \text{otherwise.} \end{cases}$$

ζ is called the *zeta function* of X. The *Moebius function* M is defined inductively by setting

$$M(x, x) = 1$$

for all $x \in X$ and

$$M(x, y) = - \sum_{\substack{z \in X: \\ x \preceq z \prec y}} M(x, z)$$

for all $x, y \in X$ with $x \prec y$; finally $M(x, y) = 0$ otherwise (that is when $x \succ y$ or x and y are not comparable).

Proposition 8.4.8 (Moebius inversion formula) *Let X be a finite poset. Then we have*

$$\sum_{z \in X} M(x, z)\zeta(z, y) = \delta_x(y) = \sum_{z \in X} \zeta(x, z)M(z, y)$$

for all $x, y \in X$.

Proof We recall that $M(x, z) = 0$ whenever $x \npreceq z$. We first establish the first equality. If $x = y$ the left hand side is

$$\sum_{z \in X} M(x, z)\zeta(z, x) = \sum_{\substack{z \in X: \\ z \preceq x}} M(x, z) = M(x, x)$$

$$+ \sum_{\substack{z \in X: \\ z \prec x}} M(x, z) = M(x, x) = 1.$$

Similarly, if $x \npreceq y$ then the left hand side is

$$\sum_{z \in X} M(x, z)\zeta(z, y) = \sum_{\substack{z \in X: \\ z \preceq y}} M(x, z) = 0$$

as there is no $z \in X$ with $x \preceq z \preceq y$.

Finally, if $x \prec y$, the first equality reduces to the definition of the Moebius function:

$$\sum_{z \in X} M(x, z)\zeta(z, y) = M(x, y) + \sum_{x \preceq z \prec y} M(x, z) = M(x, y) - M(x, y) = 0.$$

Noting that the first equality is nothing but the fact that the matrix $(M(x, y))_{x,y \in X}$ is the inverse of $(\zeta(x, y))_{x,y \in X}$, the second equality follows. $\square$

Exercise 8.4.9 Let X be a finite poset. The *incidence algebra* $\mathcal{I}(X)$ of X is the set of all matrices $(a(x, y))_{x,y \in X}$ such that $a(x, y) = 0$ whenever $x \npreceq y$.

(1) Show that $\mathcal{I}(X)$ is an algebra under matrix multiplication (which, in this setting, is called *convolution*).

(2) Show that a matrix $a \in \mathcal{I}(X)$ is invertible if and only if $a(x, x) \neq 0$ for all $x \in X$.

We now show that the inverse matrix ν' of ν may be expressed in terms of the Moebius function M.

Proposition 8.4.10 *Let* $X = \coprod_{n=0}^{N} X_n$ *be a regular semi-lattice. For* $y \in X_t$ *we have*

$$\nu'(s, t) = \sum_{z \in X_s} M(z, y).$$

In particular, the right hand side depends only on s and t.

Proof First of all note that if $z \in X_s$ then, by definition of ν,

$$\nu(r,s) = \sum_{x \in X_r} \zeta(x,z). \tag{8.25}$$

Fix $y \in X_t$ and let $x \in X_r$ with $r \le t$. If we write Moebius inversion formula in the form

$$\sum_{s=r}^{t} \sum_{z \in X_s} \zeta(x,z) M(z,y) = \delta_x(y)$$

then, summing over $x \in X_r$ we get

$$\delta_r(t) = \sum_{x \in X_r} \delta_x(y) = \sum_{x \in X_r} \sum_{s=r}^{t} \sum_{z \in X_s} \zeta(x,z) M(z,y)$$

$$\text{(by (8.25))} \quad = \sum_{s=r}^{t} \nu(r,s) \sum_{z \in X_s} M(z,y)$$

and therefore $\nu'(s,t) = \sum_{z \in X_s} M(z,y)$. $\qquad\square$

Suppose now that, for $0 \le r \le s \le N$, $M(x,y)$ is constant on each pair $(x,y) \in X_r \times X_s$ with $x \preceq y$, say $M(x,y) = m(r,s)$. For instance, this is the case when $|\{z \in X_t : x \preceq z \preceq y\}|$ is a constant depending only on r,t,s. Then the formula of Proposition 8.4.10 may be written in the form

$$\nu'(s,t) = m(s,t)\nu(s,t). \tag{8.26}$$

We end this section with a discussion on the Moebius inversion formula of the Boolean lattice. We also derive the classical inclusion–exclusion principle (it will be used in Theorem 11.7.6; see also Proposition 11.7.3 for a direct and elementary proof).

Example 8.4.11 Let Q_n be the Boolean lattice. Then

$$M(A,B) = \begin{cases} (-1)^{|B \setminus A|} & \text{if } A \subseteq B \\ 0 & \text{otherwise.} \end{cases}$$

This is just a consequence of the following elementary binomial identity: suppose that $|A| = k$, $|C| = h$ and $A \subseteq C$. Then, with M as above, we have:

$$\sum_{B \in Q_n} M(A,B)\zeta(B,C) = \sum_{\substack{B \in Q_n: \\ A \subseteq B \subseteq C}} (-1)^{|B \setminus A|} = \sum_{r=k}^{h} (-1)^{r-k} \binom{h-k}{r-k}$$

$$= \sum_{r=0}^{h-k} (-1)^r \binom{h-k}{r}$$

$$= \begin{cases} 1 & \text{if } k = h \\ 0 & \text{if } k < h. \end{cases}$$

Proposition 8.4.12 (Boolean Moebius inversion formula) *Let* $f, g \in L(Q_n)$. *Then*

$$g(A) = \sum_{\substack{B \in Q_n: \\ A \subseteq B}} f(B)$$

for all $A \in Q_n$ *if and only if*

$$f(A) = \sum_{\substack{B \in Q_n: \\ A \subseteq B}} (-1)^{|B \setminus A|} g(B)$$

for all $A \in Q_n$.

Proof This is just a consequence of the fact that M is the inverse of ζ: the first equation may be written in the form $g(A) = \sum_{B \in Q_n} \zeta(A,B) f(B)$ and the second one $f(A) = \sum_{B \in Q_n} M(A,B) g(B)$. $\square$

For instance, let $A_1, A_2, \ldots, A_n$ be n subsets of a finite set A (they are not necessarily distinct). For $T \in Q_n$ set

$$f(T) = \begin{cases} \left| \bigcap_{k \in T} A_k \setminus \bigcup_{h \in T^c} A_h \right| & \text{if } T \neq \emptyset \\ \left| A \setminus \bigcup_{k=1}^{n} A_k \right| \equiv |A_1^c \cap A_2^c \cap \cdots \cap A_n^c| & \text{otherwise.} \end{cases}$$

In other words $f(T)$ is the cardinality of the set of all $x \in A$ which belong to all A_k's with $k \in T$ and to none A_h with $h \notin T$.

Also set

$$g(T) = \left| \bigcap_{k \in T} A_k \right|.$$

Then we have

$$g(T) = \sum_{\substack{R \in Q_n: \\ T \subseteq R}} f(R) \tag{8.27}$$

for all $T \in Q_n$. Indeed, $\bigcap_{k \in T} A_k = \coprod_{\substack{R \in Q_n: \\ R \supseteq T}} \left[\bigcap_{k \in R} A_k \setminus \bigcup_{h \in R^c} A_h \right]$
because setting $R = R(x) = \{k \in \{1, 2, \ldots, n\} : x \in A_k\}$, for $x \in A$,
then $x \in \left[\bigcap_{k \in R} A_k \setminus \bigcup_{h \in R^c} A_h \right]$ and $x \notin \left[\bigcap_{k \in R'} A_k \setminus \bigcup_{h \in (R')^c} A_h \right]$ for
any $R' \neq R$. Therefore the Boolean Moebius inversion applied to (8.27)
yields

$$f(T) = \sum_{\substack{R \in Q_n: \\ T \subseteq R}} (-1)^{|R \setminus T|} g(R).$$

Taking $T = \emptyset$ we immediately obtain:

Corollary 8.4.13 (Principle of inclusion–exclusion)   *Let A_1,
$A_2, \ldots, A_n$ be n subsets of a finite set A. Then*

$$|A_1^c \cap A_2^c \cap \cdots \cap A_n^c| = \sum_{\ell=0}^{n} (-1)^\ell s_\ell \tag{8.28}$$

where $s_0 = |A|$ and $s_\ell = \sum_{\substack{T \in Q_n: \\ |T| = \ell}} |\bigcap_{h \in T} A_h|$, for all $\ell = 1, 2, \ldots, n$.

Since $|A_1 \cup A_2 \cup \cdots \cup A_n| = |A| - |A_1^c \cap A_2^c \cap \cdots \cap A_n^c|$ and $|A| = s_0$,
(8.28) may be written in the form

$$|A_1 \cup A_2 \cup \cdots \cup A_n| = \sum_{\ell=1}^{n} (-1)^{\ell-1} s_\ell.$$

For $n = 2$ this is simply $|A_1 \cup A_2| = |A_1| + |A_2| - |A_1 \cap A_2|$ and, for
$n = 3$

$$|A_1 \cup A_2 \cup A_3| = |A_1| + |A_2| + |A_3| - |A_1 \cap A_2| - |A_1 \cap A_3|$$
$$- |A_2 \cap A_3| + |A_1 \cap A_2 \cap A_3|.$$

Exercise 8.4.14 Let $X = \mathbb{T}_{q,n}$.

(1) Show that the Moebius function of X has the form

$$M(x, y) = \begin{cases} 1 & \text{if } x = y \\ -1 & \text{if } x \lessdot y \\ 0 & \text{otherwise} \end{cases}$$

and therefore

$$\nu'(s, t) = \begin{cases} 1 & \text{if } s = t \\ -1 & \text{if } s = t - 1 \\ 0 & \text{otherwise.} \end{cases}$$

(2) Show that $\rho(r, s) = \begin{cases} q^{n-s} & \text{if } s \geq r \\ 0 & \text{otherwise} \end{cases}$ in two ways: first using Lemma 8.4.5 and Example 8.2.2, and then by checking that $C_s = q^{n-s}(E_0 + E_1 + \cdots + E_s)$ (use Exercise 8.3.6).

(3) Use Theorem 8.4.6 to determine the spherical functions of the finite ultrametric space.

Exercise 8.4.15 Show that the Moebius function of $\Theta_{n,m+1}$ (see Example 8.1.10) is

$$M(\phi, \psi) = (-1)^{|\mathrm{dom}(\psi) \backslash \mathrm{dom}(\phi)|}$$

whenever $\phi \preceq \psi$. In other words, it coincides with the Moebius function of the Boolean lattice Q_n.

8.5 q-binomial coefficients and the subspaces of a finite vector space

Definition 8.5.1 (q-binomial coefficients or Gaussian numbers)
Let n be a nonnegative integer and q be a real number different from 1. We set

$$[n]_q = 1 + q + q^2 + \cdots + q^{n-1} \equiv \frac{q^n - 1}{q - 1}$$

and

$$(n)_q = [1]_q \cdot [2]_q \cdots [n]_q.$$

Moreover, for $k \in \mathbb{Z}$, $q, n \in \mathbb{R}$ and $q \neq 1$, we set (*q-binomial coefficient*)

$$\binom{n}{k}_q = \begin{cases} \prod_{i=0}^{k-1} \frac{q^{n-i}-1}{q^{k-i}-1} & \text{if } k \geq 0 \\ 0 & \text{if } k < 0. \end{cases}$$

It is easy to check, that for n a nonnegative integer, we have:

$$\binom{n}{k}_q = \frac{(n)_q}{(n-k)_q (k)_q}$$

when $0 \leq k \leq n$ and $\binom{n}{k}_q = 0$ when $k > n$.

The ordinary binomial coefficients are limiting cases of their q-analogs: if n is an integer we have

$$[n]_q \to n \text{ and } (n)_q \to n! \text{ as } q \to 1$$

and if n is real and k a nonnegative integer then

$$\binom{n}{k}_q \to \binom{n}{k} \equiv \frac{n(n-1)\cdots(n-k+1)}{k!} \text{ as } q \to 1.$$

We present now some of the basic properties of the q-binomial coefficients.

Proposition 8.5.2

(i) *For $n, k \in \mathbb{Z}$ and $n \geq 0$*

$$\binom{n}{k}_q = \binom{n}{n-k}_q \quad (symmetry);$$

(ii) *for $n \in \mathbb{R}$ and $m, k \in \mathbb{Z}$*

$$\binom{n}{m}_q \cdot \binom{m}{k}_q = \binom{n}{k}_q \cdot \binom{n-k}{m-k}_q \quad (trinomial\ reversion);$$

(iii) *for $n \in \mathbb{R}$ and $k \in \mathbb{Z}$*

$$\binom{n}{k}_q = \binom{n-1}{k}_q + q^{n-k}\binom{n-1}{k-1}_q = \binom{n-1}{k-1}_q + q^k \binom{n-1}{k}_q$$

   (*addition formula*);

(iv) *for all $x \in \mathbb{C}$ and n positive integer*

$$(1+x)(1+xq)\cdots(1+xq^{n-1}) = \sum_{j=0}^{n} q^{\binom{j}{2}} x^j \binom{n}{j}_q$$

   (*q-binomial theorem*);

(v)

$$\binom{n}{k}_q = (-1)^k q^{nk - \binom{k}{2}} \binom{k-n-1}{k}_q;$$

(vi) *for* $\ell \geq 0$ *and integer*

$$\sum_{k=-m}^{\ell-m} (-1)^k \binom{\ell}{m+k}_q \binom{s+k}{n}_q q^{\binom{k}{2}+k(m-n)}$$

$$= (-1)^{\ell+m} q^{(\ell-m)(\ell-2n+m-1)/2} \binom{s-m}{n-\ell}_q;$$

(vii)

$$\sum_{k=-m}^{\ell-m} (-1)^k \binom{\ell}{m+k}_q \binom{t-k}{n}_q q^{\binom{k}{2}+km}$$

$$= (-1)^m q^{-\binom{m}{2}+\ell(t+m-n)} \binom{t-\ell+m}{n-\ell}_q.$$

Proof (i), (ii), (iii) and (v) are easy calculations. (iv) may be proved by induction on n using the first identity in (iii). (vi) may be proved by induction on ℓ, using (iii); it is the q-analog of (6.28). (vii) may be easily deduced by applying (v) to the coefficients $\binom{s+k}{n}_q$ and $\binom{s-m}{n-\ell}_q$ in (vi) and setting $t = n - s - 1$; it is another q-analog of (6.28). $\qquad\square$

Let q be a prime power, say $q = p^m$ with p a prime number and $m \in \mathbb{N}$. Then there exists a unique (finite) field $\mathbb{F}_q$ with q elements. If $m = 1$, that is $q = p$ is a prime, this is just $\mathbb{Z}_p$, the field of integers mod p. If $m > 1$, $\mathbb{F}_q$ is usually regarded as an algebraic extension of $\mathbb{Z}_p$. We refer to the monographs by M. Artin [6], Herstein [120], Hungerford [123] and Lang [151]. We shall not use special properties of $\mathbb{F}_q$ (besides the fact that it is a field with q elements). In the following theorem we present a general counting principle for the subspaces of $\mathbb{F}_q^n$, the n-dimensional vector space over $\mathbb{F}_q$. It is taken from [74].

Theorem 8.5.3 *Let* V, W *be subspaces of* $\mathbb{F}_q^n$ *with* $\dim V = a$, $\dim W = b$ *and* $\dim(V \cap W) = x$. *Suppose that* c, y *are integers satisfying* $x \leq y \leq a$ *and* $b \leq c \leq n - a + y$. *Then the number of subspaces* $Z \subseteq \mathbb{F}_q^n$ *such that* $Z \supseteq W$, $\dim Z = c$ *and* $\dim(Z \cap V) = y$ *is*

$$\binom{a-x}{y-x}_q \binom{n-b-a+x}{c-b-y+x}_q q^{(a-y)(c-b-y+x)}.$$

Proof Any Z as in the statement may be written in the form $Z = W \oplus Z_1 \oplus Z_2$ (as usual $\oplus$ denotes the direct sum) where Z_1 is an $(y - x)$-dimensional subspace of V with $Z_1 \cap (V \cap W) = \{0\}$ and Z_2 is a $(c - b - y + x)$-dimensional subspace of $\mathbb{F}_q^n$ with $Z_2 \cap (V + W) = \{0\}$. In other words, Z may be obtained by choosing an ordered set $\mathcal{A}$ of $y - x$ linearly independent vectors in $V \setminus W \equiv V \setminus (V \cap W)$ ($\setminus$ denotes a set difference) and then an ordered set $\mathcal{B}$ of $c - b - y + x$ linearly independent vectors in $\mathbb{F}_q^n \setminus (V + W)$. The choice of $\mathcal{A}$ may be done in

$$c_1 = (q^a - q^x)(q^a - q^{x+1}) \cdots (q^a - q^{y-1})$$

different ways: the first vector v_1 may be chosen in $q^a - q^x$ = the cardinality of $V \setminus W$ ways; when V_1 has been chosen, the second vector v_2 must be taken from $V \setminus (W \oplus \langle v_1 \rangle)$, whose cardinality is $q^a - q^{x+1}$ and so on. Analogously, the set $\mathcal{B}$ may be chosen in

$$c_2 = (q^n - q^{a+b-x})(q^n - q^{a+b-x+1}) \cdots (q^n - q^{a+c-y-1})$$

different ways (note that by the *Grassmann identity* we have that $\dim(V + W) = \dim V + \dim W - \dim(V \cap W) = a + b - x$). But any Z corresponds to several different choices of $\mathcal{A}$ and $\mathcal{B}$: any Z contains

$$c_3 = (q^y - q^x)(q^y - q^{x+1}) \cdots (q^y - q^{y-1})$$

distinct $\mathcal{A}$'s and

$$c_4 = (q^c - q^{y+b-x})(q^c - q^{y+b-x+1}) \cdots (q^c - q^{c-1})$$

distinct $\mathcal{B}$'s (note that $\dim[Z \cap (V + W)] = \dim W + \dim Z_1 = b + y - x$). Therefore the number of subspaces Z is equal to

$$\frac{c_1 c_2}{c_3 c_4} = \binom{a - x}{y - x}_q \binom{n - b - a + x}{c - b - y + x}_q q^{(a-y)(c-b-y+x)}.$$

$\square$

Exercise 8.5.4 Without using the above theorem show that the number of k-dimensional subspaces in $\mathbb{F}_q^n$ is equal to $\binom{n}{k}_q$.

Corollary 8.5.5 *Let* V, W *be subspaces of* $\mathbb{F}_q^n$ *with* $\dim W = m$ *and* $\dim(V \cap W) = u$. *Then the number of subspaces* $Z \subseteq W$ *with* $\dim Z = k$ *and* $\dim(Z \cap V) = v$ *is equal to*

$$\binom{u}{v}_q \binom{m - u}{k - v}_q q^{(u-v)(k-v)}.$$

Proof This is an immediate consequence of the theorem: take $n = m$, $x = b = 0$, $a = u$, $y = v$, $c = k$ and then replace $\mathbb{F}_q^n$, V, W with W, $V \cap W$ and $\{0\}$, respectively. $\qquad\square$

Corollary 8.5.6 *If $W \subseteq V \subseteq \mathbb{F}_q^n$, $\dim W = b$, $\dim V = a$ then the number of subspaces Z with $W \subseteq Z \subseteq V$ and $\dim Z = c$ is equal to $\binom{a-b}{c-b}_q$.*

Proof In Theorem 8.5.3 take $x = b$ and $y = c$. $\qquad\square$

Exercise 8.5.7 Give a combinatorial proof of the following identities

$$\binom{n+m+1}{m+1}_q = \sum_{k=0}^{n} q^k \binom{k+m}{m}_q$$

and (q-*Vandermonde*)

$$\binom{m}{k}_q = \sum_{v=\max\{0,k-m+u\}}^{\min\{u,k\}} \binom{u}{v}_q \binom{m-u}{k-v}_q q^{(u-v)(k-v)}.$$

8.6 The q-Johnson scheme

Let q be a prime power and $n \in \mathbb{N}$. As in the previous section, we denote by $\mathbb{F}_q$ the finite field of order q, and by $\mathbb{F}_q^n$ the n-dimensional vector space over $\mathbb{F}_q$; also we denote by $GL(n, q)$ the corresponding *general linear group*. In other words, $GL(n, q)$ is the group of all invertible $n \times n$ matrices with coefficients in $\mathbb{F}_q$, equivalently, the group of all invertible linear transformations of $\mathbb{F}_q^n$. Elementary introductions to $GL(n, q)$ may be found in the monographs by Alperin and Bell [4] and Grove [109]. A more advanced treatment is in Grove's more recent book [110].

Lemma 8.6.1 *The order of $GL(n, q)$ is given by*

$$|GL(n, q)| = q^{n(n-1)/2}(q^n - 1)(q^{n-1} - 1) \cdots (q - 1).$$

Proof A matrix in $GL(n, q)$ is (uniquely) determined by its rows. The first row $a_1 \in \mathbb{F}_q^n$ can be chosen in $q^n - 1$ different ways (this corresponds to the number of nonzero vectors in $\mathbb{F}_q^n$); the second row a_2 can be chosen in $q^n - q^{n-1}$ different ways (a_2 may be any vector in $\mathbb{F}_q^n \setminus \langle a_1 \rangle$); continuing this way, the kth row a_k can be chosen in $q^n - q^{k-1}$ ways. Finally, observe that

$$(q^n - 1)(q^n - q)(q^n - q^2) \cdots (q^n - q^{n-1})$$
$$= q^{n(n-1)/2}(q^n - 1)(q^{n-1} - 1) \cdots (q - 1).$$

$\square$

Let $0 \leq m \leq n$ and denote by $\mathcal{G}_m$ the *Grassmann variety* of all m-dimensional subspaces of $\mathbb{F}_q^n$. Also set for simplicity $G = GL(n, q)$. Then G acts on $\mathcal{G}_m$ in the obvious way: if $A \in G$ and $V \in \mathcal{G}_m$, then $AV = \{Av : v \in V\}$ (note that by the nonsingularity of A the subspace AV has the same dimension of V).

Lemma 8.6.2 *The action of G on $\mathcal{G}_m$ is transitive and the stabilizer of a subspace $V_0 \in \mathcal{G}_m$ is isomorphic to the subgroup H_m consisting of all matrices of the form*

$$A = \begin{pmatrix} A_1 & A_2 \\ 0 & A_3 \end{pmatrix} \text{ such that } A_1 \in GL(m, q), A_3 \in GL(n-m, q) \quad (8.29)$$

and A_2 is an $m \times (n - m)$ matrix with coefficients in $\mathbb{F}_q$.

Proof Let $e_1, e_2, \ldots, e_n$ be the standard basis of $\mathbb{F}_q^n$ and suppose that $V_0 = \langle e_1, e_2, \ldots, e_m \rangle$. Then a matrix $A \in G$ satisfies $AV_0 = V_0$ if and only if $Ae_i \in V_0$ for all $i = 1, 2, \ldots, m$. This is clearly equivalent to A be of the form (8.29).

If V is another m-dimensional subspace and $v_1, v_2, \ldots, v_m$ is a basis of V, then denoting by $A \in G$ a matrix such that $Ae_i = v_i$ for all $i = 1, 2, \ldots, m$, then $AV_0 = V$, showing that the action is transitive.

$\square$

Let now $\mathcal{G} = \coprod_{m=0}^n \mathcal{G}_m$ the set of all subspaces of $\mathbb{F}_q^n$. It is a poset under inclusion which is also a lattice: if $V, W \in \mathcal{G}$ then their meet is their intersection $V \cap W$ and their join is their sum $V + W = \{v + w : v, \in V, w \in W\}$. The dimension $\dim V$ is a rank function so that $\mathcal{G}$ is ranked. Finally it is modular: by Corollary 8.1.6 this is equivalent to

$$\dim V + \dim W = \dim(V + W) + \dim(V \cap W)$$

for all $V, W \in \mathcal{G}$ which is nothing but the Grassmann identity.

Moreover, for the action of G in $\mathcal{G}$ the following general principle holds.

Lemma 8.6.3 *Let $V_1, V_2, W_1, W_2 \in \mathcal{G}$. Suppose that $\dim V_1 = \dim W_1$, $\dim V_2 = \dim W_2$ and $\dim(V_1 \cap V_2) = \dim(W_1 \cap W_2)$. Then there exists $A \in G$ such that*

$$AV_1 = W_1 \text{ and } AV_2 = W_2.$$

Proof Set $h = \dim V_1$, $k = \dim V_2$ and $r = \dim(V_1 \cap V_2)$. Take a basis $v_1, v_2, \ldots, v_{h+k-r}$ of $V_1 + V_2$ such that $v_1, v_2, \ldots, v_r$ is a basis of $V_1 \cap V_2$, $v_1, v_2, \ldots, v_h$ is a basis of V_1 and finally $v_1, v_2, \ldots, v_r, v_{h+1}, v_{h+2}, \ldots,$ v_{h+k-r} is a basis of V_2. Formally, replacing every v (resp. V) symbol by w (resp. W) and choosing a matrix $A \in G$ such that $Av_i = w_i$ for all $i = 1, 2, \ldots, h + k - r$, the statement follows. $\qquad\square$

Corollary 8.6.4 *Suppose that $m \leq n/2$. Then the poset $\coprod_{k=0}^{m} \mathcal{G}_k$ is spherical.*

Proof $\coprod_{k=0}^{m} \mathcal{G}_k$ is a semi-lattice and it is lower modular because the whole $\mathcal{G}$ is a modular lattice. The action of G is transitive on $\mathcal{G}_m$ and the previous lemma ensures us that if H_m is the stabilizer of a fixed $V_0 \in \mathcal{G}_m$, then for $k = 0, 1, \ldots, m$ the sets

$$\Omega_{k,j} = \{V \in \mathcal{G}_k : \dim(V \cap V_0) = k - j\}, \qquad (8.30)$$

$j = 0, 1, \ldots, k$ are the (nonempty) orbits of H_m on $\mathcal{G}_k$. $\qquad\square$

The symmetric association scheme $\mathcal{G}_m$ is usually called the *q-Johnson scheme*. On $\mathcal{G}_m$ we define the G-invariant distance $\delta(V, W) = m - \dim(V \cap W)$ that corresponds to the distance δ in Proposition 8.2.18. Let also d and d^* denote the linear operators in Section 8.3 and set $W_k = L(\mathcal{G}_k) \cap \ker d$ and $V_k = (d^*)^{m-k} W_k$.

Applying Corollary 8.2.14 and Theorem 8.3.4 we obtain:

Theorem 8.6.5

  (i) $(GL(n, q), H_m)$ *is a Gelfand pair;*

 (ii) $L(\mathcal{G}_m) = \oplus_{k=0}^{m} V_k$ *is the decomposition of $L(\mathcal{G}_m)$ into spherical representations;*

(iii) $\dim V_k = \binom{n}{k}_q - \binom{n}{k-1}_q$;

(iv) *the spherical function* $\Phi_k \in V_k$ *has the following expression:*
$\Phi_k(V_0) = 1$ *and*

$$\Phi_k(W) = \frac{1}{\binom{m}{k}_q} \sum_{i=\max\{0,h+k-m\}}^{\min\{h,k\}} \left[(-1)^i \left(\prod_{j=0}^{i-1} \frac{q^{m-k+j+1} - 1}{1 - q^{m-n+j}} \right) \right. $$
$$\left. \cdot \binom{m-h}{k-i}_q \binom{h}{i}_q q^{i(m-h-n+i-1)} \right]$$

if $W \in \Omega_{m,h}$, $h = 1, 2, \ldots, m$.

Proof (i) and (ii) are an immediate consequence of the fact that $\mathcal{G} = \coprod_{k=0}^{m} \mathcal{G}_k$ is a spherical poset (see Corollary 8.2.14 and Theorem 8.3.4). To prove (iii), first note that for $0 \leq k \leq [n/2]$ each poset $\coprod_{j=0}^{k} \mathcal{G}_j$, is spherical and therefore $L(\mathcal{G}_k) = \oplus_{j=0}^{k} (d^*)^{k-j} W_j$. In particular, $W_k = L(\mathcal{G}_k) \cap \ker d$ is irreducible, $d : L(\mathcal{G}_k) \to L(\mathcal{G}_{k-1})$ is surjective and $\dim V_k = \dim W_k = \binom{n}{k}_q - \binom{n}{k-1}_q$. We are left to prove the expression for the spherical functions in (iv). Let V_0 be the subspace stabilized by H_m. In this case the set $\Omega_{k,j}$ in the definition of spherical poset is given by (8.30)

We have just to get the expression for the coefficients $a_{k,j}$, $b_{k,j}$ and $C_{i,h}^k$ in (8.11) and (8.13). If $W \in \Omega_{k,j}$, then $a_{k,j}$ is the number of subspaces $Z \in \Omega_{k+1,j}$ such that $Z \supseteq W$. Since $\dim W = k$, $\dim(W \cap V_0) = k - j$, $\dim Z = k + 1$ and $\dim(Z \cap V_0) = k + 1 - j$, then an application of Theorem 8.5.3 with $a = m$, $b = k$, $x = k - j$, $c = k+1$ and $y = k+1-j$ yields $a_{k,j} = \frac{q^{m-k+j}-1}{q-1}$. Similarly $b_{k,j} = \frac{q^{n-k}-q^{m-k+j}}{q-1}$. Moreover, if $W \in \Omega_{m,h}$ then $C_{i,h}^k$ is the number of subspaces $Z \in \Omega_{k,i}$ such that $Z \subseteq W$ and an application of Corollary 8.5.6 with $V = V_0$, $u = m - h$ and $v = k - i$ yields

$$C_{i,h}^k = \binom{m-h}{k-i}_q \binom{h}{i}_q q^{(m-k+i-h)i}$$

and therefore the expression for the spherical functions follows from Theorem 8.3.4. $\qquad\square$

For $x \in \Omega_{n,h}$ set $\phi_k(h) = \Phi_k(x)$ (see Section 5.1).

Corollary 8.6.6

(i) *The spherical functions ϕ_k satisfy the following orthogonality relations:*

$$\sum_{h=0}^{m} \phi_k(h)\phi_t(h) \binom{m}{h}_q \binom{n-m}{h}_q q^{h^2} = \frac{\binom{n}{m}_q}{\binom{n}{k}_q - \binom{n}{k-1}_q} \delta_{k,t}.$$

(ii) *$\mathcal{G}_m$ is a distance-transitive graph with parameters*

$$b_h = \frac{1}{(q-1)^2}(q^m - q^h)(q^{n-m} - q^h)q \qquad h = 0, 1, \ldots, m-1$$

$$c_h = \frac{1}{(q-1)^2}(q^h - 1)^2 \qquad h = 1, 2, \ldots, m$$

and degree $b_0 = \frac{1}{(q-1)^2}(q^m - 1)(q^{n-m} - 1)q$.

(iii) *The spherical functions satisfy the following finite difference equation:*

$$c_h \phi_k(h-1) + (b_0 - b_h - c_h)\phi_k(h) + b_h \phi_k(h+1) = \lambda_k \phi_k(h)$$

for $h, k = 0, 1, \ldots, m$ where b_h and c_h are as above and

$$\lambda_k = \frac{1}{(q-1)^2} \cdot \left[q(1 - q^{n-m})(1 - q^m) - (1 - q^{-k})(q^{n+1} - q^k)\right].$$

Proof (i) These are the usual orthogonality relations for the spherical functions. Note that setting $x = b = 0$, $a = c = m$ and $y = m - h$ in Theorem 8.5.3 we get

$$|\Omega_{m,h}| \equiv |\{W \in \mathcal{G}_m : \delta(W, V_0) = h\}| = q^{h^2} \binom{m}{h}_q \binom{n-m}{h}_q.$$

(ii) It is easy to see that $\mathcal{G}_m$ satisfies the hypothesis in Proposition 8.3.10 and therefore that it is a distance-transitive graph (compare with Lemma 8.2.13). Now we have to compute the parameters b_h and c_h. Given $V \in \mathcal{G}_m$ with $\dim(V \cap V_0) = m - h$ then $b_h = |\{W \in \mathcal{G}_m : \dim(W \cap V) = m-1$ and $\dim(W \cap V_0) = m - h - 1\}|$. Note that by the Grassmann identity we have

$$\dim[W \cap (V \cap V_0)] = \dim W + \dim(V \cap V_0) - \dim[W + (V \cap V_0)] \geq m - h - 1.$$

Therefore a W as above may be obtained by selecting an $(m - 1)$-dimensional subspace $W_1 \subseteq V$ such that $\dim[W_1 \cap (V \cap V_0)] = m - h - 1$, then a vector $w \notin V + V_0$ and setting $W = W_1 + \langle w \rangle$. From Corollary 8.5.5 with $u = m - h$, $v = m - h - 1$ and $k = m - 1$ the number of possible W_1 is $\binom{m-h}{m-h-1}_q \binom{h}{h}_q q^h = \frac{q^{m-h}-1}{q-1} q^h$. The number of possible w is equal to $|\mathbb{F}_q^n \setminus (V + V_0)| = q^n - q^{m+h}$. But every W corresponds to the choice of $|W \setminus (W \cap V)| = q^m - q^{m-1}$ different w's. Therefore

$$b_h = \frac{(q^{m-h}-1)q^h(q^n - q^{m+h})}{(q-1)(q^m - q^{m-1})} = \frac{(q^m - q^h)(q^{n-m} - q^h)q}{(q-1)^2}.$$

Similarly, one can prove that

$$c_h = \frac{(q^h - 1)^2}{(q-1)^2}.$$

The finite difference equation is clearly that in Corollary 5.1.9 and the eigenvalues may be computed by the formula $\lambda_k = b_0 \phi_k(1)$ (Lemma 5.1.8). $\qquad\square$

Exercise 8.6.7 Show that for $n/2 \leq m \leq n$ we have $L(\mathcal{G}_m) \cong L(\mathcal{G}_{n-m}) = \bigoplus_{k=0}^{n-m} V_k$.

Exercise 8.6.8 Prove that for $q \to 1$ all the formulae for the q-Johnson scheme become the formulae for the Johnson scheme.

In the last part of this section, we apply the Moebius inversion (see Section 8.4) to the q-Johnson scheme. Denote by $P_n(q)$ the poset of all subspaces of $\mathbb{F}_q^n$. Then we have:

Proposition 8.6.9

(i) *The Moebius function in $P_n(q)$ is given by*

$$M(U, V) = (-1)^k q^{\binom{k}{2}}$$

where $U \subseteq V$ and $k = \dim V - \dim U$.

(ii) *The coefficients μ, ν, π in 8.2.1 are given by*

$$\mu(r, s) = \binom{m-r}{m-s}_q, \qquad \nu(r, s) = \binom{s}{r}_q$$

for $r \leq s$ and

$$\pi(j, r, s)$$
$$= \sum_{i=\max\{0, r+s-m\}}^{\min\{j, s\}} q^{(j-i)(s-i)} \binom{j}{i}_q \binom{m-j}{s-i}_q \binom{n-r-s+i}{n-m}_q$$

for $r - j \leq m - s$ and $j \leq r$.

(iii) *The coefficients $\nu'(r,s)$ and $\rho(r,s)$ in Section 8.4 are given by*

$$\nu'(r,s) = (-1)^{s-r} q^{\binom{s-r}{2}} \binom{s}{r}_q$$

and

$$\rho(r,s) = q^{r(m-s)} \binom{n-s-r}{n-m-r}_q \binom{m-r}{s-r}_q.$$

Proof (i) If M is as above, then given $V \subseteq W \subseteq \mathbb{F}_q^n$ and setting $M(k) = (-1)^k q^{\binom{k}{2}}$, $m = \dim W$, $u = \dim V$ we have,

$$\sum_{Z \in P_n(q)} M(V,Z) \zeta(Z,W) = \sum_{\substack{Z \in P_n(q): \\ V \subseteq Z \subseteq W}} M(\dim Z - \dim W)$$

$$\text{(by Corollary 8.5.6)} = \sum_{k=u}^{m} (-1)^{k-u} q^{\binom{k-u}{2}} \binom{m-u}{k-u}_q$$

$$= \sum_{r=0}^{m-u} (-1)^r q^{\binom{r}{2}} \binom{m-u}{r}_q$$

$$= \begin{cases} 1 & \text{if } m = u \\ 0 & \text{if } m > u \end{cases}$$

where the last equality follows from (iv) in Proposition 8.5.2 taking $x = -1$ (alternatively, it can be deduced from (vi) in the same proposition).

(ii) The formula for $\nu(r,s)$ is just the combinatorial interpretation of the q-binomial coefficients and the formula for μ is just Corollary 8.5.6. Now we prove the formula for $\pi(j,r,s)$. Take $j \leq r \leq m \leq n$, $s \leq m$ and two subspaces V, W of $\mathbb{F}_q^n$ with $\dim W = m$, $\dim V = r$ and $\dim(V \cap W) = j$. Then from Corollary 8.5.5 we get that the number of subspaces $Z \subseteq \mathbb{F}_q^n$ with $\dim Z = s$, $Z \subseteq W$ and $\dim(Z \cap V) = i$ is equal to

$$\binom{j}{i}_q \binom{m-j}{s-i}_q q^{(j-i)(s-i)}. \tag{8.31}$$

If we fix Z as above, from the Grassmann identity we have $\dim(Z+V) = r + s - i$ and therefore from Corollary 8.5.6 we get that the number of subspaces U with $\dim U = m$, $Z \subseteq U$ and $V \subseteq U$ is equal to

$$\binom{n-r-s+i}{n-m}_q. \tag{8.32}$$

From (8.31) and (8.32) the formula for $\pi(j,r,s)$ follows.

(iii) The formula for $\nu'(r,s)$ follows from (i) and (ii) and (8.26). To get the formula for $\rho(r,s)$, we must apply Lemma 8.4.5. For $0 \le r \le s \le m$ we have

$$
\begin{aligned}
\rho(r,s) &= \sum_{j=0}^{r} \nu'(j,r)\pi(j,r,s) \\
&= \sum_{j=0}^{r} (-1)^{r-j} q^{\binom{r-j}{2}} \binom{r}{j}_q \\
&\qquad \cdot \sum_{i=\max\{0,r+s-m\}}^{j} q^{(j-i)(s-i)} \binom{j}{i}_q \binom{m-j}{s-i}_q \binom{n-r-s+i}{n-m}_q \\
&=_{*} \sum_{i=\max\{0,r+s-m\}}^{r} q^{(r-i)(s-i)} \binom{r}{i}_q \binom{n-r-s+i}{n-m}_q \\
&\qquad \cdot \sum_{j=i}^{r} (-1)^{r-j} \binom{r-i}{r-j}_q \binom{m-j}{s-i}_q q^{(r-j)(i-s)+\binom{r-j}{2}} \\
&=_{**} \sum_{i=\max\{0,r+s-m\}}^{r} \binom{r}{r-i}_q \binom{n-s-(r-i)}{n-m}_q \\
&\qquad \cdot q^{\binom{r-i}{2}} (-1)^{r-i} \binom{m-r}{s-r}_q \\
&=_{***} q^{r(m-s)} \binom{n-s-r}{n-m-r}_q \binom{m-r}{s-r}_q
\end{aligned}
$$

where *), **) and ***) follow from (ii), (vi) and (vii) of Proposition 8.5.2, respectively. $\qquad\square$

Now we can apply Delsarte's formula (8.24).

Corollary 8.6.10 *For the q-Johnson scheme $\mathcal{G}_m$ the eigenvalues $p_k(\lambda_r)$ are given by*

$$
p_k(\lambda_r) = \sum_{j=0}^{\min\{k,m-r\}} (-1)^{k-j} q^{\binom{k-j}{2}+rj} \\
\cdot \binom{m-j}{m-k}_q \binom{m-r}{j}_q \binom{n-m+j-r}{j}_q, \tag{8.33}
$$

for $k,r = 0,1,\ldots,m$.

Proof From (8.24) and Proposition 8.6.9 we immediately get an expression for $p_k(\lambda_r)$ which modulo the change of variable $j = m - s$ becomes (8.33). $\qquad\square$

8.7 A q-analog of the Hamming scheme

In this section we discuss a q-analog of the Hamming scheme. If q is a prime power and n, m are positive integers, then $G^q_{n,m}$ is the set of all triples (α, β, γ) where $\alpha \in GL(n, q)$, $\beta \in GL(m, q)$ and $\gamma \in Hom(\mathbb{F}^n_q, \mathbb{F}^m_q)$.

Lemma 8.7.1 $G^q_{n,m}$ *is a group with the compososition law:*

$$(\alpha, \beta, \gamma) \cdot (\alpha_1, \beta_1, \gamma_1) = (\alpha\alpha_1, \beta\beta_1, \beta\gamma_1\alpha^{-1} + \gamma).$$

Proof $(\iota_n, \iota_m, 0)$ where ι_n and ι_m are the identity operators in $\mathbb{F}^n_q$ and $\mathbb{F}^m_q$ respectively, is the identity element. The inverse of (α, β, γ) is

$$(\alpha, \beta, \gamma)^{-1} = (\alpha^{-1}, \beta^{-1}, -\beta^{-1}\gamma\alpha).$$

The associativity is trivial and it is left as an exercise. $\qquad\square$

Exercise 8.7.2 Show that $G^q_{n,m}$ is isomorphic to the group of all matrices of the form $\begin{pmatrix} A & 0 \\ C & B \end{pmatrix}$ where $A \in GL(n, q)$, $B \in GL(m, q)$ and $C \in M_{m,n}(\mathbb{F}_q)$.

Remark 8.7.3 $G^q_{n,m}$ contains two subgroups, namely

$$GL(n, q) \times GL(m, q) \equiv \{(\alpha, \beta, \gamma) \in G^q_{n,m} : \gamma = 0\}$$

and $S^q_{n,m} \equiv \{(\alpha, \beta, \gamma) \in G^q_{n,m} : \alpha = \iota_n, \beta = \iota_m\}$. Note that $S^q_{n,m}$ is commutative: it coincides with $Hom(\mathbb{F}^n_q, \mathbb{F}^m_q)$ with the usual addition. Moreover $S^q_{n,m}$ is normal in $G^q_{n,m}$:

$$(\alpha, \beta, \gamma) \cdot (\iota_n, \iota_m, \gamma_1) \cdot (\alpha, \beta, \gamma)^{-1}$$
$$= (\alpha, \beta, \beta\gamma_1\alpha^{-1}) \cdot (\alpha^{-1}, \beta^{-1}, -\beta^{-1}\gamma\alpha)$$
$$= (\iota_n, \iota_m, \beta\gamma_1\alpha^{-1} - \gamma) \in S^q_{n,m}$$

for any choice of $(\alpha, \beta, \gamma) \in G^q_{n,m}$ and $(\iota_n, \iota_m, \gamma_1) \in S^q_{n,m}$. Finally, $G^q_{n,m} = [GL(n, q) \times GL(m, q)] \cdot S^q_{n,m}$ (since $(\alpha, \beta, \gamma) = (\alpha, \beta, 0) \cdot (\iota_n, \iota_m, \beta^{-1}\gamma\alpha)$) and $[GL(n, q) \times GL(m, q)] \cap S^q_{n,m} = \{(\iota_n, \iota_m, 0)\}$. Altogether, these properties may be expressed by saying that $G^q_{n,m}$ is the *semidirect product* of $GL(n,q) \times GL(m,q)$ with the abelian group $S_{n,m}^q$. Groups with this structure have a representation theory that can be deduced by that of their constituents; we refer to [196], [198] and [220] for more details.

We now introduce a poset that generalizes the posets in Example 8.1.10. To avoid the discussion of particular cases, we assume $n \leq m$. We set

$$\Theta_{n,m}^q = \{(\phi, V) : V \text{ is a subspace of } \mathbb{F}_q^n \text{ and}$$
$$\phi : V \to \mathbb{F}_q^m \text{ is a linear map}\}.$$

For any pair (ϕ, V) as above we set $\mathrm{dom}(\phi) = V$ and we often identify the pair (ϕ, V) just with ϕ. We introduce a partial order in $\Theta_{n,m}^q$ by setting $\phi \preceq \psi$ when $\mathrm{dom}(\phi) \subseteq \mathrm{dom}(\psi)$ and $\phi(v) = \psi(v)$ for all $v \in \mathrm{dom}(\phi)$.

The group $G_{n,m}^q$ acts on $\Theta_{n,m}^q$ by the rule

$$(\alpha, \beta, \gamma)(\phi, V) = ((\beta \phi \alpha^{-1} + \gamma)|_{\alpha V}, \alpha V) \tag{8.34}$$

for all $(\alpha, \beta, \gamma) \in G_{n,m}^q$ and $(\phi, V) \in \Theta_{n,m}^q$.

Lemma 8.7.4 *The rule in (8.34) indeed defines an action.*

Proof For $(\alpha, \beta, \gamma), (\alpha_1, \beta_1, \gamma_1) \in G_{n,m}^q$ and $(\phi, V) \in \Theta_{n,m}^q$ we have

$$(\alpha, \beta, \gamma)[(\alpha_1, \beta_1, \gamma_1)(\phi, V)] = ([\beta\beta_1\phi\alpha_1^{-1}\alpha^{-1} + \beta\gamma_1\alpha^{-1} + \gamma]|_{\alpha\alpha_1 V}$$
$$\equiv [(\alpha, \beta, \gamma) \cdot (\alpha_1, \beta_1, \gamma_1)](\phi, V).$$

Moreover, $(\iota_n, \iota_m, 0) \cdot (\phi, V) = (\phi, V)$. $\qquad\square$

Lemma 8.7.5 $\Theta_{n,m}^q$ *is a graded poset with rank function* $r : \Theta_{n,m}^q \to \{0, 1, \ldots, n\}$ *given by:* $r(\phi) = \dim[\mathrm{dom}(\phi)]$, $\phi \in \Theta_{n,m}^q$. *Moreover it is a lower semi-modular meet semi-lattice.*

Proof Adapt the arguments in Example 8.1.10. $\qquad\square$

Let $\mathcal{X}_{n,m}^q$ denote the top level of $\Theta_{n,m}^q$. Then $\mathcal{X}_{n,m}^q$ consists of those ϕ's with $\mathrm{dom}(\phi) \equiv \mathbb{F}_q^n$ and it is isomorphic to $S_{n,m}^q$ via the map

$$\begin{array}{ccc} \mathcal{X}_{n,m}^q & \to & S_{n,m}^q \\ \phi & \mapsto & (\iota_n, \iota_m, \phi). \end{array} \tag{8.35}$$

Moreover, under the identification (8.35), the action of $G^q_{n,m}$ on $\mathcal{X}^q_{n,m}$ corresponds to the conjugation on $S^q_{n,m}$:

$$(\alpha, \beta, \gamma) \cdot \phi = \beta \phi \alpha^{-1} + \gamma \mapsto (\iota_n, \iota_m, \beta \phi \alpha^{-1} + \gamma)$$
$$= (\alpha, \beta, \gamma)(\iota_n, \iota_m, \phi)(\alpha, \beta, \gamma)^{-1}.$$

In any case, if we take the trivial map $0 \equiv \phi_0 \in Hom(\mathbb{F}^n_q, \mathbb{F}^m_q)$, then the stabilizer of ϕ_0 in $G^q_{n,m}$ coincides with the subgroup $GL(n, q) \times GL(m, q)$ described in Remark 8.7.3.

Proposition 8.7.6 $\Theta^q_{n,m}$ *is a spherical poset.*

Proof We have already seen that $\Theta^q_{n,m}$ is a lower modular meet semi-lattice. The transitivity of $G^q_{n,m}$ on the top level $\mathcal{X}^q_{n,m}$ is trivial: the subgroup $S^q_{n,m}$ is transitive (its action is just the translation on the vector space $\mathcal{X}^q_{n,m}$). It remains to show that the sets $\Omega_{k,j} = \{\phi \in \Theta^q_{n,m} : \dim[\mathrm{dom}(\phi)] = k \text{ and } \dim[\mathrm{dom}(\phi \wedge \phi_0)] = k - j\}$ are the nonempty orbits of the stabilizer $GL(n, q) \times GL(m, q)$ of ϕ_0. But the condition $\dim[\mathrm{dom}(\phi \wedge \phi_0)] = k - j$ is equivalent to $\dim[\ker(\phi)] = k - j$. Clearly each $\Omega_{k,j}$ is nonempty (recall that we assume $n \leq m$). Moreover, if $\phi_1, \phi_2 \in \Omega_{k,j}$ then there exist four subspaces V_1, V_2, W_1 and W_2 of $\mathbb{F}^n_q$ such that $\mathbb{F}^n_q = \ker \phi_i \oplus V_i \oplus W_i$, $\mathrm{dom}(\phi_i) = \ker \phi_i \oplus V_i$, $\dim(\ker \phi_i) = k - j$, $\dim V_i = j$, $\dim W_i = n - k$, for $i = 1, 2$. Moreover, $\phi_i|_{V_i} : V_i \to Im(\phi_i) = \{\phi_i(v) : v \in \mathbb{F}^n_q\}$ is a linear bijection. Therefore, if we choose $\alpha \in GL(n, q)$ such that $\alpha(\ker \phi_1) = \ker \phi_2$, $\alpha(V_1) = V_2$, $\alpha(W_1) = W_2$ and $\beta \in GL(m, q)$ such that $\beta \phi_1(v) = \phi_2 \alpha(v)$ for all $v \in V_1$, that is $\beta|_{Im \phi_1} = \phi_2 \alpha(\phi_1|_{V_1})^{-1}$, then we have

$$(\alpha, \beta, 0) \cdot (\phi_1, \mathrm{dom}(\phi_1)) = (\phi_2, \mathrm{dom}(\phi_2)).$$

Indeed, if $v \in \mathrm{dom}(\phi_2)$ and $v = u + w$ with $u \in \ker \phi_2$, $w \in V_2$ then $\beta \phi_1 \alpha^{-1}(v) = \beta \phi_1 \alpha^{-1}(w) = \phi_2(w) = \phi_2(v)$ where the second equality follows from the choice of β. The proof is now complete. $\square$

Before giving the expression for the spherical functions we need a couple of lemmas.

Lemma 8.7.7 *For the spherical poset $\Theta^q_{n,m}$ the coefficients $a_{k,j}$, $b_{k,j}$ (see Section 8.3) have the expression:*

$$a_{k,j} = \frac{q^{n-k} - 1}{q - 1} q^j$$

and

$$b_{k,j} = \frac{q^{n-k} - 1}{q - 1}(q^m - q^j).$$

Proof Fix $\psi \in \Omega_{k,j}$. Thus $\dim(\mathrm{dom}(\psi)) = k$ and $\dim(\ker \psi) = k - j$. Note that

$$a_{k,j} = |\{\phi \in \Omega_{k+1,j} : \phi \succ \psi\}| \tag{8.36}$$

and

$$b_{k,j} = |\{\phi \in \Omega_{k+1,j+1} : \phi \succ \psi\}|. \tag{8.37}$$

In both cases, $\mathrm{dom}(\phi) \supseteq \mathrm{dom}(\psi)$ with $\dim(\mathrm{dom}(\phi)) = k + 1$. Therefore, by Corollary 8.5.6 (with $a = n$, $b = k$ and $c = k + 1$), the number of possible $\mathrm{dom}(\phi)$'s is equal to $\binom{n-k}{1}_q = \frac{q^{n-k}-1}{q-1}$.

If we fix $w \in \mathrm{dom}(\phi)\backslash\mathrm{dom}(\psi)$, then ϕ is determined by the value $\phi(w)$ (and different values for $\phi(w)$ clearly give different ϕ's). In the first case we want $\phi \in \Omega_{k+1,j}$ and therefore $\dim(\ker \phi) = k + 1 - j$, $\ker \phi \supsetneq \ker \psi$ or equivalently $\mathrm{Im}\phi = \mathrm{Im}\psi$. In other words $\phi(w)$ must belong to $\mathrm{Im}\psi$ and thus we have $q^j = |\mathrm{Im}\psi|$ choices for $\phi(w)$ and (8.36) follows.

In the second case we want $\phi \in \Omega_{k+1,j+1}$ and therefore $\ker \phi = \ker \psi$ or equivalently $\mathrm{Im}\phi \supsetneq \mathrm{Im}\psi$. In other words $\phi(w)$ must belong to $\mathbb{F}_q^m \backslash \mathrm{Im}\psi$ and thus we have $q^m - q^j$ choices for $\phi(w)$ and (8.37) follows as well. $\square$

Lemma 8.7.8 *For the spherical poset $\Theta_{n,m}^q$ the coefficients $C_{i,j}^k$ in (8.13) have the expression:*

$$C_{i,j}^k = \binom{n-j}{k-i}_q \binom{j}{i}_q q^{(n-j-k+i)i}. \tag{8.38}$$

Proof Fix $\psi \in \Omega_{n,j}$, that is $\psi \in Hom(\mathbb{F}_q^n, \mathbb{F}_q^m)$ and $\dim(\ker \psi) = n - j$. Then

$$C_{i,j}^k = |\{\phi \in \Omega_{k,i} : \phi \preceq \psi\}|.$$

In particular, ϕ is determined by its domain. This subspace must satisfy the condition $\dim(\mathrm{dom}(\phi) \cap \ker \psi) = k - i$ (which is equivalent to $\dim(\ker \phi) = k - i$). Then, applying Corollary 8.5.5 with $m = n$, $u = n - j$ and $v = k - i$, (8.38) follows. $\square$

As usual, we introduce the differential operators d and d^* and the subspaces $V_0, V_1, \ldots, V_n$ (see Section 8.3). Moreover, we continue to denote by ϕ_0 the trivial linear map $\phi_0 : \mathbb{F}_q^n \to \mathbb{F}_q^n$, that is $\phi_0 \equiv 0$.

Theorem 8.7.9

(i) $(G_{n,m}^q, GL(n,q) \times GL(m,q))$ *is a Gelfand pair;*

(ii) $L(\mathcal{X}_{n,m}^q) = \bigoplus_{k=0}^{n} V_k$ *is the decomposition of $L(\mathcal{X}_{n,m}^q)$ into spherical representations;*

(iii) *for $0 \leq k \leq n$, the spherical function $\Phi_k \in V_k$ has the following expression: $\Phi_k(\phi_0) = 1$ and, if $\phi \in \Omega_{n,h}$, $h = 1, 2, \ldots, n$*

$$\Phi_k(\phi) = \frac{1}{\binom{n}{k}_q} \sum_{i=\max\{0,k-n+h\}}^{\min\{h,k\}} (-1)^i \binom{n-h}{k-i}_q \binom{h}{i}_q \left(\prod_{j=0}^{i-1} \frac{1}{q^{m-j}-1} \right) q^{(n-k-h+i)i}.$$

Proof This is just an application of Theorem 8.3.4.   □

Exercise 8.7.10 Prove that $\mathcal{X}_{n,m}^q$ is a distance-transitive graph.

Let Δ be the Laplace operator on $\mathcal{X}_{n,m}^q$ (see Section 5.1). Then from Theorem 8.7.9 and Theorem 5.1.6 we get immediately the spectral analysis of Δ.

Corollary 8.7.11 V_k *is an eigenspace of Δ and the corresponding eigenvalue is*

$$\frac{(q^m - 1)(q^{n-k} - 1) - q^{n-k}(q^k - 1)}{q - 1}.$$

Proof We just note that, for $\phi \in \mathcal{X}_{n,m}^q$, $|\{\psi \in \mathcal{X}_{n,m}^q : \delta(\phi,\psi) = 1\}| = \frac{(q^n-1)(q^m-1)}{q-1}$ (see exercise below).   □

Exercise 8.7.12 Prove that

$$|\{\psi \in \mathcal{X}_{n,m}^q : \delta(\phi,\phi_0) = k\}| = \binom{n}{k}_q \prod_{j=0}^{k-1} (q^m - q^j)$$

$$\equiv \binom{m}{k}_q \prod_{j=0}^{k-1} (q^n - q^j)$$

$$\equiv q^{-k} \prod_{j=0}^{k-1} \frac{(q^{m-j} - 1)(q^{n-j} - 1)}{1 - q^{-j-1}}.$$

In the following series of exercises, we present an alternative approach to the q-Hamming scheme. It is sketched in [74] and mentioned in [206, 207]. An earlier reference is [41]. The method that we use is also an application of ideas in [44]. In what follows, we drop the assumption $n \leq m$ (i.e. $n > m$ is allowed). We will use the fact that the dual of a finite abelian group A is isomorphic to A (this will be proved in Section 9.2). First of all, for an operator $\phi \in Hom(\mathbb{F}_q^n, \mathbb{F}_q^n)$ we can define the *trace* $Tr(\phi)$ as follows: choose a basis $v_1, v_2, \ldots, v_n$ for $\mathbb{F}_q^n$ and if $\phi(v_i) = \sum_{j=1}^n a_{ij} v_j$ then $Tr(\phi) = \sum_{i=1}^n a_{ii}$.

Exercise 8.7.13 Show that $Tr(\phi)$ does not depend on the chosen basis, that $Tr : Hom(\mathbb{F}_q^n, \mathbb{F}_q^n) \to \mathbb{F}_q$ is a linear operator and that $Tr(\phi\psi) = Tr(\psi\phi)$ for all $\phi, \psi \in Hom(\mathbb{F}_q^n, \mathbb{F}_q^n)$.

Here, $\mathcal{X}_{n,m}^q \cong Hom(\mathbb{F}_q^n, \mathbb{F}_q^n)$ is an abelian group under addition and therefore is isomorphic to its dual, that is $\widehat{\mathcal{X}_{n,m}^q} \cong \mathcal{X}_{n,m}^q$. We now give an explicit parameterization.

Fix a nontrivial character χ of the additive group of $\mathbb{F}_q$. For $\psi \in \mathcal{X}_{m,n}^q$ we define $\chi_\psi : \mathcal{X}_{n,m}^q \to \mathbb{T}$ by setting

$$\chi_\psi(\phi) = \chi[Tr(\psi\phi)]$$

for any $\phi \in \mathcal{X}_{n,m}^q$.

Exercise 8.7.14 Show that the map $\psi \mapsto \chi_\psi$ is an explicit isomorphism between $\mathcal{X}_{m,n}^q$ and $\widehat{\mathcal{X}_{n,m}^q}$.

We know that $(G_{n,m}^q, GL(n,q) \times GL(m,q))$ is a symmetric Gelfand pair and that $GL(n,q) \times GL(m,q)$ has $\min\{n,m\} + 1$ orbits on $\mathcal{X}_{n,m}^q$ (compare with the proof of Proposition 8.7.6). Now we give an alternative description of the spherical representations. For $\psi \in \mathcal{X}_{n,m}^q$, the rank of ψ is defined by

$$rk(\psi) := \dim[\mathrm{Im}\psi].$$

Clearly $0 \leq rk(\psi) \leq \min\{n,m\}$. Set

$$\overline{V_k} = \langle \chi_\psi : rk(\psi) = k \rangle,$$

for $k = 0, 1, \ldots, \min\{n,m\}$.

Exercise 8.7.15 Show that $L(\mathcal{X}_{n,m}^q) = \bigoplus_{k=0}^{\min\{n,m\}} \overline{V_k}$ is the decomposition of $L(\mathcal{X}_{n,m}^q)$ into irreducible representations.

Exercise 8.7.16 Show that $\overline{V_k}$ is an eigenspace of Δ and that the corresponding eigenvalue is

$$\frac{(q^m - 1)(q^{n-k} - 1) - q^{n-k}(q^k - 1)}{q - 1}.$$

Exercise 8.7.17 From the preceding exercise and from Corollary 8.7.11 deduce that $V_k \equiv \overline{V_k}$ for $k = 0, 1, \ldots, \min\{n, m\}$. From Exercise 8.7.12 deduce that

$$\dim V_k = \binom{n}{k}_q \prod_{j=0}^{k-1} (q^n - q^j).$$

8.8 The nonbinary Johnson scheme

This section is devoted to a Gelfand pair arising from posets (it corresponds to a single level of the poset $\Theta_{n,m+1}$ in Example 8.1.10) but not covered by the theory of spherical posets developed in Section 8.2. Many details will be left as exercises.

We first need some notation. Fix two integers $n > 0$ and $m \geq 0$ and, for $k = 0, 1, \ldots, n$, denote by Ω_k the family of all k-subsets of $\{1, 2, \ldots, n\}$ (see Section 6.1) and by $\Omega_{k,m+1}$ the set of all functions whose domain is an element $A \in \Omega_k$ and whose range is $\{0, 1, \ldots, m\}$. In symbols

$$\Omega_{k,m+1} = \{\theta : A \to \{0, 1, \ldots, m\}; \ A \in \Omega_k\}.$$

It is then clear that

$$\Theta_{n,m+1} = \coprod_{k=0}^{n} \Omega_{k,m+1}.$$

Moreover $\Omega_{n,m+1}$ corresponds to the Hamming scheme $X_{n,m+1}$ (see Section 5.3). It has been shown (see Example 8.2.8) that the group $S_{m+1} \wr S_n$ acts on $\Omega_{k,m+1}$. It is also easy to see that this action is transitive.

Fix $k \in \{0, 1, \ldots, n\}$ and a function $\overline{\theta} \in \Omega_{k,m+1}$: for instance, fix $\overline{A} \in \Omega_k$ and then set $\overline{\theta}(j) = 0$ for all $j \in \overline{A}$.

Exercise 8.8.1 Show that the stabilizer of $\overline{\theta}$ is isomorphic to $(S_m \wr S_k) \times (S_{m+1} \wr S_{n-k})$, where S_m is the stabilizer of 0 in S_{m+1} and $S_k \times S_{n-k}$ is the stabilizer of $\overline{A}$ in S_n.

For $h \leq k$ define a function

$$\delta : \Omega_{k,m+1} \times \Omega_{h,m+1} \to \{0, 1, \ldots, k\} \times \{0, 1, \ldots, k\}$$

by setting, for $\theta \in \Omega_{k,m+1}$ and $\psi \in \Omega_{h,m+1}$,

$$\delta(\theta, \psi) = (|\mathrm{dom}(\theta) \cap \mathrm{dom}(\psi)|, |\{j \in \mathrm{dom}(\theta) \cap \mathrm{dom}(\psi) : \theta(j) = \psi(j)\}|).$$

Exercise 8.8.2 (1) Show that the orbits of the diagonal action of $S_{n+1} \wr S_n$ on $\Omega_{k,m+1} \times \Omega_{k,m+1}$ are precisely the sets

$$\{(\theta, \psi) : \delta(\theta, \psi) = (t, l)\} \qquad \max\{0, 2k - n\} \leq t \leq k, \ 0 \leq l \leq t.$$

(2) Deduce from (1) that $(S_{m+1} \wr S_n, (S_m \wr S_k) \times (S_{m+1} \wr S_{n-k}))$ is a symmetric Gelfand pair (use the orbit criterion (Exercise 4.3.5)).

We know that $\Theta_{n,m+1}$ is a spherical poset so that the theory in Section 8.3 can be used to recover the Hamming scheme (see also Section 8.3). But, for $k < n$, $\coprod_{h=0}^{k} \Omega_{h,m+1}$ is *not* a spherical poset as the following exercise shows.

Exercise 8.8.3 Deduce from the previous exercise and Corollary 3.13.2 that the orbits of $(S_m \wr S_k) \times (S_{m+1} \wr S_{n-k})$ on $\Omega_{h,m+1}$, with $0 \leq h \leq k$, are precisely the sets

$$\{\theta \in \Omega_{h,m+1} : \delta(\theta, \overline{\theta}) = (t, l)\} \qquad \max\{0, k + h - n\} \leq t \leq h, \ 0 \leq l \leq t.$$

From this fact deduce that $\coprod_{h=0}^{k} \Omega_{h,m+1}$ is not a spherical poset.

Indeed, the decomposition into irreducibles in Theorem 8.3.4 is not valid for $L(\Omega_{k,m+1})$ (see also Exercise 8.3.7). Nevertheless, a refinement of this decomposition holds for $L(\Omega_{k,m+1})$: it was found by Dunkl [73]. We describe a slightly different approach inspired by a generalization contained in [44]. Another different approach, based on the theory of association schemes and orthogonal polynomials, is in [219].

Remark 8.8.4 The term "nonbinary Johnson scheme" was introduced in [219]. If we consider the Hamming scheme $\Omega_{n,m+1} (\equiv X_{n,m+1}$ in the notation of Section 5.3) with the Hamming distance $d(\theta, \psi) = |\{j \in \{1, 2, \ldots, n\} : \theta(j) \neq \psi(j)\}|$ and $\overline{\theta}$ is the function defined by $\overline{\theta}(j) = 0$ for all $j = 1, 2, \ldots, n$, then the sphere $\widetilde{\Omega}_{k,m} = \{\theta \in \Omega_{n,m+1} : d(\theta, \overline{\theta}) = k\}$ coincides with $\Omega_{k,m}$: with each $\theta \in \Omega_{n,m+1}$, such that $A := \{j \in \{1, 2, \ldots, n\} : \theta(j) \neq 0\}$ has cardinality k, one associates $\theta_1 : A \to \{1, 2, \ldots, m\}$. In other words, we regard $\Omega_{k,m}$ as the set of all functions $\theta : A \to \{1, 2, \ldots, m\}$, with $A \in \Omega_k$.

In particular, for $m = 1$ (the binary case: $\Omega_{n,2}$ is the hypercube Q_n) $\widetilde{\Omega}_{k,1}$ is just the usual Johnson scheme of Chapter 6. For $m = 1$ one recovers the binary case: $\Omega_{n,2}$ is the hypercube Q_n. Finally, for $m \geq 2$ (the nonbinary case: $\Omega_{n,m+1}$ is the Hamming scheme over the nonbinary alphabet $\{0, 1, \ldots, m\}$), $\widetilde{\Omega}_{k,m}$ is what may be properly called the "nonbinary Johnson scheme".

We still need further notation. Consider the operator $D : L(\Theta_{n,m+1}) \to L(\Theta_{n,m+1})$ defined by

$$(DF)(\theta) = \sum_{\substack{\psi \in \Theta_{n,m+1} \\ \theta < \psi}} F(\psi)$$

for all $F \in L(\Theta_{n,m+1})$ and $\theta \in \Theta_{n,m+1}$, and let D^* be its adjoint. The symbols d and d^* will be limited to denote the operators introduced in Section 6.1. Clearly d, d^* and D, D^* are particular cases of the operators d, d^* introduced in Section 8.3, but we need to keep them distinct.

If $A \setminus \{1, 2, \ldots, n\}$ we denote by $[m+1]^A$ (in order to simplify the standard notation $\{0, 1, \ldots, m\}^A$) the set of all functions $\theta : A \to \{0, 1, \ldots, m\}$. Then we have the decomposition

$$\Omega_{k,m+1} = \coprod_{A \in \Omega_k} [m+1]^A$$

and

$$L(\Omega_{k,m+1}) = \bigoplus_{A \in \Omega_k} L([m+1]^A). \tag{8.39}$$

Clearly, $[m+1]^A$ is isomorphic to the Hamming scheme $X_{|A|,m+1}$.

We now introduce two kinds of products that generalize the coordinatewise product used in Section 5.3. If $A \subseteq \{1, 2, \ldots, n\}$ and $f_i \in L(\{0, 1, \ldots, m\})$, $i \in A$, we define the product $\otimes_{i \in A} f_i \in L([m+1]^A)$ by setting

$$[\otimes_{i \in A} f_i](\theta) = \prod_{i \in A} f_i[\theta(i)] \tag{8.40}$$

for all $\theta \in [m+1]^A$. This is the first kind of product.

Now recall that $L(\{0, 1, \ldots, m\}) = V_0 \oplus V_1$ with V_0 the space of constant functions, V_1 the space of functions with zero mean (cf. Example 1.8.1 and Section 5.3). If $A \subseteq \{1, 2, \ldots, n\}$, $|A| = h$ and $0 \leq t \leq h$, then we denote by $\Omega_t(A)$ the family of all t-subsets of A and set $M^{h-t,t}(A) = L(\Omega_t(A))$ (this corresponds to the space $M^{h-t,t}$ of Section 6.1 associated with the t-subsets of the H-set A).

For $0 \leq \ell \leq k \leq n$ and $B \in \Omega_\ell$, $f_i \in V_1$ for all $i \in B$ and $\gamma \in M^{n-k,k-\ell}(\complement B)$ ($\complement B = \{1, 2, \ldots, n\} \setminus B$ denotes the complement of B), we define $\gamma \otimes (\otimes_{i \in B} f_i) \in L(\Omega_{k,m+1})$ by setting

$$[\gamma \otimes (\otimes_{i \in B} f_i)](\theta) = \gamma(\mathrm{dom}(\theta) \setminus B) \cdot \prod_{i \in B} f_i[\theta(i)] \tag{8.41}$$

for all $\theta \in \Omega_{k,m+1}$ such that $\text{dom}(\theta) \supseteq B$, and

$$[\gamma \otimes (\otimes_{i \in B} f_i)](\theta) = 0$$

if $\theta \in \Omega_{k,m+1}$ but $\text{dom}(\theta) \not\supseteq B$. This is the second kind of product.

Clearly, (8.41) may be expressed in terms of (8.40): if $\phi_0 \in V_0$ is the constant function $\equiv 1$ on $\{0, 1, \ldots, m\}$, then

$$\gamma \otimes (\otimes_{i \in B} f_i) = \sum_{C \in \Omega_{k-\ell}(\complement B)} \gamma(C)[(\otimes_{i \in C} \phi_0) \otimes (\otimes_{i \in B} f_i)] \qquad (8.42)$$

where $\otimes_{i \in C} \phi_0 \in L([m+1]^C)$ is nothing but the constant function $\equiv 1$. Moreover $(\otimes_{i \in C} \phi_0) \otimes (\otimes_{i \in B} f_i)$ is a product of the first kind and

$$[(\otimes_{i \in C} \phi_0) \otimes (\otimes_{i \in B} f_i)](\theta) = \prod_{i \in B} f_i[(\theta(i)]$$

for all $\theta \in [m+1]^{C \amalg B}$.

In what follows, if G is a group acting on X, $f \in L(X)$ and $g \in G$, then the G-image of f will be simply denoted by gf, in symbols: $(gf)(x) = f(g^{-1}x)$ for all $x \in X$.

For $\gamma \in M^{n-k,k-\ell}(\complement B)$ and $\pi \in S_n$, accordingly with the notation above, we denote by $\pi\gamma \in M^{n-k,k-\ell}(\complement \pi B)$ the function defined by $(\pi\gamma)(C) = \gamma(\pi^{-1} C)$ for all $C \in \Omega_{k-\ell}(\complement \pi B)$.

In the following lemma, we study the action of $S_{m+1} \wr S_n$ on the products (8.40) and (8.41).

Lemma 8.8.5 *Let $(\sigma_1, \sigma_2, \ldots, \sigma_n; \pi) \in S_{m+1} \wr S_n$. Then*

$$(\sigma_1, \sigma_2, \ldots, \sigma_n; \pi)(\otimes_{i \in B} f_i) = \otimes_{t \in \pi B} \sigma_t f_{\pi^{-1}(t)} \qquad (8.43)$$

(here $\sigma_t \in S_{m+1}$ acts on $f_{\pi^{-1}(t)} \in L(\{0, 1, \ldots, m\})$) and

$$(\sigma_1, \sigma_2, \ldots, \sigma_n; \pi)[\gamma \otimes (\otimes_{i \in B} f_i)] = (\pi\gamma) \otimes [\otimes_{t \in \pi B} \sigma_t f_{\pi^{-1}(t)}] \qquad (8.44)$$

(here $\pi \in S_n$ acts on γ as above).

Proof Using the formula $(\sigma_1, \sigma_2, \ldots, \sigma_n; \pi)^{-1} = (\sigma_{\pi(1)}^{-1}, \sigma_{\pi(2)}^{-1}, \ldots, \sigma_{\pi(n)}^{-1};$ $\pi^{-1})$ (see Lemma 5.4.1), for $\theta \in [m+1]^{\pi B}$ we have

$$[(\sigma_1, \sigma_2, \ldots, \sigma_n; \pi)(\otimes_{i \in B} f_i)](\theta) = (\otimes_{i \in B} f_i)[(\sigma_{\pi(1)}^{-1}, \sigma_{\pi(2)}^{-1}, \ldots, \sigma_{\pi(n)}^{-1}; \pi^{-1})\theta]$$

$$= \prod_{i \in B} f_i\{\sigma_{\pi(i)}^{-1}[\theta(\pi(i))]\}$$

$$(t = \pi(i)) \quad = \prod_{t \in \pi B} [\sigma_t f_{\pi^{-1}(t)}](\theta(t))$$

$$= [\otimes_{t \in \pi B} \sigma_t f_{\pi^{-1}(t)}](\theta).$$

This proves (8.43). Then (8.44) follows from (8.43) and (8.42). □

In the following lemma we investigate the action of D and D^* on the product of type (8.41).

Lemma 8.8.6 *We have*

$$D[\gamma \otimes (\otimes_{i \in B} f_i)] = (m+1)[(d\gamma) \otimes (\otimes_{i \in B} f_i)] \tag{8.45}$$

and

$$D^*[\gamma \otimes (\otimes_{i \in B} f_i)] = (d^*\gamma) \otimes (\otimes_{i \in B} f_i) \tag{8.46}$$

where d and d^* act on $M^{n-k, k-\ell}(\complement B)$.

Proof The product $[\gamma \otimes (\otimes_{i \in B} f_i)](\theta)$ is defined for those θ such that $\mathrm{dom}(\theta) \supseteq B$. Therefore, $\{D[\gamma \otimes (\otimes_{i \in B} f_i)]\}(\xi)$ is defined for those $\xi \in \Omega_{k-1,m+1}$ satisfying the condition $|B \setminus \mathrm{dom}(\xi)| \leq 1$, that is, for the ξ's for which there exists $\theta \in \Omega_{k,m+1}$ such that $\mathrm{dom}(\theta) \supseteq B$ and $\xi \preceq \theta$.

We first consider the case $|B \setminus \mathrm{dom}(\xi)| = 0$, that is $\mathrm{dom}(\xi) \supseteq B$. If this is the case, then

$$\{D[\gamma \otimes (\otimes_{i \in B} f_i)]\}(\xi) = \sum_{\substack{\theta \in \Omega_{k,m+1}: \\ \theta \succeq \xi, \ \mathrm{dom}(\theta) \supseteq B}} [\gamma \otimes (\otimes_{i \in B} f_i)](\theta)$$

$$= \sum_{\substack{\theta \in \Omega_{k,m+1}: \\ \theta \succeq \xi}} \gamma(\mathrm{dom}(\theta \setminus B)) \cdot \prod_{i \in B} f_i(\theta(i))$$

$$(\theta(u) = v) = \sum_{u \in \complement \mathrm{dom}(\xi)} \sum_{v=0}^{m} \gamma[(\mathrm{dom}(\xi) \coprod \{u\}) \setminus B] \cdot \prod_{i \in B} f_i(\xi(i))$$

$$= (m+1)(d\gamma)(\mathrm{dom}(\xi) \setminus B) \cdot \prod_{i \in B} f_i(\xi(i))$$

$$= (m+1)[(d\gamma) \otimes (\otimes_{i \in B} f_i)](\xi).$$

If $|B \setminus \text{dom}(\xi)| = 1$ and u is the unique element in $B \setminus \text{dom}(\xi)$, then

$$\{D[\gamma \otimes (\otimes_{i \in B} f_i)]\}(\xi) = \gamma[(\text{dom}(\xi) \coprod \{u\}) \setminus B] \cdot (\sum_{v=0}^{m} f_u(v))$$

$$\cdot \prod_{i \in B \setminus \{u\}} f_i(\xi(i)) = 0$$

because $f_u \in V_1$. In particular, if $\ell = k$, then $D[\gamma \otimes (\otimes_{i \in B} f_i)] = 0$.
The proof of (8.46) is similar and it is left as an exercise. $\qquad\square$

Exercise 8.8.7 Show that for $0 \le \ell \le k \le n$, $B, B' \in \Omega_\ell$, $f_i \in V_1$
for all $i \in B$, $f'_j \in V_1$ for all $j \in B'$, $\gamma \in M^{n-k,k-\ell}(\complement B)$ and $\gamma' \in M^{n-k,k-\ell}(\complement B')$ one has

$$\langle \gamma \otimes (\otimes_{i \in B} f_i), \gamma' \otimes (\otimes_{j \in B'} f'_j) \rangle_{L(\Omega_{k,m+1})} =$$
$$= \delta_{B,B'} \cdot \langle \gamma, \gamma' \rangle_{M^{n-k,k-\ell}(\complement B)} \cdot (m+1)^{k-\ell} \cdot \langle \otimes_{i \in B} f_i, \otimes_{j \in B'} f'_j \rangle_{L([m+1]^B)}.$$
$$(8.47)$$

We recall that $M^{n-h,h-\ell}(\complement B) = \oplus_{k=0}^{\min\{n-h,h-\ell\}} S^{n-\ell-k,k}$ (see Chapter 6) as $S_{n-\ell}$-representation (think of $S_{n-\ell}$ as the group of all permutations of $\complement B$).

Definition 8.8.8 Let $0 \le \ell \le h \le n$ and $A \in \Omega_h$.

(i) We denote by $W_\ell(A)$ the subspace of $L([m+1]^A)$ spanned by all products $\otimes_{i \in A} f_i$ with f_i either in V_0 or V_1 and $|\{i \in A : f_i \in V_1\}| = \ell$.

(ii) We define $W_{h,\ell}$ as the subspace of $L(\Omega_{h,m+1})$ given by (see (8.39))

$$W_{h,\ell} = \bigoplus_{A \in \Omega_h} W_\ell(A).$$

In other words, $W_{h,\ell}$ is the subspace of $L(\Omega_{h,m+1})$ spanned by all products of the form $\gamma \otimes (\otimes_{i \in B} f_i)$ where $B \in \Omega_\ell$, $\gamma \in M^{n-h,h-\ell}(\complement B)$ (and, clearly, $f_i \in V_1$ for all $i \in B$).

(iii) For $0 \le k \le (n-\ell)/2$ we denote by $W_{h,\ell,k}$ the subspace of $W_{h,\ell}$ spanned by the products $\gamma \otimes (\otimes_{i \in B} f_i)$ with γ belonging to the subspace of $M^{n-h,h-\ell}$ isomorphic to $S^{n-\ell-k,k}$.

From Lemmas 8.8.5 and 8.8.6, Exercise 8.8.7, and the results in Sections 6.1 and 6.2, we immediately get some properties of the subspaces introduced above.

Corollary 8.8.9

(i) *The subspaces $W_{h,\ell}$ and $W_{h,\ell,k}$ are $S_{m+1} \wr S_n$-invariant.*

(ii) $\dim W_{h,\ell} = \binom{n}{\ell}\binom{n-\ell}{h-\ell}m^\ell$ *and*

$$\dim W_{h,\ell,k} = \binom{n}{\ell}\left[\binom{n-\ell-k}{k} - \binom{n-\ell-k}{k-1}\right]m^\ell.$$

(iii) $\dim W_{h,\ell} = \bigoplus_{k=0}^{\min\{n-h,h-\ell\}} W_{h,\ell,k}$ *(orthogonal direct sum).*

(iv) $W_{k+\ell,\ell,k} = W_{k+\ell,\ell} \cap \ker D.$

(v) *For $0 \leq k \leq \min\{n-h, h-\ell\}$ and $f \in W_{k+\ell,\ell,k}$ one has*

$$\|(D^*)^{h-k-\ell}f\|^2_{L(\Omega_{h,m+1})} = (m+1)^{h-k-\ell}(h-k-\ell)! \cdot$$

$$\cdot (n-k-\ell+1)_{h-k-\ell}\|f\|^2_{L(\Omega_{k+\ell,m+1})}.$$

In particular, $(D^*)^{h-k-\ell}$ is an isomorphism of $W_{k+\ell,\ell,k}$ onto $W_{h,\ell,k}$.

Exercise 8.8.10 Prove the statements in the previous corollary.

We can now state and prove the main result of this section.

Theorem 8.8.11 *For $0 \leq h \leq n$,*

$$L(\Omega_{h,m+1}) = \bigoplus_{\ell=0}^{h} \bigoplus_{k=0}^{\min\{n-h,h-\ell\}} W_{h,\ell,k} \qquad (8.48)$$

is the decomposition of $L(\Omega_{h,m+1})$ into spherical representations. In particular, the subspaces in the right hand side are irreducible pairwise inequivalent representation of $S_{m+1} \wr S_n$.

Proof First of all, note that from the results in Section 5.3 it follows that, for $A \in \Omega_h$, one has

$$L([m+1]^A) = \bigoplus_{\ell=0}^{h} W_\ell(A). \qquad (8.49)$$

From (8.39), (8.49) and the definition of $W_{h,\ell}$ it follows that $L(\Omega_{h,m+1}) = \oplus_{\ell=0}^{h} W_{h,\ell}$ with orthogonal direct sum. Therefore, from (i) and (iii) in Corollary 8.8.9 it follows that (8.48) is an orthogonal decomposition into $(S_{m+1} \wr S_n)$-invariant subspaces. It remains to prove that the $W_{h,\ell,k}$'s are irreducible.

Set

$$\mathcal{A} = \{(t,\ell) : \max\{0, 2h-n\} \leq t \leq h, \ 0 \leq \ell \leq t\}$$

and

$$\mathcal{B} = \{(k, \ell) : 0 \leq \ell \leq h, \ 0 \leq k \leq \min\{n - h, h - \ell\}\}.$$

The set $\mathcal{A}$ parameterizes the $(S_{m+1} \wr S_n)$-orbits on $\Omega_{h,m+1} \times \Omega_{h,m+1}$ and $\mathcal{B}$ parameterizes the representations in (8.48). In virtue of Proposition 4.4.4 it suffices to show that $|\mathcal{A}| = |\mathcal{B}|$. But the map $T : \mathcal{A} \to \mathcal{B}$ defined by

$$T(t, \ell) = \begin{cases} (t + n - 2h, \ell) & \text{if } n - h < h - \ell \\ (t - \ell, \ell) & \text{if } n - h \geq h - \ell \end{cases}$$

is a bijection (see exercise below). $\square$

Exercise 8.8.12 Show that the map T defined above is a bijection.

Exercise 8.8.13 Show that for $h = n$ the decomposition (8.48) coincides with the decomposition in Theorem 5.3.2. When does it coincides with the decomposition in Theorem 6.1.6?

Exercise 8.8.14 Use (8.48) to show that, in general, $L(\Omega_{k,m+1}) \cap \ker D$ is not irreducible; see also Exercise 8.3.7.

We end this section with a description of the spherical functions of the nonbinary Johnson scheme.

Let ϕ_0 and ϕ_1 be the spherical functions of the Gelfand pair (S_{m+1}, S_m) as in Exercise 5.1.15. Suppose that $0 \leq \ell \leq h \leq n$. Fix $\overline{A} \in \Omega_h$ and define $\overline{\theta} \in [m + 1]^{\overline{A}}$ by $\overline{\theta}(j) = 0$ for all $j \in \overline{A}$, as in Exercise 8.8.1. For $0 \leq u \leq \min\{n - h, h - \ell\}$ we define the function $\Psi(h, \ell, u) \in W_{h,\ell}$ by setting

$$\Psi(h, \ell, u) = \sum_{A_1 \in \Omega_\ell(\overline{A})} \sum_{\substack{A_0 \in \Omega_{h-\ell}(\complement A_1): \\ |A_0 \setminus \overline{A}| = u}} [(\otimes_{i \in A_0} \phi_0) \otimes (\otimes_{i \in A_1} \phi_1)] \qquad (8.50)$$

where $\otimes_{i \in A_j} \phi_j$ indicates the product of $|A_j|$ many ϕ_j's over A_j, in symbols $(\otimes_{i \in A_j} \phi_j)(\theta) = \prod_{i \in A_j} \phi_j[(\theta(i)]$ for all $\theta \in [m + 1]^{A_j}$, $j = 0, 1$.

Let $(S_m \wr S_h) \times (S_{m+1} \wr S_{n-h})$ be the stabilizer of $\overline{\theta}$ (see Exercise 8.8.1). From Lemma 8.8.5 it follows that each function $\Psi(h, \ell, u)$ is $[(S_m \wr S_h) \times (S_{m+1} \wr S_{n-h})]$-invariant. Moreover, the functions

$$\{\Psi(h, \ell, u) : 0 \leq u \leq \min\{n - h, h - \ell\}\}$$

constitute an orthogonal basis for the subspace of $[(S_m \wr S_h) \times (S_{m+1} \wr S_{n-h})]$-invariant functions in $W_{h,\ell}$. Indeed,

$$\Psi(h, \ell, u) \in \bigoplus_{\substack{B \in \Omega_h: \\ |B \setminus \overline{A}| = u}} L([m+1]^B)$$

and, for different values of u, the corresponding subspaces are orthogonal. Moreover, from (iii) in Corollary 8.8.9, Theorem 8.8.11 and Theorem 4.6.2, it follows that the dimension of the space of $(S_m \wr S_h) \times (S_{m+1} \wr S_{n-h})$-invariant vectors in $W_{h,\ell}$ is precisely $\min\{n - h, h - \ell\} + 1$.

In the following theorem, $\phi(n, m, k; \ell)$ denote the coefficients of the spherical functions of the Johnson scheme, as in Theorem 6.1.10. Moreover, if $B \subseteq \{1, 2, \ldots, n\}$, $|B| = t$, then the symbol $\Phi_B(t, m, k)$ denotes the spherical function of the Gelfand pair $(S_t, S_{t-m,m} \times S_m)$ which belongs to $(d^*)^{m-k} S^{t-k,k}$, where S_t is the group of all permutations of B.

Theorem 8.8.15 *The spherical function $\Psi(n, h, \ell, k)$ in $W_{h,\ell,k}$ is given by*

$$\Psi(n, h, \ell, k) = \frac{1}{\binom{h}{\ell}} \sum_{u=0}^{\min\{n-h,h-\ell\}} \phi(n - \ell, h - \ell, k; u) \cdot \Psi(h, \ell, u).$$

Proof The function $\Psi(n, h, \ell, k)$ is a linear combination of $[(S_m \wr S_h) \times (S_{m+1} \wr S_{n-h})]$-invariant functions and therefore it is invariant as well. Moreover, its value on $\overline{\theta}$ equals the value of $\frac{1}{\binom{h}{\ell}} \Psi(h, \ell, 0)$ on $\overline{\theta}$, namely

$$\frac{1}{\binom{h}{\ell}} \sum_{A_1 \in \Omega_\ell(\overline{A})} \left[(\otimes_{i \in \overline{A} \setminus A_1} \phi_0) \otimes (\otimes_{i \in A_1} \phi_1) \right] (\overline{\theta}) = 1.$$

From (8.50) and (8.42) we get

$$\Psi(n, h, \ell, k) = \frac{1}{\binom{h}{\ell}} \sum_{u=0}^{\min\{n-h,h-\ell\}} \phi(n - \ell, h - \ell, k; u) \Psi(h, \ell, u)$$

$$= \sum_{A_1 \in \Omega_\ell(\overline{A})} \sum_{u=0}^{\min\{n-h,h-\ell\}} \phi(n - \ell, h - \ell, k; u) \cdot$$

$$\cdot \sum_{\substack{A_0 \in \Omega_{h-\ell}(\complement A_1): \\ |A_0 \setminus \overline{A}| = u}} \left[(\otimes_{i \in A_0} \phi_0) \otimes (\otimes_{i \in A_1} \phi_1) \right]$$

$$= \sum_{A_1 \in \Omega_\ell(\overline{A})} \Phi_{\complement A_1}(n - \ell, h - \ell, k) \otimes (\otimes_{i \in A_1} \phi_1)$$

as

$$\Phi_{\complement A_1}(n - \ell, h - \ell, k) = \sum_{u=0}^{\min\{n-h, h-\ell\}} \phi(n - \ell, h - \ell, k; u) \cdot$$

$$\sum_{\substack{A_0 \in \Omega_{h-\ell}(\complement A_1): \\ |A_0 \setminus \overline{A}| = u}} (\otimes_{i \in A_0} \phi_0)$$

is the spherical function of the Gelfand pair $(S_{n-\ell}, S_{n-h} \times S_{h-\ell})$ belonging to the irreducible representation $S^{n-\ell-k,k}$ (here $S_{n-\ell}$ is the symmetric group on $\complement A_1$ and $S_{n-h} \times S_{h-\ell}$ is the stabilizer of $\overline{A} \setminus A_1$). Therefore, $\Psi(n, h, \ell, k)$ belongs to $W_{h,\ell,k}$ and this ends the proof. $\qquad \square$

Exercise 8.8.16 Define the coefficients

$$\psi(h, \ell, u; s) = \sum_{j=\max\{0, \ell+s+u-h\}}^{\min\{s, \ell\}} \binom{s}{j} \binom{h - u - s}{\ell - j} \left(-\frac{1}{m}\right)^j.$$

(1) Show that $\psi(h, \ell, u; s)$ is precisely the value of $\Phi(h, \ell, u)$ on those $\theta \in \Omega_{h,m+1}$ that satisfy $|\operatorname{dom}(\theta) \setminus \overline{A}| = u$ and $|\{j \in \operatorname{dom}(\theta) \cap \overline{A} : \theta(j) = 0\}| = u - s$.

(2) Show that the value of $\psi(n, h, \ell, k)$ on a $\theta \in \Omega_{h,m+1}$ as above is equal to

$$\frac{1}{\binom{h}{\ell}} \phi(n - \ell, h - \ell, k; u) \psi(h, \ell, u; s).$$

9

Complements of representation theory

In this chapter we collect several complements of representation theory. The references are the same as those for Chapter 3. Sections 9.1, 9.7 (and part of Section 9.5) are inspired by Simon's book [198]. Section 9.3 is (partially) based on the books of Sternberg [212] and Naimark and Stern [168]. Section 9.8 is inspired by A. Greenhalgh's thesis [106]. Finally, Section 9.9 is based on Behrends' monograph [18] on Markov chains.

9.1 Tensor products

In this section we intoduce the notion of tensor product of vector spaces and of representations. In several textbooks the tensor product of two spaces is defined in terms of dual spaces. Here, following Simon [198], we treat the (finite dimensional) spaces as Hilbert spaces and we make use of the scalar product.

Let V and W be two finite dimensional complex vector spaces endowed with scalar products $\langle \cdot, \cdot \rangle_V$ and $\langle \cdot, \cdot \rangle_W$, respectively. A map $B \colon V \times W \to \mathbb{C}$ is *bi-antilinear* (on $V \times W$) if

$$
\begin{aligned}
B(\alpha_1 v_1 + \alpha_2 v_2, w) &= \overline{\alpha_1} B(v_1, w) + \overline{\alpha_2} B(v_2, w) \\
B(v, \alpha_1 w_1 + \alpha_2 w_2) &= \overline{\alpha_1} B(v, w_1) + \overline{\alpha_2} B(v, w_2)
\end{aligned}
\tag{9.1}
$$

for all $v, v_1, v_2 \in V, w, w_1, w_2 \in W$ and $\alpha_1, \alpha_2 \in \mathbb{C}$.

The prefix "anti-" comes from the fact that in (9.1) on the right hand side the conjugates $\overline{\alpha_1}$ and $\overline{\alpha_2}$ of α_1 and α_2, respectively appear. Clearly, the definition of a *bilinear map* is the same as in (9.1) but with $\overline{\alpha_1}$ and $\overline{\alpha_2}$ replaced by α_1 and α_2, respectively.

The set of all bi-antilinear maps on $V \times W$ is a complex vector space in a natural way: if B_1 and B_2 are two anti-bilinear maps and $\alpha_1, \alpha_2 \in \mathbb{C}$, then their linear combination defined by setting $(\alpha_1 B_1 + \alpha_2 B_2)(v, w) = \alpha_1 B_1(v, w) + \alpha_2 B_2(v, w)$ for all $v \in V$ and $w \in W$, is again bi-antilinear.

We denote by $V \bigotimes W$ the space of all bi-antilinear maps on $V \times W$ and call it the *tensor product* of V and W.

For $v \in V$ and $w \in W$ we denote by $v \otimes w$ the element in $V \bigotimes W$ defined by

$$[v \otimes w](v', w') = \langle v, v' \rangle_V \langle w, w' \rangle_W$$

for all $v' \in V$ and $w' \in W$. Note that the map $V \times W \ni (v, w) \mapsto v \otimes w \in V \bigotimes W$ is bilinear, that is

$$(\alpha_1 v_1 + \alpha_2 v_2) \otimes (\beta_1 w_1 + \beta_2 w_2) = \sum_{i,j=1}^{2} \alpha_i \beta_j v_i \otimes w_j$$

for all $\alpha_1, \alpha_2, \beta_1, \beta_2 \in \mathbb{C}$, $v_1, v_2 \in V$ and $w_1, w_2 \in W$ (in particular, $(\alpha v) \otimes w = \alpha(v \otimes w) = v \otimes (\alpha w)$). The elements in $V \bigotimes W$ of the form $v \otimes w$ are called *simple tensors*.

We claim that the simple tensors span the whole $V \bigotimes W$. Indeed, if $\{e_i\}_{i=1}^{d_V}$ and $\{f_j\}_{j=1}^{d_W}$ denote two orthonormal bases for V and W, respectively, for $B \in V \bigotimes V$ we clearly have $B = \sum_{i=1}^{d_V} \sum_{j=1}^{d_W} B(e_i, f_j) e_i \otimes f_j$. For, if $V \ni v = \sum_{i=1}^{d_V} \alpha_i e_i$ and $W \ni w = \sum_{j=1}^{d_W} \beta_j f_j$, then $(e_i \otimes f_j)(v, w) = \overline{\alpha_i} \overline{\beta_j}$ and

$$\left[\sum_{i=1}^{d_V} \sum_{j=1}^{d_W} B(e_i, f_j) e_i \otimes f_j \right] (v, w) = \sum_{i=1}^{d_V} \sum_{j=1}^{d_W} \overline{\alpha_i} \overline{\beta_j} B(e_i, f_j) = B(v, w).$$

This, incidentally shows that $\{e_i \otimes f_j\}_{\substack{i=1,\ldots,d_V \\ j=1,\ldots,d_W}}$ is a basis for $V \bigotimes W$ so that, in particular, $\dim(V \bigotimes W) = \dim(V) \cdot \dim(W)$.

We now endow $V \bigotimes W$ with a scalar product $\langle \cdot, \cdot \rangle_{V \bigotimes W}$ by setting

$$\langle v \otimes w, v' \otimes w' \rangle_{V \bigotimes W} = \langle v, v' \rangle_V \langle w, w' \rangle_W$$

and then extending by linearity. This way, if the bases $\{e_i\}_{i=1}^{d_V}$ and $\{f_j\}_{j=1}^{d_W}$ are orthonormal in V and W, respectively, then so is $\{e_i \otimes f_j\}_{\substack{i=1,\ldots,d_V \\ j=1,\ldots,d_W}}$ in $V \bigotimes W$.

Exercise 9.1.1 (1) Show that the bilinear map

$$\phi : V \times W \quad \to \quad V \bigotimes W$$
$$(v, w) \quad \mapsto \quad v \otimes w$$

is universal in the sense that if Z is another complex vector space and $\psi : V \times W \to Z$ is bilinear, then there exists a unique linear map

$\theta : V \otimes W \to Z$ such that $\theta(v \otimes w) = \phi(v, w)$, that is such that the diagram

$$
\begin{array}{ccc}
V \times W & \xrightarrow{\phi} & V \otimes W \\
 & \searrow \psi \quad \swarrow \theta & \\
 & Z &
\end{array}
$$

is commutative (i.e. $\psi = \theta \circ \phi$).

(2) Show that the above universal property characterizes the tensor product: if U is a complex vector space and $\psi : V \times W \to U$ is a bilinear map such that

(a) $\psi(V \times W) = \{\psi(v, w) : v \in V, w \in W\}$ generates U;

(b) for any complex vector space Z with a bilinear map $\tau : V \times W \to Z$ there exists a unique linear map $\theta : U \to Z$ such that $\tau = \theta \circ \psi$,

then there exists a linear isomorphism $\varepsilon : V \otimes W \to U$ such that $\psi = \varepsilon \circ \phi$.

Lemma 9.1.2 *We have the following natural isomorphisms:*

(i)
$$
\begin{array}{ccc}
V \otimes W & \to & W \otimes V \\
v \otimes w & \mapsto & w \otimes v
\end{array}
$$

(ii)
$$
\begin{array}{ccc}
\mathbb{C} \otimes V & \to & V \\
\lambda \otimes v & \mapsto & \lambda v
\end{array}
$$

(iii)
$$
\begin{array}{ccc}
(V \otimes W) \otimes Z & \to & V \otimes (W \otimes Z) \\
(v \otimes w) \otimes z & \mapsto & v \otimes (w \otimes z)
\end{array}
$$

(iv)
$$
\begin{array}{ccc}
(V \oplus W) \otimes Z & \to & (V \otimes Z) \oplus (W \otimes Z) \\
(v + w) \otimes z & \mapsto & (v \otimes z) + (w \otimes z).
\end{array}
$$

Proof The proof is elementary and the details are left to the reader. $\quad\square$

Note that (iii), namely the associativity of the tensor product, may be recursively extended to the tensor product of k vector spaces: we then denote by $V_1 \otimes V_2 \otimes \cdots \otimes V_k$ the set of all k-antilinear maps $B : V_1 \times V_2 \times \cdots \times V_k \to \mathbb{C}$.

Let now $A \in Hom(V)$ and $B \in Hom(W)$. Define $A \otimes B \in Hom(V \otimes W)$ by setting, for $C \in V \otimes W$,

$$
\{[A \otimes B](C)\} (v', w') = C(A^* v', B^* w')
$$

where $v' \in V$ and $w' \in W$ and A^* and B^* are the adjoints of A and B, respectively. For $v, v' \in V$ and $w, w' \in W$ we then have

$$
\begin{aligned}
\{[A \otimes B](v \otimes w)\}(v', w') &= [v \otimes w](A^* v', B^* w') \\
&= \langle v, A^* v' \rangle_V \langle w, B^* w' \rangle_W \\
&= \langle Av, v' \rangle_V \langle Bw, w' \rangle_W \\
&= [(Av) \otimes (Bw)](v' \otimes w').
\end{aligned}
$$

This shows that $[A \otimes B](v \otimes w) = (Av) \otimes (Bw)$.

Lemma 9.1.3 *Let $A \in Hom(V)$ and $B \in Hom(W)$. Then $Tr(A \otimes B) = Tr(A)Tr(B)$.*

Proof Let $\{e_i\}_{i=1}^{d_V}$ and $\{f_j\}_{j=1}^{d_W}$ be two orthonormal bases in V and W, respectively. Then

$$
\begin{aligned}
Tr(A \otimes B) &= \sum_{\substack{i=1,\ldots,d_V \\ j=1,\ldots,d_W}} \langle [A \otimes B](e_i \otimes f_j), e_i \otimes f_j \rangle_{V \otimes W} \\
&= \sum_{\substack{i=1,\ldots,d_V \\ j=1,\ldots,d_W}} \langle (Ae_i) \otimes (Bf_j), e_i \otimes f_j \rangle_{V \otimes W} \\
&= \sum_{\substack{i=1,\ldots,d_V \\ j=1,\ldots,d_W}} \langle Ae_i, e_i \rangle_V \langle Bf_j, f_j \rangle_W \\
&= Tr(A)Tr(B).
\end{aligned}
$$

$\square$

Exercise 9.1.4 (Kronecker product) Given two matrices $A = (a_{i,k})_{i,k=1}^n \in M_{n,n}(\mathbb{C})$ and $B = (b_{j,h})_{j,h=1}^m \in M_{m,m}(\mathbb{C})$ we define their *Kronecker product* $A \otimes B \in M_{nm,nm}(\mathbb{C})$ as the matrix which has the following block form

$$
\begin{pmatrix}
a_{1,1}B & a_{1,2}B & \cdots & a_{1,n}B \\
a_{2,1}B & a_{2,2}B & \cdots & a_{2,n}B \\
\vdots & \vdots & & \vdots \\
a_{n,1}B & a_{n,2}B & \cdots & a_{n,n}B
\end{pmatrix}.
$$

(1) Let V and W be two complex vector spaces of dimension $\dim(V) = n$ and $\dim(W) = m$. Fix two bases $\{e_i\}_{i=1}^n \subseteq V$ and $\{f_j\}_{j=1}^m \subseteq W$. Let $L_A : V \to V$ and $L_B : W \to W$ be the linear maps induced by the matrices A and B with respect to the chosen bases. Show that the

matrix representing the operator $L_A \otimes L_B : V \otimes W \to V \otimes W$ with respect to the basis $\{e_i \otimes f_j\}_{\substack{i=1,\ldots,n \\ j=1,\ldots,m}}$ is the Kronecker product of A and B, in formulæ, $L_A \otimes L_B = L_{A \otimes B}$.

(2) Show that the Kronoecker product satisfies the following algebraic properties:

$$(A \otimes B) \otimes C = A \otimes (B \otimes C) \equiv A \otimes B \otimes C$$
$$(A + B) \otimes C = A \otimes C + B \otimes C$$
$$(A \otimes B)(C \otimes D) = (AC) \otimes (BD).$$

Let now G_1 and G_2 be two groups and (ρ_1, V_1) and (ρ_2, V_2) be two representations of G_1 and G_2, respectively. We define the *outer tensor product* of ρ_1 and ρ_2 as the representation $(\rho_1 \boxtimes \rho_2, V_1 \otimes V_2)$ of $G_1 \times G_2$ defined by setting

$$[\rho_1 \boxtimes \rho_2](g_1, g_2) = \rho_1(g_1) \otimes \rho_2(g_2) \in Hom(V_1 \bigotimes V_2)$$

for all $g_i \in G_i$, $i = 1, 2$.

If $G_1 = G_2 = G$ the *internal tensor product* of ρ_1 and ρ_2 is the G-representation $(\rho_1 \otimes \rho_2, V_1 \otimes V_2)$ defined by setting

$$[\rho_1 \otimes \rho_2](g) = \rho_1(g) \otimes \rho_2(g) \in Hom(V_1 \bigotimes V_2)$$

for all $g \in G$.

It is obvious that, modulo the isomorphism between G and $\widetilde{G} = \{(g, g) : g \in G\} \le G \times G$, the internal tensor product $\rho_1 \otimes \rho_2$ is unitarily equivalent to restriction of the outer tensor product $\rho_1 \boxtimes \rho_2$ from $G \times G$ to $\widetilde{G}$.

We have introduced the symbol "$\boxtimes$" to make a distinction between these two notions of tensor (compare with [95]). Note that, however, the space will be simply denoted by $V_1 \bigotimes V_2$ in both cases.

Lemma 9.1.5 *Let ρ_1 and ρ_2 be two representations of two groups G_1 and G_2, respectively and denote by χ_{ρ_1} and χ_{ρ_2} their characters. Then, the character of $\rho_1 \boxtimes \rho_2$ is given by*

$$\chi_{\rho_1 \boxtimes \rho_2}(g_1, g_2) = \chi_{\rho_1}(g_1) \chi_{\rho_2}(g_2)$$

for all $g_1 \in G_1$ and $g_2 \in G_2$. In particular, if both ρ_1 and ρ_2 are one dimensional, so that they coincide with their characters, then one has that $\rho_1 \boxtimes \rho_2 = \chi_{\rho_1} \boxtimes \chi_{\rho_2} = \chi_{\rho_1} \chi_{\rho_2}$, the pointwise product of the characters.

Proof This follows immediately from Lemma 9.1.3. □

Similarly, for an internal tensor product we have: $\chi_{\rho_1 \otimes \rho_2}(g) = \chi_{\rho_1}(g)\chi_{\rho_2}(g)$.

Theorem 9.1.6 *Let $\rho_1 \in \widehat{G_1}$ and $\rho_2 \in \widehat{G_2}$ two irreducible representations of two groups G_1 and G_2. Then $\rho_1 \boxtimes \rho_2$ is an irreducible representation of $G_1 \times G_2$. Moreover, if $\rho_1' \in \widehat{G_1}$ and $\rho_2' \in \widehat{G_2}$ are two other irreducible representations of G_1 and G_2, then $\rho_1 \boxtimes \rho_2 \sim \rho_1' \boxtimes \rho_2'$ if and only if $\rho_1 \sim \rho_1'$ and $\rho_2 \sim \rho_2'$.*

Proof By Proposition 3.7.4 and Corollary 3.7.8 it suffices to check that $\langle \chi_{\rho_1 \boxtimes \rho_2}, \chi_{\rho_1' \boxtimes \rho_2'} \rangle$ is either $|G_1 \times G_2| \equiv |G_1| \cdot |G_2|$ if $\rho_1' \sim \rho_1$ and $\rho_2' \sim \rho_2$, or 0 otherwise. But

$$
\begin{aligned}
\langle \chi_{\rho_1 \boxtimes \rho_2}, \chi_{\rho_1' \boxtimes \rho_2'} \rangle &= \sum_{(g_1, g_2) \in G_1 \times G_2} \chi_{\rho_1 \boxtimes \rho_2}(g_1, g_2) \overline{\chi_{\rho_1' \boxtimes \rho_2'}(g_1, g_2)} \\
&= \sum_{\substack{g_1 \in G_1 \\ g_2 \in G_2}} \chi_{\rho_1}(g_1) \chi_{\rho_2}(g_2) \overline{\chi_{\rho_1'}(g_1) \chi_{\rho_2'}(g_2)} \\
&= \sum_{g_1 \in G_1} \chi_{\rho_1}(g_1) \overline{\chi_{\rho_1'}(g_1)} \sum_{g_2 \in G_2} \chi_{\rho_2}(g_2) \overline{\chi_{\rho_2'}(g_2)} \\
&= \langle \chi_{\rho_1}, \chi_{\rho_1'} \rangle \cdot \langle \chi_{\rho_2}, \chi_{\rho_2'} \rangle \\
&= \begin{cases} |G_1| \cdot |G_2| & \text{if } \rho_1 \sim \rho_1' \text{ and } \rho_2 \sim \rho_2' \\ 0 & \text{otherwise} \end{cases}.
\end{aligned}
$$

□

Corollary 9.1.7 *Let G_1 and G_2 be two finite groups. Then $\widehat{G_1 \times G_2} \cong \widehat{G_1} \times \widehat{G_2}$.*

Proof In virtue of Theorem 3.9.10 we have that $|\widehat{G_1 \times G_2}|$ equals the number of conjugacy classes in $G_1 \times G_2$. But the latter is the product of the numbers of conjugacy classes in G_1 and G_2, which, again by Theorem 3.9.10 equals $|\widehat{G_1}| \cdot |\widehat{G_2}|$. Therefore, by the previous theorem, the map $\widehat{G_1} \times \widehat{G_2} \ni (\rho_1, \rho_2) \mapsto \rho_1 \boxtimes \rho_2 \in \widehat{G_1 \times G_2}$ is a bijection. Alternatively, it is immediate to check (exercise) that $\sum_{\rho_1 \in \widehat{G_1}} \sum_{\rho_2 \in \widehat{G_2}} (d_{\rho_1 \boxtimes \rho_2})^2 = |G_1 \times G_2|$. □

Example 9.1.8 Let G and H be two finite groups and X and Y be G- (resp. H-)homogeneous spaces. First of all note that there is a natural isomorphism
$$
L(X) \bigotimes L(Y) \cong L(X \times Y)
$$

obtained by extending by linearity the map $L(X) \bigotimes L(Y) \ni \delta_x \otimes \delta_y \mapsto \delta_{(x,y)} \in L(X \times Y)$. In other words, $F \in L(X \times Y)$ corresponds to the bi-antilinear map B_F on $L(X) \times L(Y)$ defined by setting

$$B_F(f,g) = \sum_{x \in X} \sum_{y \in Y} F(x,y)\overline{f(x)g(y)}$$

and $L(X \times Y) \ni F \mapsto B_F \in L(X) \otimes L(Y)$ is an isomorphism.

Let now λ and μ be the permutation representations of G and H on X and Y, respectively. Then $\lambda \boxtimes \mu$ corresponds to the permutation representation of $G \times H$ on $X \times Y$. If $G = H$, then the internal tensor product $\lambda \otimes \mu$ corresponds to the diagonal action of G on $X \times Y$, see Section 3.13.

Example 9.1.9 Set $X = \{1, 2, \ldots, n\}$. With the notation in Section 3.14, we investigate the decomposition into irreducibles of the representations $M^{(n-1,1)} \otimes M^{(n-1,1)}$ and $S^{(n-1,1)} \otimes S^{(n-1,1)}$. All the tensor products in this example are inner.

We first observe that S_n has exactly two orbits on $X \times X$, namely

$$\Omega_{n-2,1,1} = \{(x,y) : x \neq y\} \quad \text{and} \quad \Omega_{n-1,1} = \{(x,x) : x \in X\} \equiv X.$$

In virtue of the previous example, we thus have

$$\begin{aligned}
M^{(n-1,1)} \otimes M^{(n-1,1)} &= L(X) \otimes L(X) = L(X \times X) \\
&= L(\Omega_{n-2,1,1}) \oplus L(\Omega_{n-1,1}) \qquad (9.2) \\
&= M^{(n-2,1,1)} \oplus M^{(n-1,1)}.
\end{aligned}$$

Then, observing that $S^{(n)} \otimes S^\lambda = S^\lambda$ and using (3.28) and (3.32) we have:

$$\begin{aligned}
M^{(n-1,1)} \otimes M^{(n-1,1)} &= [S^{(n)} \oplus S^{(n-1,1)}] \otimes [S^{(n)} \oplus S^{(n-1,1)}] \\
&= S^{(n)} \oplus 2S^{(n-1,1)} \oplus [S^{(n-1,1)} \otimes S^{(n-1,1)}]
\end{aligned} \qquad (9.3)$$

and

$$\begin{aligned}
M^{(n-1,1)} \otimes M^{(n-1,1)} &= M^{(n-2,1,1)} \oplus M^{(n-1,1)} \\
&= [S^{(n)} \oplus 2S^{(n-1,1)} \oplus S^{(n-2,2)} \oplus S^{(n-2,1,1)}] \oplus \\
&\quad \oplus [S^{(n)} \oplus S^{(n-1,1)}] \\
&= 2S^{(n)} \oplus 3S^{(n-1,1)} \oplus S^{(n-2,2)} \oplus S^{(n-2,1,1)}.
\end{aligned}$$
$$(9.4)$$

Comparing (9.3) and (9.4) we deduce that

$$S^{(n-1,1)} \otimes S^{(n-1,1)} = S^{(n)} \oplus S^{(n-1,1)} \oplus S^{(n-2,2)} \oplus S^{(n-2,1,1)}. \quad (9.5)$$

Observe that this is equivalent to (3.33) in Exercise 3.14.1.

9.2 Representations of abelian groups and Pontrjagin duality

In this section A denotes a finite abelian group.

We recall the following structure theorem for finite abelian groups (see for instance the monographs by M. Artin [6], Herstein [120], Hungerford [123] and Simon [198]).

Theorem 9.2.1 *Let A be a finite abelian group. Then A is a direct sum of cyclic groups: $A = C_{d_1} \times C_{d_2} \times \cdots \times C_{d_n}$. Such a decomposition can be chosen in such a way that d_i divides d_{i+1} for $i = 1, 2, \ldots, n - 1$ and in this case it is unique.*

Remark 9.2.2 An alternative form of Theorem 9.2.1 is the following. Any finite abelian group A may be (uniquely) expressed in the form $A = C_{d_1} \times C_{d_2} \times \cdots \times C_{d_m}$, where $d_j = p_j^{\alpha_j}$ with the p_j's distinct prime numbers and α_j's positive integers, $j = 1, 2, \ldots, m$.

Corollary 9.2.3 *Let A be a finite abelian group. Then $\widehat{A}$ has a structure of an abelian group and it is isomorphic to A.*

Proof In virtue of Exercise 3.5.3, every irreducible representation of A is one dimensional and therefore coincides with its character. It is then clear that the characters form an abelian group under pointwise multiplication. In Example 3.8.1 we showed that for a cyclic group C_d one has $\widehat{C_d}$ is even isomorphic to C_d. On the other hand, by the above structure theorem, we have that $A = C_{d_1} \times C_{d_2} \times \cdots \times C_{d_n}$ and Corollary 9.1.7 yields a bijective map $\widehat{C_{d_1}} \times \widehat{C_{d_2}} \times \cdots \times \widehat{C_{d_n}} \to \widehat{A}$ which, by the second part of Lemma 9.1.5, is in fact an isomorphism of groups.

$\square$

The isomorphism in the above corollary depends on the choice of the generators for the cyclic subgroups, that is it depends on the coordinates. There is, however, a more intrinsic isomorphism between A and its *bidual* $\widehat{\widehat{A}}$. It is given by

$$A \ni g \mapsto \psi_g \in \widehat{\widehat{A}} \quad (9.6)$$

where $\psi_g(\chi) = \chi(g)$ for all $\chi \in \widehat{G}$.

Exercise 9.2.4 Prove that the map (9.6) is an isomorphism.

The situation is similar to that of finite dimensional vector spaces. If V is a finite dimensional vector space, then $V \cong V^*$ but the canonical isomorphism is with its bidual V^{**}. Moreover, the isomorphism between G and $\widehat{\widehat{G}}$ extends to locally compact abelian groups (*Pontrjagin duality*). For instance, if $\mathbb{T} = \{z \in \mathbb{C} : |z| = 1\}$ denotes the unit circle, then $\widehat{\mathbb{T}} \cong \mathbb{Z}$ but $\widehat{\widehat{\mathbb{T}}} \cong \mathbb{T}$ (this is the setting of *classical Fourier series*, see the monographs of abstract harmonic analysis by Rudin [181], Katznelson [139] or Loomis [157]).

We end this section with a characterization of the one-dimensional representations of an arbitrary (not necessarily abelian) finite group. For a group G, the *commutator* of two elements $g_1, g_2 \in G$ is the element $[g_1, g_2] := g_1^{-1} g_2^{-1} g_1 g_2 \in G$. Then, the *derived* (or *commutator*) subgroup of G is the subgroup $G' \equiv [G, G] := \langle [g_1, g_2] : g_1, g_2 \in G \rangle$ generated by the commutators. As $[g_1, g_2]^{-1} = [g_2, g_1]$, G' just consists of products of commutators.

Proposition 9.2.5 *With the above notation, we have the following:*

  (i) *G' is normal in G;*

  (ii) *if N is normal in G, then G/N is abelian if and only if $G' \leq N$;*

  (iii) *set $G_{ab} = G/G'$ (it is abelian by (ii)). For $\chi \in \widehat{G_{ab}}$ define $\widetilde{\chi} \in \widehat{G}$ as the one-dimensional representation defined by $\widetilde{\chi}(g) = \chi(gG')$. Then the map $\chi \mapsto \widetilde{\chi}$ is a one-to-one correspondence between the dual of the G_{ab} and the set of all one-dimensional representations of G.*

Proof (i) This follows immediately from the fact that $[g_1, g_2]^t = [g_1^t, g_2^t]$ for all $g_1, g_2, t \in G$, that is, the generating set of G' is conjugacy invariant.

(ii) We have $g_1 N \cdot g_2 N = g_2 N \cdot g_1 N$ if and only if $g_1 g_2 \in g_2 g_1 N$ which in turn is equivalent to the condition $[g_1, g_2] \in N$. Therefore G/N is abelian if and only if N contains all commutators.

(iii) The element $\widetilde{\chi}$ is clearly a one-dimensional representation of G. Conversely, if $\sigma \in \widehat{G}$ is one-dimensional, that is $\sigma : G \to \mathbb{T}$ is a homomorphism, then setting $\ker(\sigma) = \{g \in G : \sigma(g) = 1\}$ we have that $G/\ker(\sigma) \cong \mathbb{T}$ is abelian. By point (ii) we have that $\ker(\sigma) \geq G'$ so that $\chi(gG') := \sigma(g)$, $g \in G$, is well defined, so that χ is a character of G_{ab}. Finally one has $\widetilde{\chi} = \sigma$. $\qquad\square$

The group G_{ab} is called the *abelianization* of G. It is the largest abelian quotient of G. The point (i) in the previous proposition may be immediately extended in the following sense: G' is *fully invariant*, that is it is invariant under all endomorphisms of G.

9.3 The commutant

Let G be a finite group and denote by $M_{m,n}(\mathbb{C})$ the set of all $m \times n$ complex matrices.

Lemma 9.3.1 *Suppose that (σ, V) and (θ, U) are two G-representations with*

$$V = W_1 \oplus W_2 \oplus \cdots \oplus W_m \quad and \quad U = Z_1 \oplus Z_2 \oplus \cdots \oplus Z_n$$

as decompositions into irreducible representations.

(i) *Suppose that $W_1, W_2, \ldots, W_m, Z_1, Z_2, \ldots, Z_n$ are all equivalent. Then $Hom_G(V, U) \equiv M_{m,n}(\mathbb{C})$ as vector spaces.*
In particular, $\dim Hom_G(V, U) = mn$.

(ii) *Suppose that $W_1, W_2, \ldots, W_m$ are all equivalent, $Z_1, Z_2, \ldots, Z_n$ are all equivalent, but the W_i's are not equivalent to the Z_j's. Then $Hom_G(V, U) = 0$.*

Proof (i) Denote by $I_j : W_j \to V$ the inclusion and by $P_i : U \to Z_i$ the projection ($1 \leq j \leq m$, $1 \leq i \leq n$).

For $T \in Hom(U, V)$ set $T_{i,j} = P_i T I_j \in Hom(W_j, Z_i)$. We claim that T is G-invariant if and only if so are the $T_{i,j}$'s. If $w \in W$ and $w = w_1 + w_2 + \cdots + w_m$ with $w_j \in W_j$ then $Tw = \sum_{i=1}^{n} \sum_{j=1}^{m} T_{i,j} w_j$. Thus, by the G-invariance of the W_j's, one has $T\sigma(g)w = \sum_{i=1}^{n} \left(\sum_{j=1}^{m} T_{i,j} \sigma(g) w_j \right)$ and, by the G-invariance of the Z_i's, $\theta(g)Tw = \sum_{i=1}^{n} \left(\sum_{j=1}^{m} \theta(g) T_{i,j} w_j \right)$, for all $g \in G$, with $\theta(g) T_{i,j} w_j \in Z_i$.

Therefore, $T\sigma(g)w = \theta(g)Tw$ for all $w \in W$ and $g \in G$ if and only if $T_{i,j}\sigma(g)w_j = \theta(g)T_{i,j}w_j$ for all $w_j \in W_j$, $1 \leq j \leq m$, $1 \leq i \leq n$ and $g \in G$, proving the claim.

In other words, all I_j's commute with the $\sigma(g)$'s and the P_j's with the $\theta(g)$'s. Moreover, every $T \in Hom_G(U, V)$ is determined by the operators $T_{i,j}$'s: $T = \bigoplus_{i=1}^{n} \bigoplus_{j=1}^{m} T_{i,j}$.

By Schur's lemma, $Hom_G(W_j, Z_i)$ is one dimensional, say $Hom_G(W_j, Z_i) = \mathbb{C}L_{i,j}$, for some fixed $L_{i,j} \in Hom_G(W_j, Z_i)$. Thus, for every $T \in Hom_G(V, U)$, $T_{i,j} = a_{i,j} L_{i,j}$ (with $a_{i,j} \in \mathbb{C}$) and T is of the form $T = \sum_{i,j} a_{i,j} L_{i,j}$; the proof of (1) is now complete.

(ii) This is easy: if $W_j \not\sim Z_i$, then, with the same notation as in the previous part, $T_{i,j} = 0$ by Schur's lemma and $T = 0$. $\square$

Theorem 9.3.2 *Let (σ, V) and (θ, U) be two G-representations with*

$$V = \bigoplus_{\pi \in I} m_\pi W_\pi \quad and \quad U = \bigoplus_{\pi \in J} n_\pi W_\pi$$

as decompositions into irreducible sub-representations, with $0 < m_\pi, n_\pi$ the corresponding multiplicities (so that $I \cap J$ is the set of (irreducible) representations contained in both V and U). Then $Hom_G(V, U) \equiv \bigoplus_{\pi \in I \cap J} M_{m_\pi, n_\pi}(\mathbb{C})$ as vector spaces. In particular, $\dim Hom_G(V, U) = \sum_{\pi \in I \cap J} m_\pi n_\pi$.

Proof For $\pi \in J$ denote by $P_\pi : U \to n_\pi W_\pi$ the orthogonal projection from U onto the isotypic component of π in U. Analogously, for $\pi \in I$, denote by $I_\pi : m_\pi W_\pi \to V$ the inclusion map. Then, if $T \in Hom_G(V, W)$ one has $P_{\pi_1} T I_{\pi_2} \in Hom_G(m_{\pi_2} W_{\pi_2}, n_{\pi_1} W_{\pi_1})$ and, by (ii) in previous lemma, $P_{\pi_1} T I_{\pi_2} = 0$ if $\pi_1 \neq \pi_2$. Therefore, $T = \bigoplus_{\pi \in I \cap J} P_\pi T I_\pi$ with $P_\pi T I_\pi \in Hom_G(m_\pi W_\pi, n_\pi W_\pi)$ and the theorem follows from (i) in the previous lemma, indeed $Hom_G(V, U) = \bigoplus_{\pi \in I \cap J} Hom_G(m_\pi W_\pi, n_\pi W_\pi) \equiv \bigoplus_{\pi \in I \cap J} M_{m_\pi, n_\pi}(\mathbb{C})$. $\square$

Corollary 9.3.3 *Let V and W be two G-representations. Suppose that W is irreducible. Then*

$$\dim Hom_G(V, W) = \text{multiplicity of } W \text{ in } V = \dim Hom_G(W, V)$$

Corollary 9.3.4 *Suppose that $V = \bigoplus_{\pi \in I} m_\pi W_\pi$ is the decomposition of V into irreducible inequivalent representations. Then*

$$Hom_G(V, V) \equiv \bigoplus_{\pi \in I} M_{m_\pi, m_\pi}(\mathbb{C})$$

as algebras. In particular, $\dim Hom_G(V, V) = \sum_{\pi \in I} m_\pi^2$.

Proof For all $\pi \in I$ let $m_\pi W_\pi = W_{\pi,1} \oplus W_{\pi,2} \oplus \cdots \oplus W_{\pi, m_\pi}$ be a decomposition into irreducibles and choose $L_{i,j}^\pi : W_{\pi,j} \to W_{\pi,i}$ as in the proof of Lemma 9.3.1 in such a way that

$$L_{i,j}^\pi L_{j,k}^\pi = L_{i,k}^\pi \tag{9.7}$$

for all $i, j, k = 1, 2, \ldots, m_\pi$. This is possible because we can first choose $L_{k,1}^\pi$ for $k = 1, 2, \ldots, m_\pi$ and then define $L_{k,j}^\pi = L_{k,1}^\pi (L_{j,1}^\pi)^{-1}$. But, if $T \in Hom_G(m_\pi W_\pi, m_\pi W_\pi)$ then there exists a matrix $(a_{i,j})_{i,j=1}^{m_\pi}$ such that $T = \sum_{i,j=1}^{m_\pi} a_{i,j} L_{i,j}^\pi$. From (9.7) it follows that the map $T \mapsto (a_{i,j})_{i,j=1}^{m_\pi}$ is an isomorphism between the algebras $Hom_G(m_\pi W_\pi, m_\pi W_\pi)$ and $M_{m_\pi, m_\pi}(\mathbb{C})$. $\qquad \square$

Definition 9.3.5 The algebra $Hom_G(V, V)$ is called the *commutant* of the representation V and $m_\pi W_\pi$ is the π-*isotypic component* of V.

Suppose again that $V = \bigoplus_{\pi \in I} m_\pi W_\pi$ is the decomposition of V into irreducible representations. For any $\pi \in I$ denote by $T_\pi \in Hom_G(V, V)$ the projection onto the σ-isotypical component, that is the operator defined by setting, for all $v \in V$,

$$T_\pi v = \begin{cases} 0 & \text{if } v \in \bigoplus_{\sigma \in I \setminus \pi} m_\sigma W_\sigma \\ v & \text{if } v \in m_\pi W_\pi. \end{cases}$$

Let Z be the center of $Hom_G(V, V)$, that is $Z = \{T \in Hom_G(V, V) : ST = TS \text{ for all } S \in Hom_G(V, V)\}$. We then have:

Proposition 9.3.6

(i) *The set $\{T_\pi : \pi \in I\}$ is a basis for Z.*

(ii) *Z is isomorphic to $\mathbb{C}^{|I|} = \{(\alpha_1, \alpha_2, \ldots, \alpha_{|I|}) : \alpha_j \in \mathbb{C}\}$ under coordinatewise multiplication:*

$$(\alpha_1, \alpha_2, \ldots, \alpha_{|I|}) \cdot (\alpha_1', \alpha_2', \ldots, \alpha_{|I|}') = (\alpha_1 \alpha_1', \alpha_2 \alpha_2', \ldots, \alpha_{|I|} \alpha_{|I|}').$$

Proof The space $Hom_G(V, V)$ is isomorphic to the direct sum $\bigoplus_{\pi \in I} M_{m_\pi, m_\pi}(\mathbb{C})$. But a matrix $A \in M_{m_\pi, m_\pi}(\mathbb{C})$ commutes with any other matrix $B \in M_{m_\pi, m_\pi}(\mathbb{C})$ if and only if it is a scalar multiple of the identity: $A \in \mathbb{C}I$. This proves (ii).

Then (i) follows by observing that with the notation in the proof of Corollary 9.3.4 we have

$$T_\pi |_{m_\pi W_\pi} = \bigoplus_{k=1}^{m_\pi} L_{k,k}^\pi.$$

$\qquad \square$

Recalling the notion of multiplicity-freeness (Definition 4.4.1), we have the following generalization of Theorem 4.4.2.

Corollary 9.3.7 *A representation V is multiplicity-free if and only if $Hom_G(V, V)$ is commutative.*

9.4 Permutation representations

In this section, we examine the commutant of a permutation representation and the relative harmonic analysis.

Suppose that a finite group G acts transitively on X, that K is the stabilizer of a fixed point $x_0 \in X$ and denote by $(\lambda, L(X))$ the corresponding permutation representation. If V is any G-representation, we denote by V^K the subspace of K-invariant vectors in V. Suppose that (ρ, V) is irreducible and that V^K is nontrivial. For any $v \in V^K$ define the linear map $T_v : V \to L(X)$ defined by setting $(T_v w)(g x_0) = \sqrt{\frac{\dim V}{|X|}} \langle w, \rho(g) v \rangle_V$, for any $w \in V$ and $g \in G$.

The following proposition generalizes Corollary 4.6.4.

Proposition 9.4.1 *With the notation above one has:*

(i) *For any non–trivial $v \in V^K$ with $\|v\|_V = 1$, T_v is an isometric inversion of V into $L(X)$.*

(ii) *If $u, v \in V^K$ then $Im(T_u)$ is orthogonal to $Im(T_v)$ if and only if u is orthogonal to v.*

(iii) *The map $V^K \ni v \mapsto T_v \in Hom_G(V, L(X))$ is an antilinear isomorphism between the vector spaces V^K and $Hom_G(V, L(X))$.*

Proof (i) may be proved as (4.13) and (i) in Corollary 4.6.4. See also (9.8) below.

(ii) We first show that if $u, v \in V^K$ and $w, z \in V$, then

$$\langle T_u w, T_v z \rangle_{L(X)} = \langle w, z \rangle_V \langle v, u \rangle_V. \tag{9.8}$$

Fix $u, v \in V^K$ and define two linear maps $L, R : V \to V$ by setting: $Lw = \langle w, u \rangle v$ for all $w \in V$ and then $R = \frac{1}{|K|} \sum_{g \in G} \rho(g) L \rho(g^{-1})$. It is clear that $R \in Hom_G(V)$ and since V is irreducible, $R = a I_V$ with $a \in \mathbb{C}$ and I_V the identity map on V. But $Tr(L) = \langle v, u \rangle$ (as can be easily checked) and then $a \dim V = Tr(R) = \frac{1}{|K|} \sum_{g \in G} Tr(L) = |X| \langle v, u \rangle$. In other words, $R = \frac{|X|}{\dim V} \langle v, u \rangle I_V$.

Therefore, if $w, z \in V$, we have

$$
\begin{aligned}
\langle T_u w, T_v z \rangle_{L(X)} &= \frac{1}{|K|} \sum_{g \in G} \langle w, \rho(g)u \rangle_V \overline{\langle z, \rho(g)v \rangle_V} \cdot \frac{\dim V}{|X|} \\
&= \frac{\dim V}{|X|} \langle Rw, z \rangle_V \\
&= \langle w, z \rangle_V \langle v, u \rangle_V.
\end{aligned}
$$

Then (ii) immediatley follows from (9.8).

(iii) The map $v \mapsto T_v$ is antilinear: if $\alpha, \beta \in \mathbb{C}$, $u, v \in V^K$, then clearly $T_{\alpha u + \beta v} = \overline{\alpha} T_u + \overline{\beta} T_v$.

It remains to show that this map is a bijection. If $T \in Hom_G(V, L(X))$, then $V \ni w \mapsto (Tw)(x_0) \in \mathbb{C}$ is a linear map and therefore, by Riesz's representation theorem, there exists $v \in V$ such that, for all $w \in V$, $(Tw)(x_0) = \langle w, v \rangle_V$. It follows that for all $w \in V$ and $g \in G$

$$
\begin{aligned}
[Tw](gx_0) &= [\lambda(g^{-1})Tw](x_0) \\
\text{(because } T \in Hom_G(V, L(X))) \quad &= [T\rho(g^{-1})w](x_0) \\
&= \langle \rho(g^{-1})w, v \rangle_V \\
&= \langle w, \rho(g)v \rangle_V.
\end{aligned}
\tag{9.9}
$$

This shows that $T = T_v$. Moreover, taking $g = k \in K$ in (9.9) we get that $v \in V^K$.

It is also clear that for $T \in Hom_G(V, L(X))$ such a vector $v \in V^K$ is uniquely determined. $\qquad \square$

Corollary 9.4.2 (Frobenius reciprocity for permuation representations) *The multiplicity of V in $L(X)$ is equal to $\dim V^K$, the dimension of the subspace of K-invariant vectors in V.*

Corollary 9.4.3 *If $v_1, v_2, \ldots, v_m$ constitute an orthogonal basis for V^K, then*

$$
T_{v_1} V \bigoplus T_{v_2} V \bigoplus \cdots \bigoplus T_{v_m} V
$$

is an orthogonal decomposition of the V-isotypic component of $L(X)$.

Exercise 9.4.4 Use Corollary 9.4.2 and Corollary 9.3.4 to prove Wielandt's lemma (Theorem 3.13.3).

We can summarize the preceding facts in the following way: if $L(X) = \bigoplus_{\rho \in I} m_\rho V_\rho$ is the decomposition of $L(X)$ into irreducible G- representations, then $m_\rho = \dim V_\rho^K$ and

$$Hom_G(L(X), L(X)) \cong \bigoplus_{\rho \in I} M_{m_\rho, m_\rho}(\mathbb{C}). \qquad (9.10)$$

We already know that $Hom_G(L(X), L(X))$ is isomorphic to the algebra $L(K\backslash G/K)$ of all bi-K-invariant functions on G (see Proposition 4.2.1). In the remaining part of this section we construct an explicit isomorphism between $L(K\backslash G/K)$ and the right hand side of (9.10).

Suppose again that $L(X) = \bigoplus_{\rho \in I} m_\rho V_\rho$. For any irreducible representation $\rho \in I$, select an orthonormal basis $\{v_1^\rho, v_2^\rho, \ldots, v_{m_\rho}^\rho\}$ in V_ρ^K.

Definition 9.4.5 For $\rho \in I$, the matrix coefficients

$$\phi_{i,j}^\rho(g) = \langle v_i^\rho, \rho(g)v_j^\rho \rangle_{V_\rho} \qquad (9.11)$$

with $i, j = 1, 2, \ldots, m_\rho$, are the *spherical matrix coefficients* of (ρ, V_ρ).

Clearly, the spherical matrix coefficients generalize the spherical functions. Moreover they satisfy the usual properties of the matrix coefficients (they are the conjugates of the matrix coefficients introduced in Section 3.6).

In particular, the spherical matrix coefficients form an orthonormal basis for the vector space $L(K\backslash G/K)$ (note that from (9.10) and the isomorphism $Hom_G(L(X), L(X)) \cong L(K\backslash G/K)$ it follows that $\dim L(K\backslash G/K) = \sum_{\rho \in I} m_\rho^2$). Using the orthogonality relation in Lemma 3.6.3 we can immediately write the *spherical Fourier transform* relative to the matrix coefficients (9.11): for $F \in L(K\backslash G/K)$, $\rho \in I$ and $i, j = 1, 2, \ldots, m_\rho$

$$\widehat{f}_{i,j}(\rho) = \langle f, \phi_{i,j}^\rho \rangle_{L(G)}$$

and the relative *inversion formula* becomes

$$f(g) = \frac{1}{|G|} \sum_{\rho \in I} d_\rho \sum_{i,j=1}^{m_\rho} \phi_{i,j}^\rho(g) \widehat{f}_{i,j}(\rho).$$

From Lemma 3.9.14 we get

$$\phi_{i,j}^\rho * \phi_{h,k}^\sigma = \frac{|G|}{d_\rho} \delta_{j,h} \delta_{\rho,\sigma} \phi_{i,k}^\rho \qquad (9.12)$$

and this immediately yields the desired explicit isomorphism.

Theorem 9.4.6 *The map*

$$L(K\backslash G/K) \quad \longrightarrow \quad \bigoplus_{\rho \in I} M_{m_\rho, m_\rho}(\mathbb{C})$$
$$f \quad \mapsto \quad \bigoplus_{\rho \in I} \left(\widehat{f}_{i,j}(\rho) \right)_{i,j=1,2,\dots,m_\rho}$$

is an isomorphism of algebras.

Proof Let $f, h \in L(K\backslash G/K)$. Then, for all $\rho \in I$ and $i, j = 1, 2, \dots, m_\rho$, we have

$$\widehat{(f * h)}_{i,j}(\rho) = \sum_{k=1}^{m_\rho} \widehat{f}_{i,k}(\rho) \cdot \widehat{h}_{k,j}(\rho)$$

as it easily follows from the inversion formula and (9.12).   □

When (G, K) is not a Gelfand pair, then there are three distinct kinds of algebras that are worthwhile to study.

- The first algebra is the whole algebra $L(K\backslash G/K)$ of bi-K-invariant functions: its harmonic analysis has been described above.
- The second algebra is $\mathcal{A} = $ span $\{\phi_{i,i}^\rho : \rho \in I, \ i = 1, 2, \dots, m_\rho\}$. It depends on the choice of the bases $\{v_1^\rho, v_2^\rho, \dots, v_{m_\rho}^\rho\}$, $\rho \in I$.

Exercise 9.4.7 (1) Show that $\mathcal{A}$ is a maximal abelian subalgebra of $L(K\backslash G/K)$.
(2) Show that the operator $E_i^\rho : L(X) \to L(X)$ defined by

$$(E_i^\rho f)(gx_0) = \frac{d_\rho}{|G|} \langle \widetilde{f}, \lambda(g) \phi_{i,i}^\rho \rangle_{L(G)}$$

$f \in L(X)$, $g \in G$ (and $\widetilde{f}$ is as in (4.1)), is the projection from $L(X)$ onto $T_{v_i^\rho} V_\rho$, $\rho \in I$, $i = 1, 2, \dots, m_\rho$ ($T_{v_i^\rho} V_\rho$ as in Proposition 9.4.1).

- The last algebra is $\mathcal{B} = $ span $\{\chi_\rho^K : \rho \in I\}$ where

$$\chi_\rho^K = \sum_{i=1}^{m_\rho f} \phi_{i,i}^\rho.$$

Exercise 9.4.8 (1) Show that $\mathcal{B}$ is the center of $L(K\backslash G/K)$.
(2) Show that the linar operator $E^\rho : L(X) \to L(X)$ defined by

$$(E^\rho f)(gx_0) = \frac{d_\rho}{|G|} \langle \widetilde{f}, \lambda(g) \chi_\rho^K \rangle_{L(G)}$$

$f \in L(X)$, $g \in G$ (and $\tilde{f}$ is as in (4.1)), is the projection from $L(X)$ onto the isotypic component $m_\rho V_\rho$, $\rho \in I$.

(3) Show that if χ_ρ is the character of ρ, then $\chi_\rho^K(g) = \frac{1}{|K|}\sum_{k\in K}\overline{\chi_\rho(kg)}$.

Exercise 9.4.9 Show that $(S_{a+b+c}, S_a \times S_b \times S_c)$ (where S_n is the symmetric group on n elements and a, b, c are nonnegative integers) is a Gelfand pair if and only if $\min\{a, b, c\} = 0$.

9.5 The group algebra revisited

Let G be a finite group, (π, V) a unitary representation of G. Denote by V' the *dual* of V and let $\theta : V' \to V$ be the *Riesz map*, i.e. the bijection given by the Riesz representation theorem: if $f \in V'$ then $f(v) = \langle v, \theta(f)\rangle$ for all $v \in V$. Recall that θ is antilinear, i.e. $\theta(\alpha f_1 + \beta f_2) = \overline{\alpha}\theta(f_1) + \overline{\beta}\theta(f_2)$, for all $\alpha, \beta \in \mathbb{C}$ and $f_1, f_2 \in V^*$.

We now define the *adjoint* or *conjugate representation* (π', V') of (π, V). For $f \in V'$ and $g \in G$, $\pi'(g)f \in V'$ is defined by

$$[\pi'(g)f](v) = f[\pi(g^{-1})v].$$

for all $v \in V$.

We have

$$\langle v, \theta[\pi'(g)f]\rangle = [\pi'(g)f](v) = f[\pi(g^{-1})v]$$
$$= \langle \pi(g^{-1})v, \theta(f)\rangle = \langle v, \pi(g)[\theta(f)]\rangle$$

so that

$$\pi'(g) = \theta^{-1}\pi(g)\theta$$

for all $g \in G$. Clearly π' is a linear representation of G; in general it is not equivalent to π, but in any case it is easy to show that π' is irreducible if and only if π is irreducible.

Definition 9.5.1 *The representation π is selfconjugate (or selfadjoint) if π and π' are equivalent.*

In other words, π is selfconjugate if and only if $\chi_\pi = \chi_{\pi'}$.

Fix now an orthonormal basis $\{v_1, v_2, \ldots, v_n\}$ of V. We recall that the functions $u_{i,j} : G \to \mathbb{C}$, defined by

$$u_{i,j}(g) = \langle \pi(g)v_j, v_i\rangle \tag{9.13}$$

are the matrix coefficients of the representation π; see Section 3.6. In V' we introduce a scalar product by setting, for all f_1 and $f_2 \in V'$

$$\langle f_1, f_2 \rangle = \langle \theta(f_2), \theta(f_1) \rangle.$$

Thus, for $f \in V'$ and $v \in V$ one has

$$f(v) = \langle v, \theta(f) \rangle = \langle f, \theta^{-1}(v) \rangle$$

which shows that $(V')'$ is isometrically identified with V by mean of θ^{-1}. Moreover the double adjoint $(\pi')'$ coincides with π; indeed for all $g \in G$:

$$(\pi')'(g) = \theta\pi'(g)\theta^{-1} \equiv \pi(g).$$

Denoting by $\{f_1, f_2, \ldots, f_n\}$ the orthonormal basis in V' which is dual to $\{v_1, v_2, \ldots, v_n\}$, that is $f_i(v_j) = \delta_{i,j}$ (or, equivalently, $f_i = \theta^{-1}(v_i)$), then, for the corresponding matrix coefficients $u'_{i,j}(g)$ of π' one has

$$\begin{aligned} u'_{i,j}(g) &= \langle \pi'(g)f_j, f_i \rangle = \langle \theta(f_i), \theta[\pi'(g)f_j] \rangle \\ &= \overline{\langle v_i, \pi(g)v_j \rangle} \\ &= \overline{u_{i,j}(g)}. \end{aligned} \tag{9.14}$$

Therefore, if χ_π and $\chi_{\pi'}$ are the characters of π and π' then $\chi_\pi(g) = \overline{\chi_{\pi'}(g)}$. For instance, if χ_k is the character of the cyclic group C_n as in Section 2.1, then the adjoint character is χ_{-k}.

Denote by $\widehat{G}$ the unitary dual of G, that is a complete list of irreducible, inequivalent, unitary representations of G. It is not restrictive to suppose that $\widehat{G}$ is invariant under conjugation: for all $\sigma \in \widehat{G}$, also $\sigma' \in \widehat{G}$.

For $\sigma \in \widehat{G}$ denote by V_σ and by $d_\sigma = \dim(V_\sigma)$ the space and the dimension of σ, respectively. Also we denote by $\{v_1^\sigma, v_2^\sigma, \ldots, v_{d_\sigma}^\sigma\}$ an orthonormal basis for V_σ in such a way that $\{v_1^{\sigma'}, v_2^{\sigma'}, \ldots, v_{d_\sigma}^{\sigma'}\}$ is its dual in $V_{\sigma'} = (V_\sigma)'$; the matrix coefficients of σ are the $u_{i,j}^\sigma$'s given by $u_{i,j}^\sigma(g) = \langle \sigma(g)v_j^\sigma, v_i^\sigma \rangle$. Finally, we define three kinds of subspaces of $L(G)$ spanned by the matrix coefficients of σ, namely $M_{i,*}^\sigma = \langle u_{i,j}^\sigma : j = 1, 2, \ldots, d_\sigma \rangle$, $M_{*,j}^\sigma = \langle u_{i,j}^\sigma : i = 1, 2, \ldots, d_\sigma \rangle$ and $M^\sigma = \langle u_{i,j}^\sigma : i, j = 1, 2, \ldots, d_\sigma \rangle$.

We recall that the left-regular and right-regular representations of G are the representations $(\lambda, L(G))$ and $(\rho, L(G))$ defined by

$$[\lambda(g)F](g_1) = F(g^{-1}g_1)$$

and

$$[\rho(g)F](g_1) = F(g_1g)$$

for $F \in L(G)$ and $g, g_1 \in G$; see Section 3.4.

Theorem 9.5.2

(i) $L(G) = \oplus_{\sigma \in \widehat{G}} M^\sigma$ and each M^σ is both λ- and ρ-invariant;

(ii) $M^\sigma = \oplus_{i=1}^{d_\sigma} M_{i,*}^\sigma$; each $M_{i,*}^\sigma$ is ρ-invariant and the restriction of ρ to $M_{i,*}^\sigma$ is equivalent to σ;

(iii) $M^\sigma = \oplus_{j=1}^{d_\sigma} M_{*,j}^\sigma$; each $M_{*,j}^\sigma$ is λ-invariant and the restriction of λ to $M_{*,j}^\sigma$ is equivalent to σ'.

Proof (i) The decomposition $L(G) = \bigoplus_{\sigma \in \widehat{G}} M^\sigma$ is just the Peter–Weyl theorem (Theorem 3.7.11) and λ- and ρ-invariance are proved below.

(ii) If $g, g_1 \in G$ and $i, j \in \{1, 2, \ldots, d_\sigma\}$ then, by Lemma 3.6.4 (ii),

$$[\rho(g)u_{i,j}^\sigma](g_1) = u_{i,j}^\sigma(g_1g) = \sum_{k=1}^{d_\sigma} u_{i,k}^\sigma(g_1)u_{k,j}^\sigma(g),$$

i.e. $\rho(g)u_{i,j}^\sigma = \sum_{k=1}^{d_\sigma} u_{i,k}^\sigma u_{k,j}^\sigma(g)$ and since by (3.13) $\sigma(g)v_j^\sigma = \sum_{k=1}^{d_\sigma} v_k^\sigma$ $u_{k,j}^\sigma(g)$ this means that the map $v_j^\sigma \mapsto u_{i,j}^\sigma$, $j = 1, 2, \ldots, d_\sigma$, extends to an operator that intertwines σ with $\rho|_{M_{i,*}^\sigma}$.

(iii) If $g, g_1 \in G$ and $i, j \in \{1, 2, \ldots, d_\sigma\}$ then, by Lemma 3.6.4 (ii), (i) and by (9.14)

$$[\lambda(g)u_{i,j}^\sigma](g_1) = u_{i,j}^\sigma(g^{-1}g_1)$$

$$= \sum_{k=1}^{d_\sigma} u_{i,k}^\sigma(g^{-1})u_{k,j}^\sigma(g_1)$$

$$= \sum_{k=1}^{d_\sigma} \overline{u_{k,i}^\sigma(g)}u_{k,j}^\sigma(g_1)$$

$$= \sum_{k=1}^{d_\sigma} u_{k,i}^{\sigma'}(g)u_{k,j}^\sigma(g_1)$$

i.e. $\lambda(g)u_{i,j}^\sigma = \sum_{k=1}^{d_\sigma} u_{k,j}^\sigma u_{k,i}^{\sigma'}(g)$ and, since by (3.13) $\sigma'(g)v_i^{\sigma'} = \sum_{k=1}^{d_\sigma} v_k^{\sigma'} u_{k,i}^{\sigma'}(g)$, this shows that the map $v_i^{\sigma'} \mapsto u_{i,j}^\sigma$, $i = 1, 2, \ldots, d_\sigma$, extends to an operator that intertwines σ' with $\lambda|_{M_{*,j}^\sigma}$. $\square$

The representation M^σ is the σ-*isotypic component* of $L(G)$.

Now consider the action of $G \times G$ on G given by

$$(g_1, g_2) \cdot g = g_1 g g_2^{-1}$$

and the associated permutation representation η of $G \times G$ on $L(G)$ given by

$$[\eta(g_1, g_2)f](g) = f(g_1^{-1} g g_2)$$

for all $f \in L(G)$ and $g, g_1, g_2 \in G$. Note that $\eta(g_1, g_2) = \lambda(g_1)\rho(g_2) = \rho(g_2)\lambda(g_1)$. The stabilizer of the point 1_G is the diagonal subgroup $\widetilde{G} = \{(g, g) : g \in G\}$, clearly isomorphic to G, and we may write

$$G = (G \times G)/\widetilde{G}.$$

For $\sigma \in \widehat{G}$ let $d_\sigma = \dim(V_\sigma)$ and χ_σ denote, as usual, the dimension and the character of σ, respectively.

Lemma 9.5.3 *The restriction of η to M^σ is equivalent to $\sigma' \otimes \sigma$. In particular it is irreducible.*

Proof For $f \in V_{\sigma'}$, $v \in V_\sigma$ and $g \in G$ define

$$F_v^f(g) = f(\sigma(g)v) = [\sigma'(g^{-1})f](v) = \langle \sigma(g)v, \theta(f) \rangle$$

where $\theta : V_{\sigma'} \to V_\sigma$ is the Riesz map. Noticing that for all $i, j = 1, 2, \ldots, d_\sigma$, one has $u_{i,j}^\sigma = F_{v_j}^{\theta^{-1} v_i}$ we have that the F_v^f's span the whole of M^σ.

Moreover, if $(g_1, g_2) \in G \times G$ and $g \in G$, then

$$\begin{aligned}
[\eta(g_1, g_2)F_v^f](g) &= F_v^f(g_1^{-1} g g_2) \\
&= \langle \sigma(g_1^{-1} g g_2)v, \theta(f) \rangle \\
&= \langle \sigma(g)\sigma(g_2)v, \theta(\sigma'(g_1)f) \rangle \\
&= F_{\sigma(g_2)v}^{\sigma'(g_1)f}(g)
\end{aligned}$$

so that the surjective map $V_{\sigma'} \otimes V_\sigma \ni f \otimes v \mapsto F_v^f \in M^\sigma$ intertwines $\sigma' \otimes \sigma$ with $\eta|_{M^\sigma}$. The irreducibility of $\sigma' \otimes \sigma$ follows from Theorem 9.1.6. $\qquad\square$

A group G is said to be *ambivalent* if g^{-1} is conjugate to g for all $g \in G$: for any $g \in G$ there exists $t \in G$ such that $t^{-1} g t = g^{-1}$.

Theorem 9.5.4

  (i) $(G \times G, \widetilde{G})$ *is a Gelfand pair;*

  (ii) $L(G) = \oplus_{\sigma \in \widehat{G}} M^\sigma$ *is the decomposition of* $L(G)$ *into irreducible spherical representations of* $G \times G$;

  (iii) $\frac{1}{d_\sigma} \chi_\sigma$ *is the spherical function in* M^σ;

  (iv) *the pair is symmetric if and only if* G *is ambivalent.*

Proof We first observe that the action of $\widetilde{G}$ on G is simply the conjugation: for all $g, g_1 \in G$,

$$(g_1, g_1) \cdot g = g_1^{-1} g g_1.$$

Therefore, recalling Theorem 3.9.10,

$$
\begin{aligned}
\# \text{ orbits of } \widetilde{G} \text{ on } G \;\; &= \;\; \# \text{ conjugacy classes of } G \\
&= \;\; |\widehat{G}|.
\end{aligned}
$$

Moreover, from Lemma 9.5.3 and the fact that the matrix coefficients $\{u_{i,j}^\sigma : \sigma \in \widehat{G}, i, j = 1, 2, \ldots, d_\sigma\}$ form an orthogonal basis for $L(G)$ (see Lemma 3.6.3) we deduce that $L(G) = \oplus_{\sigma \in \widehat{G}} M^\sigma$ is the decomposition of $L(G)$ into irreducible $(G \times G)$-representations. Moreover, from Theorem 9.5.2 we deduce that this decomposition is multiplicity free (the restrictions to $\{1\} \times G$ are inequivalent for different choices of $\sigma \in \widehat{G}$). Then (i) and (ii) follow from Proposition 4.4.4.

To prove (iii), note that, in our setting, bi-$\widetilde{G}$-invariance is nothing but invariance by conjugation; in other words the algebra of bi-$\widetilde{G}$-invariant functions coincides with the algebra of central functions on G. This gives, incidentally, another proof of (i). Now (Theorem 3.9.10), the characters form a basis for the latter algebra and, in particular, χ_σ, which is a sum of matrix coefficients of σ, spans the (one-dimensional space of) central functions in M^σ, and (iii) follows from $\chi_\sigma(1_G) = d_\sigma$.

Finally, $(G \times G, \widetilde{G})$ is symmetric if and only if $(g_1^{-1}, g_2^{-1}) \in \widetilde{G}(g_1, g_2)\widetilde{G}$ for all $g_1, g_2 \in G$, that is if and only if for every pair $(g_1, g_2) \in G \times G$ there exist $g_3, g_4 \in G$ such that

$$
\begin{cases}
g_1^{-1} = g_3 g_1 g_4 \\
g_2^{-1} = g_3 g_2 g_4.
\end{cases}
\tag{9.15}
$$

Now (9.15) is equivalent to the existence of $g_3 \in G$ such that

$$
g_1^{-1} g_2 = g_3 (g_1 g_2^{-1}) g_3^{-1};
\tag{9.16}
$$

indeed, if condition (9.16) (apparently weaker) is satisfied, one can take $g_4 = g_1^{-1}g_3^{-1}g_1^{-1} \equiv g_2^{-1}g_3^{-1}g_2^{-1}$. The equivalence of (9.16) and ambivalence of G is now clear: if (9.16) is satisfied take $g_2 = 1_G$; conversely, if G is ambivalent, given g_1 and $g_2 \in G$ there exists $g_5 \in G$ such that

$$
\begin{aligned}
g_1^{-1}g_2 &= g_5(g_2^{-1}g_1)g_5^{-1} \\
&= g_5 g_2^{-1} g_1 (g_2^{-1}g_2)g_5^{-1} \\
&= (g_5 g_2^{-1})g_1 g_2^{-1}(g_5 g_2^{-1})^{-1}
\end{aligned}
$$

and taking $g_3 = g_5 g_2^{-1}$ this gives (9.16). This proves (iv). $\qquad\square$

Remark 9.5.5 It is easy to see that $(G \times G, \widetilde{G})$ is always weakly symmetric (see Exercise 4.3.3). Indeed for the automorphism $\tau : G \times G \ni (g_1, g_2) \mapsto (g_2, g_1) \in G \times G$ one always has

$$(g_1, g_2)^{-1} = (g_1^{-1}, g_2^{-1}) = (g_1^{-1}, g_1^{-1})(g_2, g_1)(g_2^{-1}, g_2^{-1}) \in \widetilde{G}\tau(g_1, g_2)\widetilde{G}$$

for all $g_1, g_2 \in G$.

Corollary 9.5.6 *The projection $E_\sigma : L(G) \to M^\sigma$ is given by*

$$E_\sigma = \frac{d_\sigma}{|G|} f * \chi_\sigma \qquad \text{for all } f \in L(G).$$

Proof This is just a reformulation in our setting of Proposition 4.7.10.

Alternatively, it also easily follows from Lemma 3.9.14: we leave to the reader to check the details. $\qquad\square$

Remark 9.5.7 Let (G, K) be a Gelfand pair and (ρ, V) a spherical representation. Take an orthonormal basis $v_1, v_2, \ldots, v_d$ in V (here $d = \dim(V)$) such that v_1 is K-invariant. Then the spherical function ϕ associated with V is given by

$$\phi(g) = \langle v_1, \rho(g)v_1 \rangle_V = \langle \rho(g^{-1})v_1, v_1 \rangle_V = \langle \rho'(g)v_1, v_1 \rangle_V$$

where ρ' denotes the conjugate representation of ρ (see Corollary 4.6.4).

In other words, the spherical function ϕ is a matrix coefficient of the conjugate representation ρ' but not of ρ (unless ρ is selfconjugate).

This happens because ϕ belongs to the sub-representation of $L(G/K)$ isomorphic to V (see Corollary 4.6.4). Note also that, by Theorem 9.5.2, the restriction of the left regular representation λ to $M_{*,j}^\sigma$ is isomorphic to σ', that is, $M_{*,1}^{\sigma'} \sim V$. $M_{*,1}^{\sigma'}$ is made up of the right K-invariant functions corresponding to V inside $L(G/K)$: this again follows from Corollary 4.6.4.

Exercise 9.5.8 Let (G, K) be a Gelfand pair and (ρ, V) a spherical representation. Denote by $\chi = \chi_\rho$ its character and by ϕ the corresponding spherical function. Also set $d = \dim(V)$. Show that

(1) $\phi(g) = \frac{1}{|K|} \sum_{k \in K} \overline{\chi(gk)}$;

(2) $\chi(g) = \frac{d}{|G|} \sum_{h \in G} \overline{\phi(h^{-1}gh)}$.

Exercise 9.5.9 Let $f^\flat$ and $f^\sharp$ be as in Section 4.8. Show that $\widehat{f^\flat}(\rho) = \widehat{f}(\rho)^*$ and $\widehat{f^\sharp}(\rho) = \widehat{f}(\rho')^*$ for all $\rho \in \widehat{G}$.

We end this section with still another view on $L(G)$ and the Fourier transform, which connects them with the material developed in Sections 9.1 and 9.3.

Example 9.5.10 Denote by $Hom(V)$ the set of all linear maps $T : V \to V$ of a finite dimensional complex vector space V. Define the *Hilbert–Schmidt scalar product* on $Hom(V)$ by setting

$$\langle T, S \rangle_{V, HS} = Tr(TS^*)$$

for all $T, S \in Hom(V)$. (Exercise: (1) show that $\langle \cdot, \cdot \rangle_{V, HS}$ is indeed a scalar product; (2) if $\{v_1, v_2, \ldots, v_n\}$ is an orthonormal basis for V, then $\langle T, S \rangle_{V, HS} = \sum_{i=1}^{n} \langle Tv_i, Sv_i \rangle$.)

Let G be again a finite group. Suppose that V is a G-representation and denote by $V = \oplus_{\sigma \in \widehat{G}} M^\sigma$ the decomposition of V into its isotypical components (thus $M^\sigma = m_\sigma V_\sigma$ with V_σ irreducible). For $T \in Hom_G(V)$ we then write $T = \oplus_{\sigma \in \widehat{G}} T_\sigma$ where $T_\sigma \in Hom_G(M^\sigma)$. For $S, T \in Hom_G(V)$ we then set

$$\langle T, S \rangle_{HS, G} = \frac{1}{|G|} \sum_{\sigma \in \widehat{G}} d_\sigma \langle T_\sigma, S_\sigma \rangle_{M^\sigma, HS}.$$

We can summarize the results in Section 3.9 in the following way.

Proposition 9.5.11 *The Fourier transform is an isometric isomorphism between the group algebra $L(G)$ and $\bigoplus_{\sigma \in \widehat{G}} Hom(V_\sigma)$.*

Recall that if V is a finite dimensional vector space and V' denotes its dual, then $Hom(V) \cong V \otimes V'$. An explicit isomorphism is given by linearly extending to the whole of $V \otimes V'$ the map

$$
\begin{aligned}
V \otimes V' &\to Hom(V) \\
v \otimes f &\mapsto T_{v,f}
\end{aligned}
\tag{9.17}
$$

where $T_{v,f}(w) = f(w)v$ for all $w \in V$.

Exercise 9.5.12 Fill in all the details relative to (9.17).

Let now (ρ, V_ρ) be a G-representation. For $f \in V'_\rho$ and $v \in V_\rho$ we define the function $F^\rho_{f,v} \in L(G)$ by setting, for all $g \in G$,

$$
F^\rho_{f,v}(g) = f[\rho(g^{-1})v]
$$

(compare with the proof of Lemma 9.5.3).

We then have two maps. The first one is the Fourier transform:

$$
\begin{aligned}
\hat{\ }: \quad L(G) &\to \bigoplus_{\rho \in \widehat{G}} [V_\rho \otimes V'_\rho] \left(\cong \bigoplus_{\rho \in \widehat{G}} Hom(V_\rho) \right) \\
F &\mapsto \widehat{F} = \bigoplus_{\rho \in \widehat{G}} \widehat{F}(\rho)
\end{aligned}
$$

where $\widehat{F}(\rho) = \sum_{g \in G} F(g)\rho(g) \in Hom(V_\rho)$. The second one (the inverse Fourier transform)

$$
\mathcal{T} : \bigoplus_{\rho \in \widehat{G}} [V_\rho \otimes V'_\rho] \to L(G)
$$

is defined by

$$
\mathcal{T}(v \otimes f) = \frac{d_\rho}{|G|} F^\rho_{f,v}
$$

for all $\rho \in \widehat{G}$, $v \in V_\rho$ and $f \in V_{rho'}$ and then extending by linearity.

Exercise 9.5.13 Show directly that the maps $\hat{\ }$ and $\mathcal{T}$ are the adjoint of each other, that is

$$
\langle F, \mathcal{T}(v \otimes f) \rangle_{L(G)} = \langle \widehat{F}, v \otimes f \rangle_{HS,G}
$$

for all $F \in L(G)$, $v \in V_\rho$ and $f \in V'_\rho$ where $\rho \in \widehat{G}$.

Exercise 9.5.14 Suppose that $(\sigma, V), (\pi, W)$ and (η, U) are three representations of a group G.
(1) Show that

$$
Hom_G(\pi' \otimes \eta, \sigma) \cong Hom_G(\eta, \pi \otimes \sigma)
$$

by constructing an explicit isomorphism.
(2) Show that $\dim Hom_G(\pi' \otimes \eta, \sigma) = \dim Hom_G(\eta, \pi \otimes \sigma)$ by computing the scalar product of the corresponding characters.

9.6 An example of a not weakly symmetric Gelfand pair

In this section we describe un example of a Gelfand pair which is not weakly symmetric. The existence of not weakly symmetric Gelfand pairs is indicated in Terras' book [220] who directly refers to Krieg's monograph [148] but in both cases no details are given. We thank Ernst Vinberg [225] for addressing our attention to such an interesting example.

Example 9.6.1 Let p be a prime number. Denote by $\mathbb{F}_p$ the field with p elements and by

$$G = GL(2, p) = \left\{ \begin{pmatrix} a & b \\ c & d \end{pmatrix} : ad - bc \neq 0 \right\}$$

the *general linear group* of 2×2 matrices over $\mathbb{F}_p$. Let

$$K = Aff(2, \mathbb{F}_p) := \left\{ \begin{pmatrix} a & b \\ 0 & 1 \end{pmatrix} : a \neq 0 \right\}$$

the subgroup of *affine transformations* of $\mathbb{F}_p$ (that is $\begin{pmatrix} a & b \\ 0 & 1 \end{pmatrix} \cdot x = ax + b$ for all $x \in \mathbb{F}_p$). Then $(GL(2, p), Aff(2, \mathbb{F}_p))$ is a Gelfand pair. Moreover, if $p \geq 5$ it is not weakly symmetric.

It is a simple computation to show that the number of K-double cosets is p. Indeed, let $\begin{pmatrix} a & b \\ c & d \end{pmatrix} \in GL(2, p)$. If $c \neq 0$ then

$$\begin{pmatrix} a & b \\ c & d \end{pmatrix} = \begin{pmatrix} \frac{bc - ad}{c} & \frac{a}{c} \\ 0 & 1 \end{pmatrix} \begin{pmatrix} 0 & 1 \\ 1 & 0 \end{pmatrix} \begin{pmatrix} c & d \\ 0 & 1 \end{pmatrix}$$

while, if $c = 0$ (and therefore $a, d \in \mathbb{F}_p^*$),

$$\begin{pmatrix} a & b \\ 0 & d \end{pmatrix} = \begin{pmatrix} \frac{a}{d} & 0 \\ 0 & 1 \end{pmatrix} \begin{pmatrix} d & 0 \\ 0 & d \end{pmatrix} \begin{pmatrix} 1 & \frac{b}{a} \\ 0 & 1 \end{pmatrix}.$$

Setting $w = \begin{pmatrix} 0 & 1 \\ 1 & 0 \end{pmatrix}$ and $I_d = \begin{pmatrix} d & 0 \\ 0 & d \end{pmatrix}$, an instant of thought shows that

$$G = KwK \coprod \left(\coprod_{d \in \mathbb{F}_p^*} KI_d K \right). \tag{9.18}$$

Note that if h is in the center of G, then one has

$$\mathbf{1}_{KgK} * \mathbf{1}_{KhK} = \left(\sum_{k_1,k_2 \in K} \delta_{k_1 g k_2} \right) * \left(\sum_{k_3,k_4 \in K} \delta_{k_3 h k_4} \right)$$

$$= \sum_{k_1,k_2,k_3,k_4 \in K} \delta_{k_1 g k_2 k_3 h k_4}$$

$$= \sum_{k_1,k_2,k_3,k_4 \in K} \delta_{k_1 h g k_2 k_3 k_4}$$

$$= \left(\sum_{k_1,k_2 \in k} \delta_{k_1 h k_2} \right) * \left(\sum_{k_3,k_4 \in k} \delta_{k_3 g k_4} \right)$$

$$= \mathbf{1}_{KhK} * \mathbf{1}_{KgK}.$$

Therefore, as the diagonal matrices I_d belong to the center of G and the algebra $L(K\backslash G/K)$ of bi-K-invariant functions is linearly spanned by the characteristic functions $\mathbf{1}_{KwK}$ and $\mathbf{1}_{KI_dK}$, $d \in \mathbb{F}_p^*$, one has that $L(K\backslash G/K)$ is commutative. In other words, (G, K) is a Gelfand pair.

To show that (G, K) is not weakly symmetric we recall the description of the group $Aut(GL(2, p))$ of automorphisms of $GL(2, p)$. There are two kinds of automorphisms in $Aut(GL(2, p))$ (see [124]):

- $\tau_B : A \mapsto BAB^{-1}$, where $B \in GL(2, p)$ (inner automorphisms);
- $\tau_\chi : A \mapsto \chi(A)A$ where $\chi : G \to \mathbb{F}_p^*$ is a group homomorphism (radial automorphisms).

Moreover, any automorphism $\tau \in Aut(GL(2, p))$ can be expressed as a product of an inner and a radial automorphism. Our goal is to show that the condition

$$\tau(g) \in Kg^{-1}K \tag{9.19}$$

is verified by no automorphism $\tau \in Aut(GL(2, p))$ if $p \geq 5$.

Claim 1. *Suppose that $\tau \in Aut(GL(2, p))$ satisfies (9.19). Then necessarily*

$$\tau(I_d) = I_{d^{-1}} \tag{9.20}$$

and

$$\tau(K) = K. \tag{9.21}$$

The proof is straightforward. (9.20) follows from (9.19) with $g = I_d$ and the fact that an automorphism leaves the center invariant (also recall (9.18)), while (9.21) follows from (9.19) with $g \in K$.   □

Claim 2. *There is no radial automorphism such that both (9.20) and (9.21) hold.*

Indeed, let $\tau = \tau_\chi$ be a radial automorphism. Looking at the (2,2)-entries, formula (9.21) has the stronger formulation: $\tau(k) = k$ for all $k \in K$. Moreover, $\tau(w) = \lambda w$ with $\lambda = \chi(w) \in \mathbb{F}_p$ such that $\lambda^2 = 1$. Suppose that $a, d \in \mathbb{F}_p^*$. Then

$$\tau \left[\begin{pmatrix} a & 0 \\ 0 & d \end{pmatrix} \right] = \tau \left[\begin{pmatrix} \frac{a}{d} & 0 \\ 0 & 1 \end{pmatrix} \right] \tau \left[\begin{pmatrix} d & 0 \\ 0 & d \end{pmatrix} \right]$$

$$= \begin{pmatrix} \frac{a}{d} & 0 \\ 0 & 1 \end{pmatrix} \begin{pmatrix} d^{-1} & 0 \\ 0 & d^{-1} \end{pmatrix}$$

$$= \begin{pmatrix} \frac{a}{d^2} & 0 \\ 0 & \frac{1}{d} \end{pmatrix}.$$

On the other hand,

$$w \begin{pmatrix} a & 0 \\ 0 & d \end{pmatrix} w = \begin{pmatrix} d & 0 \\ 0 & a \end{pmatrix}$$

so that, if $a^2 \neq d^2$,

$$\tau \left[w \begin{pmatrix} a & 0 \\ 0 & d \end{pmatrix} w \right] = \tau \left[\begin{pmatrix} d & 0 \\ 0 & a \end{pmatrix} \right] = \begin{pmatrix} \frac{d}{a^2} & 0 \\ 0 & \frac{1}{a} \end{pmatrix}$$

differs from

$$\tau(w) \tau \left[\begin{pmatrix} a & 0 \\ 0 & d \end{pmatrix} \right] \tau(w) = \lambda^2 w \tau \left[\begin{pmatrix} a & 0 \\ 0 & d \end{pmatrix} \right] w = \begin{pmatrix} \frac{1}{d} & 0 \\ 0 & \frac{a}{d^2} \end{pmatrix}.$$

Thus τ fails to be multiplicative, a contradiction.   □

We now show that there is no automorphism $\tau \in Aut(GL(2, p))$ verifying (9.19). This will be done via a case-by-case analysis.

- We first observe that (9.20) rules out all inner automorphisms (in the case $p \geq 5$).
- τ cannot be radial: this immediately follows from Claims 1 and 2.
- Suppose now that τ is a composition of an inner and a radial automorphism, say $\tau = \tau_\chi \tau_B$, and that τ satisfies (9.19). We distinguish two cases: $B \in K I_d K$, for some $d \in \mathbb{F}_p^*$, and $B \in KwK$. In the first case, as I_d is central, $\tau_B = \tau_k$ for some $k \in K$. From (9.19), as $\tau_B(K) = K$, we deduce that $\tau_\chi(K) = K$. Also,

as $\tau_B(I_d) = I_d$, from (9.20) we deduce that $\tau_\chi(I_d) = I_{d-1}$. These two facts contradict Claim 2.

Suppose now that $B \in KwK$, say $B = k_1 w k_2$. Let $a, b \neq 0$ and set $k = k_2^{-1} \begin{pmatrix} a & b \\ 0 & 1 \end{pmatrix} k_2 \in K$. Then

$$\tau_B(k) = k_1 w \begin{pmatrix} a & b \\ 0 & 1 \end{pmatrix} w k_1^{-1} = k_1 \begin{pmatrix} 1 & 0 \\ b & a \end{pmatrix} k_1^{-1} = \begin{pmatrix} * & * \\ \frac{b}{x} & * \end{pmatrix}$$

where x denotes the $(1,1)$-entry of k_1. But then

$$\tau(k) = \tau_\chi \left[\begin{pmatrix} * & * \\ \frac{b}{x} & * \end{pmatrix} \right] = \begin{pmatrix} * & * \\ \lambda \frac{b}{x} & * \end{pmatrix} \notin K.$$

This contradicts (9.21).

9.7 Real, complex and quaternionic representations: the theorem of Frobenius and Schur

Let (π, V) be an irreducible representation of a finite group G and let (π', V') denote its conjugate representation as in Section 9.5. We recall that $\chi_{\pi'} = \overline{\chi_\pi}$ and therefore π is selfconjugate (Definition 9.5.1) if and only if $\chi_\pi(g) \in \mathbb{R}$ for all $g \in G$, that is its character is a real valued function.

Definition 9.7.1 The representation π is *complex* when it is not selfconjugate.

Equivalently, π is complex if and only if π and π' are not unitarily equivalent (i.e. $\chi_\pi \neq \chi_{\pi'}$).

Further, we split the class of selfconjugate representations into two subclasses.

Definition 9.7.2 Let π be selfconjugate and suppose that there exists an orthonormal basis $\{v_1, v_2, \ldots, v_d\}$ in V such that the corresponding matrix coefficients are real valued: $u_{i,j}(g) = \langle \pi(g)v_j, v_i \rangle \in \mathbb{R}$ for any $g \in G$ and $i, j = 1, 2, \ldots, d$. Then we say that π is *real*. Otherwise we say that π is *quaternionic*.

The etymological reason for the last definition becomes apparent from the following basic example.

Example 9.7.3 Let $Q = \{1, -1, i, -i, j, -j, k, -k\}$ denote the *quaternionic* group. We recall that the multiplication law is given by

$$ij = k \qquad jk = i \qquad ki = j$$

$$ji = -k \qquad kj = -i \qquad ik = -j$$

$$i^2 = -1 \qquad j^2 = -1 \qquad k^2 = -1$$

together with the sign rule

$$(-x)y = -(xy) \text{ and } (-x)(-y) = xy$$

for all $x, y \in Q$.

Define the two-dimensional unitary representation ρ of Q by setting

$$\rho(\pm 1) = \pm \begin{pmatrix} 1 & 0 \\ 0 & 1 \end{pmatrix} \quad \rho(\pm i) = \pm \begin{pmatrix} i & 0 \\ 0 & -i \end{pmatrix}$$

$$\rho(\pm j) = \pm \begin{pmatrix} 0 & i \\ i & 0 \end{pmatrix} \quad \rho(\pm k) = \pm \begin{pmatrix} 0 & -1 \\ 1 & 0 \end{pmatrix}. \tag{9.22}$$

It is easy to check that ρ is indeed a representation. In fact Q is sometimes defined as the group consisting of the eight matrices above.

Its character is given by

$$\chi_\rho(g) = \begin{cases} 2 & \text{if } g = 1 \\ -2 & \text{if } g = -1 \\ 0 & \text{if } g \neq \pm 1. \end{cases}$$

As $\sum_{g \in Q} |\chi_\rho(g)|^2 = 4 + 4 = |Q|$ we have that π is irreducible (Corollary 3.7.8), moreover χ_ρ being real valued, π is selfconjugate.

We now show that π is quaternionic: more explicitly we show that there is no basis of $\mathbb{C}^2$ for which ρ is represented by real matrices.

Suppose, by contradiction, that U is a unitary matrix such that $U\rho(g)U^*$ is real valued for any $g \in Q$. Since the matrices $A = \begin{pmatrix} 1 & 0 \\ 0 & 1 \end{pmatrix}$, $B = \begin{pmatrix} 0 & 1 \\ 1 & 0 \end{pmatrix}$ $C = \begin{pmatrix} 0 & -1 \\ 1 & 0 \end{pmatrix}$ and $D = \begin{pmatrix} 1 & 0 \\ 0 & -1 \end{pmatrix}$ form a basis of the vector space $M_{2,2}(\mathbb{R})$, then there exist $\alpha, \beta, \gamma, \delta : Q \to \mathbb{R}$ such that

$$U\rho(g)U^* = \alpha(g)A + \beta(g)B + \gamma(g)C + \delta(g)D.$$

As $\chi_\rho(g) = tr(U\rho(g)U^*) = \alpha(g)tr(A) = 2\alpha(g)$ vanishes for $g \neq \pm 1$, we conclude that

$$\alpha(1) = 1, \quad \alpha(-1) = -1 \tag{9.23}$$

and, for all $g \neq \pm 1$,

$$\alpha(g) = 0. \tag{9.24}$$

Moreover direct computations show that, for all $g, h \in Q$,

$$\alpha(gh) = \alpha(g)\alpha(h) + \beta(g)\beta(h) - \gamma(g)\gamma(h) + \delta(g)\delta(h). \tag{9.25}$$

Suppose now that $g, h \neq \pm 1$. Setting $g = h$ in (9.25) we deduce from (9.23) and (9.24) that

$$\beta(g)^2 - \gamma(g)^2 + \delta(g)^2 = -1. \tag{9.26}$$

On the other hand from (9.25) and (9.24) one deduces that, for $h \neq g$,

$$\beta(g)\beta(h) - \gamma(g)\gamma(h) + \delta(g)\delta(h) = 0. \tag{9.27}$$

Set $b_1 = \beta(g), b_2 = \beta(h), c_1 = \gamma(g), c_2 = \gamma(h), d_1 = \delta(g)$ and $d_2 = \delta(h)$. With this position and supposing that $h \neq g$ equations (9.26) and (9.27) altogether give

$$\begin{cases} b_1^2 - c_1^2 + d_1^2 = -1 \\ b_1^2 - c_1^2 + d_1^2 = -1 \\ b_1 b_2 - c_1 c_2 + d_1 d_2 = 0. \end{cases}$$

From the above system we have

$$(1 + b_1^2 + d_1^2)(1 + b_2^2 + d_2^2) = (c_1 c_2)^2 = (b_1 b_2 + d_1 d_2)^2.$$

This yields

$$1 + b_1^2 + b_2^2 + d_1^2 + d_2^2 = -(b_1 d_2 + b_2 d_1)^2$$

a contradiction. Therefore ρ is quaternionic.

Exercise 9.7.4 Determine all the irreducible representations of Q.

Exercise 9.7.5 Let (ρ, V) be a selfadjoint representation of a group G. Suppose that V is endowed with a scalar product $\langle \cdot, \cdot \rangle$ and that there exists a basis (not necessarily orthonormal) $\{v_1, v_2, \ldots, v_d\}$ in V such that the corresponding matrix coefficients are real valued.

(1) Show that there exists a ρ-invariant scalar product $(\cdot, \cdot)$ with an orthonormal basis $\{w_1, w_2, \ldots, w_d\}$ in V such that the corresponding matrix coefficients are real valued: $u_{i,j}(g) = (\rho(g)w_j, w_i) \in \mathbb{R}$ for any $g \in G$ and $i, j = 1, 2, \ldots, d$.

(2) Show that if $\widetilde{\rho}(g)$ is the matrix representing $\rho(g)$ in a fixed $\langle \cdot, \cdot \rangle$-orthonormal basis $\{u_1, u_2, \ldots, u_d\}$ in V, then there exists a unitary matrix U such that $U\widetilde{\rho}(g)U^*$ is real for all $g \in G$.

This shows that it is not restrictive in Definition 9.7.2 to impose orthonormality for the basis in the representation space V.

Suppose now that ρ is selfconjugate. This means that there exists a unitary operator $W : V \to V'$ which intertwines ρ and ρ'. Let $\{v_1, v_2, \ldots, v_d\}$ be a basis in V and denote by $\{f_1, f_2, \ldots, f_d\}$ the dual basis in V'. Then condition $W\rho(g)W^* = \rho'(g)$ becomes

$$\overline{u_{i,j}(g)} = \sum_{t,k=1}^{d} w_{i,k} u_{k,t}(g) \overline{w_{j,t}} \tag{9.28}$$

for all $g \in G$ and $i, j = 1, 2, \ldots, d$, where $W = (w_{i,k})_{i,k=1,2,\ldots,d}$ is the matrix representing W.

Denoting by $U(g) = (u_{i,j}(g))_{i,j=1,2,\ldots,d}$ the matrix representing $\rho(g)$ (with respect to the basis $\{v_1, v_2, \ldots, v_d\}$) and, for a general matrix A, denoting $\overline{A}$ its conjugate (obtained by conjugating all coefficients) then (9.28) may be written

$$\overline{U(g)} = WU(g)W^*. \tag{9.29}$$

Lemma 9.7.6 *Suppose that ρ is selfconjugate. Then, with W as above, we have*

$$W\overline{W} = \overline{W}W = \pm I.$$

More precisely, $\overline{W}W = I$ if ρ is real and $\overline{W}W = -I$ if ρ is quaternionic.

Proof Conjugating (9.29) we obtain

$$U(g) = \overline{W}\,\overline{U(g)}(\overline{W})^* \equiv \overline{W}\,\overline{U(g)}W^T$$

and therefore

$$U(g) = \overline{W}WU(g)W^*(\overline{W})^* = (\overline{W}W)U(g)(\overline{W}W)^*.$$

This means that the linear operator $V \to V$ represented by the matrix $\overline{W}W$ intertwines ρ with itself. By Schur's lemma,

$$\overline{W}W = cI \tag{9.30}$$

with $|c| = 1$ as $\overline{W}W$ is unitary. Moreover, from $(\overline{W}W)W = W(\overline{W}W) = WcI$ we deduce that $W\overline{W} = cI$, while conjugating (9.30) gives $W\overline{W} = \overline{c}I$. Thus c is real and in fact $c = \pm 1$.

Suppose that $c = 1$, that is $\overline{W}W = I$ or, equivalently, $W^* = \overline{W}$. We want to prove that ρ is real by showing that there exists a unitary matrix B such that $B^*U(g)B$ is a real matrix for all $g \in G$. As W is unitary, there exists a unitary matrix A such that

$$A^*WA = \Lambda \tag{9.31}$$

with Λ unitary and diagonal:

$$\Lambda = \begin{pmatrix} \lambda_1 & 0 & 0 & \cdots & 0 \\ 0 & \lambda_2 & 0 & \cdots & 0 \\ \vdots & \vdots & \vdots & \ddots & \vdots \\ 0 & 0 & \cdots & \lambda_{d-1} & 0 \\ 0 & 0 & \cdots & 0 & \lambda_d \end{pmatrix}.$$

Conjugating both sides of (9.31) we get

$$A^T\overline{W}A = \overline{\Lambda}$$

while from $\overline{W}W = I$ we get

$$I = A^T\overline{W}WA = A^T\overline{W}A \cdot A^TWA = \overline{\Lambda}A^TWA$$

which, in virtue of $\Lambda^{-1} = \overline{\Lambda}$ becomes

$$A^TW\overline{A} = \Lambda.$$

Therefore we conclude that $W(A + \overline{A}) = A\Lambda + \overline{A}\Lambda = (A + \overline{A})\Lambda$. In particular, W has a set of <u>real</u> eigenvectors and we can say that there exists a <u>real</u> unitary matrix A_1 with the same properties of A, that is

$$A_1^*WA_1 \equiv A_1^TWA_1 = \Lambda$$

and

$$A_1^T\overline{W}A_1 = \overline{\Lambda}$$

(A_1 may be obtained by an application of the Gram–Schmidt process to the real eigenvectors of W). Let Θ be a unitary and diagonal matrix such that $\Theta^2 = \Lambda$ and define

$$B = A_1\overline{\Theta}.$$

Then B is unitary and we have

$$\overline{B^*U(g)B} = \overline{\Theta}A_1^T\overline{U(g)}A_1\Theta$$

from (9.29) $= \overline{\Theta}A_1^T W U(g) W^* A_1\Theta$

from $W^* = \overline{W}$ $= \overline{\Theta}A_1^T W A_1 A_1^T U(g) A_1 A_1^T \overline{W} A_1\Theta$

$$= \overline{\Theta}\Theta^2 A_1^T U(g) A_1 \overline{\Theta}^2\Theta$$

$$= \Theta A_1^T U(g) A_1 \overline{\Theta}$$

$$= B^*U(g)B$$

and therefore ρ is real.

Conversely suppose that ρ is real so that there exists a unitary matrix B such that $\overline{B^*U(g)B} = B^*U(g)B$ for all g in G. This is equivalent to

$$\overline{U(g)} = \overline{B}B^*U(g)BB^T$$

and by Schur's lemma (see also Section 9.3) we have that $\overline{B}B^*$ must be a multiple of W in (9.29):

$$W = c\overline{B}B^*$$

with $|c| = 1$. Therefore

$$\overline{W}W = \overline{c}BB^T c\overline{B}B^* = I.$$

$$\square$$

The following frundamental theorem shows how it is possible to distinguish among real, complex and quaternionic representations by looking at their characters.

Theorem 9.7.7 (Frobenius and Schur) *Let χ be the character of an irreducible representation (ρ, V) of G. Then*

$$\frac{1}{|G|}\sum_{g\in G}\chi(g^2) = \begin{cases} 1 & \text{if} \quad \rho \text{ is real} \\ 0 & \text{if} \quad \rho \text{ is complex} \\ -1 & \text{if} \quad \rho \text{ is quaternionic.} \end{cases} \qquad (9.32)$$

Proof Fix an orthonormal basis $\{v_1, v_2, \ldots, v_d\}$ in V and set, as usual, $u_{i,j} = \langle \rho(g)v_j, v_i \rangle$. Then

$$\frac{1}{|G|}\sum_{g\in G}\chi(g^2) = \frac{1}{|G|}\sum_{g\in G}\sum_{i=1}^d u_{i,i}(g^2) = \frac{1}{|G|}\sum_{g\in G}\sum_{i,j=1}^d u_{i,j}(g)u_{j,i}(g)$$

$$(9.33)$$

where the last equality comes from Lemma 3.6.4. If ρ is real we can choose the basis $\{v_1, v_2, \ldots, v_d\}$ in such a way that the matrix coefficients $u_{i,j}(g)$ are all real that is $u_{i,j}(g) = \overline{u_{i,j}(g)}$. By the orthogonality relations (Proposition 3.6.3) we get from (9.33):

$$\frac{1}{|G|} \sum_{g \in G} \chi(g^2) = \frac{1}{|G|} \sum_{i,j=1}^{d} \frac{|G|}{d} \delta_{i,j} = 1.$$

If ρ is complex then from (9.14) we get $u_{j,i}(g) = \overline{u'_{j,i}(g)}$ where $u'_{j,i}(g)$ is the matrix coefficients of the conjugate representation ρ' and, since ρ and ρ' are not equivalent, again from the orthogonality relations we obtain

$$\frac{1}{|G|} \sum_{g \in G} \chi(g^2) = 0.$$

Finally suppose that ρ is quaternionic. Then from (9.29) written in the form $U(g) = \overline{W U(g)} W^T$ (by Lemma 9.7.6), (9.33) becomes:

$$\begin{aligned}
\frac{1}{|G|} \sum_{g \in G} \chi(g^2) &= \sum_{i,j=1}^{d} \sum_{k,t=1}^{d} \overline{w_{j,k}} w_{i,t} \cdot \frac{1}{|G|} \sum_{g \in G} u_{i,j}(g) \overline{u_{k,t}(g)} \\
&= \sum_{i,j=1}^{d} \sum_{k,t=1}^{d} \overline{w_{j,k}} w_{i,t} \delta_{i,k} \delta_{j,t} \frac{1}{d} \\
&= \frac{1}{d} tr[\overline{W} W] \\
\text{(by Lemma 9.7.6)} &= \frac{1}{d} tr[-I] = -1.
\end{aligned}$$

$\square$

For any $h \in G$ set $\mathcal{S}(h) = |\{g \in G : g^2 = h\}|$ that is $\mathcal{S}(h)$ is the number of square roots of h in G. For any $\rho \in \widehat{G}$ set

$$c(\rho) = \begin{cases} 1 & \text{if } \rho \text{ is real;} \\ 0 & \text{if } \rho \text{ is complex;} \\ -1 & \text{if } \rho \text{ is quaternionic.} \end{cases} \tag{9.34}$$

Then we have

Corollary 9.7.8

$$\mathcal{S}(h) = \sum_{\rho \in \widehat{G}} c(\rho) \chi_\rho(h)$$

and in particular

$$\mathcal{S}(1_G) = \sum_{\substack{\rho \in \widehat{G} \\ \rho \ real}} d_\rho - \sum_{\substack{\rho \in \widehat{G} \\ \rho \ quaternionic}} d_\rho.$$

Proof From the Theorem of Frobenius and Schur (9.7.7) we have

$$\frac{1}{|G|} \sum_{h \in G} \overline{\chi_\rho(h)} \mathcal{S}(h) = c(\rho)$$

(note that $\mathcal{S}(h)$ and $c(\rho)$ are real valued and we may conjugate formula (9.32)).

Now, $\mathcal{S}(h)$ is a central function and the characters constitute an orthogonal basis of the central functions (Theorem 3.9.10). Therefore the statement follows from the orthogonality relations of the characters (Proposition 3.7.4).   □

Definition 9.7.9 Let G be a group. A conjugacy class C in G is *ambivalent* if $g^{-1} \in C$ for all $g \in C$.

Therefore G is ambivalent (see Section 9.5) if and only if all conjugacy classes are ambivalent.

The following theorem relates the number of ambivalent classes with the number of selfconjugate representations.

Theorem 9.7.10 *Let h be the number of selfconjugate representations of G. Then we have*

(i) $h = \frac{1}{|G|} \sum_{g \in G} \mathcal{S}(g)^2$

(ii) h *is equal to the number of ambivalent classes in G.*

Proof (i) From the previous corollary and the orthogonality relations of the characters we get

$$\frac{1}{|G|} \sum_{g \in G} \mathcal{S}(g)^2 = \frac{1}{|G|} \sum_{g \in G} \mathcal{S}(g)\overline{\mathcal{S}(g)}$$

$$= \sum_{\rho, \pi \in \widehat{G}} c(\rho)c(\pi) \frac{1}{|G|} \sum_{g \in G} \chi_\rho(g)\overline{\chi_\pi(g)}$$

$$= \sum_{\rho, \pi \in \widehat{G}} c(\rho)c(\pi)\delta_{\rho,\pi} = h.$$

(ii) Note that if ρ' is the conjugate of ρ we have $\chi_\rho(g) = \overline{\chi_{\rho'}(g)}$ (see Section 9.5) and therefore

$$\frac{1}{|G|} \sum_{g \in G} \chi_\rho(g)^2 = \begin{cases} 1 \text{ if } \rho \sim \rho' \\ 0 \text{ if } \rho \not\sim \rho'. \end{cases}$$

Let $C_1, C_2, \ldots, C_m$ be the conjugacy classes of G. Then

$$h = \frac{1}{|G|} \sum_{g \in G} \sum_{\rho \in \widehat{G}} \chi_\rho(g)^2 = \sum_{i=0}^m \frac{|C_i|}{|G|} \sum_{\rho \in \widehat{G}} \chi_\rho(c_i)^2 \qquad (9.35)$$

where $c_i \in C_i$. Since $\overline{\chi_\rho}(c_i) = \chi_\rho(c_i^{-1})$ by Proposition 3.7.2, we get from the dual orthogonality relations for the characters (see Exercise 3.9.13),

$$\sum_{\rho \in \widehat{G}} \chi_\rho(c_i)^2 = \sum_{\rho \in \widehat{G}} \chi_\rho(c_i)\overline{\chi_\rho(c_i^{-1})} = \begin{cases} \frac{|G|}{|C_i|} & \text{if } C_i \text{ is ambivalent;} \\ 0 & \text{otherwise.} \end{cases}$$

Thus, from (9.35) we deduce (ii). $\qquad\qquad\qquad\qquad\qquad\qquad\square$

As outlined in the following exercise, one can directly prove that

$$h = \frac{1}{|G|} \sum_{g \in G} p(g)^2$$

(see [166] and [229]).

Exercise 9.7.11 Set $p(g) = |\{u \in G : gug^{-1} = u^{-1}\}|$.

(1) Show that $p(g) = \rho(g^2)$.
(2) Deduce that $\sum_{g \in G} p(g) = \sum_{t \in G} \rho(g)^2$.
(3) Show that the number of ambivalent classes of G equals $\frac{1}{|G|} \sum_{g \in G} p(g)$.

Corollary 9.7.12 *G is ambivalent if and only if any irreducible representation is selfconjugate.*

Corollary 9.7.13 *Suppose that $|G|$ is odd. Then any non-trivial irreducible representation is complex.*

Proof Set $n = 2k-1 = |G|$. Clearly, for all $g \in G$ one has $g^{n+1} = g^n \cdot g = 1_G \cdot g = g$. From this we deduce that $(g^k)^2 = g$, that is g is a square and $\mathcal{S}(g) \geq 1$ for all $g \in G$. On the other hand, as $\sum_{g \in G} \mathcal{S}(g) = |G|$ we necessarily have $\mathcal{S}(g) = 1$ for all $g \in G$. Therefore from (i) in the previous theorem one deduces

$$h = \frac{1}{|G|} \sum_{g \in G} \mathcal{S}(g)^2 = 1$$

that is there is only one selfconjugate representation, namely the trivial one. □

We end this section by illustrating the Theorem of Frobenius and Schur for a Gelfand pair (G, K). Denote, as usual, by $X = G/K$ the corresponding homogeneous space and by $x_0 \in X$ the point stabilized by K. Also let $X = \coprod_{j=0}^{N} \Omega_j$ be the decomposition of X into K-orbits with $\Omega_0 = \{x_0\}$. Finally let $g_0 = 1_G, g_1, \ldots, g_N$ denote representatives for the double cosets KgK in G so that $G = \coprod_{j=0}^{N} Kg_jK$ and $\Omega_j = Kg_jx_0$ (see Sections 3.13 and 4.1).

Define a map $\theta : \{0, 1, \ldots, N\} \to \{0, 1, \ldots, N\}$ in such a way that

$$g_j^{-1} \in Kg_{\theta(j)}K.$$

It is obvious that θ is an involution.

Definition 9.7.14 A K-orbit Ω_j is *symmetric* if $gx_0 \in \Omega_j$ implies $g^{-1}x_0 \in \Omega_j$ for all $g \in G$.

In other words, Ω_j is symmetric if the corresponding double coset is symmetric, that is $g \in Kg_jK$ implies $g^{-1} \in Kg_jK$, equivalently if $\theta(j) = j$.

Let now (ρ_i, V_i), ϕ_i and $d_i = \dim(V_i)$ be as in Section 4.7 and $c(\rho_i)$ as in (9.34).

Exercise 9.7.15 (1) (**Frobenius–Schur Theorem for Gelfand pairs**) Show that

$$\frac{1}{|G|} \sum_{g \in G} \phi_i(g^2x_0) = \frac{c(\rho_i)}{d_i}.$$

(2) For any $x \in X$ set $\mathcal{S}(x) = |\{g \in G : g^2x_0 = x\}|$. Show that

$$\mathcal{S}(x) = |K| \sum_{i=0}^{N} c(\rho_i)\phi_i(x).$$

(3) Set $A = \{i : \rho_i$ is selfconjugate$\}$. Show that

$$\frac{1}{|G|} \sum_{x \in X} \mathcal{S}(x)^2 = |K| \sum_{i \in A} \frac{1}{d_i}.$$

(4) Show that the number $|A|$ of selfconjugate spherical representations of G equals the number of symmetric K-orbits on X, in other words

$$|A| = |\{j : \theta(j) = j\}|.$$

(5) Prove that indeed every selfconjugate spherical representation is real.

For variations and generalizations on this exercise and, more generally, on the whole Section 9.7 we refer to the following papers: [38], [141] and [166].

9.8 Greenhalgebras

This section contains an exposition of a theory that generalizes the construction in Theorem 9.5.4. We introduce a class of algebras that after Diaconis' paper [57] we call Greenhalgebras.

Actually, A. Greenhalgh in his thesis [106] developed a more general theory while the construction in this section was already considered by Brender [34]. Particular cases were also considered in [137], [33], [223] and [108].

If $H \leq G$ and $f \in L(G)$, we say that f is H-conjugacy invariant when

$$f(h^{-1}gh) = f(g)$$

foa all $g \in G$ and $h \in H$.

Definition 9.8.1 Let G be a finite group and $K \trianglelefteq H \leq G$. The *Greenhalgebra* associated with the triple (G, H, K) is

$$\mathcal{G}(G, H, K) = \{f \in L(G) : f(h^{-1}gh) = f(g) \text{ and } f(k_1gk_2) = f(g),$$
$$\forall h \in H, g \in G, k_1, k_2 \in K\},$$

that is $\mathcal{G}(G, H, K)$ is the set of all $f \in L(G)$ that are both H-conjugacy invariant and bi-K-invariant. It is easy to see that $\mathcal{G}(G, H, K) = \{f \in L(G) : f(h^{-1}gh) = f(g) \text{ and } f(gk) = f(g), \forall h \in H, g \in G, k \in K\}$ since $f(k_1gk_2) = f(k_1gk_2k_1k_1^{-1})$.

Example 9.8.2 (1) For $K = \{1_G\}$ and $H = G$ we get the algebra of central functions on G.

(2) For $K = H \leq G$ it is the algebra of bi-K-invariant functions on G.

(3) For $H = K = \{1_G\}$ it coincides with $L(G)$.

(4) For $K = \{1_G\}$, $H \leq G$ it is the algebra of all H-conjugacy invariant functions on G. It is also called *the centralizer of H (or of $L(H)$) in $L(G)$*. It coincides with $\{f \in L(G) : f * \delta_h = \delta_h * f, \ \forall h \in H\}$ and it is usually denoted by $L(G)^H$ (or $\mathbb{C}[G]^H$).

In what follow, we fix G with $K \trianglelefteq H \leq G$. We will think of G/K as the set of right K-cosets in G.

Lemma 9.8.3 $G \times H$ *acts on* G/K *by:*

$$(g, h) \cdot g_0 K = g g_0 h^{-1} K$$

and the stabilizer of K is the subgroup $B = \{(kh, h) : k \in K, h \in H\}$.

Proof The coset $(g, h) \cdot g_0 K$ is well defined: $(g, h) \cdot g_0 k K = g g_0 k h^{-1} K = g g_0 h^{-1} \cdot h k h^{-1} K = g g_0 h^{-1} K$, for any $k \in K$. It is easy to check that indeed it is an action. Moreover, $(g, h)K = K$ if and only if $g h^{-1} \in K$ and this is equivalent to $(g, h) \in B$. $\qquad\square$

Remark 9.8.4 It is easy to see that $B = (K \times \{1_H\})\widetilde{H}$ where $\widetilde{H} = \{(h, h) : h \in H\} \cong H$, $(K \times \{1_H\}) \cap \widetilde{H} = \{1_{G \times H}\}$ and $K \times \{1_H\}$ is normal in B. In other words, $B \cong K \rtimes H$, where $\rtimes$ denotes the semidirect product (see Section 8.7).

In other words, $G \times H$ acts on G by $(g, h) \cdot g_0 = g g_0 h^{-1}$ and permutes among themselves the classes of G/K. The action of $G \times H$ on $L(G/K)$ (identified with the subspace of right K-invariant functions on G) will be denoted by η:

$$[\eta(g, h)f](g_0) = f(g^{-1} g_0 h)$$

(compare with the η in Lemma 9.5.3).

Lemma 9.8.5 *Let B as in the previous lemma. Then we have the following isomorphism of algebras:*

$$\mathcal{G}(G, H, K) \cong \mathcal{G}(G \times H, B, B)$$

that is $\mathcal{G}(G, H, K)$ is isomorphic to the algebra of bi-B-invariant functions on $G \times H$.

Proof From Proposition 4.2.1 and Lemma 9.8.3 we get

$$\mathcal{G}(G \times H, B, B) \cong Hom_{G \times H}\left(L(G/K), L(G/K)\right).$$

But the latter is a subalgebra of $Hom_G\left(L(G/K), L(G/K)\right) \cong \mathcal{G}(G, K, K)$ (the algebra of bi-K-invariant functions on G.) Suppose that $Tf = f * \phi$ with ϕ bi-K-invariant. Then

$$[\eta(g_1, h)Tf](g_0) = Tf(g_1^{-1}g_0 h)$$
$$= \sum_{g \in G} f(g_1^{-1}g_0 h g^{-1})\phi(g)$$
$$= \sum_{t \in G} f(g_1^{-1}g_0 t^{-1})\phi(th),$$

while

$$\{T[\eta(g_1, h)f]\}(g_0) = \sum_{g \in G} f(g_1^{-1}g_0 g^{-1}h)\phi(g)$$
$$= \sum_{t \in G} f(g_1^{-1}g_0 t^{-1})\phi(ht)$$

and therefore $T\eta(g_1, h) = \eta(g_1, h)T$, that is $T \in Hom_{G \times H}(L(G/K), L(G/K))$, if and only if the bi-$K$-invariant function ϕ satisfies $\phi(ht) = \phi(th)$, that is if and only if $\phi \in \mathcal{G}(G, H, K)$. $\qquad\square$

Exercise 9.8.6 Prove directly Lemma 9.8.5 by showing that

$$\Phi : \mathcal{G}(G \times H, B, B) \to \mathcal{G}(G, H, K),$$

defined by setting for $F \in \mathcal{G}(G \times H, B, B)$ and $g \in G$, $\Phi(F)(g) = \sqrt{|H|}F(g, 1)$, is an isomorphism of algebras. Moreover $\|\Phi(F)\|_{L(G)} = \|F\|_{L(G \times H)}$.

We recall that any irreducible $G \times H$-representation is of the form $\sigma \boxtimes \theta$ with $\sigma \in \widehat{G}$, $\theta \in \widehat{H}$ (see Section 9.1). Now we want to determine the multiplicity of $\sigma \boxtimes \theta$ in η. With the symbol $Res_H^G \sigma$ we will denote the restriction of σ to H, that is the representation of H given by $\sigma(h)$, $h \in H$. Moreover, θ' will denote the adjoint of θ (see Section 9.5)

Theorem 9.8.7 *Let $\sigma \in \widehat{G}$ and $\theta \in \widehat{H}$. If $\sigma \boxtimes \theta$ is contained in η then $Res_K^H \theta = (\dim\theta)\iota_K$, where ι_K is the trivial representation of K.*

Moreover, the multiplicity of $\sigma \boxtimes \theta$ in η is equal to the multiplicity of θ' in $Res_H^G \sigma$.

Proof Denote by V and W the representation spaces of σ and θ, respectively.

Suppose that $\sigma \boxtimes \theta$ is contained in η. Note that the linear map $T :$ $V \otimes W \rightarrow L(G/K)$ belongs to $Hom_{G \times H}(\sigma \boxtimes \theta, \eta)$ if and only if, for $g, g_0 \in G$, $h \in H$, $v \in V$, $w \in W$, we have:

$$\{T[\sigma(g)v \otimes \theta(h)w]\}(g_0) = \{T(v \otimes w)\}(g^{-1}g_0 h). \tag{9.36}$$

In particular, setting $g = 1_G$, $h = k \in K$, (9.36) becomes

$$\{T[v \otimes \theta(k)w]\}(g_0) = \{T(v \otimes w)\}(g_0), \quad \forall g_0 \in G.$$

If T is nontrivial, $\ker(T) = \{0\}$ and this forces $\theta(k)w = w$. That is if $\sigma \boxtimes \theta$ is contained in η then $Res_K^H \theta = (\dim\theta)\iota_K$

Now suppose that this condition is satisfied. By Corollary 9.3.3 we end the proof if we construct a linear isomorphism

$$Hom_{G \times H}(\sigma \boxtimes \theta, \eta) \cong Hom_H(Res_H^G \sigma, \theta').$$

Let W' be the dual space of W (i.e. the representation space of θ'). If $T \in Hom_{G \times H}(\sigma \boxtimes \theta, \eta)$ we define $\widetilde{T} : V \rightarrow W'$ by setting $(\widetilde{T}v)(w) = [T(v \otimes w)](1_G)$ for all $v \in V$ and $w \in W$. Clearly $\widetilde{T}v \in W'$ and $\widetilde{T}$ is a linear map.

Moreover, if $h \in H$ we have

$$\left[\widetilde{T}\sigma(h)v\right](w) = \{T[\sigma(h)v \otimes w]\}(1_G)$$

$$\text{by } (9.36) = [T(v \otimes w)](h^{-1})$$

$$\text{again by } (9.36) = \{T[v \otimes \theta(h^{-1})w]\}(1_G)$$

$$= \left(\widetilde{T}v\right)[\theta(h^{-1})w]$$

$$= \left[\theta'(h)\widetilde{T}v\right](w)$$

that is $\widetilde{T}\sigma(h) = \theta'(h)\widetilde{T}$ and $\widetilde{T} \in Hom_H(Res_H^G \sigma, \theta')$. It remains to show that the map $T \mapsto \widetilde{T}$ is a bijection. First of all, observe that T is determined by $\widetilde{T}$:

$$[T(v \otimes w)](g) = [T(\sigma(g^{-1})v \otimes w)](1_G)$$

$$= \left\{\widetilde{T}[\sigma(g^{-1})v]\right\}(w). \tag{9.37}$$

Therefore the map is injective. It is also surjective and (9.37) is its inversion. Indeed, given $\widetilde{T}$ we can define T by (9.37). Then we have (recall that we are supposing $Res_K^H\theta = (\dim\theta)i_K$)

$$
\begin{aligned}
\left[T(v \otimes w)\right](gk) &= \left[\widetilde{T}\sigma(k^{-1}g^{-1})v\right](w) \\
&= \left[\theta'(k^{-1})\widetilde{T}\sigma(g^{-1})v\right](w) \\
&= \left[\widetilde{T}\sigma(g^{-1})v\right](\theta(k)w) \\
&= \left[\widetilde{T}\sigma(g^{-1})v\right](w) \\
&= \left[T(v \otimes w)\right](g)
\end{aligned}
$$

that is $T(v \otimes w) \in L(G/K)$. Moreover

$$
\begin{aligned}
\left\{T\left[\sigma(g)v \otimes \theta(h)w\right]\right\}(g_0) &= \left[\widetilde{T}\sigma(g_0^{-1}g)v\right](\theta(h)w) \\
&= \left[\theta'(h^{-1})\widetilde{T}\sigma(g_0^{-1}g)v\right](w) \\
&= \left[\widetilde{T}\sigma(h^{-1}g_0^{-1}g)v\right](w) \\
&= \left[T(v \otimes w)\right](g^{-1}g_0h)
\end{aligned}
$$

and by (9.36) T belongs to $Hom_{G \times H}(\sigma \boxtimes \theta, \eta)$. □

Theorem 9.8.8 *The following conditions are equivalent*

(i) *The algebra $\mathcal{G}(G, H, K)$ is commutative.*
(ii) *$(G \times H, B)$ is a Gelfand pair.*
(iii) *For any $\sigma \in \widehat{G}$ and any $\theta \in \widehat{H}$ with $Res_K^H\theta \cong (\dim\theta)\iota_K$, ι_K the trivial representation of K, the multiplicity of θ' in $Res_H^G\sigma$ is ≤ 1.*

Proof The equivalence between (i) and (ii) follows from Lemma 9.8.5 and the definition of Gelfand pair. The equivalence between (ii) and (iii) follows from Theorem 4.4.2 and Theorem 9.8.7. □

Example 9.8.9 (1) If $K = \{1_G\}$ and $H = G$ Theorems 9.8.7 and 9.8.8 yield (i) and (ii) in Theorem 9.5.4.

(2) If $K = H \leq G$ they generalize Theorems 4.4.2 and 4.6.2.

(3) If $K = H = \{1_G\}$ we recover Theorem 3.7.11.

The fourth example deserves more evidence. Now $K = \{1_G\}$ and $H \leq G$.

Corollary 9.8.10 *Suppose $H \leq G$ and let $G \times H$ acts on G by $(g, h) \cdot g_0 = gg_0h^{-1}$ Let η be the corresponding permutation representation. For $\sigma \in \widehat{G}$ and $\theta \in \widehat{H}$ let $m_{\theta\sigma}$ be the multiplicity of θ' in $\mathrm{Res}_H^G \sigma$ Then*

$$\eta = \oplus_{\sigma \in \widehat{G}} \oplus_{\theta \in \widehat{H}} m_{\theta\sigma} \sigma \boxtimes \theta.$$

Now we want to determine when the pair $(G \times H, B)$ is symmetric, and therefore a sufficient condition for the commutativity of the algebra $\mathcal{G}(G, H, K)$.

Proposition 9.8.11 *The pair $(G \times H, B)$ is symmetric if and only if the following condition is satisfied: for any $g \in G$ there exist $k_1, k_2 \in K$ and $h_1 \in H$ such that:*

$$g^{-1} = k_1 h_1 g h_1^{-1} k_2. \tag{9.38}$$

Proof It is just an easy exercise: the pair is symmetric if and only if for any $(g, h) \in G \times H$ there exist $(k_1 h_1, h_1), (k_2 h_2, h_2) \in B$ such that

$$\begin{cases} h^{-1} = h_1 h h_2 \\ g^{-1} = k_1 h_1 g k_2 h_2. \end{cases} \tag{9.39}$$

Taking $h = 1_G$ we get $h_2 = h_1^{-1}$ and $g^{-1} = k_1 h_1 g h_1^{-1} \cdot (h_1 k_2 h_1^{-1})$.

Conversely, if (9.38) is satisfied then taking $(g, h) \in G \times H$ and $\bar{k}_1, \bar{k}_2 \in K$, $\bar{h}_1 \in H$ such that

$$(gh^{-1})^{-1} = \bar{k}_1 \bar{h}_1 (gh^{-1}) \bar{h}_1^{-1} \bar{k}_2$$

we get (9.39), by setting $h_1 = h^{-1} \bar{h}_1$, $k_1 = h^{-1} \bar{k}_1 h$, $k_2 = h^{-1} \bar{h}_1^{-1} \bar{k}_2 \bar{h}_1 h$ and $h_2 = h^{-1} \bar{h}_1^{-1}$ $\qquad \square$

Corollary 9.8.12 *If (9.38) is satisfied then the algebra $\mathcal{G}(G, H, K)$ is commutative.*

Corollary 9.8.13 *Suppose that $H \leq G$. If for any $g \in G$ there exists $h \in H$ such that $g^{-1} = hgh^{-1}$ then the algebra of H-conjugacy invariant functions in G is commutative and for any $\sigma \in \widehat{G}$ the restriction $\mathrm{Res}_H^G \sigma$ is multiplicity-free.*

Proof This is the case $K = \{1_G\}$ in Corollary 9.8.12 and in Theorem 9.8.8. $\qquad \square$

Exercise 9.8.14 Suppose that $\mathcal{G}(G, H, K)$ is commutative. For $\sigma \in \widehat{G}$ and $\theta \in \widehat{H}$ with $Res_K^H \theta = (\dim\theta)\iota_K$ and θ' contained in $Res_H^G \sigma$, define the function $\phi_{\sigma,\theta} \in L(G)$ by setting:

$$\phi_{\sigma,\theta}(g) = \frac{1}{|H| \cdot |K|} \sum_{h \in H} \sum_{k \in K} \overline{\chi_\sigma(gkh)} \; \overline{\chi_\theta(h)}$$

where χ_σ and χ_θ are the characters of σ and θ, rispectively.

Show that these functions form an orthonormal basis for $\mathcal{G}(G, H, K)$ and that $\|\phi_{\sigma,\theta}\|_{L(G)}^2 = \frac{|G|}{(\dim\sigma)(\dim\theta)}$.

Exercise 9.8.15 A *Schur algebra* on a finite group G is a subalgebra $\mathcal{A}$ of $L(G)$ such that there exists a partition $G = A_0 \coprod A_1 \coprod \cdots \coprod A_m$ that satisfies:

  (i) The functions $\mathbf{1}_{A_0}, \mathbf{1}_{A_1}, \ldots, \mathbf{1}_{A_m}$ form a basis of $\mathcal{A}$ (as a vector space).
  (ii) $A_0 = \{1_G\}$.
  (iii) For any $i \in \{0, 1, \ldots, m\}$ there exists $j \in \{0, 1, \ldots, m\}$ such that:
    $g \in A_i \Rightarrow g^{-1} \in A_j$.

Show that $\mathcal{G}(G, H, K)$ is a Schur algebra.

A classical reference on Schur algebras is the monograph by Wielandt [228]. See also [122] and [217, 218]. The survey article by Takacs [216] contains many applications of Schur algebras to probability.

We now present our main example of Greenhalgebra. It is taken from Greenhalgh's thesis [106]. We take $G = S_n$, $H = S_{n-k} \times S_k$ and $K = S_{n-k}$. First of all, the homogeneous space S_n/S_{n-k} may be identified with the set

$$\{(i_1, i_2, \ldots, i_k) : i_1, i_2, \ldots, i_k \in \{1, 2, \ldots, n\}, i_h \neq i_j \text{ if } h \neq j\},$$

that is with the set of all *ordered* sequences of k *distinct* elements of $\{1, 2, \ldots, n\}$. Indeed S_n acts on this space and the stabilizer of $(1, 2, \ldots, k)$ is just S_{n-k} (seen as the group of all permutations of $\{k+1, k+2, \ldots, n\}$). If we identify S_k with the group of all permutations of $\{1, 2, \ldots, k\}$, then $S_n \times (S_{n-k} \times S_k)$ acts on S_n/S_{n-k} by setting

$$(\pi, \gamma) \cdot (i_1, i_2, \ldots, i_k) = \left(\pi(i_{\gamma^{-1}(1)}), \pi(i_{\gamma^{-1}(2)}), \ldots, \pi(i_{\gamma^{-1}(k)})\right) \quad (9.40)$$

for all $\gamma \in S_{n-k} \times S_k$ and $\pi \in S_n$.

It is easy to verify that this is an action. Moreover, the stabilizer of $(1, 2, \ldots, k)$ in $S_n \times (S_{n-k} \times S_k)$ is exactly

$$B = \{(\nu\gamma, \gamma) : \nu \in S_{n-k}, \gamma \in S_{n-k} \times S_k\}. \quad (9.41)$$

Clearly, any element of B stabilizes $(1, 2, \ldots, k)$. Conversely, set $(i_1, i_2, \ldots, i_k) \equiv (1, 2, \ldots, k)$ and suppose that $(\pi, \gamma)(i_1, i_2, \ldots, i_k) = (i_1, i_2, \ldots, i_k)$. Since $i_{\gamma^{-1}(j)} = \gamma^{-1}(j)$ (simply because $i_j = j$) we get $j = \pi(i_{\gamma^{-1}(j)}) = \pi(\gamma^{-1}(j))$ for $j = 1, 2, \ldots, k$, that is $\nu := \pi\gamma^{-1} \in S_{n-k}$ and $(\pi, \gamma) = (\nu\gamma, \gamma) \in B$. We have determined the action of Lemma 9.8.3 in the present setting and thus we have:

Lemma 9.8.16 *The homogeneous space* $[S_n \times (S_{n-k} \times S_k)]/B$ *is exactly* S_n/S_{n-k} *with the action given by (9.40).*

Exercise 9.8.17 Show that $B \cong (S_{n-k} \rtimes S_{n-k}) \times S_k$.

Our next goal is to prove that $(S_n \times (S_{n-k} \times S_k), B)$ is a symmetric Gelfand pair and therefore that the algebra $\mathcal{G}(S_n, S_{n-k} \times S_k, S_{n-k})$ is commutative. First of all we need some definitions. An *oriented graph* is a couple $\mathfrak{G} = (V, E)$ where V is the vertex set and the edge set E is a subset of $V \times V$, that is an edge is an *ordered* pair of vertices (v, w), $v, w \in V$ ($v = w$ is allowed), v is the *initial* vertex and w is the *final* vertex. It is just a slightly modification of the definition given in Section 1.6. When we draw an oriented graph, we put arrows instead of segments (for an example see Exercise 1.10.8).

If $\mathfrak{G}_1 = (V_1, E_1)$, $\mathfrak{G}_2 = (V_2, E_2)$ are two oriented graphs, we say that are isomorphic and we write $\mathfrak{G}_1 \cong \mathfrak{G}_2$ if there exists a bijection $\phi : V_1 \to V_2$ such that $(v, w) \in E_1 \Leftrightarrow (\phi(v), \phi(w)) \in E_2$, for all $v, w \in V_1$.

We will use a particular construction with oriented graphs to parameterize the orbits of $S_n \times (S_{n-k} \times S_k)$ on $(S_n/S_{n-k}) \times (S_n/S_{n-k})$. Given a pair $x, y \in S_n/S_{n-k}$, $x = (i_1, i_2, \ldots, i_k)$, $y = (j_1, j_2, \ldots, j_k)$, we associate with x, y the oriented graph $\mathfrak{G}(x, y) = (V, E)$ defined as follows. The vertex set is $V = \{i_1, i_2, \ldots, i_k\} \cup \{j_1, j_2, \ldots, j_k\}$ while the edge set is $E = \{(i_1, j_1), (i_2, j_2), \ldots, (i_k, j_k)\}$.

Note that the degree of a vertex $v \in V$ is ≤ 2:

- if $i_h = j_h$ they form a loop that is not connected with any other vertex;
- if $v \in \{i_1, i_2, \ldots, i_k\} \cap \{j_1, j_2, \ldots, j_k\}$ and $v = i_h = j_t$ for $h \neq t$ then the degree of v is 2 and it is the initial vertex of an edge and the terminal vertex of another edge;
- if $v \in \{i_1, i_2, \ldots, i_k\} \triangle \{j_1, j_2, \ldots, j_k\}$ then the degree of v is 1 and it is the terminal (resp. initial) vertex of an edge if $v \in \{j_1, j_2, \ldots, j_k\}$ (resp. if $v \in \{i_1, i_2, \ldots, i_k\}$).

The following is an example: $(n = 15, k = 8)$

$$x = (1, 7, 14, 15, 10, 3, 13, 9)$$
$$y = (3, 1, \ 6, 15, 11, 9, 14, 7).$$

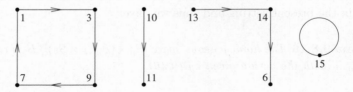

Figure 9.1.

Note that an oriented graph $\mathfrak{G}(x, y)$ is composed by (isolated) loops, (oriented) closed paths and (oriented) nonclosed paths. Moreover, if $v_1 \to v_2 \to \cdots \to v_m$ is a nonclosed path, then $v_1 \in \{i_1, i_2, \ldots, i_k\}$ and $v_m \in \{j_1, j_2, \ldots, j_k\}$. Clearly, $\mathcal{G}(y, x)$ is obtained from $\mathcal{G}(x, y)$ just by reversing the orientation of the edges. From these considerations it follows easily that:

Lemma 9.8.18 $\mathfrak{G}(x, y) \cong \mathfrak{G}(y, x)$.

Moreover,

Lemma 9.8.19 $(x, y), (x', y') \in (S_n/S_{n-k}) \times (S_n/S_{n-k})$ *belong to the same* $S_n \times (S_{n-k} \times S_k)$ *orbit if and only if* $\mathfrak{G}(x, y) \cong \mathfrak{G}(x', y')$.

Proof Suppose that $x = (i_1, i_2, \ldots, i_k)$, $y = (j_1, j_2, \ldots, j_k)$, $x' = (i'_1, i'_2, \ldots, i'_k)$, $y' = (j'_1, j'_2, \ldots, j'_k)$, $\gamma \in S_{n-k} \times S_k$, $\pi \in S_n$ and

$$i'_h = \pi(i_{\gamma^{-1}(h)}), \quad j'_h = \pi(j_{\gamma^{-1}(h)}), \quad h = 1, 2, \ldots, k.$$

Then the graphs $\mathfrak{G}(x, y)$ and $\mathfrak{G}(x', y')$ are isomorphic: the isomorphism ϕ is given by:

$$\phi(v) = \pi(v) \quad \forall v \in \{i_1, i_2, \ldots, i_k\} \cup \{j_1, j_2, \ldots, j_k\}.$$

This proves the "only if" part. Note that the effect of γ is just a permutation of the edges, while π changes the vertices but not the structure of the graph.

Conversely suppose that $\mathfrak{G}(x, y) \cong \mathfrak{G}(x', y')$ and denote by ϕ the corresponding isomorphism. Then we can take $\gamma \in S_{n-k} \times S_k$ such that the ϕ-image of the edge $(i_{\gamma^{-1}(h)}, j_{\gamma^{-1}(h)})$ is (i'_h, j'_h) for $h = 1, 2, \ldots, k$. Then setting $\pi(i_{\gamma^{-1}(h)}) = i'_h$ and $\pi(j_{\gamma^{-1}(h)}) = j'_h$, $h = 1, 2, \ldots, k$ and extent-ing π to a permutation of the whole $\{1, 2, \ldots, n\}$ we have $(\pi, \gamma)(x, y) = (x', y')$. $\qquad \square$

For instance, if x, y are as in Figure 9.1 and

$$x' = (2,\ 3, 1,\ 7, 11,\ 4, 14, 12)$$
$$y' = (2, 11, 5, 13,\ 4, 12,\ 7,\ 3),$$

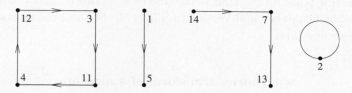

Figure 9.2.

then we can take $\gamma = (4, 1, 8, 5, 3)(6, 2)(7)$ and

$$\pi = (1, 12, 10)(15, 2)(9, 11, 5)(3)(6, 13, 14, 7, 4)(8).$$

Corollary 9.8.20 $(S_n \times (S_{n-k} \times S_k), B)$ *is a symmetric Gelfand pair.*

Corollary 9.8.21 *The algebra* $\mathcal{G}(S_n, S_{n-k} \times S_k, S_{n-k})$ *is commutative.*

Corollary 9.8.22 S_n *is ambivalent. In particular, any irreducible* S_n*-representation is selfconjugate.*

Proof The ambivalence of S_n is just Corollary 9.8.20 for $k = n$ combined with Theorem 9.5.4. Then we can apply Corollary 9.7.12. $\qquad \square$

Denote by $S^{(n-k)}$ the trivial representation of S_{n-k} (as in Section 3.14).

Corollary 9.8.23 *Suppose that* S^μ *is an irreducible representation of* S_n *and that* $S^{(n-k)} \otimes S^\lambda$ *is an irreducible representation of* $S_{n-k} \times S_k$ *which is trivial on* S_{n-k} *(and therefore* S^λ *is an irreducible representation of* S_k*). Then the multiplicity of* $S^{(n-k)} \otimes S^\lambda$ *in* $\mathrm{Res}^{S_n}_{S_{n-k} \times S_k} S^\mu$ *is* ≤ 1.

For $k = n - 1$, $S_{n-1} \times S_1 \cong S_{n-1}$ (the stabilizer of a point, say n) and we get

Corollary 9.8.24 *The algebra of S_{n-1}-conjugacy invariant functions in S_n is commutative. For any irreducible representation S^μ of S_n, $Res_{S_{n-1}}^{S_n} S^\mu$ is multiplicity-free.*

Note that in Corollaries 9.8.23 and 9.8.24 we do not need to know an explicit parameterization of the irreducible representations of S_n, as it will be developed in Chapter 10.

We refer to Koornwinder's paper [147] for a result similar to Corollary 9.8.24 in the setting of classical compact groups. Corollary 9.8.24 is the starting point for the approach to the representation theory of S_n developed by Okounkov and Vershik in [173, 174]. See also the monograph by Kleshchev [144].

9.9 Fourier transform of a measure

In this section we present a characterization of ergodicity for a random walk on a finite group. It is taken from Behrends' book [18].

Let G be a finite group. For a subset $\Delta \subseteq G$ set $\Delta^{-1} = \{g^{-1} : g \in \Delta\}$. Denote by G_Δ the set of all $g \in G$ that can be expressed as a product

$$g = g_1 g_2 \cdots g_{2r} \tag{9.42}$$

with r-many g_i belonging to Δ and with the remaining r in Δ^{-1}. For instance, if $g_1, g_2, g_3, g_4 \in \Delta$, then the element $g = g_1 g_2^{-1} g_3^{-1} g_4 \in G_\Delta$, while the element $g' = g_2 g_1^{-1} g_4$ in general does not belong to G_Δ. Moreover, for any $\Delta \subseteq G$, we denote by $G(\Delta)$ (resp. $G^N(\Delta)$) the subgroup (resp. the normal subgroup) generated by Δ.

Exercise 9.9.1 (Discrete intermediate value theorem) Let $a < b \in \mathbb{Z}$ and set $[a, b] = \{a, a+1, \ldots, b-1, b\} \subset \mathbb{Z}$. Suppose that $f : [a, b] \to \mathbb{Z}$ is such that $|f(t) - f(t+1)| = 1$ for all $t \in [a, b-1]$ and $f(a)f(b) < 0$. Then there exists $u \in [a+1, b-1]$ such that $f(u) = 0$.

Lemma 9.9.2

(i) G_Δ *is a subgroup of G. Moreover, setting $\Delta\Delta^{-1} = \{gh^{-1} : g, h \in \Delta\}$ we have*

$$G(\Delta\Delta^{-1}) \leq G_\Delta \leq G^N(\Delta\Delta^{-1}). \tag{9.43}$$

(ii) *If $1_G \in \Delta$, then $G_\Delta = G(\Delta)$.*

(iii) $G_\Delta = G$ *if and only if there exists $k \in \mathbb{N}$ such that every element $g \in G$ can be written as a product of k elements in Δ.*

Proof First recall that in a finite group, a subset is a subgroup if and only if it is closed under multiplication.

(i) The first part of the statement immediately follows from the preceding observation.

The first inclusion in (9.43) follows immediately from the fact that $\Delta\Delta^{-1} \subseteq G_\Delta$ and the fact that G_Δ is a subgroup.

We now prove the second inclusion (in 9.43) by induction on r. The identity $g^{-1}h = (h^{-1}h)g^{-1}h = h^{-1}(hg^{-1})h$ implies that $\Delta^{-1}\Delta \subseteq G^N(\Delta\Delta^{-1})$ and this is the case $r = 1$.

For $h \in \Delta \cup \Delta^{-1}$ we define the element $\overline{h} \in \Delta$ by setting

$$\overline{h} = \begin{cases} h & \text{if } h \in \Delta \\ h^{-1} & \text{if } h \in \Delta^{-1}. \end{cases} \tag{9.44}$$

With this notation, if $g \in G_\Delta$, then (9.42) becomes

$$g = (\overline{g_1})^{\epsilon_1}(\overline{g_2})^{\epsilon_2} \cdots (\overline{g_{2r}})^{\epsilon_{2r}}$$

with $\epsilon_i = \pm 1$. Obviously one has $\sum_{i=0}^{2r} \epsilon_i = 0$.

Suppose by induction that the assertion is true for all $\ell \leq r - 1$.

We distinguish two cases:

- $\epsilon_1 = \epsilon_{2r}$. Applying Exercise 9.9.1 with $a = 0, b = 2r - 1$ and $f(t) = \sum_{i=0}^{t} \epsilon_i$, we have that there exists u (necessarily an even integer, say $u = 2i$) such that $f(u) = 0$. This infers that $g = g'g''$ with $g' = g_1g_2 \cdots g_{2i}$ and $g'' = g_{2i+1}g_{2i+2} \cdots g_{2r}$ both belonging to G_Δ. By recurrence g' and g'' belong to $G^N(\Delta\Delta^{-1})$ and so does g.

- $\epsilon_1 \neq \epsilon_{2r}$. In this case we have g_1g_{2r} and $g' = g_2g_3 \cdots g_{2r-1}$ both belong to G_Δ (and therefore to $G^N(\Delta\Delta^{-1})$, by recurrence). But then by normality $g_1g'g_1^{-1} \in G^N(\Delta\Delta^{-1})$ and the same holds for $g = (g_1g'g_1^{-1})(g_1g_{2r})$.

(ii) Suppose that $1_G \in \Delta$. Every element in $G(\Delta)$, the subgroup generated by Δ, is of the form $g = g_1g_2 \cdots g_k$ with $g_i \in \Delta$ (recall the observation at the beginning of the proof). But then $g = g_1g_2 \cdots g_k 1_G^{-1} 1_G^{-1} \cdots 1_G^{-1} \in G_\Delta$.

(iii) Suppose first that $G = G_\Delta$ and denote by $N = |G|$ the order of this group. Recall that $g^N = 1_G$ for all $g \in G$.

Fix $g \in G$ and express it as $g = g_1g_2 \cdots g_{2r}$ where $r = r(g)$ of the g_i's are in Δ and the other r in Δ^{-1}. With the notation in (9.44), replace each $g_i \in \Delta^{-1}$ with $\overline{g_i}^{N-1}$. This shows that g can be expressed as a product of $k(g) = r + r(N - 1) = rN$ elements in Δ. The same holds for 1_G which can be expressed as $1_G = g^{k(g)}$ for any element $g \in \Delta$. It follows that $G = \Delta\Delta \cdots \Delta$ with $k = \sum_{g \in G} k(g)$ factors.

Conversely, if $G = \Delta\Delta \cdots \Delta$ with k factors, then for any $g \in G$ one can write $g = g_1 g_2 \cdots g_k$ and $1_G = h_1 h_2 \cdots h_k$ with all g_i's and h_j's in Δ. But then

$$g = g_1 g_2 \cdots g_k h_k^{-1} h_{k-1}^{-1} \cdots h_2^{-1} h_1^{-1} \in G_\Delta$$

and the proof is complete. □

Exercise 9.9.3 Show that in (9.43) both inclusions may be proper.

Exercise 9.9.4 Let μ be a probability measure on a finite group G and denote by $\Delta = \{g \in G : \mu(g) > 0\}$ its support. Show that Δ^k is the support of μ^{*k} for all $k = 1, 2, \ldots$.

Recall that for a linear operator A on a finite dimensional vector space, the Hilbert-Schmidt norm is $\|A\|_{HS} := \sqrt{tr(AA^*)}$ (see Example 9.5.10).

Lemma 9.9.5 (Upper bound lemma) *Let G be a finite group, μ a probability measure and denote by U the uniform distribution on G. Then*

$$\|\mu^{*k} - U\|_{TV} \leq \frac{1}{4} \sum_{\rho \in \widehat{G} \setminus \{\iota_G\}} d_\rho tr\left([\widehat{\mu}(\rho)]^k ([\widehat{\mu}(\rho)]^*)^k\right). \tag{9.45}$$

Proof From the Cauchy–Schwarz inequality we deduce that $\|\mu^{*k} - U\|_{TV} \leq \frac{1}{4}|G|\|\mu^{*k} - U\|^2$. By Plancherel's formula (3.20), and the fact that $\widehat{U}(\rho) = 1$ if $\rho = \iota_G$, the trivial representation and vanishes otherwise, the last term equals $\frac{1}{4} \sum_{\rho \in \widehat{G} \setminus \{\iota_G\}} d_\rho tr\left([\widehat{\mu}(\rho)]^k ([\widehat{\mu}(\rho)]^*)^k\right)$. □

Proposition 9.9.6 *Let G be a finite group and μ a probability measure with support $\Delta = \{g \in G : \mu(g) > 0\}$. Then the following conditions are equivalent:*

(a) *$G_\Delta = G$.*

(b) *There is $k \in \mathbb{N}$ such that $G = \Delta\Delta \cdots \Delta$ with k factors.*

(c) *For any nontrivial irreducible representation ρ one has*

$$\|\widehat{\mu}(\rho)^k\|_{HS} \to 0 \text{ as } k \to \infty.$$

(d) *Denoting by U the uniform distribution on G one has*

$$\|\mu^{*k} - U\|_{TV} \to 0 \text{ as } k \to \infty.$$

(e) *The corresponding Markov chain is ergodic.*

Proof The equivalence (a)$\Leftrightarrow$(b) follows from Lemma 9.9.2.

(b) is equivalent to (e) by definition of ergodicity: indeed we have (P is the stochastic matrix associated with the measure μ, namely $p(g,h) = \mu(g^{-1}h)$)

$$P \text{ is } k\text{-ergodic} \iff p^{(k)}(g,h) > 0 \text{ for all } g, h \in G$$
$$\text{by the } G\text{-invariance of } P \iff p^{(k)}(1_G, h) > 0 \text{ for all } h \in G$$
$$\iff \mu^{*k}(h) > 0 \text{ for all } h \in G$$
$$\iff \text{supp } \mu^{*k} = G$$
$$\text{by the previous exercise} \iff \Delta^k = G.$$

(e) is equivalent to (d) in virtue of the ergodic theorem (Theorem 1.4.1); observe that in this case the (unique) stationary distribution is given by the uniform distribution $U \equiv 1/|G|$.

(c) $\Leftrightarrow$ (d): recall that the norms $\|\cdot\|_{TV}$ and $\|\cdot\|_{L(G)}$ are equivalent (G is finite and $\|\cdot\|_{TV} = \frac{1}{2}\|\cdot\|_{L^1(G)} \leq \frac{\sqrt{|G|}}{2}\|\cdot\|_{L(G)} \leq \frac{\sqrt{|G|}}{2}\|\cdot\|_{L^1(G)}$). Therefore $\|\mu^{*k} - U\|_{TV} \to 0$ if and only if $\|\mu^{*k} - U\|_{L(G)} \to 0$, as $k \to \infty$. With the same arguments as in the upper bound lemma (Lemma 9.9.5) we have $\|\mu^{*k} - U\|_{L(G)} \to 0$, if and only if $\|\widehat{\mu}(\rho)^k\|_{HS}^2 = tr\left([\widehat{\mu}(\rho)]^k([\widehat{\mu}(\rho)]^*)^k\right) \to 0$ as $k \to \infty$, for all nontrivial representation $\rho \in \widehat{G}$. $\qquad\square$

Exercise 9.9.7 Suppose that G is a finite abelian group. Let μ be a probability measure on G and denote by Δ its support. Prove that the following assertions are equivalent (we use additive notation for the operation on G).

(a) $G(\Delta - \Delta) = G$ (in particular, $G_\Delta = G$).

(b) There is a $k \in \mathbb{N}$ such that $G = \Delta + \Delta + \cdots + \Delta$ with k summands.

(b′) There are no proper subgroups H of G and $g_0 \in G$ such that $\Delta \subseteq g_0 + H$.

(c) For any nontrivial character χ, one has $|\widehat{\mu}(\chi)| < 1$.

(d) Denoting by U the uniform distribution on G one has

$$\|\mu^{*k} - U\|_{TV} \to 0 \text{ as } k \to \infty.$$

(e) The corresponding Markov chain is ergodic.

10

Basic representation theory of
the symmetric group

This chapter contains a standard exposition on James' approach to the representation theory of the symmetric group. We follow the monographs by James [132] and Sagan [182]. See also Diaconis [55] and Sternberg's [212] books. The most complete reference on the subject is the book by James and Kerber [134]. Sections 10.6 and 10.7 are based on Diaconis's book [55]. Section 10.8 is an introduction to Stanley's paper [201]. See also Sagan's book [182]. In Section 10.9 we prove a characterization of the irreducible representations of the symmetric group due to James; we follow the elementary proof in [187].

10.1 Preliminaries on the symmetric group

In this section we recall some elementary facts on the symmetric group (see, for instance, the algebra books by Herstein [120], Hungerford [123] and M. Artin [6]).

For $n \geq 2$, we denote by S_n the *symmetric group* of degree n, that is the group of all permutations of the set $\{1, 2, \ldots, n\}$.

A permutation $\gamma \in S_n$ is called a *cycle* of length d, and we denote it as $\gamma = (a_1, a_2, \ldots, a_d)$, with $1 \leq a_i \neq a_j \leq n$ for $1 \leq i \neq j \leq d$, if

$$\gamma(a_1) = a_2, \ \gamma(a_2) = a_3, \ \ldots, \gamma(a_{d-1}) = a_d, \ \gamma(a_d) = a_1$$

and

$$\gamma(b) = b \ \text{ if } b \neq a_i \text{ for all } i\text{'s}.$$

Two cycles $\gamma = (a_1, a_2, \ldots, a_d)$ and $\theta = (b_1, b_2, \ldots, b_{d'})$ with $\{a_1, a_2, \ldots, a_d\} \cap \{b_1, b_2, \ldots, b_{d'}\} = \emptyset$ are said to be *disjoint*. It is clear that any two disjoint cycles commute ($\gamma\theta = \theta\gamma$) and that any permutation can be uniquely expressed as product of pairwise disjoint cycles. We call the

319

type or the *cycle structure* of a permutation the sequence of the lengths (in decreasing order) of its cycles. In other words, if

$$\pi = (a_1, a_2, \ldots, a_{\mu_1})(b_1, b_2, \ldots, b_{\mu_2}) \cdots (c_1, c_2, \ldots, c_{\mu_k})$$

with $\mu_1 \geq \mu_2 \geq \cdots \geq \mu_k$ and disjoint cycles, the type of π is $(\mu_1, \mu_2, \ldots, \mu_k)$ (and we suppose that $\mu_1 + \mu_2 + \cdots + \mu_k = n$, that is also the trivial cycles, i.e. the points fixed by π, are included in the cycle structure).

The following problem was presented to us by Adriano Garsia who, in turn, heard it from Richard Stanley. We thank Richard Stanley for addressing Peter Winkler's home page and to the latter for kindly giving us the permission to reproduce it here.

Exercise 10.1.1 (The prisoners' problem [230]) There are $2n$ prisoners senteced to death. The director of the prison decides to give them a "last chance" of survival. Each prisoner has a "registration number" between 1 and $2n$ assigned. These numbers are placed into $2n$ boxes, one number inside each box, and the boxes, numbered from 1 to $2n$, are placed in a room. The prisoners (all together) have their lives saved if and only if each of them finds his own registration number according to the following rule. Each prisoner, one at a time, is led into the room. He may open at most n boxes (i.e. the half of them) and has to leave the room exactly as he found it. He cannot communicate with the other prisoners after he has visited the room. The prisoners, however, have the possibility to plot together a strategy in advance. Find a strategy which would ensure them a probability of survival exceeding 30%.

Comment. Note that if no strategy is established, so that each prisoner opens at random n boxes, then their probability of survival is exactly $(1/2)^{2n}$ which is almost zero. On the other hand, if $n = 1$, so that there are exactly two prisoners and two boxes, the strategy for which prisoner i opens the box i, $i = 1, 2$, gives a $1/2$ probability of survival.

As usual in group theory, two elements η and η' in S_n are *conjugate* if there exists $\pi \in S_n$ such that $\pi\theta\pi^{-1} = \theta'$. It is clear that conjugacy is an equivalence relation; we call the *conjugacy class* of an element $\eta \in S_n$ the class $\{\pi\eta\pi^{-1} : \pi \in S_n\}$ containing η.

Before stating the following lemma we need the following definition.

Definition 10.1.2 Let n be a positive integer. A *partition of length h* of n is an ordered sequence of positive integers $\lambda = (\lambda_1, \lambda_2 \ldots, \lambda_h)$ such that $\lambda_1 + \lambda_2 + \cdots \lambda_h = n$ and $\lambda_1 \geq \lambda_2 \geq \cdots \geq \lambda_h$. We use the expression $\lambda \vdash n$ to say that λ is a partition of n.

For instance, the cycle structure of a permutation $\pi \in S_n$ is a partition of n.

We have the following remarkable fact.

Lemma 10.1.3 *If* $\theta = (a_1, a_2, \ldots, a_d) \cdots (b_1, b_2, \ldots, b_{d'})$ *is the decomposition into disjoint cycles of* $\theta \in S_n$, *then, for all* $\pi \in S_n$ *one has that*

$$\pi\theta\pi^{-1} = (\pi(a_1), \pi(a_2), \ldots, \pi(a_d)) \cdots (\pi(b_1), \pi(b_2), \ldots, \pi(b_{d'})) \quad (10.1)$$

is the decomposition into disjoint cycles of the conjugate $\pi\theta\pi^{-1}$.

Proof Observe that $\pi\theta\pi^{-1}(\pi(x)) = \pi(\theta(x))$ for all $x \in \{1, 2, \ldots, n\}$. Therefore, the cycle $(a_1, a_2, \ldots, a_d) \equiv (a_1, \theta(a_1), \ldots, \theta^{d-1}(a_1))$ of the permutation θ corresponds to the cycle $(\pi(a_1), \pi(a_2), \ldots, \pi(a_d))$ of $\pi\theta\pi^{-1}$. $\square$

Corollary 10.1.4 *Two permutations are conjugate if and only if they have the same cycle structure. In particular, if* $\mathcal{C}_\mu$ *denotes the set of all permutations that have cycle structure* $\mu \vdash n$, *then the map* $\mu \mapsto \mathcal{C}_\mu$ *is a bijection between the set of all partitions of n and the conjugacy classes in S_n.*

There is another way to write a partition of n. If $\mu = (\mu_1, \mu_2, \ldots, \mu_k) \vdash n$, we can represent it in the form $(1^{r_1}, 2^{r_2}, \ldots, n^{r_n})$ where r_j denotes the number of $i \in \{1, 2, \ldots, k\}$ with $\mu_i = j$, $j = 1, 2, \ldots, n$. Thus, $r_1 + 2r_2 + \cdots + nr_n = n$. Usually one omits the terms of the form i^{r_i} with $r_i = 0$ and simply writes i instead of i^1. For instance, the partition $(7, 5, 5, 4, 4, 4, 1) \vdash 30$ may be also represented in the form $(1, 4^3, 5^2, 7)$. Analogously $(1, 1, \cdots, 1) \vdash n$ is represented by (1^n).

Proposition 10.1.5 *Let* $\mu \vdash n$ *with* $\mu = (1^{r_1}, 2^{r_2}, \ldots, n^{r_n})$ *and denote by* $\mathcal{C}_\mu$ *the corresponding conjugacy class. Then*

$$|\mathcal{C}_\mu| = \frac{n!}{r_1! 1^{r_1} r_2! 2^{r_2} \cdots r_n! n^{r_n}}.$$

Proof S_n acts transitively on $\mathcal{C}_\mu$ by conjugation: $\mathcal{C}_\mu \ni \sigma \mapsto \pi\sigma\pi^{-1} \in \mathcal{C}_\mu$ for all $\pi \in S_n$. If $\sigma \in \mathcal{C}_\mu$, then the stabilizer of σ in S_n is the subgroup $Z_\sigma = \{\pi \in S_n : \pi\sigma\pi^{-1} = \sigma\}$, which is nothing but the centralizer of σ in S_n. From Lemma 10.1.3 it follows that $\pi \in Z_\sigma$ if and only if: (a) for $j = 1, 2, \ldots, n$ it permutes among themselves the cycles of length j and/or (b) it cyclically permutes each cycle of σ.

Now, for each j, (a) may be performed in $r_j!$ different ways, while, for each cycle of length j, (b) can be performed in j different ways. Altogether we get $|Z_\sigma| = r_1!1^{r_1}r_2!2^{r_2} \cdots r_n!n^{r_n}$. Finally, $|\mathcal{C}_\mu| = |S_n|/|Z_\sigma|$ is as we wished to prove. $\qquad\square$

With $\lambda = (\lambda_1, \lambda_2, \cdots, \lambda_m) \vdash n$ we associate the group

$$S_\lambda = S_{\lambda_1} \times S_{\lambda_2} \times \cdots \times S_{\lambda_m}$$

which can be naturally embedded into S_n: if $\{1, 2, \ldots, n\} = \coprod_{i=1}^m A_i$ with $|A_i| = \lambda_i$, is a partition of $\{1, 2, \ldots, n\}$, then we may regard S_λ as the subgroup $S_{A_1} \times S_{A_2} \times S_{A_m}$, where S_{A_i} denotes the group of all permutations of A_i. This is clearly a subgroup of S_n: $(\pi_1, \pi_2, \ldots, \pi_m) \in S_{A_1} \times S_{A_2} \times S_{A_m}$ is the same thing as the permutation $\pi_1\pi_2 \cdots \pi_m \in S_n$.

A subgroup of this type is called a *Young subgroup* of S_n corresponding to $\lambda \vdash n$.

Remark 10.1.6 Note that we write the permutations on the left of the permuted object, that is, if $\gamma \in S_n$ and $i \in \{1, 2, \ldots, n\}$, the γ-image of i is $\gamma(i)$. As a consequence, the product of nondisjoint cycles is as in the following example: $(1, 2, 5) \cdot (1, 2, 4) = (1, 5)(2, 4)$, that is, the second cycle on the left hand side sends $1 \to 2$ and the first one $2 \to 5$, and so on. This agrees with the convention in Hungerford [123]. On the other hand, Herstein [120] writes the permutations on the right and therefore, in that book, one finds $(1, 2, 5) \cdot (1, 2, 4) = (1, 4)(2, 5)$.

Exercise 10.1.7 (1) Show that S_n is generated by the set of all transpositions.

(2) Show that S_n is generated by the transpositions $(i, i + 1)$ for $i = 1, 2, \ldots, n - 1$.

10.2 Partitions and Young diagrams

With a partition $\lambda = (\lambda_1, \lambda_2, \ldots, \lambda_k) \vdash n$ one associates its *Young diagram*, namely an arrangement (from left to right and from the top to the bottom) of n squared boxes in k left-justified rows, the ith one containing exactly λ_i boxes. In other words, the first row contains λ_1 boxes, the second row λ_2 boxes, with the ith box of the second row below the ith box of the first row and so on. For instance, the Young diagram corresponding to the partition $(4, 4, 3, 1, 1) \vdash 13$ is

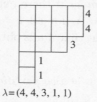

$$\lambda = (4, 4, 3, 1, 1)$$

Figure 10.1.

Also, rows and columns are numbered as rows and column of a matrix.

Note that in the diagram of λ there are $t := \lambda_1$ columns and the jth column contains exactly $\lambda'_j = |\{i : \lambda_i \geq j\}|$ boxes. The *conjugate* partition λ' of λ is defined by $\lambda' = (\lambda'_1, \lambda'_2, \ldots, \lambda'_t)$. Clearly its associated Young diagram is obtained by transposing the Young diagram of λ.

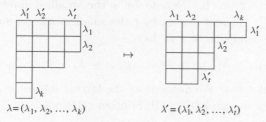

Figure 10.2.

We now introduce a partial order on the set of all partitions of n.

Definition 10.2.1 Let $\lambda = (\lambda_1, \lambda_2, \ldots, \lambda_h)$ and $\mu = (\mu_1, \mu_2, \ldots, \mu_k)$ be two partitions of n. We say that μ *dominates* λ and we write

$$\lambda \trianglelefteq \mu$$

if $k \leq h$ and

$$\sum_{i=1}^{\ell} \lambda_i \leq \sum_{i=1}^{\ell} \mu_i$$

for $\ell = 1, 2, \ldots, k$.

For instance one has $(3, 2, 2, 1) \trianglelefteq (4, 3, 1)$. It is clear that $\trianglelefteq$ is a partial order relation and the partitions $(1, 1, \ldots, 1)$ and (n) are the minimal and maximal elements, respectively. The order is not total: there exist elements that are not comparable. This is the case, for instance, for the partitions $\lambda = (3,3)$ and $\mu = (4,1,1)$.

Suppose that λ and μ are partitions of n. We say that μ is obtained from λ by a *single-box up-move* if there exist positive integers a and b with $a < b$ such that $\mu_i = \lambda_i$ for all $i \neq a, b$ and $\mu_a = \lambda_a + 1$ and $\lambda_b = \mu_b - 1$. It is also obvious that if μ is obtained from λ by a single-box up-move, then $\lambda \trianglelefteq \mu$.

Lemma 10.2.2 *Let λ and μ be two partitions of n. Then μ dominates λ if and only if there exists a chain*

$$\lambda^0 \trianglelefteq \lambda^1 \trianglelefteq \cdots \trianglelefteq \lambda^{s-1} \trianglelefteq \lambda^s$$

where $\lambda^0 = \lambda$, $\lambda^s = \mu$ and λ^{i+1} is obtained from λ^i by a single-box up-move, $i = 1, 2, \ldots, s-1$.

Proof We only prove the nontrivial implication. Suppose $\lambda = (\lambda_1, \lambda_2, \ldots, \lambda_h)$ is dominated by $\mu = (\mu_1, \mu_2, \ldots, \mu_k)$ and $\lambda \neq \mu$. Set $\mu_j = 0$ for $j = k+1, k+2, \ldots, h$. Denote by s the smallest integer such that $\sum_{i=1}^{s} \lambda_i < \sum_{i=1}^{s} \mu_i$ and denote by t the smallest integer such that $\mu_t < \lambda_t$. From the definition of s we have

$$\mu_1 = \lambda_1, \ \mu_2 = \lambda_2, \ldots, \mu_{s-1} = \lambda_{s-1}, \mu_s > \lambda_s. \tag{10.2}$$

It follows that $t > s$. We define v as the largest integer $v \geq t$ such that $\lambda_v = \lambda_t$. Then we can add to (10.2) other relations:

$$\begin{cases} \mu_i \geq \lambda_i & i = s+1, s+2, \ldots, t-1 \\ \lambda_i = \lambda_t > \mu_t \geq \mu_i & i = t+1, t+2, \ldots, v. \end{cases} \tag{10.3}$$

We define $\overline{\lambda}$ as the partition of n obtained from λ by the following single-box up-move:

$$\overline{\lambda} = (\lambda_1, \lambda_2, \ldots, \lambda_{s-1}, \lambda_s + 1, \lambda_{s+1}, \ldots, \lambda_{v-1}, \lambda_v - 1, \lambda_{v+1}, \ldots, \lambda_h).$$

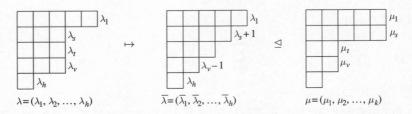

Figure 10.3.

Clearly, $\overline{\lambda}$ is still a partition: $\lambda_{s-1} = \mu_{s-1} \geq \mu_s > \lambda_s$ yields $\lambda_{s-1} \geq \lambda_s + 1$ and $\lambda_v > \lambda_{v+1}$ yields $\lambda_v \geq \lambda_{v+1} + 1$.

We now show that $\overline{\lambda} \trianglelefteq \mu$. Clearly, it suffices to prove that $\sum_{i=1}^z \mu_i \geq \sum_{i=1}^z \overline{\lambda}_i$, for $z = s, s+1, \ldots, v-1$, equivalently that $\sum_{i=1}^z \mu_i > \sum_{i=1}^z \lambda_i$, for the same values of z. From $\sum_{i=1}^z \mu_i \geq \sum_{i=1}^z \lambda_i$ and from (10.2) and (10.3) it follows that the quantity $\sum_{i=1}^z \mu_i - \sum_{i=1}^z \lambda_i$ is positive for $z = s$ and $z = v - 1$, not decreasing for $s \leq z \leq t - 1$, decreasing for $t \leq z \leq v-1$. Therefore it cannot be $\sum_{i=1}^z \mu_i = \sum_{i=1}^z \lambda_i$ for $s \leq z \leq v-1$ and $\overline{\lambda} \trianglelefteq \mu$.

Finally, one can take $\lambda^1 = \overline{\lambda}$ and iterate the procedure until one reaches μ. $\qquad\square$

10.3 Young tableaux and the Specht modules

Definition 10.3.1 Let λ be a partition of n and consider the corresponding Young diagram D_λ. A *Young tableau* of *shape* λ (or, simply, a λ-*tableau*) is a bijection t of the set of n boxes of D_λ into $\{1, 2, \ldots, n\}$. We denote by $t_{i,j}$ the number corresponding to the box in the ith row and jth column and by T_λ the set of all λ-tableaux.

Example 10.3.2 *If $n = 3$ and $\lambda = (2,1)$ we have the following λ-tableaux.*

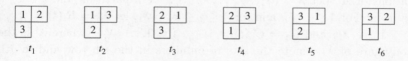

Figure 10.4.

It is clear that for any partition $\lambda \vdash n$ the number of λ-tableaux is $|T_\lambda| = n!$.

For $\lambda \vdash n$ we now define an action of S_n on T_λ by setting, for all $\pi \in S_n$ and $t \in T_\lambda$,

$$(\pi[t])_{i,j} = \pi(t_{i,j}).$$

Example 10.3.3 For example, if $n = 3$, $\lambda = (2,1)$ as above and with $\pi = (1,2)$ we have:

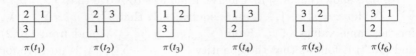

$$\quad \pi(t_1) \qquad\quad \pi(t_2) \qquad\quad \pi(t_3) \qquad\quad \pi(t_4) \qquad\quad \pi(t_5) \qquad\quad \pi(t_6)$$

Figure 10.5.

We note that this action is transitive.

Lemma 10.3.4 *Let $\lambda, \mu \vdash n$ be two partitions of n and $t \in T_\lambda$ and $s \in T_\mu$ be two Young tableaux. Suppose that, for each row i of t, the elements $\{t_{i,1}, t_{i,2}, \ldots, t_{i,\lambda_i}\}$ belong to different columns of s. Then $\lambda \trianglelefteq \mu$.*

Proof As the elements of the first row of t belong to different columns of s we necessarily have $\lambda_1 \leq \mu_1$. Now, given three elements from the first two rows of t, as two of them are necessarily in the same row and therefore belong to different columns of s, we have that these three elements cannot belong to a single column of s. In other words, every column of s contains at most two elements from the first two rows of t. Up to permuting the elements in each single column of s, we may suppose that the elements corresponding to the first two rows of t belong to the top two rows of s. This shows that $\lambda_1 + \lambda_2 \leq \mu_1 + \mu_2$. Analogously, iterating this argument, each column of s contains at most i elements from the first i rows of t showing that $\lambda_1 + \lambda_2 + \cdots + \lambda_i \leq \mu_1 + \mu_2 + \cdots + \mu_i$ and this completes the proof. $\qquad\square$

Definition 10.3.5 For any set A denote by S_A the group of all permutations of A. Let $\lambda = (\lambda_1, \lambda_2, \ldots, \lambda_h) \vdash n$ be a partition and $t \in T_\lambda$ a tableau. For $1 \leq i \leq h$ and for $1 \leq j \leq k := \lambda_1$, let $R_i = R_i(t) = \{t_{i,j} : j = 1, \ldots, \lambda_i\}$ and $C_j = C_j(t) = \{t_{i,j} : i = 1, \ldots, \lambda'_j\}$ (where λ' is the conjugate of λ) denote the sets of numbers in the ith row and in jth column of t, respectively. Then the Young subgroups

$$R_t = S_{R_1} \times S_{R_2} \times \cdots \times S_{R_h} \qquad \text{and} \qquad C_t = S_{C_1} \times S_{C_2} \times \cdots \times S_{C_k}$$

are called the *row-stabilizer* and *column-stabilizer* of t, respectively. Moreover, given $t_1, t_2 \in T_\lambda$ we say that they are *equivalent* and we write $t_1 \sim t_2$ if $R_i(t_1) = R_i(t_2)$ for all $1 \leq i \leq h$, that is if each row of t_1 contains the same elements of the corresponding row in t_2. This is clearly an equivalence relation; the equivalence class of t is called a λ-*tabloid* and it is denoted by $\{t\}$. We denote by $\mathfrak{T}_\lambda$ the set of all λ-tabloids.

We can also think of a tabloid as a tableaux with no order within the rows. That is,

1	2	3	5	8
4	6	7		

1	8	2	5	3
7	4	6		

$t \in T_{(5,3)}$ $\{t\} \in \mathfrak{T}_{(5,3)}$

Figure 10.6.

where in the diagram of t the sets $\{1, 2, 3, 6, 8\}$ and $\{4, 5, 7\}$ are unordered (they are viewed just as subsets of $\{1, 2, \ldots, 8\}$).

Observe that the action of S_n on T_λ induces a transitive action on the λ-tabloids, namely $\pi\{t\} = \{\pi t\}$, as clearly $t_1 \sim t_2$ infers $\pi(t_1) \sim \pi(t_2)$ for all $t_1, t_2 \in T_\lambda$ and $\pi \in S_n$. It is also clear that the stabilizer in S_n of a tabloid $\{t\}$ is nothing but the subgroup R_t. Thus the number of all λ-tabloids is

$$|\mathfrak{T}_\lambda| = \frac{|S_n|}{|R_t|} = \frac{n!}{\lambda!}$$

where for $\lambda = (\lambda_1, \lambda_2, \ldots, \lambda_h) \vdash n$ we use the notation $\lambda! = \lambda_1! \lambda_2! \cdots \lambda_h!$.

Definition 10.3.6 For $\lambda \vdash n$ denote by $M^\lambda = L(\mathfrak{T}_\lambda)$ the permutation representation of S_n induced by the action on the λ-tabloids.

Example 10.3.7 Consider the partition $(n - 1, 1) \vdash n$. Clearly there are $\frac{n!}{(n-1)!} = n$ distinct $(n-1, 1)$-tabloids: these are uniquely determined by the unique element in the second row.

2	3	$\cdots$	n
1			

1	3	$\cdots$	n
2			

$\cdots$

1	2	$\cdots$	$n-1$
n			

Figure 10.7.

Thus, $M^{n-1,1}$ is isomorphic to the space $L(\{1, 2, \ldots, n\})$ of all complex valued functions on $\{1, 2, \ldots, n\}$.

More generally, for $k \leq n/2$, a $(n - k, k)$-tabloid is uniquely characterized by the k numbers appearing in its second row. This way, $M^{n-k,k}$ is isomorphic to the space of all complex valued functions on the space of all k-subsets of $\{1, 2, \ldots, n\}$; see Chapter 6 on the Johnson scheme. Analogously, for $k < n$, a $(n - k, 1, \ldots, 1)$-tabloid is determined by the ordered k-tuple consisting of the numbers appearing in the last k rows.

This way, $M^{(n-k,1,\ldots,1)}$ is isomorphic to the space of all complex valued functions defined on the set of all ordered k-tuples of distinct elements in $\{1, 2, \ldots, n\}$; see also Section 3.14. In general, $\mathfrak{T}_\lambda$ is isomorphic to the S_n-homogeneous space $S_n/(S_{\lambda_1} \times S_{\lambda_2} \times \cdots \times S_{\lambda_h})$ and M^λ is isomorphic to the corresponding permutation representations.

With a λ-tabloid $\{t\}$ we denote by $e_{\{t\}} \in M^\lambda$ the Dirac function at $\{t\}$. Clearly, $\{\{t\} : t \in \mathfrak{T}_\lambda\}$ is an orthonormal basis for M^λ. Here we adopt a slight change of notation (more similarly to that adopted in the literature relative to the representation theory of S_n; see for instance [132, 182]). Indeed, a function $f \in M^\lambda$ will be seen as a linear combination $\sum_{\{t\} \in \mathfrak{T}_\lambda} f(\{t\}) e_{\{t\}}$. The permutation representation of S_n on M^λ will be written in the form: $\pi e_{\{t\}} = e_{\{\pi t\}}$ and $\pi f \equiv \pi(\sum_{\{t\} \in \mathfrak{T}_\lambda} f(\{t\}) e_{\{t\}}) = \sum_{\{t\} \in \mathfrak{T}_\lambda} f(\{t\}) e_{\{\pi t\}}$, which is cleary equivalent to saying that $(\pi f)(\{t\}) = f(\pi^{-1}\{t\})$.

Let now $\mu \vdash n$ and $t \in T_\mu$. We set $A_t = \sum_{\pi \in C_t} \text{sign}(\pi)\pi$, where sign is the character of S_n given by $\text{sign}(\pi) = 1$ if π is an even permutation and $\text{sign}(\pi) = -1$ otherwise (see Exercise 3.4.5). If λ is another partition of n we can think of A_t as a linear operator $A_t : M^\lambda \to M^\lambda$ by setting

$$A_t e_{\{s\}} = \sum_{\pi \in C_t} \text{sign}(\pi) e_{\{\pi s\}} \tag{10.4}$$

for all $s \in T_\lambda$. In other words, A_t corresponds to the element f of the group algebra $L(S_n)$ given by $f(\pi) = \text{sign}(\pi)$ if $\pi \in C_t$ and $f(\pi) = 0$ otherwise. Moreover (10.4) is its natural action on M^λ.

The linear operator $A_t : M^\lambda \to M^\lambda$ is symmetric: indeed,

$$\langle A_t u, v \rangle = \langle \sum_{\pi \in C_t} \text{sign}(\pi)\pi u, v \rangle = \sum_{\pi \in C_t} \text{sign}(\pi) \langle u, \pi^{-1} v \rangle = \langle u, A_t v \rangle$$

for all $u, v \in M^\lambda$, where the last equality follows from the fact that $\text{sign}(\pi^{-1}) = \text{sign}(\pi)$ and that $\langle \pi e_{\{u\}}, e_{\{v\}} \rangle = \langle e_{\{u\}}, \pi^{-1} e_{\{v\}} \rangle$.

Definition 10.3.8 Let t be a λ-tableau. The *polytabloid* associated with t is the vector e_t in M^λ defined by

$$e_t = A_t e_{\{t\}}. \tag{10.5}$$

Example 10.3.9 Let $\lambda = (n-2, 2) \vdash n$. Consider the λ-tableau

$$t = \begin{array}{|c|c|c|c|} \hline 1 & 2 & \cdots & n-2 \\ \hline n-1 & n \\ \hline \end{array}$$

Figure 10.8.

Then, its column stabilizer is $C_t = S_{\{1,n-1\}} \times S_{\{2,n\}}$ and therefore

$$A_t = 1 - (1, n-1) + (1, n-1)(2, n) - (2, n).$$

Correspondingly, the associated polytabloid is

$$e_t = e_{\{t\}} - e_{\{t_1\}} + e_{\{t_2\}} - e_{\{t_3\}}$$

where

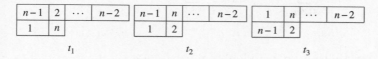

$$t_1 \qquad\qquad t_2 \qquad\qquad t_3$$

Figure 10.9.

Lemma 10.3.10 *For a polytabloid e_t and a permutation $\pi \in S_n$ one has*

$$\pi e_t = e_{\pi t}. \tag{10.6}$$

Proof Let A be a subset of $\{1, 2, \ldots, n\}$ and denote by $S_n(A)$ its stabilizer in S_n. Since every element of $S_n(A)$ is a product of its cycles we have, in virtue of (10.1), that $\pi S_n(A)\pi^{-1} = S_n(\pi A)$ for all $\pi \in S_n$. Thus for every $\pi \in S_n$ we have $\pi C_t \pi^{-1} = C_{\pi t}$ and

$$\pi e_t = \pi \left(\sum_{\pi' \in C_t} \text{sign}(\pi') e_{\pi'(t)} \right) = \sum_{\pi' \in C_t} \text{sign}(\pi') \pi e_{\pi'\{t\}}$$

$$= \sum_{\pi' \in C_t} \text{sign}(\pi') e_{\pi\pi'\{t\}} = \sum_{\sigma \in C_{\pi t}} \text{sign}(\pi^{-1}\sigma\pi) e_{\pi\pi^{-1}\sigma\pi\{t\}})$$

$$= e_{\pi\{t\}}.$$

$\square$

Definition 10.3.11 For $\lambda \vdash n$ we define S^λ as the subspace of M^λ spanned by all the λ-polytabloids. S^λ is called the *Specht module* associated with λ.

Note that, in virtue of the previous lemma, any Specht module S^λ, $\lambda \vdash n$, is S_n-invariant.

Lemma 10.3.12 *Let* $\lambda, \mu \vdash n$ *be two partitions; let* $t \in T_\lambda$ *and* $s \in T_\mu$ *be two tableaux. Suppose that there are two numbers belonging simultaneously to the same row of* s *and to the same column of* t. *Then* $A_t e_{\{s\}} = 0$.

Proof Let $a, b \in \{1, 2, \ldots, n\}$ denote the two numbers as in the statement. Then we have

$$(1 - (a, b)) e_{\{s\}} = e_{\{s\}} - e_{\{s\}} = 0.$$

On the other hand, $\{1, (a, b)\}$ is a subgroup of C_t: denote by $\{\gamma_1, \ldots, \gamma_z\}$ a set of representatives for the corresponding left-cosets. We then have

$$A_t e_{\{s\}} = \sum_{\pi \in C_t} \text{sign}(\pi) e_{\pi\{s\}} = \sum_{i=1}^{z} \text{sign}(\gamma_i) \gamma_i (1 - (a, b)) e_{\{s\}} = 0.$$

$\square$

Corollary 10.3.13 *Let* $\lambda, \mu \vdash n$ *be two partitions; let* $t \in T_\lambda$ *and* $s \in T_\mu$ *be two tableaux. Suppose that* $A_t e_{\{s\}} \neq 0$. *Then* $\lambda \trianglerighteq \mu$. *In particular, if in addition,* $\lambda = \mu$ *one has* $A_t e_{\{s\}} = \pm e_t$.

Proof From the previous lemma we immediately have that the elements belonging to a given row of s necessarily belong to different columns of t. In virtue of Lemma 10.3.4 we deduce that $\lambda \trianglerighteq \mu$.

Now suppose, in addition, that $\lambda = \mu$. We show that there exist two elements $\sigma \in R_s$ and $\theta \in C_t$ such that $\sigma s = \theta t$. Let $s_{1,1}, s_{1,2}, \ldots, s_{1,\lambda_1}$ be the elements of the first row in s. As they belong to different columns in t, there is $\sigma_1 \in C_t$ such that $\{s_{1,1}, s_{1,2}, \ldots, s_{1,\lambda_1}\} = \{(\sigma_1 t)_{1,1}, (\sigma_1 t)_{1,2}, \ldots, (\sigma_1 t)_{1,\lambda_1}\}$. Let then $\theta_1 \in S_{\lambda_1} \leq R_s$ be such that $(\theta_1 s)_{1,j} = (\sigma_1 t)_{1,j}$ for all j's. Repeating the same argument for the part of the tableaux $\theta_1 s$ and $\sigma_1 t$ below the first rows, respectively, we obtain elements $\theta_i \in S_{\lambda_i} \leq R_s$ and $\theta_i \in C_t$ (fixing the first $i - 1$ rows of t), $i = 2, 3, \ldots, h$, so that $\sigma = \sigma_h \cdots \sigma_2 \sigma_1 \in R_s$ and $\theta = \theta_h \cdots \theta_2 \theta_1 \in C_t$ are the right permutations.

Thus, $\{s\} = \{\sigma s\} = \{\theta t\} = \theta\{t\}$ and $e_{\{s\}} = e_{\theta\{t\}} = \theta e_{\{t\}}$ so that

$$A_t e_{\{s\}} = \sum_{\pi \in C_t} \text{sign}(\pi) \pi \theta e_{\{t\}} = \text{sign}(\theta) A_t e_{\{t\}} = \text{sign}(\theta) e_t.$$

$\square$

Lemma 10.3.14 *Let* t *be a* λ-*tableau and* $f \in M^\lambda$. *Then there exists* $c \in \mathbb{C}$ *such that* $A_t f = c e_t$.

Proof By linearity it suffices to prove the statement in the case $f = e_{\{s\}}$ where $s \in T_\lambda$; but then, in virtue of the previous corollary (with $\lambda = \mu$), we have that

$$A_t f = c e_t, \qquad \text{with } c \in \{1, -1\}.$$

$\square$

Theorem 10.3.15 (James' submodule theorem) *Let W be an S_n-invariant subspace of M^λ. Then S^λ is contained either in W or in its orthogonal complement $W^\perp$. In particular, S^λ is an irreducible representation of S_n.*

Proof Suppose that for all $f \in W$ and for all $s \in T_\lambda$ one has $A_s f = 0$. Then

$$0 = \langle A_s f, e_{\{s\}} \rangle = \langle f, A_s e_{\{s\}} \rangle = \langle f, e_s \rangle$$

so that $W \leq (S^\lambda)^\perp$.

On the other hand if there exist $f \in W$ and $s \in T_\lambda$ such that $A_s f \neq 0$ then, as $A_s f$ clearly belongs to W, from the previous lemma we deduce that also $e_s \in W$. As for every tableau $t \in T_\lambda$ there exists $\pi \in S_n$ such that $\pi s = t$ and thus, by Lemma 10.3.10, $\pi e_s = e_t$, one deduces that $e_t \in W$ and therefore $S^\lambda \leq W$. $\square$

Proposition 10.3.16 *Let $\lambda, \mu \vdash n$ be two partitions and $F \in \text{Hom}_{S_n}(M^\lambda, M^\mu)$, $F \neq 0$. If $S^\lambda \not\subset \ker F$ then $\lambda \unrhd \mu$. Moreover if $\lambda = \mu$ then $F|_{S^\lambda} = c\mathbf{1}|_{S^\lambda}$.*

Proof Noticing that $\ker F$ is S_n-invariant, the condition $S^\lambda \not\subset \ker F$ infers, by previous lemma, that $\ker F \leq (S_\lambda)^\perp$. It follows that for $t \in T_\lambda$ one has

$$A_t F e_{\{t\}} = F A_t e_{\{t\}} = F e_t \neq 0. \tag{10.7}$$

Now, $F e_{\{t\}}$ is a linear combination of μ-tabloids and therefore there exists $s \in T_\mu$ such that $A_t e_{\{s\}} \neq 0$. In virtue of Corollary 10.3.13 we deduce that $\lambda \unrhd \mu$.

If $\lambda = \mu$, then, by (10.7) and Lemma 10.3.14, we have $F e_t = c e_t$ with $c \in \mathbb{C}$ so that, for $\pi \in S_n$

$$F(e_{\pi t}) = F(\pi e_t) = \pi F(e_t) = \pi c e_t = c e_{\pi t}$$

showing that $F|_{S^\lambda} = c\mathbf{1}|_{S^\lambda}$. $\square$

Corollary 10.3.17 *If* $Hom_{S_n}(S^\lambda, S^\mu) \neq \{0\}$ *then* $\lambda \trianglerighteq \mu$.

Proof Let $0 \neq F' \in Hom_{S_n}(S^\lambda, S^\mu)$. We denote by F its extension $F : M^\lambda \to M^\mu$ obtained by setting $F|_{(S^\lambda)^\perp} = 0$. Then $0 \neq F \in Hom_{S_n}(M^\lambda, M^\mu)$ and the previous proposition yields the proof. $\square$

Exercise 10.3.18 Prove that $\dim Hom_{S_n}(S^\lambda, M^\lambda) = 1$.

Theorem 10.3.19 *Let* $\lambda, \mu \vdash n$ *be two partitions. If* $\lambda \neq \mu$ *then* $S^\lambda \not\sim S^\mu$.

Proof Suppose $S^\lambda \sim S^\mu$. Then $Hom_{S_n}(S^\lambda, S^\mu) \neq \{0\}$ and therefore, by the previous corollary $\lambda \trianglerighteq \mu$. By symmetry we also have $\mu \trianglelefteq \lambda$ and therefore $\lambda = \mu$. $\square$

We are now able to state the main result of this section.

Theorem 10.3.20 *The set* $\{S^\lambda : \lambda \vdash n\}$ *is a complete list of all irreducible representations of* S_n.

Proof Observe that

$$|\text{partitions of } n| = |\text{conjugacy classes of } S_n| = |\widehat{S_n}|$$

and, since by the previous theorem the S^λ's are pairwise nonequivalent, the statement follows. $\square$

Remark 10.3.21 As the polytabloids are linear combinations of tabloids with integer and therefore real coefficients, we have that S^λ is a *real* representation (cf. Corollary 9.8.22).

Corollary 10.3.22 *Let* $\mathcal{P}(n)$ *be the set of all partitions of* n *and* $\widehat{S_n}$ *the dual of* S_n. *Then the map*

$$\mathcal{P}(n) \ni \lambda \mapsto S^\lambda \in \widehat{S_n} \tag{10.8}$$

has the following properties:

  (i) $S^\lambda \leq M^\lambda$;
  (ii) *if* $S^\lambda \leq M^\mu$ *then* $\lambda \trianglerighteq \mu$.

Moreover, (10.8) is uniquely determined by the above properties.

Proof Clearly $S^\lambda \leq M^\lambda$ (with multiplicity one by Exercise 10.3.18). If S^λ is isomorphic to a sub-representation of M^μ, then there exists $F \in \mathrm{Hom}_{S_n}(M^\lambda, M^\mu)$ with $S^\lambda \not\leq \ker F$ and, by Proposition 10.3.16, $\lambda \trianglerighteq \mu$.

If $\mathcal{P}(n) \ni \lambda \mapsto V^\lambda \in \widehat{S_n}$ is another map satisfying (i) and (ii) then, clearly, $V^{(n)} = M^{(n)} = S^{(n)}$. Suppose that there exists $\mu \in \mathcal{P}(n)$ such that $S^\mu \neq V^\mu$. We can suppose that μ is maximal with respect to $\trianglerighteq$ so that $S^\lambda = V^\lambda$ for $\lambda \trianglerighteq \mu$. Decomposing M^μ and keeping in mind property (ii) we get a contradiction. $\qquad\square$

Exercise 10.3.23 With the notation of Chapter 6, directly verify that the polytabloids $e_t \in S^{(n-k,k)}$ are in the kernel of the operator d defined in (6.2), namely $de_t = 0$.

10.4 Representations corresponding to transposed tableaux

Consider the partition $(1, 1, \ldots, 1) \vdash n$ which we shall simply denote by (1^n). For each (1^n)-tableau t one clearly has $R_t = \{1\}$ and $C_t = S_n$. Moreover, given any two tableaux $t, s \in T_{(1^n)}$ there exists $\pi_t \in S_n$ such that $\pi_t s = t$ and therefore for the corresponding polytabloids we have

$$e_t = \sum_{\pi \in S_n} \mathrm{sign}(\pi)\pi e_{\{t\}} = \sum_{\pi \in S_n} \mathrm{sign}(\pi)\pi\pi_t e_{\{s\}} = \mathrm{sign}(\pi_t)e_s.$$

In other words the irreducible representation $S^{(1^n)}$ is one-dimensional and it is equivalent to the sign representation: $\pi e_s = \mathrm{sign}(\pi)e_s$.

Theorem 10.4.1 *Let $\lambda \vdash n$ and denote by λ' its transposed partition. Then*

$$S^{\lambda'} = S^\lambda \otimes S^{(1^n)}.$$

Proof Let $t \in T_\lambda$ and denote by $t' \in T_{\lambda'}$ the corresponding *transposed* tableau.

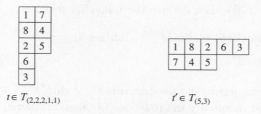

$t \in T_{(2,2,2,1,1)}$ $\qquad\qquad\qquad\qquad t' \in T_{(5,3)}$

Figure 10.10.

Clearly $R_{t'} = C_t$ and $C_{t'} = R_t$. Consider the operators

$$A_{t'} = \sum_{\pi \in C_{t'}} \text{sign}(\pi)\pi \quad \text{and} \quad B_{t'} = \sum_{\pi \in R_{t'}} \pi$$

and denote by u a vector spanning the one-dimensional space of the representation $S^{(1^n)}$. Consider the map $\theta : M^{\lambda'} \to S^\lambda \otimes S^{(1^n)}$ given by

$$\theta(\pi e_{\{t'\}}) := \pi B_{t'}(e_{\{t\}} \otimes u) = \pi \left(\sum_{\sigma \in C_t} \sigma e_{\{t\}} \otimes \sigma u \right)$$

$$= \pi \left(\sum_{\sigma \in C_t} \text{sign}(\sigma)\sigma e_{\{t\}} \otimes u \right) = \pi(e_t \otimes u) = (\pi e_t \otimes u)\text{sign}(\pi)$$

for all $\pi \in S_n$ and extending by linearity. We observe that θ is well defined. Indeed if $\pi_1 e_{\{t'\}} = \pi_2 e_{\{t'\}}$ with $\pi_1, \pi_2 \in S_n$ then on has $\pi_2^{-1}\pi_1 \in R_{t'}$ and therefore $\pi_1 B_{t'} = \pi_2 B_{t'}$. Moreover θ is surjective and commutes with the action of S_n. Indeed, for all $\pi, \sigma \in S_n$ one has

$$\theta(\sigma\pi e_{\{t'\}}) = (\sigma\pi e_t \otimes u)\text{sign}(\sigma)\text{sign}(\pi)$$

$$= \sigma((\pi e_t \otimes u)\text{sign}(\pi)) = \sigma(\theta(\pi e_{\{t'\}})).$$

The image of $e_{t'}$ under θ is given by

$$\theta(e_{t'}) = \theta(\sum_{\pi \in C_{t'}} \text{sign}(\pi)\pi e_{\{t'\}})$$

$$= \sum_{\pi \in R_t} \text{sign}(\pi)\theta(\pi e_{\{t'\}})$$

$$= \sum_{\pi \in R_t} \pi e_t \otimes u$$

$$= (B_t e_t) \otimes u.$$

Since $\langle B_t e_t, e_{\{t'\}} \rangle = \langle e_t, B_t e_{\{t'\}} \rangle = \langle e_t, |R_t| e_{\{t'\}} \rangle = |R_t| \neq 0$ one has $\theta(e_{t'}) \neq 0$. Therefore $S^{\lambda'} \not\subset \ker\theta$ so that, in virtue of Theorem 10.3.15, $\ker\theta \leq (S^{\lambda'})^\perp$. We then deduce the following relation

$$\dim S^\lambda = \dim(\text{Im}\theta) = \dim M^{\lambda'} - \dim(\ker\theta) \geq \dim M^{\lambda'} - \dim(S^{\lambda'})^\perp$$

$$= \dim S^{\lambda'}. \tag{10.9}$$

With analogous arguments we have $\dim S^{\lambda'} \geq \dim(S^{(\lambda')'}) = S^\lambda$. Substituting this last inequality in (10.9) one obtains $\dim(\ker\theta) = \dim(S^{\lambda'})^\perp$. Thus $\ker\theta = (S^{\lambda'})^\perp$ and θ is an isomorphism. This ends the proof. $\qquad\square$

Corollary 10.4.2 *Under the same hypotheses of the preceding theorem one has* $\chi_{\lambda'}(\pi) = \text{sign}(\pi)\chi_\lambda(\pi)$ *for all* $\pi \in S_n$. *In particular* $d_\lambda = d_{\lambda'}$ *and, for all transpositions* $\tau \in S_n$, $\chi_\lambda(\tau) = -\chi_{\lambda'}(\tau)$.

Proof In virtue of Lemma 9.1.5 for all $\pi \in S_n$ we have

$$\chi_{\lambda'}(\pi) = \chi_{(1^n)}(\pi)\chi_\lambda(\pi) = \text{sign}(\pi)\chi_\lambda(\pi).$$

In particular, for $\pi = 1$ one has $d_{\lambda'} = d_\lambda$, while if τ is a transposition $\chi_\lambda(\tau) = -\chi_{\lambda'}(\tau)$. □

10.5 Standard tableaux

Definition 10.5.1 Let $\lambda \vdash n$ be a partition. A tableau $t \in T_\lambda$ is *standard* if the numbers in each row and in each column are growing from left to right and from the top to the bottom, respectively. The polytabloid e_t corresponding to a standard tableau t is also called a *standard* polytabloid.

Example 10.5.2 For $n = 8$ and $\lambda = (5,3)$ consider the following λ-tabloids.

$$t \in T_{(5,3)} \qquad\qquad s \in T_{(5,3)}$$

Figure 10.11.

Then t is standard while s is not.

Theorem 10.5.3 *The standard λ-polytabloids constitute a basis for S^λ. Thus, the dimension of S^λ equals the number of all standard λ-polytabloids.*

The proof of the theorem is divided into two parts: in the first part we show that λ-polytabloids are independent (Corollary 10.5.9) while in the second one we show that they span S^λ (Theorem 10.5.12).

Definition 10.5.4 Let n be a positive integer. A *composition* of n is an ordered sequence $a = (a_1, a_2, \ldots, a_k)$ of nonnegative integers such that $a_1 + a_2 + \cdots + a_k = n$. The numbers a_i are called the *parts* of a.

Note that a partition $\lambda \vdash n$ is therefore a composition (with the weakly decreasing and positivity conditions on its parts) but, clearly, not vice versa. For instance, $(5, 4, 1)$ is a partition of 10, but $(5, 2, 3)$ is simply a composition of 10.

With a composition one associates, in analogy with the partitions, the *Young diagram* and the tableaux and one defines the *dominance order* of two compositions by comparing the first i parts, $i = 1, 2, \ldots$ In other words, if $a = (a_1, a_2, \ldots, a_k)$ and $b = (b_1, b_2, \ldots, b_h)$ are compositions of n, then $a \trianglerighteq b$ means that $k \leq h$ and $\sum_{i=1}^{z} a_i \geq \sum_{i=1}^{z} b_i$ for all $z = 1, 2, \ldots, k$. One can also extend the notion of a tabloid: if $a = (a_1, a_2, \ldots, a_k)$ is a composition of n, then an *a-tabloid* or a *tabloid of shape a*, is an arrangement of the numbers $1, 2, \ldots, n$ into k rows, where the ith row contains a_i numbers and there is no order within each row. For instance, given the composition $a = (2, 3)$ of 5, then

$$\begin{array}{|ccc|}\hline 1 & 5 & \\ \hline 2 & 4 & 3 \\ \hline \end{array}$$

Figure 10.12.

is an *a*-tabloid. We denote by $\mathfrak{T}_a$ the S_n-homogeneous space of all *a*-tabloids and by M^a the corresponding permutation module. It is clear that if λ is the partition of n obtained by rearranging the parts of a in a decreasing order, then $\mathfrak{T}_\lambda \cong \mathfrak{T}_a$ and $M^\lambda \sim M^a$ as S_n-homogeneous space and as permutation module respectively. For instance $M^{(1,7,5)} \cong M^{(7,5,1)}$ as S_{13}-modules.

From now on, unless otherwise specified, we shall consider tabloids associated with compositions, that is with a more general shape than those associated with partition.

Given a partition $\lambda \vdash n$ and a λ-tabloid $\{t\}$, for each $i \in \{1, 2, \ldots, n\}$ one denotes by $\{t^i\}$ the tabloid formed by all elements $\{1, 2, \ldots, i\}$ in $\{t\}$ and by λ^i the composition which corresponds to $\{t^i\}$. For example, for $n = 5$, $\lambda = (3, 2)$ and the λ-tabloid

$$\{t\} = \begin{array}{|ccc|}\hline 2 & 5 & 4 \\ \hline 1 & 3 & \\ \hline \end{array}$$

Figure 10.13.

one has

$$\{t^1\} = \frac{\overline{}}{1} \qquad \{t^2\} = \frac{\overline{2}}{1} \qquad \{t^3\} = \frac{\overline{2}}{1\ \ 3}$$

$$\lambda^1 = (0,1), \qquad\qquad \lambda^2 = (1,1), \qquad\qquad \lambda^3 = (1,2),$$

$$\{t^4\} = \frac{\overline{2\ \ 4}}{1\ \ 3} \qquad\qquad \{t^5\} = \frac{\overline{2\ \ 5\ \ 4}}{1\ \ 3}$$

$$\lambda^4 = (2,2), \qquad\qquad\qquad \lambda^5 = (3,2)$$

Figure 10.14.

The dominance order on generalized tabloids is determined by the dominance ordering on the corresponding compositions.

Definition 10.5.5 Let $\{s\}$ and $\{t\}$ be two tabloids with composition sequences $(\lambda^i)_i$ and $(\mu^i)_i$, respectively. Then $\{s\}$ *dominates* $\{t\}$ (and we write $\{s\} \trianglerighteq \{t\}$) if $\lambda^i \trianglerighteq \mu^i$ for all i's.

Lemma 10.5.6 *Let* $\{t\}$ *be a tabloid and suppose that in* $\{t\}$ *there exist* $a < b$ *with* a *in a lower row than* b. *Then* $\{t\} \lhd (a,b)\{t\}$.

Proof Suppose that a appears in row r and b in row q with $q < r$. Let $\lambda^i = (\lambda^i_1, \lambda^i_2, \ldots)$ and $\mu^i = (\mu^i_1, \mu^i_2, \ldots)$ be the composition series of $\{t\}$ and $(a,b)\{t\}$, respectively. Then we have:

$$\lambda^i = \mu^i \qquad \text{for } i < a \text{ or } i \geq b$$

and

$$\mu^i = (\lambda^i_1, \lambda^i_2, \ldots, \lambda^i_q + 1, \ldots, \lambda^i_r - 1, \ldots) \qquad \text{for } a \leq i < b.$$

Therefore we have $\mu^i \trianglerighteq \lambda^i$ with strict inequality for $a \leq i < b$. $\qquad\square$

Suppose that $f \in M^\mu$ and $f = \sum_{\{t\} \in \mathcal{I}_\mu} f(\{t\})e_{\{t\}}$. We say that $e_{\{t\}}$ (or $\{t\}$) appears in f if $f(\{t\}) \neq 0$.

Corollary 10.5.7 *If* t *is standard and* $e_{\{s\}}$ *appears in* e_t, *then* $\{t\} \trianglerighteq \{s\}$. *In other words* $\{t\}$ *is the maximum tabloid that appears in* e_t.

Proof If $e_{\{s\}}$ appears in $e_t = \sum_{\pi \in C_t} \text{sign}(\pi)\pi e_{\{t\}}$ then there exists $\pi \in C_t$ such that $\{s\} = \pi\{t\}$, and we may suppose that $s = \pi t$. A column inversion in s is a pair a, b in the same column such that $a < b$ and a is in a lower row than b. If no column inversion exists, then s is standard and necessarily $s = t$. Otherwise, applying the previous lemma one has $(a, b)\{s\} \rhd \{s\}$. This way, the number of column inversion decreases, and after a finite number of steps we get $\{t\} \trianglerighteq \{s\}$. $\square$

Lemma 10.5.8 *Let $v_1, v_2, \ldots, v_m$ be elements in M^μ. Suppose that for each v_i one can choose a tabloid $\{t_i\}$ appearing in v_i such that $\{t_i\}$ is maximum in v_i and the $\{t_i\}$'s are all distinct. Then $v_1, v_2, \ldots, v_m$ are independent.*

Proof Up to a permutation of the indices we can assume that $\{t_1\}$ is maximal among the $\{t_i\}$'s. This ensures that $\{t_1\}$ appears only in v_1. Suppose now that $c_1 v_1 + c_2 v_2 + \cdots + c_m v_m = 0$. Then necessarily $c_1 = 0$ and $c_2 v_2 + \cdots + c_m v_m = 0$. Recursively repeating the previous argument one gets $c_2 = \cdots = c_m = 0$ and the statement follows. $\square$

Corollary 10.5.9 *Let $\lambda \vdash n$ be a partition. Then the standard λ-polytabloids are linearly independent.*

Proof The statement follows immediately from Corollary 10.5.7 and Lemma 10.5.8 applied to the standard tabloids $\{t\}$'s and their associated polytabloids e_t's. $\square$

We now show that the standard polytabloids of shape λ span S^λ.

Suppose that A and B are disjoint subsets of $\{1, 2, \ldots, n\}$ and denote by S_A, S_B and $S_{A \cup B}$ the corresponding symmetric groups. Then $S_A \times S_B \leq S_{A \cup B}$. Moreover, if t is a λ-tableau, $\lambda \vdash n$, and A and B are contained in different columns of t, then we can always choose a transversal Π for $S_A \times S_B$ in $S_{A \cup B}$ (that is, $S_{A \cup B} = \coprod_{\pi \in \Pi} \pi(S_A \times S_B)$) in such a way that for any $\pi \in \Pi$ the elements of $A \cup B$ are increasing down the columns of πt. Indeed, if Π is any transversal, for any $\pi \in \Pi$ we can choose $\sigma_\pi \in S_A \times S_B$ such that $\sigma_\pi \pi \in \pi(S_A \times S_B)$ and the above property holds for $\sigma_\pi \pi t$. Then we can replace Π with $\Pi' = \{\sigma_\pi \pi : \pi \in \Pi\}$.

Definition 10.5.10 Let t be a tableau and let A and B be subsets of two consecutive columns, say the jth and the $(j + 1)$st, respectively. Let Π be the set of transversals for $S_A \times S_B$ in $S_{A \cup B}$ such that the elements of

$A \cup B$ are increasing down the columns of πt. Then the *Garnir element* associated with t and A, B is $g_{A,B} = \sum_{\pi \in \Pi} \mathrm{sign}(\pi)\pi$.

For example, for the tableau

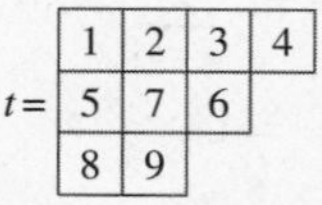

$$t = \begin{array}{|c|c|c|c|} \hline 1 & 2 & 3 & 4 \\ \hline 5 & 7 & 6 \\ \cline{1-3} 8 & 9 \\ \cline{1-2} \end{array}$$

Figure 10.15.

choose $A = \{7,9\}$ and $B = \{6,3\}$ in the second and third column, respectively. In this case we can take

$$\Pi = \{1, (6,7), (6,9,7), (3,6,7), (3,6,9,7), (3,7)(6,9)\},$$

so that the Garnir element is

$$g_{A,B} = 1 - (6,7) + (6,9,7) + (3,6,7) - (3,6,9,7) + (3,7)(6,9),$$

as the following picture shows:

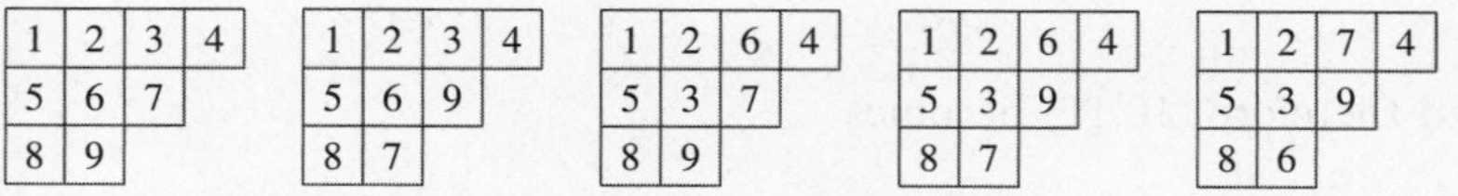

$(6,7)t$ $(6,9,7)t$ $(3,6,7)t$ $(3,6,9,7)t$ $(3,7)(6,9)t$

Figure 10.16.

Proposition 10.5.11 *Let t be a λ-tableau and A and B as in the definition of a Garnir element. If $|A \cup B|$ is greater than λ'_j (i.e. the number of elements in the jth column of t), then $g_{A,B}e_t = 0$.*

Proof We first show that

$$\sum_{\pi \in S_{A \cup B}} \mathrm{sign}(\pi)\pi e_t = 0. \tag{10.10}$$

Let $\sigma \in C_t$. As $|A \cup B| > \lambda_j'$, there exist $a \in A$ and $b \in B$ such that a and b belong to the same row of σt. Thus $(a, b) \in S_{A \cup B}$ so that

$$\sum_{\pi \in S_{A \cup B}} \text{sign}(\pi) \pi e_{\{\sigma t\}} = \sum_{\pi' \in \Pi'} \text{sign}(\pi') \pi'[1 - (a, b)] e_{\{\sigma t\}}$$

$$= \sum_{\pi' \in \Pi'} \text{sign}(\pi') \pi'[e_{\{\sigma t\}} - e_{\{\sigma t\}}]$$

$$= 0$$

where Π' is a set of transversals in $S_{A \cup B}$ for the subgroup $\{1, (a, b)\}$. Summing up over all elements in C_t (with the corresponding sign), we obtain (10.10).

By definition of the Garnir element we clearly have

$$\sum_{\pi \in S_{A \cup B}} \text{sign}(\pi) \pi = g_{A,B} \sum_{\pi' \in (S_A \times S_B)} \text{sign}(\pi') \pi'$$

so that (10.10) becomes

$$g_{A,B} \sum_{\pi' \in (S_A \times S_B)} \text{sign}(\pi') \pi' e_t = 0. \tag{10.11}$$

As $S_A \times S_B \subseteq C_t$, for $\sigma \in S_A \times S_B$ we have:

$$\text{sign}(\sigma) \sigma e_t = \text{sign}(\sigma) \sigma [\sum_{\pi \in C_t} \text{sign}(\pi) \pi]\{t\} =_{\theta = \sigma \pi} [\sum_{\theta \in C_t} \text{sign}(\theta) \theta]\{t\} = e_t]$$

and therefore (10.11) becomes

$$g_{A,B}[|S_A \times S_B| e_t] \equiv |S_A \times S_B|[g_{A,B} e_t] = 0$$

and the statement follows. $\qquad\qquad\square$

Given a tableau t, the same way as the associated tabloid is defined by $\{t\} = \{\pi t : \pi \in R_t\}$, we define the *column*-tabloid as $[t] = \{\sigma t : \sigma \in C_t\}$. Replacing "row" with "column" everywhere in Definition 10.5.5 we obtain a notion of dominance for column-tabloids. It will be denoted with the same symbol $\trianglerighteq$ used before: it will be clear from the context which ordering the symbol will refer to.

We are now in position to prove that the standard λ-polytabloids span S^λ.

Theorem 10.5.12 *The set of all standard λ-polytabloids spans S^λ.*

Proof Recalling that for $\sigma \in C_t$ and $s = \sigma t$ one has

$$e_s = \sigma e_t = \text{sign}(\sigma)e_t,$$

it suffices to prove that e_t is in the span of the standard λ-polytabloids whenever t has increasing columns. We proceed by induction with respect to the dominance order for column-tabloids. It is clear that the maximum element is standard: indeed it is given by numbering the cells of each column from the top to the bottom and from the leftmost to the rightmost column.

For example, for $\lambda = (4, 3, 2) \vdash 9$, the maximum element is

Figure 10.17.

Let t be a λ-tableau with increasing columns. By induction, we may assume that all tableaux s with $[s] \rhd [t]$ are in the span of the standard λ-polytabloids. If t is standard there is nothing to prove. If it is not standard, let j be the first column such that there is an entry $t_{i,j}$ greater than $t_{i,j+1}$. By assumption on t we have $t_{i,j} < t_{i+1,j} < \cdots < t_{\lambda'_j, j}$ and $t_{1,j+1} < t_{2,j+1} < \cdots < t_{i-1,j+1} < t_{i,j+1}$. Setting $A = \{t_{i,j}, t_{i+1,j}, \ldots, t_{\lambda'_j, j}\}$ and $B = \{t_{1,j+1}, t_{2,j+1}, \ldots, t_{i-1,j+1}, t_{i,j+1}\}$ denote by $g_{A,B} = \sum_{\pi \in \Pi} \text{sign}(\pi)\pi$ the associated Garnir element. Observing that $|A \cup B| = \lambda'_j + 1$, by Proposition 10.5.11 we have $g_{A,B}e_t = 0$ which we rewrite as

$$e_t = - \sum_{\pi \in \Pi \setminus \{1\}} \text{sign}(\pi)e_{\pi t}. \tag{10.12}$$

Applying the column-analog of Lemma 10.5.6, one gets $[\pi t] \rhd [t]$. Indeed, $[\pi t] = \sigma[t]$ where $\sigma = \sigma_1 \sigma_2 \cdots \sigma_k$ with the σ_i's (commuting) transpositions of the form $\sigma_i = (a_i, b_i)$ where $a_i \in A$ and $b_i \in B$. This ensures, in particular, that $a_i > b_i$ and that b_i belongs to the column at the (immediate) right of that of a_i.

By induction, all terms in the right side of (10.12), and hence e_t as well, are in the span of the standard λ-tabloids, and the proof is complete.

$\square$

Corollary 10.5.13 *The set* $\{e_t : t \text{ standard } \lambda\text{-tableau}\}$ *is a basis for* S^λ. *In particular the dimension* d_λ *of* S_λ *equals the number of standard* λ-*tableaux.*

Remark 10.5.14 If t is a standard λ-tabloid then its conjugate $t' \in T_{\lambda'}$ is also standard. As a consequence $\dim S^\lambda = \dim S^{\lambda'}$. We have already proved this fact in Corollary 10.4.2.

Example 10.5.15 We have the following identities (see also Exercises 3.14.1, 3.14.2 and Theorem 6.1.6).

(1) $\dim S^{(n-k,k)} = \binom{n}{k} - \binom{n}{k-1}$, for $k \leq n/2$;

(2) $\dim S^{(n-k,1,1,\dots,1)} = \binom{n-1}{k}$.

Proof (1) It is clear that an $(n-k,k)$-standard tableau is determined by the elements in the second row. These form a k-subset of $X = \{1,2,\dots,n\}$. Note that, however, not every k-subset is admissible: for instance, one cannot take a subset containing either 1 or $\{2,3\}$. More precisely, it is easy to see that a k-subset $\{i_1, i_2, \dots, i_k\}$, with $i_1 < i_2 < \cdots < i_k$, is admissible if and only if $i_1 \geq 2, i_2 \geq 4, \dots, i_k \geq 2k$. Then the following is a bijection between the nonadmissible k-subsets and the $(k-1)$-subsets of X:

$$\{i_1, i_2, \dots, i_k\} \mapsto [\{1, 2, \dots, 2h-1\} \backslash \{i_1, i_2, \dots, i_h\}] \cup \{i_{h+1}, i_{h+2}, \dots, i_k\} \tag{10.13}$$

where, still supposing $i_1 < i_2 < \cdots < i_k$, $h = \min\{j : i_j \leq 2j - 1\}$ (and therefore $i_{h-1} = 2h - 2$ and $i_h = 2h - 1$). For instance, if $k = 3$, then $\{1, i, j\} \mapsto \{i, j\}$, $\{2, 3, j\} \mapsto \{1, j\}$ (for $j > 3$), $\{2, 4, 5\} \mapsto \{1, 3\}$ and $\{3, 4, 5\} \mapsto \{1, 2\}$. It follows that there are precisely $\binom{n-1}{k}$ nonadmissible subsets.

(2) This is easy: any standard $(n-k, 1, 1, \dots, 1)$-tableau is determined by the elements in the first row: 1 is the first element, the other $n-k-1$ are arbitrary (in $\{2, 3, \dots, n\}$). $\qquad\square$

Exercise 10.5.16 Show that (10.13) is a bijection.

As regarding an estimate of the dimension d_λ of each S^λ we have the trivial upper bound $d_\lambda \leq \sqrt{n!}$. Indeed as for any finite group G one has $\sum_{\pi \in \hat{G}} d_\pi^2 = |G|$ (cf. Theorem 3.7.11) we have $\sum_{\lambda \vdash n} d_\lambda^2 = |S_n| = n!$ and therefore for each fixed λ one has the asserted estimate. One can improve this estimate as follows.

Proposition 10.5.17 *Let $\lambda = (\lambda_1, \lambda_2, \ldots, \lambda_k) \vdash n$ be a partition and set $\overline{\lambda} = (\lambda_2, \ldots, \lambda_k) \vdash n - \lambda_1$. Then*

$$d_\lambda \leq \binom{n}{\lambda_1} d_{\overline{\lambda}} \leq \binom{n}{\lambda_1} \sqrt{(n - \lambda_1)!}. \tag{10.14}$$

Proof It follows immediately that there are $\binom{n}{\lambda_1}$ different ways to choose the elements in the first row and than once the first row is settled one has at most $d_{\overline{\lambda}}$ independent ways to fill up the remaining boxes of the Young diagram of λ (which indeed constitute the Young diagram of $\overline{\lambda}$).

Combining the previous two estimates, we get that $d_\lambda \leq \binom{n}{\lambda_1} \sqrt{(n - \lambda_1)!}$. $\square$

Observe that it is not the case, in general, that a standard $\overline{\lambda}$-tableau extends to a standard λ-tableau.

10.6 Computation of a Fourier transform on the symmetric group

Let T denote the set of all transpositions of S_n.

Consider the probability distribution P on S_n defined by

$$P(\pi) = \begin{cases} \frac{1}{n} & \text{if } \pi = 1 \\ \frac{2}{n^2} & \text{if } \pi \in T \\ 0 & \text{otherwise.} \end{cases} \tag{10.15}$$

Observe that P is central, that is, it is constant on the conjugacy classes: $P = \frac{1}{n}\delta_1 + \frac{2}{n^2}\mathbf{1}_T$, where δ_1 is the Dirac function at the identity element of S_n.

In virtue of Proposition 3.9.7, one easily computes its Fourier transform with respect to any representation ρ of S_n, namely

$$\widehat{P}(\rho) = \left(\frac{1}{d_\rho} \sum_{\pi \in S_n} P(\pi)\chi_\rho(\pi)\right) I = \left(\frac{1}{d_\rho}\chi_\rho(1)\frac{1}{n} + \sum_{\tau \in T} \frac{2}{n^2}\frac{\chi_\rho(\tau)}{d_\rho}\right) I$$

$$= \left(\frac{1}{n} + \frac{n-1}{n}r(\rho)\right) I \tag{10.16}$$

where $r(\rho) = \frac{\chi_\rho(\tau)}{d_\rho}$ (which does not depend on $\tau \in T$) and we have used the facts that $\chi_\rho(1) = d_\rho$ and the number of transpositions in S_n equals $n(n-1)/2$.

Moreover if U is the uniform distribution on S_n, that is $U(\pi) = \frac{1}{n!}$ for all $\pi \in S_n$, we have

$$\widehat{U}(\rho) = \begin{cases} I & \text{if } \rho \text{ is trivial} \\ 0 & \text{otherwise.} \end{cases} \tag{10.17}$$

As $r(\rho)$ plays a prominent role in our further developments, we shall establish an explicit formula for it. We first need an easy but useful lemma.

Lemma 10.6.1 *Let* $\lambda = (\lambda_1, \lambda_2, \ldots, \lambda_k) \vdash n$ *be a partition and denote by* $\lambda' = (\lambda'_1, \lambda'_2, \ldots, \lambda'_{\lambda_1})$ *its transpose. Then we have*

$$\sum_{i=1}^{k}(i-1)\lambda_i = \sum_{j=1}^{\lambda_1}\binom{\lambda'_j}{2}. \tag{10.18}$$

Proof In the Young diagram of λ fill up the first row with zeroes, the second one with ones, and so on, so that in the last row is filled-up with $(k-1)$'s. Summing up these numbers by rows and by columns one obtains the left hand side and the right hand side of (10.18), respectively. $\square$

From now on, given a partition $\lambda \vdash n$, we denote by χ_λ and d_λ, the character and the dimension of S^λ, respectively. Moreover we set $r(\lambda) = \frac{\chi_\lambda(\tau)}{d_\lambda}$ where τ is any transposition in S_n.

The aim of the next theorem is to compute $r(\lambda)$ for any partition $\lambda \vdash n$.

Theorem 10.6.2 *Let* $\lambda = (\lambda_1, \lambda_2, \ldots, \lambda_k) \vdash n$. *Then*

$$r(\lambda) = \frac{1}{n(n-1)} \sum_{i=1}^{k} \left[\lambda_i^2 - (2i-1)\lambda_i\right]. \tag{10.19}$$

Proof Let t be a λ-tableau. Note that the characteristic function $\mathbf{1}_T$ is a central element of the group algebra of S_n; in particular, it commutes with A_t. In the notation of the present chapter, convolution becomes a formal product: $\mathbf{1}_T A_t = \sum_{\tau \in T} \sum_{\pi \in C_t} \tau\pi\text{sign}(\pi)$ (see also Remark 10.6.6).

We split T as follows: $T = T_1 \coprod T_2 \coprod [T \setminus (T_1 \cup T_2)]$ where $T_1 = T \cap C_t$ and $T_2 = T \cap R_t$. For $\tau \in T_1$ one has

$$A_t \tau = \sum_{\pi \in C_t} \text{sign}(\pi)\pi\tau = \sum_{\gamma \in C_t} -\text{sign}(\gamma)\gamma = -A_t.$$

Thus

$$A_t \tau e_{\{t\}} = -A_t e_{\{t\}} = -e_t.$$

Obviously if $\tau \in T_2$ on has $A_t \tau e_{\{t\}} = e_t$. Finally, if $\tau \in T \setminus (T_1 \cap T_2)$ there exist two elements $x, y \in \{1, 2, \ldots, n\}$ that belong to the same row of τt and to the same column of t. Indeed $\tau = (a, b)$ where, necessarily, $a, b \in \{1, 2, \ldots, n\}$ belong to different rows and different columns of t; therefore there exists an element c which belongs to the same row of a and the same column of b (or vice versa): then take $x = b$ and $y = c$ (or $x = a$ and $y = c$, respectively). Therefore, by Lemma 10.3.12 we deduce that $A_t \tau e_{\{t\}} = 0$.

We thus have

$$\mathbf{1}_T e_t = \mathbf{1}_T A_t e_{\{t\}} = A_t \mathbf{1}_T e_{\{t\}}$$

$$= \sum_{\tau \in T} A_t \tau e_{\{t\}}$$

$$= \sum_{\tau \in T_1} A_t \tau e_{\{t\}} + \sum_{\tau \in T_2} A_t \tau e_{\{t\}} + \sum_{\tau \in T \setminus (T_1 \cup T_2)} A_t \tau e_{\{t\}}$$

$$= -|T_1| e_t + |T_2| e_t$$

$$= \left[\sum_{i=1}^{k} \binom{\lambda_i}{2} - \sum_{j=1}^{\lambda_1} \binom{\lambda'_j}{2} \right] e_t$$

$$= \left[\sum_{i=1}^{k} \binom{\lambda_i}{2} - \sum_{j=1}^{k} (i-1)\lambda_i \right] e_t$$

$$= \frac{1}{2} \sum_{i=1}^{k} \left[\lambda_i^2 - (2i-1)\lambda_i \right] e_t$$

where in the last but one identity we used the previous lemma.

We have already seen in (10.16) that

$$\widehat{\mathbf{1}_T}(\rho_\lambda) = \left(\frac{1}{d_\lambda} \sum_{\pi \in S_n} \mathbf{1}_T \chi_\lambda(\pi) \right) I = \binom{n}{2} \frac{\chi_\lambda(\tau)}{d_\lambda} I.$$

But the Fourier transform $\widehat{\mathbf{1}_T}(\rho_\lambda)$ coincides with the eigenvalue of the operator $\mathbf{1}_T = \sum_{\tau \in T} \tau$ restricted to S^λ and the formula of the theorem holds. $\qquad\square$

Corollary 10.6.3 *Let* $\lambda, \mu \vdash n$ *be two partitions. If* $\lambda \triangleright \mu$ *then* $r(\lambda) > r(\mu)$.

Proof In virtue of Lemma 10.2.2 it suffices to prove the statement in the case where λ is obtained by μ by a single-box up-move. Thus suppose that there exist a, b with $b > a$ such that

$$\lambda_a = \mu_a + 1, \quad \lambda_b = \mu_b - 1 \text{ and } \lambda_c = \mu_c$$

for all $c \neq a, b$. Therefore one has

$$
\begin{aligned}
r(\lambda) - r(\mu) &= \frac{1}{n(n-1)}[\lambda_a^2 - (2a-1)\lambda_a + \lambda_b^2 - (2b-1)\lambda_b - \mu_a^2 \\
&\quad + (2a-1)\mu_a - \mu_b^2 + (2b-1)\mu_b] \\
&= \frac{2}{n(n-1)}[\mu_a - \mu_b + b - a + 1] \\
&\geq \frac{2}{n(n-1)} > 0
\end{aligned}
$$

where the last but one inequality is due to the fact that $\mu_a \geq \mu_b$. Note that these facts continue to hold true even in the limit case $\lambda_b = 0$. $\quad\square$

Lemma 10.6.4 *For $\lambda = (\lambda_1, \lambda_2, \ldots, \lambda_k) \vdash n$ one has*

(i) $r(\lambda) \leq \frac{\lambda_1 - 1}{n-1}$;

(ii) $r(\lambda) \leq 1 - \frac{2(n-\lambda_1)(\lambda_1+1)}{n(n-1)}$ *if $\lambda_1 \geq n/2$.*

Proof (i) We have:

$$
\begin{aligned}
r(\lambda) &= \frac{1}{n(n-1)} \sum_{i=1}^{k} [\lambda_i^2 - (2i-1)\lambda_i] \leq \frac{1}{n(n-1)} \sum_{i=1}^{k} [\lambda_i(\lambda_i - 1)] \\
&\leq \frac{\lambda_1 - 1}{n(n-1)} \sum_{i=1}^{k} \lambda_i = \frac{\lambda_1 - 1}{n-1}.
\end{aligned}
$$

(ii) We trivially have $(\lambda_1, n - \lambda_1) \trianglerighteq \lambda$. Thus, from the previous corollary we have

$$
\begin{aligned}
r(\lambda) &\leq r(\lambda_1, n - \lambda_1) \\
&= \frac{1}{n(n-1)} \left(\lambda_1^2 - \lambda_1 + n^2 - 2n\lambda_1 + \lambda_1^2 - 3n + 3\lambda_1 \right) \\
&= \frac{n(n-1) + 2(\lambda_1^2 + \lambda_1 - n\lambda_1 - n)}{n(n-1)} = 1 - \frac{2(n-\lambda_1)(\lambda_1+1)}{n(n-1)}.
\end{aligned}
$$

$\quad\square$

Corollary 10.6.5 *Suppose that for $\lambda \vdash n$ one has $r(\lambda) \geq 0$. Then*

$$
\left| \frac{1}{n} + \frac{n-1}{n} r(\lambda) \right| \leq
\begin{cases}
\frac{\lambda_1}{n} & \text{for all } \lambda \vdash n \\
1 - \frac{2(n-\lambda_1)(\lambda_1+1)}{n^2} & \text{if } \lambda_1 \geq \frac{n}{2}.
\end{cases}
$$

Remark 10.6.6 In the following exercise, an element f of the group algebra $L(S_n)$ will be identified with the formal linear combination $\sum_{\pi \in S_n} f(\pi)\pi$; compare with the definition of A_t in Section 10.3. This way, the convolution of two elements $\sum_{\pi \in S_n} a_\pi \pi$ and $\sum_{\sigma \in S_n} b_\sigma \sigma$ is just the formal product $\sum_{\gamma \in S_n} \left(\sum_{\substack{\pi, \sigma \in S_n: \\ \pi\sigma = \gamma}} a_\pi b_\sigma \right) \gamma$, and the right (resp. left) regular representation of S_n on $L(S_n)$ is simply: $\rho(\sigma)[\sum_{\pi \in S_n} a_\pi \pi] = \sum_{\pi \in S_n} a_\pi \pi \sigma^{-1}$ (resp. $\lambda(\sigma)[\sum_{\pi \in S_n} a_\pi \pi] = \sum_{\pi \in S_n} a_\pi \sigma \pi$) for all $\sigma \in S_n$. The exercise requires also the knowledge of Section 9.5.

Exercise 10.6.7 (1) Let $\lambda \vdash n$ and t be a λ-tableau. Set $B_t = \sum_{\sigma \in R_t} \sigma$. Show that the map $M^\lambda \ni e_{\{t\}} \mapsto \pi B_t \in L(S_n)$ with $\pi \in S_n$, is a homomorphism of group representations $(L(S_n)$ with the left regular representation).

(2) Prove the following formula for the character χ_λ of S^λ:

$$\chi_\lambda = \frac{d_\lambda}{n!} \sum_{\pi \in S_n} \sum_{\gamma \in C_{\pi t}} \sum_{\sigma \in R_{\pi t}} \operatorname{sign}(\gamma)\gamma\sigma$$

(a similar formula can be found in Section VI.6 of Simon's book [198]).

(3) For $\lambda, \mu \vdash n$, set $\phi_\lambda(\mu) = \frac{\chi_\lambda(\theta)}{d_\lambda}$, where θ is any element in $\mathcal{C}_\mu$ (the set of all permutations of S_n of type μ). Thus ϕ_λ is a spherical function of the Gelfand pair $(S_n \times S_n, \widetilde{S_n})$. Show that, for any λ-tableau t,

$$\phi_\lambda(\mu) = \sum_{\sigma \in C_t} \frac{|\{\gamma \in R_t : \sigma\gamma \in \mathcal{C}_\mu\}|}{|\mathcal{C}_\mu|} \operatorname{sign}(\sigma).$$

(4) Use (3) to prove that for $\lambda = (\lambda_1, \lambda_2, \ldots, \lambda_k)$ one has

$$\phi_\lambda(1^{n-2}, 2) = \frac{1}{n(n-1)} \sum_{i=1}^{k} [\lambda_i^2 - (2i-1)\lambda_i]$$

(cf. with Theorem 10.6.2) and

$$\phi_\lambda(1^{n-3}, 3) = \frac{1}{2n(n-1)(n-2)} \cdot$$
$$\sum_{i=1}^{k} [2\lambda_i^3 + 3\lambda_i^2(1-2i) + \lambda_i(6i^2 - 6i + 1)] - \frac{3}{2(n-2)}.$$

(5) Using (3) prove the final formulæ (apart the formula d_λ) in Ingram's paper [128].

10.7 Random transpositions

Now we are in a position to analyze the random transposition model from Section 1.2. From the discussion at the end of Section 1.3 and in Section 3.9 we know that this model is described by the repeated convolution powers of the probability measure P in (10.15).

Lemma 10.7.1 *For $n > 2$ and $k > 0$ one has*

$$\left|1 - \frac{2}{n}\right|^{2k} + (n-1)^2 \left|1 - \frac{4}{n}\right|^{2k} \le (n-1)^2 \left|1 - \frac{2}{n}\right|^{2k}.$$

Proof Dividing by $\left|1 - \frac{2}{n}\right|^{2k}$ one obtains that the statement is equivalent to the relation $\left|\frac{n-4}{n-2}\right|^{2k} \le \frac{n^2-2n}{(n-1)^2}$. The latter holds true as

$$\left|\frac{n-4}{n-2}\right|^{2k} \le \frac{n-4}{n-2} \le \frac{n^2-2n}{(n-1)^2}$$

where the last inequality can be checked directly for $n > 2$. $\square$

We now show that the random transposition model has a sharp cutoff after $\frac{1}{2}n \log n$ steps.

Theorem 10.7.2 (Diaconis and Shahshahani [69]) *There exist universal (that is independent of n) positive constants a and b such that*

(i) *if $k = \frac{1}{2}n \log n + cn$ with $c > 0$, then $\|P^{*k} - U\| \le ae^{-2c}$;*
(ii) *if $k = \frac{1}{2}n \log n - cn$ with $0 < c < \frac{1}{8}\log n$, then $\|P^{*k} - U\| \ge 1 - be^{-2c}$.*

Proof We keep the notation of the previous sections, namely, for a partition $\lambda \vdash n$ the corresponding irreducible representation is $(\rho_\lambda, S^\lambda)$ with dimension d_λ and character χ_λ. Also denote $r(\lambda) = \frac{\chi_\lambda(\tau)}{d_\lambda}$ where τ is any transposition of S_n. We recall the expression for the Fourier transform of the distribution P in (10.16), namely

$$\widehat{P}(\rho_\lambda) = \left(\frac{1}{n} + \frac{n-1}{n} r(\lambda)\right) I.$$

In virtue of the upper bound lemma 9.9.5 we have

$$\|P^{*k} - U\|^2 \leq \frac{1}{4} \sum_{\substack{\lambda \vdash n \\ \lambda \neq (n)}} d_\lambda^2 \left(\frac{1}{n} + \frac{n-1}{n} r(\lambda) \right)^{2k}.$$

In virtue of Theorem 10.6.2 we have

$$r\left((1^n)\right) = \frac{1}{n(n-1)} \sum_{j=1}^n (1 - (2j-1)) = \frac{1}{n(n-1)} \left(2n - 2\sum_{j=1}^n j \right) = -1$$

and

$$r\left((n-1,1)\right) = \frac{1}{n(n-1)} \left((n-1)^2 - (n-1) + 1 - 3 \right) = \frac{n-3}{n-1}.$$

Moreover, observing that $(2, 1^{n-2})$ is the transpose of $(n-1, 1)$, in virtue of Corollary 10.4.2 we have

$$r\left((2, 1^{n-2})\right) = -r\left((n-1,1)\right) = -\frac{n-3}{n-1}.$$

From these relations one deduces

$$d_{(1^n)}^2 \left| \frac{1}{n} + \frac{n-1}{n} r((1^n)) \right|^{2k} + d_{(2,1^{n-2})}^2 \left| \frac{1}{n} + \frac{n-1}{n} r((2, 1^{n-2})) \right|^{2k}$$

$$= \left| \frac{1}{n} - \frac{n-1}{n} r((1^n)) \right|^{2k} + (n-1)^2 \left| \frac{1}{n} - \frac{n-1}{n} \frac{n-3}{n-1} \right|^{2k}$$

$$= \left| 1 - \frac{2}{n} \right|^{2k} + (n-1)^2 \left| 1 - \frac{4}{n} \right|^{2k}$$

(from Lemma 10.7.1)

$$\leq (n-1)^2 \left| 1 - \frac{2}{n} \right|^{2k}$$

$$= d_{(n-1,1)}^2 \left| \frac{1}{n} + \frac{n-1}{n} r((n-1,1)) \right|^{2k}. \tag{10.20}$$

Note that $d_{(1^n)} = d_{(n)} = 1$ and $d_{(2,1^{n-2})} = d_{(n-1,1)} = n - 1$ (Corollary 10.4.2 and Example 10.5.15).

Again, from Corollary 10.4.2 it follows that, for any permutation $\lambda \vdash n$ one has either $r(\lambda) \geq 0$ or $r(\lambda') > 0$. Therefore

$$\sum_{\substack{\lambda : r(\lambda) < 0 \\ \lambda \neq (1^n), (2,1^{n-2})}} d_\lambda^2 \left| \frac{1}{n} + \frac{n-1}{n} r(\lambda) \right|^{2k}$$

$$= \sum_{\substack{\lambda : r(\lambda) < 0 \\ \lambda \neq (1^n), (2,1^{n-2})}} d_\lambda^2 \left| \frac{1}{n} - \frac{n-1}{n} r(\lambda') \right|^{2k}$$

$$= \sum_{\substack{\lambda' : r(\lambda') > 0 \\ \lambda' \neq (n), (n-1,1)}} d_\lambda^2 \left| \frac{1}{n} - \frac{n-1}{n} r(\lambda') \right|^{2k}$$

$$\leq \sum_{\substack{\lambda : r(\lambda) > 0 \\ \lambda \neq (n), (n-1,1)}} d_\lambda^2 \left| \frac{1}{n} + \frac{n-1}{n} r(\lambda) \right|^{2k}.$$

Now

$$\frac{1}{4} \sum_{\substack{\lambda \vdash n \\ \lambda \neq (n)}} = \frac{1}{4} \left(\sum_{\substack{\lambda : r(\lambda) < 0 \\ \lambda \neq (1^n), (2,1^{n-2})}} + \sum_{\lambda = (1^n), (2,1^{n-2})} + \sum_{\lambda : r(\lambda) = 0} + \sum_{\substack{\lambda : r(\lambda) > 0 \\ \lambda \neq (n)}} \right)$$

$$\leq \frac{1}{4} \left(\sum_{\substack{\lambda : r(\lambda) > 0 \\ \lambda \neq (n), (n-1,1)}} + \sum_{\lambda = (n-1,1)} + \sum_{\lambda : r(\lambda) = 0} + \sum_{\substack{\lambda : r(\lambda) > 0 \\ \lambda \neq (n)}} \right)$$

$$= \frac{1}{4} \left(2 \sum_{\substack{\lambda : r(\lambda) > 0 \\ \lambda \neq (n)}} + \sum_{\lambda : r(\lambda) = 0} \right)$$

$$\leq \frac{1}{2} \sum_{\substack{\lambda : r(\lambda) \geq 0 \\ \lambda \neq (n)}}$$

and therefore, recalling (10.20),

$$\frac{1}{4} \sum_{\substack{\lambda \vdash n \\ \lambda \neq (n)}} d_\lambda^2 \left(\frac{1}{n} + \frac{n-1}{n} r(\lambda) \right)^{2k} \leq \frac{1}{2} \sum_{\substack{\lambda : r(\lambda) \geq 0 \\ \lambda \neq (n)}} d_\lambda^2 \left(\frac{1}{n} + \frac{n-1}{n} r(\lambda) \right)^{2k}.$$

The first term of the last sum, namely the one corresponding to $\lambda = (n-1,1)$, equals $(n-1)^2(1-\frac{2}{n})^{2k}$ which, in virtue of the basic inequality $1 - x \le e^{-x}$, is bounded above by $n^2 \exp(-\frac{4k}{n}) \equiv \exp(-\frac{4k}{n} + 2\log n)$. Thus, for $k = \frac{1}{2}n(\log n + c)$, with $c > 0$, the latter equals e^{-2c} which tends to zero as $c \to \infty$. As usual, the idea of the proof consists in showing that the other terms are geometrically smaller than this first term so that the same bound holds for the whole sum.

Fix $\alpha \in (0,\frac{1}{4})$ and split the above sum in two parts according to whether λ_1 is greater than $(1-\alpha)n$ or not. We then have

$$\sum_{\substack{\lambda:r(\lambda)\ge 0 \\ \lambda\ne(n)}} d_\lambda^2 \left(\frac{1}{n} + \frac{n-1}{n}r(\lambda)\right)^{2k} = \sum_{j=1}^{n-1} \sum_{\substack{\lambda:r(\lambda)\ge 0 \\ \lambda_1=n-j}} d_\lambda^2 \left(\frac{1}{n} + \frac{n-1}{n}r(\lambda)\right)^{2k}$$

$$= \sum_{j=1}^{\alpha n} \sum_{\substack{\lambda:r(\lambda)\ge 0 \\ \lambda_1=n-j}} d_\lambda^2 \left(\frac{1}{n} + \frac{n-1}{n}r(\lambda)\right)^{2k}$$

$$+ \sum_{j>\alpha n}^{n-1} \sum_{\substack{\lambda:r(\lambda)\ge 0 \\ \lambda_1=n-j}} d_\lambda^2 \left(\frac{1}{n} + \frac{n-1}{n}r(\lambda)\right)^{2k}$$

$$\le \sum_{j=1}^{\alpha n} \sum_{\substack{\lambda:r(\lambda)\ge 0 \\ \lambda_1=n-j}} d_\lambda^2 \left(1 - \frac{2j(n-j+1)}{n^2}\right)^{2k}$$

$$+ \sum_{j>\alpha n}^{n-1} \sum_{\substack{\lambda:r(\lambda)\ge 0 \\ \lambda_1=n-j}} d_\lambda^2 \left(1 - \frac{j}{n}\right)^{2k}$$

$$\le \sum_{j=1}^{\alpha n} \binom{n}{j} \frac{n!}{(n-j)!} \left(1 - \frac{2j(n-j+1)}{n^2}\right)^{2k} \qquad (*)$$

$$+ \sum_{j>\alpha n}^{n-1} \binom{n}{j} \frac{n!}{(n-j)!} \left(1 - \frac{j}{n}\right)^{2k}. \qquad (**)$$

Here, we have used Corollary 10.6.5 in the first inequality, while the last one follows from

$$\sum_{\lambda:\lambda_1=n-j} d_\lambda^2 \le \binom{n}{j}^2 \sum_{\lambda:\lambda_1=n-j} d_{(\lambda_2,\ldots,\lambda_k)}^2 = \binom{n}{j}^2 \sum_{\bar\lambda\vdash j} d_{\bar\lambda}^2 = \binom{n}{j} \frac{n!}{(n-j)!}$$

(here the inequality follows from equation (10.14), while the last equality is a consequence of (iii) in Theorem 3.7.11).

In order to bound the sum $(*)$, we choose $k = \frac{1}{2}n\log n + cn$ with $c > 0$ as in the statement. Using the estimate $1 - x \le e^x$ we obtain

$$\sum_{j=1}^{\alpha n} \binom{n}{j} \frac{n!}{(n-j)!} \left(1 - \frac{2j(n-j+1)}{n^2}\right)^{2k}$$

$$\le \sum_{j=1}^{\alpha n} \frac{(n!)^2}{j!((n-j)!)^2} \exp\left(-\frac{4kj(n-j+1)}{n^2}\right)$$

$$\le n^2 \exp\left(-\frac{4k}{n}\right) \sum_{j=1}^{\alpha n} \frac{((n-1)!)^2}{j!((n-j)!)^2} \exp\left(-\frac{4k(j-1)(n-j)}{n^2}\right)$$

$$= n^2 \exp\left(-2\log n - 4c\right) \cdot$$

$$\cdot \sum_{j=1}^{\alpha n} \frac{((n-1)!)^2}{j!((n-j)!)^2} \exp\left(-4\frac{(j-1)(n-j)}{n^2}\left(\frac{1}{2}n\log n + cn\right)\right)$$

$$\le e^{-4c} \sum_{j=1}^{\alpha n} \frac{((n-1)!)^2}{j!((n-j)!)^2} \exp\left(-2\frac{(j-1)(n-j)}{n}\log n\right)$$

$$= e^{-c} \sum_{j=1}^{\alpha n} \frac{((n-1)!)^2}{j!((n-j)!)^2} n^{-2\frac{(j-1)(n-j)}{n}}$$

$$\le e^{-4c} \sum_{j=1}^{\alpha n} \frac{n^{2(j-1)} n^{-2(j-1)(1-j/n)}}{j!}$$

$$= e^{-4c} \sum_{j=1}^{\alpha n} \frac{n^{(2/n)j(j-1)}}{j!}.$$

In the lust sum the ratio between the $(j+1)$st and the jth term is $\frac{1}{j+1}n^{\frac{4j}{n}}$. In order to study this quotient consider the function $h(x) = \frac{1}{x+1}n^{\frac{4x}{n}}$ whose derivative is $h'(x) = \frac{n^{4x/n}}{x+1}\left(\frac{4\log n}{n} - \frac{1}{x+1}\right)$. Thus $h(x)$ is decreasing for $x \le 4\frac{n}{\log n} - 1$, increasing for $x \ge 4\frac{n}{\log n} - 1$ and attains the minimum at $x = \frac{n}{4\log n} - 1$. Therefore the absolute maximum of the sequence $\{h(j)\}_{j=1,\ldots,\alpha n - 1}$ is attained at either $j = 1$ or $j = \alpha n - 1$. The values at these extrema are less than a constant $q < 1$: indeed $h(1) = \frac{1}{2}n^{4/n}$ which is decreasing with respect to n and it is less than 1 as soon as $n > 17$; the second quantity $h(\alpha n - 1) = \frac{1}{\alpha n}n^{\frac{4(\alpha n - 1)}{n}} \le \frac{n^{4\alpha - 1}}{\alpha n}$ is also decreasing with respect to n and (because $\alpha < 1/4$) we have $\frac{1}{\alpha}n^{4\alpha - 1} < 1$ as soon as $n > (1/\alpha)^{1/(1-4\alpha)}$. Setting

$$q = \max\left\{\frac{1}{2}n^{4/n}, \frac{1}{\alpha}n^{4\alpha - 1} : n \ge \max\{17, (1/\alpha)^{1/(1-4\alpha)}\}\right\}$$

we obtain

$$e^{-4c} \sum_{j=1}^{\alpha n} \frac{n^{(2/n)j(j-1)}}{j!} \le e^{-4c} \sum_{j=0}^{\alpha n - 1} q^j = e^{-4c} \frac{1 - q^{\alpha n}}{1 - q} \le \frac{e^{-4c}}{1 - q}.$$

As far the sum in (∗∗) is concerned, for $k = \frac{1}{2} n \log n + cn$ (with $c > 0$), we have

$$\sum_{j > \alpha n}^{n-1} \binom{n}{j} \frac{n!}{(n-j)!} \left(1 - \frac{j}{n}\right)^{2k} = \sum_{j > \alpha n}^{n-1} \binom{n}{j} \frac{n!}{(n-j)!} \left(1 - \frac{j}{n}\right)^{n \log n + 2cn}$$

$$\le e^{-4c} \sum_{j > \alpha n}^{n-1} \binom{n}{j} \frac{n!}{(n-j)!} \left(1 - \frac{j}{n}\right)^{n \log n}$$

$$\tag{10.21}$$

as, for $n \ge \frac{-2}{\log(1-\alpha)}$ (in fact if α is sufficiently close to $1/4$ one can take $n \ge 7$), one has $2cn \log(1 - \alpha) \le -4c$ which infers $(1 - \frac{j}{n})^{2cn} \le (1 - \alpha)^{2cn} \le e^{-4c}$.

In order to bound the last sum in (10.21) we consider the ratio between the $(j + 1)$st and the jth term. This is

$$\frac{(n-j)^2}{j+1} \left(\frac{n-j-1}{n-j}\right)^{n \log n} = \frac{(n-j)^2}{j+1} \left(1 - \frac{1}{n-j}\right)^{n \log n}$$

and it is decreasing in j. The first ratio equals:

$$\frac{(n-\alpha n)^2}{\alpha n + 1} \left(1 - \frac{1}{n-\alpha n}\right)^{n \log n} \le \frac{n(1-\alpha)^2}{\alpha} \exp\left[n \log n \log\left(1 - \frac{1}{n - \alpha n}\right)\right]$$

$$\le \frac{n(1-\alpha)^2}{\alpha} \exp\left(\frac{-n \log n}{n - \alpha n}\right)$$

$$= \frac{(1-\alpha)^2}{\alpha} n^{-\frac{\alpha}{1-\alpha}}$$

where the last inequality follows from the trivial estimate $\log(1-x) \le -x$ and it holds for $n > 2 > \frac{1}{1-\alpha}$.

For $n > \left(\frac{(1-\alpha)^2}{\alpha}\right)^{\frac{1-\alpha}{\alpha}}$ the above first ratio is smaller than 1. Therefore the sum (10.21) is dominated by the first summand times n, that is, it is bounded by $e^{-4c} n \binom{n}{\alpha n} \frac{n!}{(n - \alpha n)!} (1 - \alpha)^{n \log n}$. We recall (see also Exercise 10.7.4) that we can deduce Stirling's formula from

$$n^n e^{-n} \sqrt{2\pi n} \le n! \le n^n e^{-n} \sqrt{2\pi n} \left(1 + \frac{1}{12n - 1}\right)$$

which also implies that

$$n^n e^{-n} \sqrt{2\pi n} \leq n! \leq 2n^n e^{-n} \sqrt{2\pi n}.$$

In virtue of this inequality we conclude that (10.21) is smaller than

$$e^{-4c} n \binom{n}{\alpha n} \frac{n!}{(n - \alpha n)!} (1 - \alpha)^{n \log n}$$

$$= n e^{-4c} \frac{(n!)^2}{(\alpha n)! [(n - \alpha n)!]^2} (1 - \alpha)^{n \log n}$$

$$\leq 4 e^{-4c} \frac{\sqrt{n}}{\sqrt{2\pi}} n^{\alpha n} e^{-\alpha n} \frac{(1 - \alpha)^{n \log n - 2n + 2\alpha n}}{(1 - \alpha) \alpha^{\alpha n} \sqrt{\alpha}}$$

$$= 4 e^{-4c} \frac{\sqrt{n}}{\sqrt{2\pi}} \frac{(1 - \alpha)^{-2n + 2\alpha n} n^{n(\alpha + \log(1 - \alpha))}}{(1 - \alpha) e^{\alpha n} \alpha^{\alpha n} \sqrt{\alpha}}.$$

In the last term the fraction tends to 0 as n tends to $+\infty$ (since the dominant term is $n^{n(\alpha + \log(1 - \alpha))}$ and we have that $\alpha + \log(1 - \alpha) < 0$). Therefore there exists a constant a^2 such that the last term and thus (10.21) is bounded by $a^2 e^{-4c}$. This completes the proof of the upper estimate.

We now start proving the lower bound. First of all, we look at the summand in

$$\frac{1}{4} \sum_{\substack{\lambda \vdash n \\ \lambda \neq (n)}} d_\lambda^2 \left(\frac{1}{n} + \frac{n - 1}{n} r(\lambda) \right)^{2k}$$

which tends to zero the most slowly, that is the one that corresponds to the partition $(n - 1, 1)$. We recall that if ρ is a representation of the symmetric group and χ_ρ the corresponding character we have for every probability distribution Q

$$E_Q(\chi_\rho) = \sum_{g \in S_n} Q(g) tr(\rho(g)) = tr \left(\sum_{g \in S_n} Q(g) \rho(g) \right) = tr \left(\widehat{Q}(\rho) \right).$$

In particular, from (10.16) and Theorem 10.6.2, we have:

$$E_{P*k}(\chi_\lambda) = tr \left(\widehat{P^{*k}}(\rho_\lambda) \right) = d_\lambda \left(\frac{1}{n} + \frac{n - 1}{n} r(\lambda) \right)^k$$

which for $\lambda = (n - 1, 1)$ becomes

$$E_{P*k}(\chi_{(n-1,1)}) = (n - 1) \left(1 - \frac{2}{n} \right)^k.$$

By definition of the total variation distance it follows that $\|P^{*k} - U\|_{TV} \geq |P^{*k}(A) - U(A)|$ for every subset $A \subseteq S_n$. To obtain the lower estimate we consider the family of subsets

$$A_\alpha = \{g \in S_n : |\chi_{(n-1,1)}(g)| \leq \alpha\} \quad \text{with} \quad 0 < \alpha \leq E_{P^{*k}}(\chi_{(n-1,1)})$$

which is well defined since $E_{P^{*k}}(\chi_{(n-1,1)}) > 0$.

Observe that for $g \in A_\alpha$ we have

$$|E_{P^{*k}}(\chi_{(n-1,1)}) - \chi_{(n-1,1)}(g)| \geq E_{P^{*k}}(\chi_{(n-1,1)}) - \alpha \geq 0.$$

It follows that

$$P^{*k}(A_\alpha) \leq \frac{Var_{P^{*k}}(\chi_{(n-1,1)})}{(E_{P^{*k}} - \alpha)^2}. \tag{10.22}$$

Indeed, the left hand side of (10.22) is majorized by

$$P^{*k}\left(\{g \in S_n : |E_{P^{*k}}(\chi_{(n-1,1)}) - \chi_{(n-1,1)}(g)| \geq E_{P^{*k}}(\chi_{(n-1,1)}) - \alpha\}\right)$$

which, in turn is majorized, by an immediate consequence of Chebyshev's inequality (Corollary 1.9.6), by the right hand side of 10.22.

To compute $Var_{P^{*k}}(\chi_{(n-1,1)}) = E_{P^{*k}}(\chi^2_{(n-1,1)}) - E_{P^{*k}}(\chi_{(n-1,1)})^2$, we observe that $\chi^2_{(n-1,1)}$ is the character of the representation of S_n given by the tensor product $S^{(n-1,1)} \otimes S^{(n-1,1)}$ (see Lemma 9.1.5). By (9.5) this decomposes as

$$S^{(n-1,1)} \otimes S^{(n-1,1)} = S^{(n)} \oplus S^{(n-1,1)} \oplus S^{(n-2,2)} \oplus S^{(n-2,1,1)}$$

and therefore $\chi^2_{(n-1,1)} = \chi_{(n)} + \chi_{(n-1,1)} + \chi_{(n-2,2)} + \chi_{(n-2,1,1)}$ which in turn implies

$$E_{P^{*k}}(\chi^2_{(n-1,1)}) = E_{P^{*k}}(\chi_{(n)}) + E_{P^{*k}}(\chi_{(n-1,1)})$$
$$+ E_{P^{*k}}(\chi_{(n-2,2)}) + E_{P^{*k}}(\chi_{(n-2,1,1)}).$$

Clearly $E_{P^{*k}}(\chi_{(n)}) = \sum_{g \in S_n} P^{*k}(g) = 1$, because $\rho_{(n)}$ is the trivial representation.

Moreover by Theorem 10.6.2 we have

$$r((n-2,2)) = \frac{n-4}{n} \quad \text{and} \quad r((n-2,1,1)) = \frac{n-5}{n-1}$$

which coupled with Example 10.5.15 give

$$E_{P^{*k}}(\chi_{(n-2,2)}) = \frac{n(n-3)}{2}\left(\frac{n^2 - 4n + 4}{n^2}\right)^k$$

and

$$E_{P^{*k}}(\chi_{(n-2,1,1)}) = \frac{(n-1)(n-2)}{2}\left(\frac{n-4}{n}\right)^k.$$

Simple computations show that

$$E_{P^{*k}}(\chi_{(n-2,2)}) + E_{P^{*k}}(\chi_{(n-2,1,1)}) - E_{P^{*k}}(\chi_{(n-1,1)})^2 \leq 0$$

and therefore

$$\begin{aligned} Var_{P^{*k}}(\chi_{(n-1,1)}) &= E_{P^{*k}}(\chi_{(n-1,1)}^2) - E_{P^{*k}}(\chi_{(n-1,1)})^2 \\ &\leq 1 + E_{P^{*k}}(\chi_{(n-1,1)}), \end{aligned} \tag{10.23}$$

so that, to bound $Var_{P^{*k}}(\chi_{(n-1,1)})$ we are left to estimate $E_{P^{*k}}(\chi_{(n-1,1)})$. To this end observe that

$$\log(1-x) = -x - \frac{x^2}{2}\omega(x)$$

where $\omega(x) \geq 0$ and $\lim_{x \to 0}\omega(x) = 1$

Set $k = \frac{1}{2}n\log n - cn$ with $0 \leq c \leq \frac{1}{8}\log n$. Then we have

$$\begin{aligned} E_{P^{*k}}(\chi_{(n-1,1)}) &= (n-1)\exp\left[k\log\left(1 - \frac{2}{n}\right)\right] \\ &= (n-1)\exp\left[\left(\frac{1}{2}n\log n - cn\right)\left(-\frac{2}{n} - \frac{2}{n^2}\omega\left(\frac{2}{n}\right)\right)\right] \\ &= \frac{n-1}{n}e^{2c}\exp\left[\frac{-\log n + 2c}{n}\omega\left(\frac{2}{n}\right)\right] \end{aligned}$$

which, for n sufficiently large, gives the two sided inequality

$$\frac{3}{4}e^{2c} \leq E_{P^{*k}}(\chi_{(n-1,1)}) \leq e^{2c}. \tag{10.24}$$

From (10.23) and (10.24) we get

$$Var_{P^{*k}}(\chi_{(n-1,1)}) \leq 1 + E_{P^{*k}}(\chi_{(n-1,1)}) \leq 1 + e^{2c}$$

Choose $\alpha = \frac{e^{2c}}{2}$: the first inequality in (10.24) ensures that $\alpha \leq E_{P^{*k}}(\chi_{(n-1,1)})$. In particular (for n large),

$$(E_{P^{*k}}(\chi_{(n-1,1)}) - \alpha)^2 = \left(E_{P^{*k}}(\chi_{(n-1,1)}) - \frac{e^{2c}}{2}\right)^2 \geq \frac{1}{16}e^{4c}.$$

From these estimates and (10.22) we obtain

$$P^{*k}(A_\alpha) \leq \frac{Var_{P^{*k}}(\chi_{(n-1,1)})}{(E_{P^{*k}} - \alpha)^2} \leq \frac{1 + e^{2c}}{\frac{1}{16}e^{4c}} = 16e^{-2c} + 16e^{-4c}. \tag{10.25}$$

On the other hand it is easy to prove (cf. (10.17)) that

$$E_U(\chi_{(n-1,1)}) = 0 \quad \text{and} \quad Var_U(\chi_{(n-1,1)}) = \frac{1}{|S_n|} \sum_{g \in S_N} \chi^2_{(n-1,1)}(g) = 1$$

which, in virtue of Markov's inequality (Proposition 1.9.5), gives

$$U(A_\alpha) = 1 - U(A_\alpha^c) = 1 - U\left(\{g \in S_n : \chi_{(n-1,1)}(g)^2 > \alpha^2\}\right)$$
$$\geq 1 - \frac{Var_U(\chi_{(n-1,1)})}{\alpha^2} = 1 - \frac{1}{\alpha^2}.$$

The latter, coupled with (10.25), gives

$$\|P^{*k} - U\|_{TV} \geq U(A_\alpha) - P^{*k}(A_\alpha) \geq 1 - 4e^{-4c} - 16e^{-2c} - 16e^{-4c}$$
$$\geq 1 - 36e^{-2c}.$$

$\square$

In the following exercises we sketch a proof of Wallis' formula and of Stirling's formula. We set $(2n)!! = 2n(2n-2)(2n-4)\cdots 2$ and $(2n+1)!! = (2n+1)(2n-1)(2n-3)\cdots 3\cdot 1$.

Exercise 10.7.3 (Wallis formula) (1) Prove that

$$\int_0^{\pi/2} \sin^n x\, dx = \begin{cases} \frac{(n-1)!!}{n!!}\cdot\frac{\pi}{2} & \text{if } n \text{ is even} \\ \frac{(n-1)!!}{n!!} & \text{if } n \text{ is odd.} \end{cases}$$

(2) Integrating the inequalities $\sin^{2n+1}(x) < \sin^{2n}(x) < \sin^{2n-1}(x)$ for $0 < x < \frac{\pi}{2}$ and using (1), prove Wallis' formula:

$$\frac{\pi}{2} = \lim_{n\to\infty} \frac{(2n-2)!!(2n)!!}{[(2n-1)!!]^2}.$$

Exercise 10.7.4 (Stirling formula) (1) Prove that

$$e < \left(1 + \frac{1}{n}\right)^{n+1/2} < e^{1+1/[12n(n+1)]}.$$

(2) Set $a_n = \frac{n!}{n^n e^{-n}\sqrt{2\pi n}}$ and $b_n = a_n e^{-1/12n}$. Prove that there exists $\ell \in \mathbb{R}$ such that $b_n < \ell < a_n$ and $\lim_{n\to\infty} a_n = \ell = \lim_{n\to\infty} b_n$.

(3) From Wallis' formula deduce that $\ell = 1$.

(4) Use (2) and (3) to prove that

$$n^n e^{-n} \sqrt{2\pi n} < n! < n^n e^{-n} \sqrt{2\pi n} \left(1 + \frac{1}{12n - 1}\right)$$

and therefore $n! \sim n^n e^{-n} \sqrt{2\pi n}$ for $n \to \infty$.

An elementary proof of Stirling's formula for the Gamma function can be found in [61].

10.8 Differential posets

We know (see Section 10.5) that for $\lambda \vdash n$ the dimension d_λ of the irreducible representation S^λ of the symmetric group S_n is equal to the number of standard tableaux of shape λ. Therefore, from the general theory, we get the following identity:

$$\sum_{\lambda \vdash n} d_\lambda^2 = n! \tag{10.26}$$

which has a purely combinatorial meaning: the sets

$$\bigcup_{\lambda \vdash n} \{(s, t) : s, t \text{ are standard tableaux of shape } \lambda\}$$

and S_n have the same cardinality. From this point of view, (10.26) is the starting point of a very deep and interesting field of modern combinatorics, the so-called Robinson–Schensted–Knuth correspondence; beautiful accounts may be found in [94], [182] and [205].

On the other hand, in [201] R. Stanley published a combinatorial proof of (10.26) that made use of ideas similar to those developed in Chapter 8. In particular, in that paper Stanley formalized a remarkable class of posets where the operators d and d^* satisfy a simple commutation relation. The present section requires knowledge of Sections 8.1 and 8.3 (just the definition of d and d^*). Let P be a (possibly infinite) poset. We say that P is *locally finite* if for any choice of $x, y \in P$ with $x \preceq y$ one has $|\{z \in P : x \preceq z \preceq y\}| < \infty$. We use the following notation: if $x \in P$ we set $C^+(x) = \{z \in P : z \gtrdot x\}$ and $C^-(x) = \{y \in P : x \gtrdot y\}$. Note that $C^+(x) \cap C^+(y)$ is just the set of all $z \in P$ that cover both x and y; there is a similar interpretation for $C^-(x) \cap C^-(y)$.

We are now in a position to give the definition of differential poset.

Definition 10.8.1 Let P be a poset and r a positive integer. We say that P is *r-differential* when

(i) P has a 0-element, it is graded and locally finite;
(ii) if $x, y \in P$ with $x \neq y$ then $|C^+(x) \cap C^+(y)| = |C^-(x) \cap C^-(y)|$;
(iii) for any $x \in P$ we have $|C^+(x)| = |C^-(x)| + r$.

Note that if P has a 0-element and it is locally finite, then $|C^-(x)|$ is finite for any $x \in P$. More is true if also (iii) is satisfied. To avoid misunderstanding, in this section we denote the rank of an element $x \in P$ by $rk(x)$.

Lemma 10.8.2 *If a poset P satisfies (i) and (iii) of the previous definition and $P_n = \{x \in P : rk(x) = n\}$ then $|P_n|$ is finite for any $n \in \mathbb{N}$.*

Proof By induction. Clearly $|P_0| = 1$. Suppose that $|P_n|$ and $|P_{n-1}|$ are finite. If $x \in P_n$ then $|C^-(x)| \leq |P_{n-1}|$ and therefore, by (iii), $|C^+(x)| \leq |P_{n-1}| + r$. But then

$$|P_{n+1}| \leq |P_n|(|P_{n-1}| + r).$$

$\square$

Now we show that if (ii) is satisfied then the sets $C^+(x) \cap C^+(y)$ and $C^-(x) \cap C^-(y)$ have cardinality ≤ 1.

Lemma 10.8.3 *If P satisfies (i) and (ii) then $|C^-(x) \cap C^-(y)| \leq 1$.*

Proof Suppose that this is not true. Then there exists a counterexample, that is a couple $x, y \in P$, $x \neq y$ with $|C^-(x) \cap C^-(y)| > 1$, and we may suppose that this counterexample is chosen in such a way that $rk(x) = rk(y)$ is minimum (it is clear that if $rk(x) \neq rk(y)$ then $C^-(x) \cap C^-(y) = \emptyset$). But then there exist $u, v \in P$, $u \neq v$, with $u, v \in C^-(x) \cap C^-(y)$. Therefore $|C^+(u) \cap C^+(v)| \geq 2$ and by (ii), $|C^-(u) \cap C^-(v)| \geq 2$. Since $rk(u) = rk(v) = rk(x) - 1$, this is a contradiction with the minimality assumption. $\square$

Lemma 10.8.4 *Suppose that P is a lattice and satisfies (i) and (iii). Then P satisfies (ii) (and therefore it is r-differential) if and only if it is modular.*

Proof A locally finite lattice is modular if and only if for any $x, y \in P$, $x \neq y$ we have:

$$x \wedge y \in C^-(x) \cap C^-(y) \text{ is equivalent to } x \vee y \in C^+(x) \cap C^+(y).$$

$\square$

Consider the vector space $L(P) = \{f : P \to \mathbb{C}\}$. Since P may be infinite, the dimension of $L(P)$ may be infinite. Indeed $L(P)$ may be endowed with a topology (see [201]) but we prefer to restrict our attention to subspaces of the form $\oplus_{k=0}^{n}L(P_k)$ (for a suitable n) that are finite dimensional by Lemma 10.8.2. On $L(P)$ we may introduce the operators d and d^* of Section 8.3:

$$d\delta_x = \sum_{y \in C^-(x)} \delta_y \quad \text{and} \quad d^*\delta_x = \sum_{z \in C^+(x)} \delta_z, \ \forall x \in P.$$

Clearly, for any $n \in \mathbb{N}$ we may think of d as a linear operator $d : \oplus_{k=0}^{n}L(P_k) \to \oplus_{k=0}^{n-1}L(P_k)$ and of d^* as a linear operator $d^* : \oplus_{k=0}^{n-1}L(P_k) \to \oplus_{k=0}^{n}L(P_k)$.

Now we prove the commutation rule satisfied by d and d^* in a differential poset.

Theorem 10.8.5 *Suppose that the poset P has 0-element, is graded, and $|P_n| < \infty, \forall n \in \mathbb{N}$. Then P is r-differential if and only if $dd^* - d^*d = rI$ (I is the identity operator on $L(P)$).*

Proof If $x \in P_n$ then

$$dd^*\delta_x = \sum_{y \in P_n} |C^+(x) \cap C^+(y)|\delta_y$$

and

$$d^*d\delta_x = \sum_{y \in P_n} |C^-(x) \cap C^-(y)|\delta_y$$

and therefore

$$(dd^* - d^*d)\delta_x = \sum_{\substack{y \in P_n \\ y \neq x}} \left[|C^+(x) \cap C^+(y)| - |C^-(x) \cap C^-(y)|\right]\delta_y$$

$$+ \left[|C^+(x)| - |C^-(x)|\right]\delta_x.$$

Thus we deduce that $(dd^* - d^*d)\delta_x = r\delta_x, \forall x \in P$ is equivalent to (ii) and (iii) in the Definition 10.8.1. $\qquad\square$

Corollary 10.8.6 *In an r-differential poset, the operators d and d^* satisfies the following general commutation rule:*

$$d(d^*)^n = rn(d^*)^{n-1} + (d^*)^nd. \tag{10.27}$$

Proof Straightforward by induction. $\qquad\square$

More generally, from the above corollary, immediately one has that if p is a polynomial in one variable and p' is its derivative then

$$dp(d^*) = rp'(d^*) + p(d)d.$$

The commutation relation (10.27) is similar to that satisfied in the Boolean lattice (see Section 6.1) but it is simpler: there is no dependence on the level of the poset. In [203] Stanley has considered several modifications of the axioms of a differential poset and in one case he covers the Boolean lattice (and also the lattice of subspaces of a vector space over a finite field). Some of the results in [203] should be compared to those in Chapters 6 and 8 (see, for instance, Proposition 2.10 and Corollary 2.11 in [203]).

Now we are in a position to state and prove the main result of this section, from which (10.26) will follow as an application. For $x \in P$, $SC(0, x)$ will denote the set of all saturated chain from 0 to x; in symbols

$$SC(0, x) = \{x_0, x_1, \ldots, x_n : 0 = x_0 \lessdot x_1 \lessdot \cdots \lessdot x_n = x\}.$$

Theorem 10.8.7 *If P is an r-differential poset, then for all $n \in \mathbb{N}$ and $x \in P$, we have*

$$\sum_{x \in P_n} |SC(0, x)|^2 = n! r^n.$$

Proof First, we will prove that

$$d^n (d^*)^n \delta_0 = \left[\sum_{x \in P_n} |SC(0, x)|^2 \right] \delta_0. \tag{10.28}$$

Indeed, arguing as in (8.15) we can say that the value of $d^n(d^*)^n \delta_0$ at 0 itself is equal to the number of chains of the form

$$0 = x_0 \lessdot x_1 \lessdot \cdots \lessdot x_n \gtrdot y_{n-1} \gtrdot y_{n-2} \gtrdot \cdots \gtrdot y_0 = 0$$

with $x_n \in P_n$, that is to $\sum_{x \in P_n} |SC(0, x)|^2$. On the other hand, we can compute $d^n(d^*)^n \delta_0$ by repeatedly using Corollary 10.8.6: we have $d\delta_0 = 0$ and therefore

$$\begin{aligned}
d^n (d^*)^n \delta_0 &= d^{n-1}[d(d^*)^n \delta_0] \\
&= rn d^{n-1}(d^*)^{n-1} \delta_0 \\
&= \cdots = \\
&= r^{n-1} n(n-1) \cdots 2(dd^* \delta_0) \\
&= r^n n! \delta_0.
\end{aligned} \tag{10.29}$$

From (10.28) and (10.29) the theorem follows. □

In order to get (10.26) from Theorem 10.8.7, we must study more closely the Young lattice $\mathbb{Y}$ introduced in Exercise 8.1.13. First of all, there is a nice pictorial way to describe $\mathbb{Y}$: we may think of it as the set of all Young diagrams ordered by inclusion: $\lambda \preceq \mu$ if and only if the Young diagram of λ is contained in the Young diagram of μ.

For instance, to say that $\lambda = (4, 3, 3, 1) \prec \mu = (5, 3, 3, 2)$ is the same thing as to say that the Young diagram of λ is contained in the Young diagram of μ (see Figure 10.18).

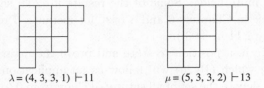

$\lambda = (4, 3, 3, 1) \vdash 11$ $\mu = (5, 3, 3, 2) \vdash 13$

Figure 10.18. $\lambda \preceq \mu$.

The following illustrates the bottom of the Young lattice.

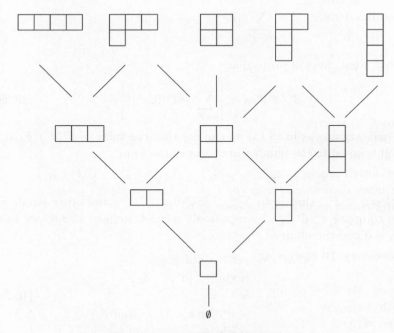

$\emptyset$

Figure 10.19. The Young lattice

Definition 10.8.8 Let $(\lambda_1, \lambda_2, \ldots, \lambda_k)$ be a partition. We say that the box (i, λ_i) is an *inner corner* of the Young diagram of λ if $\lambda_i > \lambda_{i+1}$ or it coincides with the box (k, λ_k). We say that in position $(j, \lambda_j + 1)$ the Young diagram of λ has an *outer corner* if $\lambda_{j-1} > \lambda_j$; moreover we always assume that in position $(1, \lambda_1 + 1)$ or $(k+1, 1)$ there is an outer corner.

In other words, an inner corner is a box that can be *removed*: (i, λ_i) is an inner corner if and only if $(\lambda_1, \lambda_2, \ldots, \lambda_i - 1, \ldots, \lambda_k)$ is still a partition. An outer corner is a box that can be added: (j, λ_j) is an outer corner if and only if $(\lambda_1, \lambda_2, \ldots, \lambda_j + 1, \ldots, \lambda_k)$ is still a partition.

Figure 10.20. The symbols $+$ and $-$ indicate the outer and inner corners.

Lemma 10.8.9

(i) $\mu \gtrdot \lambda$ *if and only if λ may be obtained from μ by removing one box (equivalently, μ may be obtained from λ by adding one box);*

(ii) *if $\lambda = (\lambda_1, \lambda_2, \ldots, \lambda_k)$ covers h elements in $\mathbb{Y}$ then it is covered by $h + 1$ elements.*

Proof (i) is trivial.

(ii) Observe that (i, λ_i) is an inner corner if and only if $(i+1, \lambda_{i+1}+1)$ is an outer corner. Therefore, recalling that $(1, \lambda_1 + 1)$ and $(k+1, 1)$ are outer corners and (k, λ_k) is an inner corner, we have that if λ has an inner corner exactly in the rows $i_1 < i_2 < \cdots < i_h = k$ it has an outer corner exactly in the rows $j_0 = 1 < j_1 = i_1 + 1 < \cdots < j_{h-1} = i_{h-1} + 1 < j_h = k + 1$. $\qquad\square$

Corollary 10.8.10 $\mathbb{Y}$ *is a 1-differential poset.*

Proof We have already seen that $\mathbb{Y}$ is a locally finite modular lattice with 0-element (see Exercise 8.1.13). Moreover it is graded: if $\lambda \vdash n$ then $rk(\lambda) = n$. Then, applying Lemma 10.8.4 and ii) in Lemma 10.8.9, we conclude immediately that $\mathbb{Y}$ is 1-differential. $\qquad\square$

In order to apply Theorem 10.8.7 to the Young lattice, we need a description of the saturated chain $SC(\emptyset, \lambda)$ (in this setting, the zero element is the trivial partition (0) and it is denoted by $\emptyset$). Let t be a standard λ-tableau, with λ a partition of n. We can consider the following chain of partitions:

$\lambda^n = \lambda$;

λ^{n-1} is the partition whose Young diagram is obtained by the Young diagram of λ by removing the box containing n;

λ^{n-2} is obtained by λ^{n-1} by removing the box containing $n-1$;

$\dots \dots \dots \dots$

$\lambda^1 = (1)$;

$\lambda^0 = (0)$.

Clearly, each λ^j is a partition: n is always in a removable box in a standard tableau of shape λ and the same holds for $n-k$ in λ^{n-k}. Moreover, $\lambda \gtrdot \lambda^{n-1} \gtrdot \lambda^{n-2} \gtrdot \cdots \gtrdot \lambda^1 \gtrdot \lambda^0$ (this construction has a close resemblance with that used in Section 10.5 immediately after Definition 10.5.4).

The following lemma is obvious.

Lemma 10.8.11 *Let $ST(\lambda)$ the set of all standard tableaux of shape λ. The map*

$$ST(\lambda) \ni t \mapsto (\lambda^0, \lambda^1, \dots, \lambda^n) \in SC(\emptyset, \lambda)$$

is a bijection.

Example 10.8.12 If

$$t = \begin{array}{|c|c|c|c|} \hline 1 & 2 & 4 & 8 \\ \hline 3 & 5 & 6 \\ \cline{1-3} 7 & 9 \\ \cline{1-2} \end{array}$$

Figure 10.21.

then the corresponding saturated chain is

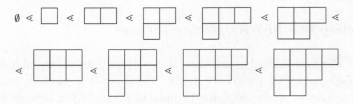

Figure 10.22.

Finally we can use Theorem 10.8.7 to prove (10.26).

Theorem 10.8.13 *For a partition λ of n, let d_λ denote the number of standard tableaux of shape λ. Then*

$$\sum_{\lambda \vdash n} d_\lambda^2 = n!.$$

Proof $\mathbb{Y}$ is a 1-differential poset and $d_\lambda = |SC(\emptyset, \lambda)|$. $\qquad\square$

The theory of differetial posets has been generalized by S. Fomin to a class of graded graphs; see the papers by Fomin [92] and [93]. See also Sagan's book [182], Section 5.2. In recent years, several papers on differential posets and related subjects have been written by J.D. Farley (see for instance [83, 84, 85]).

10.9 James' intersection kernels theorem

In this section we present a short elementary proof of a theorem of G.D. James that generalizes, to all the irreducible representations of S_n, the characterization of the irreducible representations $S^{n-k,k} = M^{n-k,k+} \cap \ker d$. Actually, James used a characteristic–free approach to the representation theory of S_n, namely he considered representations of S_n on vector spaces over arbitrary fields (for instance, the representations of finite groups over fields of positive characteristic, the so-called *modular representations*, are well studied: see Huppert's book on character theory [125] and the monograph by Curtis and Reiner [48]). We are interested only in unitary representations on vector spaces over the complex numbers (these are also called *ordinary representations*). We follow the proof in [187] which is based on a generalization of Theorem 10.6.2; see also [186, 189]. The original sources are James' paper [131] and monograph [132]. See also the exposition in Appendix C of Sternberg's book [212].

Let $a = (a_1, a_2, \ldots, a_k)$ be a composition of n (see Definition 10.5.4). For $i, j \in \{1, 2, \ldots, k\}$, $i \neq j$, set

$$b = (a_1, a_2, \ldots, a_{i-1}, a_i + 1, a_{i+1}, \ldots a_{j-1}, a_j - 1, a_{j+1}, \ldots, a_k)$$

and define a linear map

$$d_{i,j} : M^a \to M^b$$

by setting, for each a-tabloid $\{t\}$,

$$d_{i,j}e_{\{t\}} = \sum_{\{s\}} e_{\{s\}}$$

where the sum runs over all b-tabloids $\{s\}$ such that: the jth row of $\{s\}$ is contained in the jth row of $\{t\}$, the ith row of $\{t\}$ is contained in the ith row of $\{s\}$, and, for $k \neq i, j$, the kth row of $\{s\}$ coincides with the kth row of $\{t\}$. In other words, $d_{i,j}e_{\{t\}}$ is the characterisitic function of the set of all b-tabloids $\{s\}$ that can be obtained from $\{t\}$ by moving a single number from the jth to the ith row.

The operator $d_{i,j}$ commutes with the action of S_n on M^a and M^b. The goal of this section is to present a characterization of the irreducible representations of S_n as intersections of the kernels of the operators $d_{i,i+1}$.

We first prove the easy (assuming the results from Section 10.3):

Lemma 10.9.1 *Let* $\lambda = (\lambda_1, \lambda_2, \ldots, \lambda_k)$ *be a partition of* n *and* S^λ *be the corresponding irreducible representation (see Definition 10.3.11). Then,*

$$S^\lambda \subseteq M^\lambda \bigcap \left(\cap_{i=1}^{k-1} \ker d_{i,i+1} \right).$$

Proof Let t be a λ-tableau and $e_t = A_t e_{\{t\}}$ as in Definition 10.3.8. Then $d_{i,i+1}e_t = A_t d_{i,i+1}e_{\{t\}}$. But $d_{i,i+1}e_{\{t\}} = \sum_{\{s\}} e_{\{s\}}$, where the sum runs over all tabloids $\{s\}$ obtained by moving a single number from the $(i+1)$st to the ith row. Therefore, there exist two numbers on the same row of s and on the same column of t, so that, by Lemma 10.3.12, $A_t e_{\{s\}} = 0$. Note that

$$(\lambda_1, \lambda_2, \ldots, \lambda_{i-1}, \lambda_i + 1, \lambda_{i+1} - 1, \lambda_{i+2}, \ldots, \lambda_k)$$

is not necessarily a partition; but this is not a problem in the proof of Lemma 10.3.12. $\qquad\square$

For $a = (a_1, a_2, \ldots, a_n)$ a composition of n we now introduce a Laplace operator $\Delta_a \in Hom_{S_n}(M^a, M^a)$ that generalizes, in the present setting, the operator Δ in Section 6.1 and the operator Δ_n in Section 8.3.

If $\{t\}$ is an a-tabloid, we put

$$\Delta e_{\{t\}} = \sum_{\{u\}} e_{\{u\}}$$

where the sum runs over all a-tabloids $\{u\}$ that can be obtained from $\{t\}$ by picking two numbers x, y that belong to different rows and switching them. Clearly, Δ is S_n-invariant and it may be represented in the following way: if T is the set of all transpositions in S_n and $T_{\{t\}} := T \cap R_t$ is the set of all transpositions that move two elements belonging to the same row of t (R_t is as in Definition 10.3.5) then

$$\Delta e_{\{t\}} = \sum_{\sigma \in T \setminus T_{\{t\}}} \sigma e_{\{t\}}.$$

Before presenting the spectral analysis of Δ, we need the following lemma which is just a reformulation of the computation of the Fourier transform in Section 10.6.

Lemma 10.9.2 *Let $\mu \vdash n$ be a partition of n and V be an irreducible representation of S_n isomorphic to S^μ. Then, for any $f \in V$ we have*

$$\left(\sum_{\sigma \in T} \sigma \right) f = \frac{n(n-1)}{2} r(\mu) f$$

where $r(\mu)$ is as in Theorem 10.6.2 (with λ replaced by μ).

Proof The element $\sum_{\sigma \in T} \sigma$ belongs to the center of S_n and therefore its restriction to V is a multiple of the identity: $\left(\sum_{\sigma \in T} \sigma \right) |_V = c I_V$. Taking the traces from both sides of the last identity we get

$$|T| \chi_\mu(\sigma) = c d_\mu$$

from which the lemma follows. $\qquad\qquad\qquad\qquad\qquad\qquad\square$

Lemma 10.9.3 *Let $\mu = (\mu_1, \mu_2, \ldots, \mu_k)$ and $a = (a_1, a_2, \ldots, a_k)$ be a partition and a composition of n, respectively. Then the S^μ-isotypic component (see Definition 9.3.5) of M^a is an eigenspace of Δ_a and the corresponding eigenvalue is*

$$\frac{n(n-1)}{2} r(\mu) - \sum_{i=1}^{k} \frac{a_i(a_i - 1)}{2}.$$

Proof Suppose that V is an irreducible sub-representation of M^a isomorphic to S^μ and that

$$f = \sum_{\{t\} \in \mathfrak{T}_a} f(\{t\}) e_{\{t\}} \in V$$

where $\mathfrak{T}_a$ is the set of all a-tabloids (as in Section 10.5).

Then

$$\Delta f = \sum_{\{t\}\in\mathfrak{T}_{\{t\}}} f(\{t\})\Delta e_{\{t\}}$$

$$= \sum_{\{t\}\in\mathfrak{T}_{\{t\}}} f(\{t\}) \sum_{\sigma\in T\backslash T_{\{t\}}} \sigma e_{\{t\}}$$

$$= \sum_{\{t\}\in\mathfrak{T}_{\{t\}}} f(\{t\}) \left[\sum_{\sigma\in T} \sigma e_{\{t\}} - \sum_{\sigma\in T_{\{t\}}} \sigma e_{\{t\}}\right]$$

$$=_{(*)} \sum_{\sigma\in T} \sigma f - \left(\sum_{i=1}^{k} \frac{a_i(a_i-1)}{2}\right) f$$

$$=_{(**)} \left[\frac{n(n-1)}{2} r(\mu) - \sum_{i=1}^{k} \frac{a_i(a_i-1)}{2}\right] f$$

where $=_{(*)}$ follows from the fact that $\sigma e_{\{t\}} = e_{\{t\}}$ for any $\sigma \in T_{\{t\}}$ and $=_{(**)}$ follows from Lemma 10.9.2.   □

We now prove some formulæ for the operators $d_{i,j}$ and Δ_a. Note that $d_{j,i}$ is the adjoint of $d_{i,j}$ and the formula below for Δ_a is a generalization of the analogous formula for the Johnson scheme given in Lemma 6.1.3.

Lemma 10.9.4 *Let $a = (a_1, a_2, \ldots, a_h)$ be a composition of n.*

(i) *If $i, j, k \in \{1, 2, \ldots, h\}$ are distinct, then, for all $f \in M^a$,*

$$d_{i,k}d_{k,j}f = d_{i,j}f + d_{k,j}d_{i,k}f$$

(ii)

$$M^a \bigcap \left(\bigcap_{1\leq i<j\leq h} \ker d_{i,j}\right) = M^a \bigcap \left(\bigcap_{i=1}^{h-1} \ker d_{i,i+1}\right)$$

(iii)

$$\Delta_a f = \sum_{1\leq i<j\leq h} d_{j,i}d_{i,j}f - \sum_{j=1}^{k}(j-1)a_j f.$$

Proof (i) If $\{t\}$ is an a-tabloid, then $d_{i,k}d_{k,j}e_{\{t\}} = \sum_{\{s\}} e_{\{s\}}$ where the sum runs over all a-tabloids $\{s\}$ that can be obtained from $\{t\}$ by moving a number x from row j to row k and then a number y from row k to row i. Summing up the cases $x \neq y$ we get exactly $d_{i,k}d_{k,j}e_{\{t\}}$; summing up the cases $x = y$ we obtain exactly $d_{i,j}e_{\{t\}}$.

(ii) If $f \in M^a \cap \left(\bigcap_{i=1}^{h-1} \ker d_{i,i+1} \right)$ then using (i) one has, for all $i \in \{1, 2, \ldots, h-1\}$ and $1 \leq t \leq h-i$,

$$d_{i,i+t}f = (d_{i,i+1}d_{i+1,i+t} - d_{i+1,i+t}d_{i,i+1})f$$
$$= d_{i,i+1}d_{i+1,i+t}f.$$

In particular, if $d_{i,j}f = 0$ holds for $1 \leq i < j \leq i+t-1$ then it also holds for $1 \leq i < j \leq i+t$ and, by iterating, we get $f \in M^a \cap \left(\bigcap_{1 \leq i \leq j \leq h} \ker d_{i,j} \right)$.

(iii) If $\{t\}$ is an a-tabloid, then $d_{i,k}d_{k,j}e_{\{t\}} = \sum_{\{s\}} e_{\{s\}}$ where the sum runs over all a-tabloids $\{s\}$ that can be obtained from $\{t\}$ by choosing two different rows i, j, with $i < j$, moving a number x from row j to row i and then moving a number y from row i to row j. Summing up the cases $x \neq y$ we get exactly $\Delta_a e_{\{t\}}$; summing up the cases $x = y$ we obtain exactly $\sum_{1 \leq i < j \leq h} a_j = \sum_{j=1}^{k}(j-1)a_j$ times $e_{\{t\}}$. $\square$

We now prove that $M^\lambda \cap \left(\bigcap_{i=1}^{h-1} \ker d_{i,i+1} \right)$ is an eigenspace of Δ_λ.

Lemma 10.9.5 *If* $\lambda = (\lambda_1, \lambda_2, \ldots, \lambda_k) \vdash n$ *then* $M^\lambda \cap \left(\bigcap_{i=1}^{h-1} \ker d_{i,i+1} \right)$ *is an eigenspace of* Δ_λ *and the corresponding eigenvalue is* $-\sum_{j=1}^{k}(j-1)\lambda_j$.

Proof Using (ii) and (iii) in the preceding lemma, if $f \in M^\lambda$ and $d_{i,i+1}f = 0$ for all $i = 1, 2, \ldots, k-1$, then also $d_{i,j}f = 0$ for $i < j$. Thus, $\Delta_\lambda f = -\sum_{j=1}^{k}(j-1)\lambda_j f$. $\square$

From Lemma 10.9.1, Lemma 10.9.3 and Lemma 10.9.5 we get another proof of (10.19) in Theorem 10.6.2.

Corollary 10.9.6

$$r(\lambda) = \frac{1}{n(n-1)} \sum_{i=1}^{k} \left[\lambda_i^2 - (2i-1)\lambda_i \right].$$

Proof S^λ is an eigenspace of Δ_λ. From Lemma 10.9.3 the corresponding eigenvalue is $\frac{n(n-1)}{2}r(\lambda) - \sum_{i=1}^{k} \frac{\lambda_i(\lambda_i-1)}{2}$; from Lemma 10.9.1 and Lemma 10.9.5 it is equal to $-\sum_{j=1}^{k}(j-1)\lambda_j$. Equating these two expressions we get the result. $\square$

We are now in a position to prove James' intersection kernel theorem for the ordinary irreducible representations of S_n.

Theorem 10.9.7 *For any partition* $\lambda = (\lambda_1, \lambda_2, \ldots, \lambda_k) \vdash n$ *we have*

$$S^\lambda = M^\lambda \cap \left(\bigcap_{i=1}^{h-1} \ker d_{i,i+1} \right).$$

Proof We show that S^λ is the eigenspace of Δ_λ with smallest eigenvalue. Indeed, from Lemma 10.9.3 (with $a = \lambda$) and Corollary 10.6.3 we have that if $\mu \rhd \lambda$, then the eigenvalue corresponding to the S^μ-isotypic component is strictly greater than the eigenvalue corresponding to S^λ. Moreover from Proposition 10.3.16 and Corollary 10.3.22 we have that

$$M^\lambda = S^\lambda \bigoplus \left[\bigoplus_{\mu \rhd \lambda} m_\mu S^\mu \right].$$

Finally, Lemmas 10.9.1 and 10.9.5 force $M^\lambda \cap \left(\bigcap_{i=1}^{h-1} \ker d_{i,i+1} \right)$ to be equal to S^λ. □

James' intersection kernel theorem may also be proved for a class of irreducible representations of $GL(n, q)$. See James' book on the representation theory of linear groups [133] and, for a proof similar to the one in this section, [175].

11

The Gelfand pair $(S_{2n}, S_2 \wr S_n)$ and random matchings

11.1 The Gelfand pair $(S_{2n}, S_2 \wr S_n)$

Let X denote the set of all partitions of $\{1, 2, \ldots, 2n\}$ consisting of two-elements subsets: such partitions are thus of the form

$$\{\{i_1, i_2\}, \{i_3, i_4\}, \ldots, \{i_{2n-1}, i_{2n}\}\},$$

where $\{i_1, i_2, \ldots, i_{2n}\} = \{1, 2, \ldots, 2n\}$. The subsets $\{i_1, i_2\}$, $\{i_3, i_4\}, \ldots$, $\{i_{2n-1}, i_{2n}\}$ are the *parts* of the partition; note that there is no order between or within the parts. The symmetric group S_{2n} acts on X in a natural way: if $\pi \in S_{2n}$ and $x = \{\{i_1, i_2\}, \{i_3, i_4\}, \ldots, \{i_{2n-1}, i_{2n}\}\} \in X$, then

$$\pi x = \{\{\pi i_1, \pi i_2\}, \{\pi i_3, \pi i_4\}, \ldots, \{\pi i_{2n-1}, \pi i_{2n}\}\}.$$

The permutation representation $L(X)$ associated with this action will be the subject of the present section.

Fix an element $x_0 = \{\{i_1, i_2\}, \{i_3, i_4\}, \ldots, \{i_{2n-1}, i_{2n}\}\} \in X$ and denote by $K = \{g \in S_{2n} : gx_0 = x_0\}$ its stabilizer. If $k \in K$ and $1 \le h \le n$, then $k\{i_{2h-1}, i_{2h}\}$ is again one of the subsets $\{i_1, i_2\}, \ldots, \{i_{2n-1}, i_{2n}\}$, that is K permutes among themselves these n two-elements subsets. We thus have a homomorphism $\phi : K \to S_n$, where S_n is viewed as the group of all permutations of the above n two-elements subsets, and whose kernel is $\ker(\phi) = S_2 \times \cdots \times S_2$ (n-times); it is easy to see (exercise) that K is indeed the wreath product of S_2 with S_n: $K = S_2 \wr S_n$; see Section 5.3. In particular,

$$|X| = \frac{(2n)!}{n!2^n}. \tag{11.1}$$

Let now $x, y \in X$, say $x = \{\{i_1, i_2\}, \{i_3, i_4\}, \ldots, \{i_{2n-1}, i_{2n}\}\}$, and, respectively, $y = \{\{j_1, j_2\}, \{j_3, j_4\}, \ldots, \{j_{2n-1}, j_{2n}\}\}$. With the pair x, y

371

we associate the graph $\mathcal{C}(x, y)$ whose vertex set is $\{1, 2, \ldots, 2n\}$ and whose edges are $\{i_1, i_2\}, \{i_3, i_4\}, \ldots, \{i_{2n-1}, i_{2n}\}$ and $\{j_1, j_2\}, \{j_3, j_4\}, \ldots, \{j_{2n-1}, j_{2n}\}$.

For instance, if $2n = 10$ and $x = \{\{1, 3\}, \{2, 4\}, \{5, 7\}, \{8, 6\}, \{9, 10\}\}$ and $y = \{\{1, 2\}, \{3, 4\}, \{5, 6\}, \{7, 8\}, \{9, 10\}\}$, then $\mathcal{C}(x, y)$ is the following graph

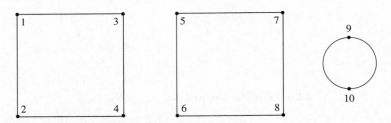

Figure 11.1.

It is immediate that $\mathcal{C}(x, y)$ is a disjoint union of cycles of even length: any vertex has degree two as it is incident to exactly one edge in x and one edge in y. If $2\lambda_1, 2\lambda_2, \ldots, 2\lambda_k$, with $\lambda_1 \geq \lambda_2 \geq \cdots \geq \lambda_k$ are the lengths of the cycles of $\mathcal{C}(x, y)$, then $\sum_{i=1}^{k} \lambda_i = n$ so that $\lambda = (\lambda_1, \lambda_2, \ldots, \lambda_k)$ is a partition of n. Following [130, 64] we define a function from $X \times X$ to the set of all partitions of n by setting, for $x, y \in X$, $d(x, y) = \lambda$, where λ is as above. The following facts are easy:

(i) $d(x, y) = (1, 1, \ldots, 1)$ if and only if $x = y$;

(ii) $d(x, y) = d(y, x)$ for all $x, y \in X$;

(iii) $d(x, y) = d(x', y')$ if and only if there exists $\pi \in S_{2n}$ such that $\pi(x) = x'$ and $\pi(y) = y'$.

Indeed, if $d(x, y) = d(x', y')$ then the graphs $\mathcal{C}(x, y)$ and $\mathcal{C}(x', y')$ are isomorphic and an explicit isomorphism (that sends edges of $\mathcal{C}(x, y)$ to edges of $\mathcal{C}(x', y')$) yields the desired $\pi \in S_{2n}$.

Theorem 11.1.1 $(S_{2n}, S_2 \wr S_n)$ *is a symmetric Gelfand pair and the number of its spherical representations equals the number of partitions of* n.

Proof (x, y) and (y, x) belong to the same orbit of S_{2n} on $X \times X$ and therefore, by Lemma 4.3.4, $(S_{2n}, S_2 \wr S_n)$ is a symmetric Gelfand pair. In addition, (iii) implies that the orbits of $S_2 \wr S_n$ on X may be indexed by the partitions of n: such orbits correspond to the "spheres" $A_\lambda = \{x \in X : d(x_0, x) = \lambda\}$. Thus, the number of irreducible representations that appear in the decomposition of $L(X)$ is the same as the number of partitions of n. $\square$

11.2 The decomposition of $L(X)$ into irreducible components

Here we use the notation and terminology for the representation theory of the symmetric group from Chapter 10.

For $\lambda \vdash n$, $\lambda = (\lambda_1, \lambda_2, \ldots, \lambda_s)$ we denote by 2λ the partition of $2n$ given by $2\lambda = (2\lambda_1, 2\lambda_2, \ldots, 2\lambda_s)$. If t is a 2λ-tableau, $\{t\}$ its associated 2λ-tabloid and

$$x = \{\{i_1, i_2\}, \{i_3, i_4\}, \ldots, \{i_{2n-1}, i_{2n}\}\}$$

is an element of X, we say that $\{t\}$ *covers* x if i_{2h-1} and i_{2h} belong to the same row of $\{t\}$, for all $h = 1, 2 \ldots, n$.

For a 2λ-tabloid $\{t\}$, let $\mathcal{C}_t$ denote the set of all $x \in X$ covered by $\{t\}$ and $\mathbf{1}_{\mathcal{C}_t}$ be the characteristic function of $\mathcal{C}_t$.

Then, for every $\lambda \vdash n$ we may define a map D_λ from the set $\{e_{\{t\}} : \{t\}$ is a 2λ-tabloid$\}$ into $L(X)$ by setting $D_\lambda(e_{\{t\}}) = \mathbf{1}_{\mathcal{C}_t}$. It is clear that D_λ extends to a linear map (still denoted by D_λ) from $M^{2\lambda}$ to $L(X)$. Moreover as x is covered by $\{t\}$ if and only if πx is covered by $\pi\{t\}$, $x \in X$, $\{t\}$ a 2λ-tabloid and $\pi \in S_{2n}$, one has $D_\lambda(\pi e_{\{t\}}) = \pi D_\lambda(e_{\{t\}})$, that is, D_λ intertwines the representations of S_{2n} on $M^{2\lambda}$ and $L(X)$.

Now we are in a position to obtain the decomposition of $L(X)$ into irreducible representations:

Theorem 11.2.1

$$L(X) = \oplus_{\lambda \vdash n} S^{2\lambda}. \tag{11.2}$$

Proof First we show that, for $\lambda \vdash n$, $S^{2\lambda}$ is contained as a sub-representation in $L(X)$. Since D_λ intertwines the representations of S_{2n} on $M^{2\lambda}$ and $L(X)$, and $S^{2\lambda}$ is an irreducible subspace of $M^{2\lambda}$, we only have to show that there exists $v \in S^{2\lambda}$ such that $D_\lambda v \neq 0$. For this purpose we introduce the following definition.

If t is a 2λ-tableau and $x_0 = \{\{i_1, i_2\}, \{i_3, i_4\}, \ldots, \{i_{2n-1}, i_{2n}\}\}$ is a fixed element of X, we say that t is *good with respect to* x_0, or simply, x_0-*good*, if, for $h = 1, 2, \ldots, n$, i_{2h-1} and i_{2h} are on the same row in t, one on an odd column and the other on the successive (from left to right) column of the diagram of 2λ (in the diagram of 2λ every odd column is followed by a column of the same length). For instance, if $x_0 = \{\{1,3\}, \{2,5\}, \{4,6\}, \{7,9\}, \{8,10\}\}$ then the tableau is x_0-good.

4	6	10	8
9	7	1	3
2	5		

Figure 11.2.

We now show that if t is x_0-good, then $D_\lambda e_t$ takes the value $\lambda'_1! \lambda'_2! \cdots \lambda'_r!$ on x_0, where $\lambda' = (\lambda'_1, \lambda'_2, \ldots, \lambda'_r) \vdash n$ is the conjugate partition of λ.

We have $D_\lambda e_t = \sum_{\sigma \in C_t} \mathrm{sign}(\sigma) D_\lambda e_{\{\sigma t\}}$. But if t is x_0-good and $\sigma \in C_t$, then $\{\sigma t\}$ covers x_0 if and only if σt is x_0-good. Indeed these are the cases if and only if, for each odd column of t, the restriction of σ to the numbers on the column has the same action of its restriction to the even column immediately on the right: since this happens if and only if $\sigma \in C_t \cap K$, where K is the stabilizer of x_0, then $D_\lambda e_{\{\sigma t\}}(x_0) \neq 0$ if and only if $\sigma \in C_t \cap K$. Since $|C_t \cap K| = \lambda'_1! \lambda'_2! \cdots \lambda'_r!$ and for every $\sigma \in K \cap C_t$ one has $\mathrm{sign}(\sigma) = 1$, the assertion follows.

So $D_\lambda e_t \neq 0$ and this proves that $L(X)$ contains an invariant subspace S_{2n}-isomorphic to $S^{2\lambda}$. Since the number of irreducible subrepresentations of $L(X)$ is the same as the number of partitions of n we have proved the decomposition (11.2). $\qquad\square$

Other proofs of this decomposition may be found in [134, 138, 165, 183, 221, 222] and the recent paper [103]. See also [24]. q-analogs of this Gelfand pair are considered in [13, 14, 101, 119, 126, 127, 140, 145, 158].

11.3 Computing the spherical functions

Let x_0 be a fixed element of X and K its stabilizer. The spherical function in $D_\lambda S^{2\lambda}$ is its (unique) K-invariant element $\phi_\lambda \in D_\lambda S^{2\lambda}$ such that $\phi_\lambda(x_0) = 1$.

If t is an x_0-good 2λ-tableau, from the previous section we immediately have that

$$\phi_\lambda = \frac{1}{\lambda'_1! \lambda'_2! \cdots \lambda'_r!} P D_\lambda e_t \qquad (11.3)$$

where $P = 1/|K| \sum_{k \in K} k$ is the projection to the subspace of K-invariant functions. But

$$PD_\lambda e_t = \frac{1}{|K|} \sum_{k \in K} \sum_{\sigma \in C_t} \text{sign}(\sigma) D_\lambda e_{\{k\sigma t\}};$$

setting $K_t := K \cap C_t$, a system of representatives for the cosets of C_t/K_t is given by the subgroup C'_t of C_t consisting of those elements that fix every number belonging to an even column of t (in other words, C'_t is the group of all permutations of the numbers on the odd columns of t that fix such columns setwise, multiplied by the identity permutation of the set of numbers on the even columns of t). Since $|K_t| = \lambda'_1! \cdots \lambda'_r!$ and $\text{sign}(k') = 1$ for all $k \in K_t$, we have:

$$
\begin{aligned}
PD_\lambda e_t &= \frac{1}{|K|} \sum_{k \in K} \sum_{k' \in K_t} \sum_{\sigma \in C'_t} D_\lambda e_{\{kk'\sigma t\}} \text{sign}(k'\sigma) \\
&= \lambda'_1! \cdots \lambda'_r! \frac{1}{|K|} \sum_{k \in K} \sum_{\sigma \in C'_t} D_\lambda e_{\{k\sigma t\}} \text{sign}(\sigma).
\end{aligned}
\tag{11.4}
$$

(Here is another proof of the decomposition of $L(X)$: (11.4) when written without D_λ shows that Pe_t is a nonzero K-invariant vector in $S^{2\lambda}$.) If $\mu \vdash n$ and, as in the previous section, $A_\mu = \{x \in X : d(x, x_0) = \mu\}$, $\mathbf{1}_{A_\mu}$ is the characteristic function of A_μ and f complex valued function on A_μ, then

$$Pf = \frac{1}{|A_\mu|} \left(\sum_{x \in A_\mu} f(x) \right) \mathbf{1}_{A_\mu}.$$

Moreover, if $x \in X$ then the value of $\sum_{\sigma \in C'_t} \text{sign}(\sigma) D_\lambda e_{\{\sigma t\}}$ on x is equal to $\sum_\sigma \text{sign}(\sigma)$ where the sum runs over all $\sigma \in C'_t$ such that $\{\sigma t\}$ covers x. Thus, defining $a^\lambda_\mu = \Sigma_{\sigma \in C'_t} \text{sign}(\sigma)|\{x \in A_\mu : \{\sigma t\}$ covers $x\}|$, which does not depend on t, we get

$$\phi_\lambda = \sum_{\mu \vdash n} \frac{a^\lambda_\mu}{|A_\mu|} \mathbf{1}_{A_\mu}. \tag{11.5}$$

11.4 The first nontrivial value of a spherical function

In this section we want to use the above results to give an explicit formula for the first nontrivial value of the spherical functions ϕ_λ, that is, for $\lambda \vdash n$, $\mu = (1^{n-1}, 2)$ and $x \in A_\mu$, we want to compute $\phi_\lambda(x)$. Since the values of ϕ_λ depend only on $d(x, x_0)$, such a quantity will be denoted by $\phi_\lambda(\mu)$.

From the preceding section, we have to compute the constants $|A_\mu|$ and a_μ^λ. If t is an x_0-good tableau, a moment of reflection shows that if $\sigma \in C_t'$, $\{\sigma t\}$ covers some $x \in A_\mu$ if and only if σ is the identity or a transposition. In fact, the elements of A_μ are exactly those of the form πx_0, with π a transposition of two numbers in two different parts of the partition x_0 (following [55], such a transposition will be called a *switch*), and then an $x \in A_\mu$ may be covered by a tabloid $\{\sigma t\}$ with $\sigma \in C_t'$, iff $x = \pi x_0$, where π switches two numbers on the same row in t (but in different parts of x_0 so that $\{t\}$ covers x and we may take $\sigma =$ identity), or two numbers on the same odd column in t (and so $\{\pi t\}$ covers x and we may take $\sigma = \pi$).

The number of $x \in A_\mu$ covered by $\{t\}$ is $\sum_{i=1}^k \lambda_i(\lambda_i - 1)$. In fact there are exactly $2\binom{\lambda_i}{2} = \lambda_i(\lambda_i - 1)$ transpositions that switch two numbers on the ith row that belong to differents parts of x_0. For every transposition $\sigma \in C_t'$, there exists exactly one $x \in A_\mu$ covered by $\{\sigma t\}$ (such an x is given by σx_0) and the number of transpositions in C_t' is $\sum_{j=1}^s \binom{\lambda_j'}{2}$, where λ' is the partition conjugate to λ. By Lemma 10.6.1, this quantity equals $\sum_{i=1}^k \lambda_i(i - 1)$. Since $|A_\mu| = n(n-1)$, applying (11.5) we obtain

$$
\begin{aligned}
\phi_\lambda(\mu) &= \frac{1}{n(n-1)} \left[\sum_{i=1}^n \lambda_i(\lambda_i - 1) - \sum_{i=1}^n \lambda_i(i - 1) \right] \\
&= \frac{1}{n(n-1)} \sum_{i=1}^n \lambda_i(\lambda_i - i).
\end{aligned}
\tag{11.6}
$$

This formula has already been obtained using the theory of zonal polynomials (see [59, 202]) or the character theory of the symmetric group [64].

11.5 The first nontrivial spherical function

In this section we compute $\phi_\lambda(\mu)$ for $\mu \vdash n$ and $\lambda = (n-1, 1)$. First some notation. If $j_1, \ldots, j_{2m}$, $m \leq n$, are $2m$ distinct numbers in $\{1, \ldots, 2n\}$, we denote by $\mathcal{P}(\{j_1, j_2\}, \ldots, \{j_{2m-1}, j_{2m}\})$ the set of all partitions $x \in X$ such that $\{j_1, j_2\}, \ldots, \{j_{2m-1}, j_{2m}\}$ are parts of x, and by $\mathbf{1}(\{j_1, j_2\}, \ldots, \{j_{2m-1}, j_{2m}\})$ the characteristic function of such a set.

We observe that if $x_0 = \{\{j_1, j_2\}, \ldots, \{j_{2m-1}, j_{2m}\}\}$, $\mu \vdash n$ and the multiplicity of 1 in μ is denoted by $a_1(\mu)$, then any $x \in A_\mu$ is contained in exactly $a_1(\mu)$ sets of the family $\{\mathcal{P}(\{i_{2r-1}, i_{2r}\}) : r = 1, 2, \ldots, n\}$.

We then have, for $h = 1, 2, \ldots, n$,

$$
P\mathbf{1}(\{i_{2h-1}, i_{2h}\}) = \frac{1}{|K|} \sum_{k \in K} \mathbf{1}(k\{i_{2h-1}, i_{2h}\})
$$

$$
= P\left[\frac{1}{n} \sum_{r=1}^{n} \mathbf{1}(\{i_{2r-1}, i_{2r}\})\right]
$$

$$
= P\left[\frac{1}{n} \sum_{\mu \vdash n} a_1(\mu) \mathbf{1}_{A_\mu}\right]
$$

$$
= \sum_{\mu \vdash n} \frac{a_1(\mu)}{n} \mathbf{1}_{A_\mu}.
$$

In a similar way, if $\{i, j\}$ is not one of the parts of x_0, then

$$
P\mathbf{1}(\{i, j\}) = \frac{1}{|K|} \sum_{k \in K} \mathbf{1}(k\{i, j\})
$$

$$
= P\left[\frac{1}{\binom{2n}{2} - n} \sum_{\{s,t\} \notin x_0} \mathbf{1}(\{s, t\})\right]
$$

$$
= P\left[\sum_{\mu \vdash n} \frac{n - a_1(\mu)}{2n(n-1)} \mathbf{1}_{A_\mu}\right]
$$

$$
= \sum_{\mu \vdash n} \frac{n - a_1(\mu)}{2n(n-1)} \mathbf{1}_{A_\mu}.
$$

If t is an x_0-good 2λ-tableau then, $D_\lambda e_{\{t\}} = \mathbf{1}(\{i_{2h-1}, i_{2h}\})$, where i_{2h-1} and i_{2h} are the numbers on the second row of t. Moreover C_t' is formed by two elements: the identity and the transposition (i_{2h-1}, i_{2m-1}), if i_{2m-1} and i_{2m} are the numbers on the first two cells of the first row of t, while $C_t = C_t' \times \{id, (i_{2h}, i_{2m})\}$. Since t is x_0-good, $\{i_{2h-1}, i_{2h}\}$ and $\{i_{2m-1}, i_{2m}\}$ are two of the parts of x_0 and so we have

$$
D_\lambda e_t = \mathbf{1}(\{i_{2h-1}, i_{2h}\}) - \mathbf{1}(\{i_{2m-1}, i_{2h-1}\})
$$
$$
+ \mathbf{1}(\{i_{2m-1}, i_{2m}\}) - \mathbf{1}(\{i_{2h-1}, i_{2m}\}).
$$

But then

$$
\phi_\lambda(\mu) = \frac{1}{2!} P D_\lambda e_t = \sum_{\mu \vdash n} \left[\frac{a_1(\mu)}{n} - \frac{1}{2n(n-1)}(n - a_1(\mu))\right] \mathbf{1}_{A_\mu}
$$

$$
= \sum_{\mu \vdash n} \left[1 - \frac{(2n-1)(n - a_1(\mu))}{2n(n-1)}\right] \mathbf{1}_{A_\mu}. \tag{11.7}
$$

This formula has already been obtained with a similar but more particular argument in [59].

11.6 Phylogenetic trees and matchings

A *rooted binary tree* is a tree T (that is a finite connected simple graph without loops or circuits) with a unique vertex R (*the root*) of degree 2 and all the remaining vertices of degree 3 (*internal vertices*) or 1 (*the leaves*: in the pictures below these are indicated with •).

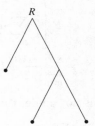

Figure 11.3.

A *labeling* of the set L of the leaves of T is a function $\sigma : L \to \{1, 2, \ldots, |L|\}$. A *phylogenetic tree* is a rooted binary tree where the leaves carry different labels (in other words σ is a bijection). We identify two phylogenetic trees if there exists a graph isomorphism (necessarily carrying root to root and leaves to leaves) which preserves the labeling. For example, there is only one phylogenetic tree with two leaves, namely:

Figure 11.4.

and there are exactly three distinct phylogenetic trees with three leaves:

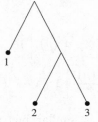
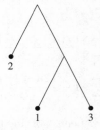
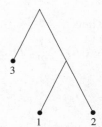

Figure 11.5.

Schröder [194] showed that there are $(2n)!/2^n n!$ distinct phylogenetic trees with $n+1$ leaves. The goal of this section is to give a combinatorial proof of this formula by explicitily giving a bijection between phylogenetic tree and matchings. Note that T is not necessarily homogeneous in the sense of Chapter 7 (see example below). Note that a tree with ℓ leaves has necessarily $2\ell - 1$ vertices (including the root): this can be easily proved by induction (exercise).

Notation: given two adjacent vertices x and y in a rooted binary tree, their distances $d'(x, R), d'(y, R)$ from the root R differ by one. Suppose $d'(x, R) < d'(y, R)$: the vertex x minimizing this distance is called the *father* of y and, reciprocally, y is the *child* of x. This way all vertices distinct form the root R have exactly one father, and all vertices which are not leaves have exactly two children. Two children of the same father are called *siblings*. Note that a phylogenetic tree is uniquely determined once the root and the parental relations are given.

A *perfect matching* (or simply a *matching*) on $2n$ points is a partition of $\{1, 2, \ldots, 2n\}$ into n two-element subsets, that is an element of the homogeneous space $S_{2n}/(S_2 \wr S_n)$ studied in the previous section. From a graph theoretical point of view, given $2n$ vertices, a matching is a set of edges which is incident to all vertices and minimal, or, equivalently, such that each vertex has degree one. In this setting, although from the graph theoretical point of view all matching are graph-isomorphic, no identification occurs.

We have already seen (see (11.1)) that there are $(2n)!/2^n n!$ distinct matchings on $2n$ elements. For example, there is only one matching on two $(n = 1)$ points and there are exactly three distinct matchings on four $(n = 2)$ points:

$$\{\{1, 2\}\{3, 4\}\}, \quad \{\{1, 3\}\{2, 4\}\} \text{ and } \{\{1, 4\}\{2, 3\}\}.$$

We now present a proof, due to Diaconis and Holmes [63],[64], the set of phylogenetic trees on $n+1$ leaves and the set of perfect matchings on $2n$ points are equipotent.

Algorithm. Start with a phylogenetical tree T with ℓ leaves which are labeled with the symbols $1, 2, \ldots, \ell$. Extend the labeling to all internal vertices (root excluded) with the symbols $\ell+1, \ell+2, \ldots, 2(\ell-1)$, giving, at each step, the minimal (not yet used) label to the father (not yet labeled) which has both children already labeled and one of these two labels is the minimal possible available (among the children whose father is not yet labeled).

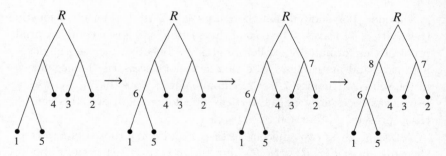

Figure 11.6.

When all vertices (except the root) are labeled, create a matching on $2n = 2(\ell - 1)$ points by grouping siblings. For instance, from the example above we get the matching

$$\{1,5\}, \ \{6,4\}, \ \{8,7\}, \ \{3,2\}.$$

Conversely, suppose we are given a perfect matching on $2n$ points. We now construct a phylogenetic tree that corresponds to the matching, that is we invert the correspondence described above.

The vertex set of the tree is $\{1, 2, \ldots, 2n, R\}$, R is the root, the set of leaves is $L = \{1, 2, \ldots, n+1\}$ and the pairs of siblings are exactly the matched pairs. We only have to assign a father with each matched pair. Consider those matched pairs in $\{1, 2, \ldots, n+1\}$: by the *pigeonhole principle* at least one such a pair exists. Take the pair with the smallest integer and choose $n+2$ as their father. Next among the pairs in $\{1, 2, \ldots, n+2\}$ with no father assigned choose the pair with the smallest integer and let $n+3$ be its father. Continuing this way, this procedure ends when assigning to the last pair the root R as a father.

This way one obtains the desired phylogenetic tree (note that the number of fathers is n and equals the number of pairs in the matching).

11.7 Random matchings

We now study a natural invariant random walk on $X = S_{2n}/(S_2 \wr S_n)$. First we give an informal description due to Diaconis.

Consider n couples at a party each talking only to the person they arrived with. The host, seeking to let the people know each other, picks two of the couples at random and switches a random person between them. Identifying each person with a number from 1 to $2n$ we can model the above mixing procedure with a Markov chain on the set X of all partitions in two elements subsets of $\{1, 2, \ldots, 2n\}$ and transition matrix P

$$p(x, y) = \begin{cases} \frac{1}{n(n-1)} & \text{if } d(x, y) = (1^{n-1}, 2) \\ 0 & \text{otherwise} \end{cases}$$

where d is the function defined in Section 11.1. Indeed given an element x the number of distinct couples of pairs in x is $\binom{n}{2}$ and for each of this couple we have two different element y with $d(x, y) = (1^{n-1}, 2)$.

Exercise 11.7.1 Translate this random walk into the setting of phylogenetic trees.

As usual our main goal is to study the rate of convergence to the stationary distibution. This random walk is clearly a procedure that generates a *random matching*.

Theorem 11.7.2 (Diaconis and Holmes [64]) *Let π be the uniform distribution on X and $\nu_0^{(k)}$ the distribution probability after k steps if the initial distribution is the Dirac function centered at $\{\{1, 2\}, \{3, 4\}, \ldots, \{2n-1, 2n\}\}$. There exists a positive constant a such that if $k = \frac{1}{2}n(\log n + c)$ then*

$$\|\nu_0^{(k)} - \pi\|_{TV} \leq ae^{-2c}.$$

Proof In virtue of the upper bound lemma (4.22) we must estimate from above the quantity

$$\sum_{\substack{\lambda \vdash n \\ \lambda \neq (n)}} d_{2\lambda} \beta_\lambda^{2k} \tag{11.8}$$

where $\lambda = (\lambda_1, \lambda_2, \ldots, \lambda_r) \vdash n$, $d_{2\lambda} = \dim(S^{2\lambda})$ and (recall (11.6)) $\beta_\lambda = \frac{1}{n(n-1)} \sum_{i=1}^r \lambda_i(\lambda_i - i)$. The bound will be achieved in several steps. Step 1 and Step 3 are the analogs of Lemma 10.6.4, while Step 2 is an analog of Corollary 10.6.3. The rest of the proof is similar to the proof of Theorem 10.7.2 (upper bound for random transpositions) but with some more technicalities. □

Step 1 $\beta_\lambda \leq \lambda_1/n$.

Proof $\beta_\lambda \leq \frac{\lambda_1 - 1}{n(n-1)} \sum_{i=1}^r \lambda_i = \frac{\lambda_1 - 1}{n-1} \leq \lambda_1/n$, where the first inequality follows immediately from the condition that $\lambda_1 \geq \lambda_2 \geq \cdots \geq \lambda_r$. □

Step 2 *If $\lambda \trianglerighteq \mu$ then $\beta_\lambda \geq \beta_\mu$.*

Proof It clearly suffices to show the inequality in the special case:

$$\lambda_a = \mu_a + 1$$
$$\lambda_b = \mu_b - 1$$
$$\lambda_c = \mu_c$$

for some $a < b$ and all $c \neq a, b$. Then

$$\beta_\lambda - \beta_\mu = \frac{1}{n(n-1)} \left[\lambda_a(\lambda_a - a) + \lambda_b(\lambda_b - b) - \mu_a(\mu_a - a) - \mu_b(\mu_b - b) \right]$$

$$= \frac{1}{n(n-1)} \left[2(\mu_a - \mu_b) + b - a + 2 \right] > 0.$$

$\square$

Step 3 *If $\lambda_1 \geq \frac{n}{2}$ then $\beta_\lambda \leq 1 - \frac{(n-\lambda_1)(2\lambda_1+1)}{n(n-1)}$.*

Proof Using the previous step we get, for $\lambda_1 \geq \frac{n}{2}$:

$$\beta_\lambda \leq \beta_{(\lambda_1, n-\lambda_1)} = \frac{1}{n(n-1)} \left[\lambda_1(\lambda_1 - 1) + (n - \lambda_1)(n - \lambda_1 - 2) \right]$$

$$= 1 - \frac{(n - \lambda_1)(2\lambda_1 + 1)}{n(n-1)}.$$

$\square$

Step 4

$$\sum_{\substack{\lambda \vdash n \\ \lambda_1 = n-j}} d_{2\lambda} \leq \frac{(2n)!}{(2n-2j)! j! 2^j} \leq \frac{n^{2j} 2^j}{j!}.$$

Proof Recall that, by Proposition 10.5.17 for the dimension d_λ of the representation S^λ, we have the estimate

$$d_\lambda \leq \binom{n}{\lambda_1} d_{\overline{\lambda}}$$

where $\overline{\lambda} = (\lambda_2, \lambda_3, \ldots, \lambda_k) \vdash (n - \lambda_1)$. We also recall that as $L(X) = \oplus_{\lambda \vdash n} S^{2\lambda}$, one has $\frac{(2n)!}{2^n n!} = \sum_{\lambda \vdash n} d_{2\lambda}$. Combining these two facts we deduce

$$\sum_{\substack{\lambda \vdash n \\ \lambda_1 = n-j}} d_{2\lambda} \leq \binom{2n}{2\lambda_1} \sum_{\overline{\lambda} \vdash j} d_{2\overline{\lambda}}$$

$$= \binom{2n}{2\lambda_1} \frac{(2j)!}{2^j j!} = \frac{(2n)!}{(2n-2j)! j! 2^j} \leq \frac{n^{2j} 2^j}{j!}.$$

$\square$

Step 5 *If $\lambda_1 \geq n/2 \geq 4$ then the eigenvalue β_λ is nonnegative.*

Proof For any partition $\lambda = (\lambda_1, \lambda_2, \ldots, \lambda_k)$ we have $\lambda \unrhd (\lambda_1, 1, 1, \ldots, 1)$. Using Step 2 we get:

$$2n(n-1)\beta_\lambda \geq 2n(n-1)\beta_{(\lambda_1, 1, 1, \ldots, 1)}$$
$$= 2\lambda_1(\lambda_1 - 1) + 2\sum_{i=2}^{n-\lambda_1}(1 - i)$$
$$= \lambda_1^2 - n^2 + 2\lambda_1 n + (n - 3\lambda_1)$$
$$\geq \frac{n^2}{4} - 2n \geq 0.$$

$\square$

Set $R_j = \{\lambda \vdash n : \lambda_1 = n - j, \ \beta_\lambda \geq 0\}$.

Step 6 *For every $\alpha \in (0, \frac{1}{4})$ there exists a positive constant a_1 (not depending on n) such that if $k = \frac{1}{2}n(\log n + c)$ with $c > 0$ then*

$$\sum_{j=1}^{\alpha n}\sum_{\lambda \in R_j} d_{2\lambda}\beta_\lambda^{2k} \leq a_1 e^{-2c}.$$

Proof Using Step 3 and Step 4 we can estimate the term corresponding to $j = 1$ in the sum of the statement. Indeed for $k = \frac{1}{2}n(\log n + c)$ and $\lambda = (n - 1, 1)$, we have that

$$d_{2\lambda}\beta_\lambda^{2k} \leq 2n^2\left[1 - \frac{2n-1}{n(n-1)}\right]^{2k}$$
$$\leq 2n^2\exp(-2k\frac{2n-1}{n(n-1)})$$
$$= 2n^2\exp\left[-(\log n + c)\frac{2n-1}{n-1}\right]$$
$$\leq 2e^{-2c}.$$

To bound the remaing part in the sum we notice that, for $j \leq \alpha n \leq \frac{n}{4}$,

$$\frac{j(2n - 2j + 1)}{n(n-1)} \geq \frac{2}{n} + \frac{2(j-1)}{n} - \frac{2j^2}{n^2}.$$

This, coupled with Step 3 and Step 4, gives

$$\sum_{j=2}^{\alpha n} \sum_{\lambda \in R_j} d_{2\lambda} \beta_\lambda^{2k} \leq \sum_{j=2}^{\alpha n} \frac{n^{2j} 2^j}{j!} \left[1 - \frac{j(2n - 2j + 1)}{n(n-1)} \right]^{2k}$$

$$\leq \sum_{j=2}^{\alpha n} \frac{n^{2j} 2^j}{j!} \exp\left[-2k \frac{j(2n - 2j + 1)}{n(n-1)} \right]$$

$$\leq \exp\left(\frac{-4k}{n} \right) \sum_{j=2}^{\alpha n} \frac{n^{2j} 2^j}{j!} \exp\left\{ \frac{-2k}{n} \left[2(j-1) - \frac{2j^2}{n} \right] \right\}$$

$$= e^{-2c} \sum_{j=2}^{\alpha n} \frac{n^{2j} 2^j}{n^2 j!} \exp\left\{ -(\log n + c) \left[2(j-1) - \frac{2j^2}{n} \right] \right\}$$

$$\leq e^{-2c} \sum_{j=2}^{\alpha n} \frac{n^{\frac{2j^2}{n}} 2^j}{j!}.$$

The remaining part of the proof consists in bounding the last sum. Setting $t_j = \frac{n^{\frac{2j^2}{n}} 2^j}{j!}$ we observe that the ratio between two consecutive t_j's is

$$\frac{t_{j+1}}{t_j} = \frac{2}{j+1} n^{(4j+2)/n} =: h(j+1).$$

The function $h(x) = \frac{2}{x} n^{\frac{4x-2}{n}}$, $x \in [3, \alpha n]$ is decreasing for $x \leq \bar{x} = \frac{n}{4 \log n}$ and increasing for $x \geq \bar{x}$ and thus attains its maximum value at the extremes. We have

$$h(3) \quad = \quad \tfrac{2}{3} n^{10/n} \qquad \rightarrow \quad \tfrac{2}{3} \text{ as } n \to \infty$$

$$h(\alpha n) \quad = \quad \tfrac{2}{\alpha n} n^{(4\alpha n - 2)/n} \quad \rightarrow \quad 0 \text{ as } n \to \infty$$

where in the second estimate we have used the fact that $\alpha < \frac{1}{4}$. We deduce that, for n sufficiently large, the ratio $\frac{t_{j+1}}{t_j}$ is smaller than $q < 1$ and thus

$$\sum_{j=2}^{\alpha n} t_j \leq t_2 \sum_{j=2}^{\alpha n} q^{j-2} \leq t_2 \frac{1}{1-q}.$$

Since $t_2 = 2n^{8/n} \to 2$ as $n \to \infty$ the proof is complete. $\qquad \square$

Step 7 *For every $\alpha \in (0, \frac{1}{4})$ there exists a positive constant a_2 (not depending on n) such that if $k = \frac{1}{2} n (\log n + c)$ with $c > 0$ then*

$$\sum_{j=\alpha n}^{n} \sum_{\lambda \in R_j} d_{2\lambda} \beta_\lambda^{2k} \leq a_2 e^{-2c}.$$

Proof By Steps 1 and 4 we have

$$\sum_{j=\alpha n}^{n} \sum_{\lambda \in R_j} d_{2\lambda} \beta_\lambda^{2k} \leq \sum_{j=\alpha n}^{n} \left(1 - \frac{j}{n}\right)^{2k} \frac{(2n)!}{(2n - 2j)! 2^j j!}.$$

If $k = \frac{1}{2} n (\log n + c)$, $c > 0$ the term which depends on k becomes

$$\left(1 - \frac{j}{n}\right)^{2k} = \left(1 - \frac{j}{n}\right)^{nc} \left(1 - \frac{j}{n}\right)^{n \log n};$$

as $j \geq \alpha n$ if n is sufficiently large we have

$$\left(1 - \frac{j}{n}\right)^{nc} \leq (1 - \alpha)^{nc} \leq e^{-2c}.$$

It remains to bound uniformly

$$\sum_{j=\alpha n}^{n} \left(1 - \frac{j}{n}\right)^{n \log n} \frac{(2n)!}{(2n - 2j)! 2^j j!}. \tag{11.9}$$

The ratio between two consecutive terms of the above sum is given by

$$\frac{1}{(j+1)} \left[1 - \frac{1}{n - j}\right]^{n \log n} (n - j)(2n - 2j - 1)$$

which is a decreasing function of j. For $j = \alpha n$ it equals

$$\left[1 - \frac{1}{n(1 - \alpha)}\right]^{n \log n} \frac{1}{\alpha n + 1} (n - j)(2n - 2j - 1) \leq \frac{2n^2}{\alpha n + 1} n^{-\frac{1}{1-\alpha}},$$

which goes to zero as n goes to $+\infty$ because $0 < \alpha < 1/4$. Thus we can estimate from above the sum (11.9) by n times the first term i.e. by

$$n \frac{(2n)!}{(2n - 2\alpha n)! 2^{\alpha n} (\alpha n)!} (1 - \alpha)^{n \log n};$$

that, by Stirling's formula, is bounded by

$$\gamma^n n^{\alpha n + \frac{1}{2}} n^{n \log(1 - \alpha)} \tag{11.10}$$

where γ is a suitable constant. Since $\alpha + \log(1 - \alpha) < 0$, the quantity (11.10) goes to zero as n goes to $+\infty$ and this ends the proof. $\square$

We consider now the contribution in the upper bound lemma given by the negative eigenvalues. We recall that if $\lambda = (\lambda_1, \lambda_2, \ldots, \lambda_r) \vdash n$, then $\lambda' = (\lambda_1', \lambda_2', \ldots, \lambda_s') \vdash n$ denotes the conjugate partition of λ. We denote by Q_h the set of all partitions λ of n such that $\lambda_1' = h$ and the corresponding eigenvalue β_λ is negative.

Step 8 If $\lambda \in \cup_{h \leq \frac{2}{3}n} Q_h$ and n is large, then $|\beta_\lambda| \leq \frac{1}{3}$.

Proof The smallest element in $\cup_{h \leq \frac{2}{3}n} Q_h$ with respect to $\trianglelefteq$, is the partition $\theta = \left(2^{\frac{n}{3}}, 1^{\frac{n}{3}}\right)$. We compute the corresponding eigenvalue:

$$
\begin{aligned}
\beta_\theta &= \frac{1}{n(n-1)} \left[\sum_{i=1}^{\frac{n}{3}} 2(2-i) + \sum_{i=\frac{n}{3}+1}^{\frac{2n}{3}} 1(1-i) \right] \\
&= \frac{1}{n(n-1)} \left[\frac{4n}{3} + \frac{4n}{3} - n - \frac{n(n+3)}{18} - \frac{2n(2n+3)}{18} \right] \\
&= \frac{-5n+21}{18(n-1)}.
\end{aligned}
$$

For n sufficiently large we deduce that $|\beta_{\theta_{2/3n}}| \leq \frac{1}{3}$. $\qquad$ $\square$

Step 9 *There exists a positive constant a_3 (not depending on n) such that if $k = \frac{1}{2}n(\log n + c)$ with $c > 0$ then*

$$
\sum_{h=1}^{\frac{2n}{3}} \sum_{\lambda \in Q_h} d_{2\lambda} \beta_\lambda^{2k} \leq a_3 e^{-2c}.
$$

Proof Using the previous step we have

$$
\begin{aligned}
\sum_{h=1}^{\frac{2n}{3}} \sum_{\lambda \in Q_h} d_{2\lambda} \beta_\lambda^{2k} &\leq \left(\frac{1}{3}\right)^{2k} \sum_{\lambda \vdash n} d_{2\lambda} \\
&= \left(\frac{1}{3}\right)^{2k} \frac{(2n)!}{n! 2^n} \\
&\qquad \text{by Stirling formula} \\
&\leq A \left(\frac{1}{3}\right)^{2k} \left(\frac{2n}{e}\right)^n \\
&\qquad \text{for } k = \frac{1}{2}n(\log n + c) \\
&\leq B \frac{n^n}{n^{n \log 3}} \left(\frac{2}{e}\right)^n e^{-2c} \\
&\leq a_3 e^{-2c}.
\end{aligned}
$$

$\square$

To establish the upper bound we are left to consider the remaining terms with negative eigenvalues.

Step 10 *There exists a positive constant a_4 (not depending on n) such that if $k = \frac{1}{2}n(\log n + c)$ with $c > 0$ then*

$$\sum_{h=\frac{2n}{3}}^{n} \sum_{\lambda \in Q_h} d_{2\lambda} \beta_\lambda^{2k} \leq a_4 e^{-2c}.$$

Proof With respect to $\trianglelefteq$ the smallest partition is (1^n). The corresponding eigenvalue is

$$\beta_{(1^n)} = \frac{1}{n(n-1)} \sum_{i=1}^{n}(1-i) = -\frac{1}{2}$$

which, by Step 2, implies

$$|\beta_\lambda| \leq 1/2, \quad \forall \lambda \in Q_h. \tag{11.11}$$

On the other hand we have, always for $\lambda \in Q_h$, that

$$d_{2\lambda} \leq \binom{2n}{2h}\binom{2h}{h} d_{2\overline{\lambda}} \tag{11.12}$$

where $\overline{\lambda} \vdash (n-h)$ is obtaining by deleting the first column of the diagram of λ. Indeed forming a 2λ standard tableau, we have $\binom{2n}{2h}$ ways to choose the elements in the first two columns and when we have selected these $2h$ elements we have $\binom{2h}{h}$ ways to choose among them the elements in the first column. Note that we do not guarantee that all the remaining h elements respect the tableau order and this explains the inequality. From (11.12) we deduce

$$\sum_{\lambda \in Q_h} d_{2\lambda} \leq \frac{(2n)!}{(2n-2h)!(h!)^2} \sum_{\overline{\lambda} \vdash n-h} d_{2\overline{\lambda}}$$

$$= \frac{(2n)!}{(2n-2h)!(h!)^2} \cdot \frac{(2n-2h)!}{2^{n-h}(n-h)!}$$

$$= \frac{(2n)!}{2^{n-h}(n-h)!(h!)^2}.$$

Using Stirling's formula we have that, if $h \geq 2/3n$, the latter term is bounded above by $\gamma^n n^{\frac{2}{3}n}$, for a suitable constant γ. Thus, taking into account (11.11) we deduce that for $k = \frac{1}{2}n(\log n + c)$

$$\sum_{h=2/3n}^{n} \sum_{\lambda \in Q_h} \beta_\lambda^{2k} d_{2\lambda} \leq \frac{n}{3} \cdot \left(\frac{1}{2}\right)^{2k} \sum_{\lambda \in Q_h} d_{2\lambda} \leq a_4 e^{-2c}.$$

It is clear that Step 6, Step 7, Step 9 and Step 10 give the proof of the theorem. $\square$

Now we proceed by giving the lower bound. The techniques we use are different from those used in the previous lower bounds. We adapt to this case the original method developed by Diaconis and Shahshahani [69]; see also the treatment of the random transposition model in Diaconis' book [55]. Our treatment follows quite closely Chapter 4 of Feller's monograph [87]. A variation of this method can be found in [64]. First we need the so-called principle of inclusion–exclusion. We have already given a proof of it in Section 8.4; in order to make this section selfcontained, we sketch another (less conceptual) proof.

Proposition 11.7.3 *Suppose that $A_1, A_2, \ldots, A_n$ are finite subsets of a finite set Ω. Then*

$$|A_1 \cup A_2 \cup \cdots \cup A_n| = \sum_{r=1}^{n} (-1)^{r-1} \sum_{1 \leq i_1 < i_2 < \cdots < i_r \leq n} |A_{i_1} \cap A_{i_2} \cap \cdots \cap A_{i_r}|.$$

$$(11.13)$$

Proof We prove the result by induction. For $n = 2$ it is just the trivial equality $|A_1 \cup A_2| = |A_1| + |A_2| - |A_1 \cap A_2|$. Suppose that (11.13) holds and let us prove it for $n + 1$. We have

$$|A_1 \cup A_2 \cup \cdots \cup A_n \cup A_{n+1}| = |A_1 \cup A_2 \cup \cdots \cup A_n| + |A_{n+1}|$$
$$- |(A_1 \cap A_{n+1}) \cup (A_2 \cap A_{n+1}) \cup \cdots \cup (A_n \cap A_{n+1})|$$

$$= \sum_{r=1}^{n} (-1)^{r-1} \sum_{1 \leq i_1 < i_2 < \cdots < i_r \leq n} |A_{i_1} \cap A_{i_2} \cap \cdots \cap A_{i_r}| + |A_{n+1}|$$

$$+ \sum_{r=1}^{n} (-1)^{r} \sum_{1 \leq i_1 < i_2 < \cdots < i_r \leq n} |A_{i_1} \cap A_{i_2} \cap \cdots \cap A_{i_r} \cap A_{n+1}|$$

$$= \sum_{r=1}^{n+1} (-1)^{r-1} \sum_{1 \leq i_1 < i_2 < \cdots < i_r \leq n+1} |A_{i_1} \cap A_{i_2} \cap \cdots \cap A_{i_r}|.$$

$\square$

We now apply the principle of inclusion–exclusion in our setting.

Lemma 11.7.4 *Fix the element $x_0 = \{\{1,2\}, \ldots \{2n-1, 2n\}\} \in X$ and let*

$$A = \{x \in X : \exists k \in \{1, 2, \ldots, n\} \text{ s.t. } \{2k-1, 2k\} \in x\}.$$

Then

$$|A| = \sum_{r=1}^{n} (-1)^{r-1} \frac{[2(n-r)]!}{2^{n-r}(n-r)!} \binom{n}{r}.$$

Proof For each $i \in \{1, 2, \ldots, n\}$ define $A_i = \{x \in X : \{2i-1, 2i\} \in x\}$ and for $1 \le i_1 < i_2 < \cdots < i_r \le n$ define

$$A_{i_1, i_2, \ldots, i_r} = \{x \in X : \{2i_1-1, 2i_1\}, \{2i_2-1, 2i_2\}, \ldots, \{2i_r-1, 2i_r\} \in x\}.$$

It is immediate to check that the cardinality of $A_{i_1, i_2, \ldots, i_r}$ depends only on the number r and equals the number of partitions of a $(2n - 2r)$-set consisting of two-elements subsets, namely $\frac{[2(n-r)]!}{2^{n-r}(n-r)!}$. Since $A_{i_1} \cap \ldots \cap A_{i_r} = A_{i_1, i_2, \ldots, i_r}$ and $A = A_1 \cup A_2 \cup \ldots \cup A_n$ by the method of inclusion–exclusion we have

$$|A| = \sum_{r=1}^{n} (-1)^{r-1} \sum_{i_1 < i_2 < \cdots < i_r} |A_{i_1, i_2, \ldots, i_r}| = \sum_{r=1}^{n} (-1)^{r-1} \frac{[2(n-r)]!}{2^{n-r}(n-r)!} \binom{n}{r}.$$

$\square$

Lemma 11.7.5 *Let π be the uniform distribution on X and A as in the previous lemma. Then*

$$\lim_{n \to \infty} \pi(A) = 1 - \frac{1}{\sqrt{e}}.$$

Proof By the previous lemma we have

$$1 - \pi(A) = 1 - \frac{|A|}{|X|} = \sum_{r=0}^{n} (-1)^r \frac{[2(n-r)]! 2^r (n!)^2}{[(n-r)!]^2 (2n)! r!}.$$

By the Stirling formula, for r fixed and $n \to \infty$ the term $\frac{[2(n-r)]!(n!)^2}{[(n-r)!]^2(2n)!}$ is asymptotical to $\frac{1}{2^{2r}}$. On the other hand

$$\frac{[2(n-r)]!(n!)^2}{[(n-r)!]^2(2n)!} \le 1 \tag{11.14}$$

thus, if we want to compute the following limit

$$\lim_{n \to \infty} \sum_{r=0}^{n} (-1)^r \frac{[2(n-r)]! 2^r (n!)^2}{[(n-r)!]^2 (2n)! r!}$$

it suffices to choose N sufficiently large such that $\sum_{r=N+1}^{\infty} \frac{2^r}{r!}$ is small and then take $n > N$ in such a manner that $\sum_{r=0}^{N} (-1)^r \frac{[2(n-r)]! 2^r (n!)^2}{[(n-r)!]^2 (2n)! r!}$ is close to $\sum_{r=0}^{N} \frac{(-1)^r}{2^r} \frac{1}{r!}$ which, for $N \to \infty$, tends to $\frac{1}{\sqrt{e}}$. $\square$

Theorem 11.7.6 *Let* π *and* $\nu_0^{(k)}$ *be as in Theorem 11.7.2. If* $k = \frac{1}{2}n(\log n - 2c)$ *and* $c > 0$ *we have*

$$\|\nu_0^{(k)} - \pi\|_{TV} \geq e^{-\frac{1}{2}} - \exp(-e^{2c}) + o(1) \text{ as } n \to \infty.$$

Proof First of all, if A is as in the previous lemmas, then $\nu_0^{(k)}(A)$ is larger than the probability that after k random switching there is a pair that has not been exchanged and this probability is equal to the probability p_k that, after having chosen k pairs of elements in $\{1, \ldots, n\}$ there exists an element that is not contained in any of the pairs chosen. Such a probability p_k can be computed using the inclusion–exclusion method: fixed r numbers between 1 and n the number of couples that one can form from the remaining $n - r$ numbers is $\frac{(n-r)(n-r-1)}{2}$. This implies that

$$
\begin{aligned}
p_k &= \frac{1}{\binom{n}{2}^k} \sum_{r=1}^{n} (-1)^r \binom{n}{r} \left[\frac{(n-r)(n-r-1)}{2} \right]^k \\
&= \sum_{r=1}^{n} (-1)^{r-1} \binom{n}{r} \left(1 - \frac{r}{n}\right)^k \left(1 - \frac{r}{n-1}\right)^k.
\end{aligned}
$$

To estimate from below this quantity we follow quite closely Chapter 4 in [87]. Set $[n]_r = n(n-1)\cdots(n-r+1)$. Clearly, $[n]_r \leq n^r$ and therefore, using $1 - x \leq e^{-x}$, we get

$$[n]_r \left(1 - \frac{r}{n-1}\right)^k \left(1 - \frac{r}{n}\right)^k \leq \left(ne^{-\frac{2k}{n}}\right)^r.$$

Moreover, using $x\exp(1 - x) \leq 1$, one obtains that (for $x = \frac{n-1}{n-r-1}$)

$$
\begin{aligned}
[n]_r \left(1 - \frac{r}{n-1}\right)^k \left(1 - \frac{r}{n}\right)^k &\geq (n-1-r)^r \left(\frac{n-1-r}{n-1}\right)^{2k} \\
&= (n-1)^r \left(1 - \frac{r}{n-1}\right)^{2k+r} \\
&\geq \left[(n-1)e^{-\frac{2k+r}{n-1-r}}\right]^r.
\end{aligned}
$$

The ratio between the extremes of these inequalities is equal to

$$
\frac{\left(ne^{\frac{-2k}{n}}\right)^r}{\left((n-1)e^{-\frac{2k+r}{n-1-r}}\right)^r} = \left\{\frac{n}{n-1} \exp\left[\frac{2kr + rn + 2k}{n(n-1-r)}\right]\right\}^r
$$

which, for $k = \frac{1}{2}n \log n - cn$ with $c > 0$, becomes

$$\left\{ \frac{n}{n-1} \exp \left[\frac{r + \log n - 2c + r \log n - 2cr}{(n-1-r)} \right] \right\}^r.$$

This quantity, for r fixed, tends to 1 as $n \to \infty$. Moreover, always for $k = \frac{1}{2}n \log n - cn$

$$[n]_{\underline{r}} \left(1 - \frac{r}{n-1}\right)^k \left(1 - \frac{r}{n}\right)^k \le \left(ne^{-\frac{2k}{n}}\right)^r = e^{2rc}.$$

Thus, if n is sufficiently large we have

$$\sum_{r=0}^{n} (-1)^r \binom{n}{r} \left(1 - \frac{r}{n}\right)^k \left(1 - \frac{r}{n-1}\right)^k \approx \sum_{r=0}^{n} (-1)^r \frac{e^{2rc}}{r!} \approx \exp(-e^{2c}).$$

More generally, setting $\lambda = ne^{-\frac{2k}{n}}$ if $k, n \to \infty$ in such a manner that λ remains bounded, then $p_k \to e^{-\lambda}$. Thus $\nu_0^{(k)}(A) \ge 1 - \exp(-e^{2c}) + o(1)$ where $o(1)$ tends to zero as $n \to \infty$. Finally, we have

$$|\pi(A) - \nu_0^{(k)}(A)| = \nu_0^{(k)}(A)) - \pi(A) \ge e^{-\frac{1}{2}} - \exp(-e^{2c}) + o(1)$$

which ends the proof. $\qquad\square$

Exercise 11.7.7 Use the same techniques to derive the lower bound for the random transposition model (Theorem 10.7.2).

Appendix 1
The discrete trigonometric transforms

Let (X, π) be a finite probability space. Suppose that there is an involution on X, that is a bijection $\sigma : X \to X$ such that $\sigma^2 = id$. Define an action of σ on $L(X)$ by setting

$$(\sigma f)(x) = f(\sigma x)$$

for all $x \in X$ and $f \in L(X)$. Suppose that the probability measure π is σ-invariant, that is $\pi(\sigma x) = \pi(x)$ for all $x \in X$.

We can decompose the set X in the following way: set $X_0 = \{x \in X : \sigma(x) = x\}$, then choose $X_1 \subset X \setminus X_0$ in such a way that $X = X_{-1} \coprod X_0 \coprod X_1$ and X_{-1} is the image of X_1 under σ. Clearly X_1 and X_{-1} are not uniquely determined.

We can define

$$V = \{f \in L(X) : \sigma f = f\} \equiv \{f \in L(X) : f(\sigma x) = f(x), \ \forall x \in X\}$$

the space of σ-even functions and

$$W = \{f \in L(X) : \sigma f = -f\} \equiv \{f \in L(X) : f(\sigma x) = -f(x), \ \forall x \in X\}$$

the space of σ-odd functions. Clearly, for any $f \in L(X)$ we have

$$f = \frac{f + \sigma f}{2} + \frac{f - \sigma f}{2} \in V + W.$$

It is easy to show that the decomposition $L(X) = V \oplus W$ is orthogonal with respect to $\langle \cdot, \cdot \rangle_\pi$ (recall that we have assumed the σ-invariance of π).

Moreover the maps

$$\begin{array}{ccc} V & \to & L(X_0 \coprod X_1) \\ f & \mapsto & f_{|X_0 \coprod X_1} \end{array}$$

and

$$W \rightarrow L(X_1)$$
$$f \mapsto f_{|X_1},$$

where the symbol $_|$ denotes the restriction, are natural isomorphisms of vector spaces (a σ-even (resp. -odd) function is determined by its values on $X_0 \coprod X_1$ (resp. X_1)). If $\widetilde{\pi}$ is defined on $X_0 \coprod X_1$ by setting

$$\widetilde{\pi}(x) = \begin{cases} \pi(x) & \text{if} \quad x \in X_0 \\ 2\pi(x) & \text{if} \quad x \in X_1 \end{cases}$$

and $\overline{\pi}$ is defined on X_1 by setting $\overline{\pi}(x) = 2\pi(x)$ for $x \in X_1$, then we have

$$\langle f_1, f_2 \rangle_\pi = \langle f_{1|X_0 \coprod X_1}, f_{2|X_0 \coprod X_1} \rangle_{\widetilde{\pi}} \quad \forall f_1, f_2 \in V$$

and

$$\langle f_1, f_2 \rangle_\pi = \langle f_{1|X_1}, f_{2|X_1} \rangle_{\overline{\pi}} \quad \forall f_1, f_2 \in W.$$

Proposition A1.0.8 *Let T be a linear operator on $L(X)$ represented by the matrix $(t(x,y))_{x,y \in X}$. Then*

(i) *T is selfadjoint with respect to $\langle \cdot, \cdot \rangle_\pi$ if and only if $\pi(x)t(x,y) = \pi(y)\overline{t(y,x)}$ for all $x, y \in X$;*

(ii) *T is σ-invariant, that is $T\sigma = \sigma T$, if and only if $t(\sigma x, \sigma y) = t(x,y)$ for all $x, y \in X$; moreover this condition is equivalent to the T-invariance of the space V and W.*

Proof (i) is an easy exercise (compare also with the proof of Proposition 1.5.3).

(ii) Assume that V and W are T-invariant. Then, for $f \in L(X)$, we have $T(f + \sigma f) \in V$ and $T(f - \sigma f) \in W$, that is

$$\begin{cases} T(f + \sigma f) = \sigma T(f + \sigma f) \\ T(f - \sigma f) = -\sigma T(f - \sigma f). \end{cases}$$

Subtracting the second equation from the first, one gets that $\sigma T f = T\sigma f$. $\qquad \square$

In the following proposition, under the assumption of the symmetry with respect to σ, the spectral analysis of T is reduced to the spectral analysis of two operators on $L(X_0 \coprod X_1)$ and $L(X_1)$. The easy proof is left as an exercise.

Proposition A1.0.9 *Let T be a linear operator on $L(X)$. Suppose that T is selfadjoint and σ-invariant. Then we have*

(i) *A function $f \in V$ satisfies $Tf = \lambda f$ if and only if $\widetilde{T}f(x) = \lambda f(x)$ for all $x \in X_0 \coprod X_1$, where $\widetilde{T}$ is the linear operator on $L(X_0 \coprod X_1)$ associated with the matrix*

$$\widetilde{t}(x,y) = \begin{cases} t(x,y) & \text{if } y \in X_0 \\ t(x,y) + t(x,\sigma y) & \text{if } y \in X_1; \end{cases}$$

moreover $\widetilde{T}$ is selfadjoint with respect to $\widetilde{\pi}$.

(ii) *A function $f \in W$ satisfies $Tf = \lambda f$ if and only if $\overline{T}f(x) = \lambda f(x)$ for all $x \in X_1$, where $\overline{T}$ is the linear operator on $L(X_1)$ associated with the matrix*

$$\overline{t}(x,y) = t(x,y) - t(x,\sigma y), \quad x,y \in X_1;$$

moreover $\overline{T}$ is selfadjoint with respect to $\overline{\pi}$.

We want to apply Proposition A1.0.9 to get the so-called *discrete trigonometric transforms* (DTT). These transforms are defined on the space $L(P_n)$ where P_n is the path $\{0,1,\ldots,n-1\}$. For each transform we have a positive function π defined on P_n and a set of functions $S_0, S_1, \ldots, S_{n-1}$ that are orthogonal with respect to $\langle \cdot, \cdot \rangle_\pi$. Therefore each of these transforms has the form, for $f \in L(P_n)$,

$$\widehat{f}(h) = \sum_{k=0}^{n-1} f(k)S_h(k)\pi(k)$$

associated with an inversion formula

$$f = \sum_{h=0}^{n-1} \frac{1}{\|S_h\|_\pi^2} \widehat{f}(h)S_h.$$

The set $S_0, S_1, \ldots, S_{n-1}$ has the remarkable property to be a set of eigenvectors of the linear operator on $L(P_n)$ represented by the matrix

$$\frac{1}{2}\begin{pmatrix} \beta_0 & \beta_1 & & & & \\ 1 & 0 & 1 & & & \\ & 1 & 0 & 1 & & \\ & & \ddots & \ddots & \ddots & \\ & & & 1 & 0 & 1 \\ & & & & \beta_2 & \beta_3 \end{pmatrix} \quad (1.1)$$

where $\beta_0, \beta_1, \beta_2$ and β_3 are coefficients that depend on the particular transform.

Altogether there are 16 matrices (and the corresponding transforms) and in the following table we name them and we specify the corresponding choice of the parameters β's (DCT means *discrete cosine transform*; DST *discrete sine transform*).

Table A1.1.

(β_2, β_3) / (β_0, β_1)	(2,0)	(1,0)	(1,1)	(1,−1)
(0, 2)	DCT-1	DCT-3	DCT-5	DCT-7
(0, 1)	DST-3	DST-1	DST-7	DST-5
(1, 1)	DCT-6	DCT-8	DCT-2	DCT-4
(−1, 1)	DST-8	DST-6	DST-4	DST-2

For instance, the DCT-1 is obtained by considering an orthonormal basis for the matrix

$$\frac{1}{2}\begin{pmatrix} 0 & 2 & & & & \\ 1 & 0 & 1 & & & \\ & 1 & 0 & 1 & & \\ & & \ddots & \ddots & \ddots & \\ & & & 1 & 0 & 1 \\ & & & & 2 & 0 \end{pmatrix}.$$

The eigenvalue problem for the matrix (1.1) may be also written in the form

$$\begin{cases} \frac{1}{2}\left(\beta_0 f(0) + \beta_1 f(1)\right) = \lambda f(0) \\ \frac{1}{2}\left(f(k-1) + f(k+1)\right) = \lambda f(k) \text{ for } k = 1, 2, \ldots, n-2 \\ \frac{1}{2}\left(\beta_0 f(0) + \beta_1 f(1)\right) = \lambda f(n-1). \end{cases} \quad (1.2)$$

In the system (1.2) the central row represents the eigenvalue problem for the *discrete Laplace operator* $(\Delta f)(k) = \frac{f(k-1)+f(k+1)}{2}$ while the first and the last row are *boundary conditions*.

In the following tables we give the value of $S_k(h)$ for each trigonometric transform, together with the corresponding eigenvalue λ_h.

The fact that the functions S_h, shown in the above tables, solve the eigenvalue problem (1.2) may be proved directly by means of elementary trigonometry (see Exercise A1.0.10). In what follows, we will show how

Table A1.2.

	$S_h(k)$	λ_h
DCT-1	$\cos hk \dfrac{\pi}{n-1}$	$\cos \dfrac{h\pi}{n-1}$
DCT-2	$\cos h\left(k+\dfrac{1}{2}\right)\dfrac{\pi}{n}$	$\cos \dfrac{h\pi}{n}$
DCT-3	$\cos\left(h+\dfrac{1}{2}\right)k\dfrac{\pi}{n}$	$\cos \dfrac{\left(h+\frac{1}{2}\right)\pi}{n}$
DCT-4	$\cos\left(h+\dfrac{1}{2}\right)\left(k+\dfrac{1}{2}\right)\dfrac{\pi}{n}$	$\cos \dfrac{\left(h+\frac{1}{2}\right)\pi}{n}$
DCT-5	$\cos hk \dfrac{\pi}{n-\frac{1}{2}}$	$\cos \dfrac{h\pi}{n-\frac{1}{2}}$
DCT-6	$\cos h\left(k+\dfrac{1}{2}\right)\dfrac{\pi}{n-\frac{1}{2}}$	$\cos \dfrac{h\pi}{n-\frac{1}{2}}$
DCT-7	$\cos\left(h+\dfrac{1}{2}\right)k\dfrac{\pi}{n-\frac{1}{2}}$	$\cos \dfrac{\left(h+\frac{1}{2}\right)\pi}{n-\frac{1}{2}}$
DCT-8	$\cos\left(h+\dfrac{1}{2}\right)\left(k+\dfrac{1}{2}\right)\dfrac{\pi}{n+\frac{1}{2}}$	$\cos \dfrac{\left(h+\frac{1}{2}\right)\pi}{n+\frac{1}{2}}$

to obtain the formulæ for the S_h's by a repeated application of Proposition A1.0.9. We start by applying it to the simple random walk on C_n, that will be seen as the set of integers mod n.

DST-1 Consider the involution

$$\sigma: \quad \begin{aligned} C_{2n+2} &\rightarrow C_{2n+2} \\ k &\mapsto 2n-k, \end{aligned}$$

which fixes n and $2n+1$ and switches $n-k$ with $n+k$, for $k=1,2,\ldots,n$ (that is, it is the reflection of C_{2n} through the axis from n and $2n+1$). Therefore $X_0 = \{n, 2n+1\}$ and we can take $X_1 = \{0, 1, \ldots, n-1\}$. If we apply (ii) in Proposition A1.0.9 to the simple random walk on C_{2n+2} we get the following eigenvalue problem on X_1:

$$\begin{cases} \frac{1}{2}f(1) = \lambda f(0) \\ \frac{1}{2}\left(f(k-1)+f(k+1)\right) = \lambda f(k) \text{ for } k=1,2,\ldots,n-2 \\ \frac{1}{2}f(n-2) = \lambda f(n-1). \end{cases} \quad (1.3)$$

Table A1.3.

	$S_h(k)$	λ_h
DST-1	$\sin(h+1)(k+1)\dfrac{\pi}{n+1}$	$\cos\dfrac{(h+1)\pi}{n+1}$
DST-2	$\sin(h+1)\left(k+\dfrac{1}{2}\right)\dfrac{\pi}{n}$	$\cos\dfrac{(h+1)\pi}{n}$
DST-3	$\sin\left(h+\dfrac{1}{2}\right)(k+1)\dfrac{\pi}{n}$	$\cos\dfrac{\left(h+\frac{1}{2}\right)\pi}{n}$
DST-4	$\sin\left(h+\dfrac{1}{2}\right)(k+\dfrac{1}{2})\dfrac{\pi}{n}$	$\cos\dfrac{\left(h+\frac{1}{2}\right)\pi}{n}$
DST-5	$\sin(h+1)(k+1)\dfrac{\pi}{n+\frac{1}{2}}$	$\cos\dfrac{(h+1)\pi}{n+\frac{1}{2}}$
DST-6	$\sin(h+1)\left(k+\dfrac{1}{2}\right)\dfrac{\pi}{n+\frac{1}{2}}$	$\cos\dfrac{(h+1)\pi}{n+\frac{1}{2}}$
DST-7	$\sin\left(h+\dfrac{1}{2}\right)(k+1)\dfrac{\pi}{n+\frac{1}{2}}$	$\cos\dfrac{\left(h+\frac{1}{2}\right)\pi}{n+\frac{1}{2}}$
DST-8	$\sin\left(h+\dfrac{1}{2}\right)\left(k+\dfrac{1}{2}\right)\dfrac{\pi}{n-\frac{1}{2}}$	$\cos\dfrac{\left(h+\frac{1}{2}\right)\pi}{n-\frac{1}{2}}$

Indeed the matrix $\bar{t}(x,y)$ in (ii) Proposition A1.0.9 now reduces simply to $t(x,y)$ with $x,y \in X_1$, because in this case $t(x,\sigma y) = 0$ if $x,y \in X_1$. Note that $\overline{T}f(0) = \frac{1}{2}f(1)$ because $2n+1 \notin X_1$.

Observe that (1.3) is nothing but the eigenvalue problem for DST-1. To solve it, write the eigenfunctions of the simple random walk Q on C_{2n+2} in the form $e_h(k) = \omega^{(h+1)(k+1)}$ with $\omega = \exp\left(\frac{\pi i}{n+1}\right)$. Then

$$L(C_{2n+2}) = \langle e_{-1}\rangle \oplus \langle e_0, e_{2n}\rangle \oplus \cdots \oplus \langle e_{n-1}, e_{n+1}\rangle \oplus \langle e_n\rangle.$$

where $\langle v_0, v_1, \ldots, v_k\rangle$ denotes the linear span of the vectors $v_0, v_1, \ldots, v_k$ and should not be confused with the scalar product $\langle \cdot, \cdot \rangle$.

The functions e_{-1} and e_n are even with respect to σ; moreover $\langle e_h,$ $e_{2n-h}\rangle$ for $h = 0, 1, \ldots, n-1$, is the eigenspace of Q corresponding to the eigenvalue $\lambda_h = \cos\frac{(h+1)\pi}{n+1}$ and contains the σ-odd function

$$e_h(k) - e_h(2n-k) = \omega^{(h+1)(k+1)} - \omega^{-(h+1)(k+1)} = 2i\sin\frac{(h+1)(k+1)\pi}{n+1}.$$

Omitting the factor $2i$, we obtain the formula for the DST-1 in Table A1.3 that is, the coefficients of DST-1 is just the imaginary part of e_h.

<u>DCT-1</u> Consider the involution

$$\sigma: \begin{array}{ccc} C_{2n-2} & \to & C_{2n-2} \\ k & \mapsto & -k. \end{array}$$

Then applying (i) in Proposition A1.0.9 to the simple random walk on C_{2n-2} one gets the DCT-1, by taking $X_1 = \{1, 2, \ldots, n-2\}$ and therefore $X_0 \coprod X_1 = \{0, 1, \ldots, n-1\}$. This has already been done in Section 2.6 (but on $\{0, 1, \ldots, n\}$).

In other words the formulæ for the DCT-1 and DST-1 are obtained from the DFT by restriction to the space of odd and even functions, respectively (recall that even and odd refer to a symmetry of C_n that fixes two opposite points).

<u>DCT-2</u> Consider the involution

$$\sigma: \begin{array}{ccc} C_{2n} & \to & C_{2n} \\ k & \mapsto & 2n-1-k, \end{array}$$

which does not have fixed points (but it fixes the edges $\{2n-1, 0\}$ and $\{n-1, n\}$).

Taking $X_1 = \{0, 1, \ldots, n-1\}$ and applying (i) in Proposition A1.0.9 to the simple random walk on C_{2n}, we get the following eigenvalue problem on X_1:

$$\begin{cases} \frac{1}{2}\left(f(0) + f(1)\right) = \lambda f(0) \\ \frac{1}{2}\left(f(k-1) + f(k+1)\right) = \lambda f(k) \text{ for } k = 1, 2, \ldots, n-2 \\ \frac{1}{2}\left(f(n-2) + f(n-1)\right) = \lambda f(n-1) \end{cases}$$

which is the problem corresponding to DCT-2. The problem may be solved by considering the eigenfunction of the simple random walk on C_{2n} that are even with respect to σ. But if $\omega = \exp\left(\frac{\pi i}{n}\right)$ then

$$\omega^{hk} + \omega^{h(2n-1-k)} = 2\omega^{-\frac{h}{2}} \cos\left[\left(k + \frac{1}{2}\right) h \frac{\pi}{n}\right]$$

and one gets the formula for DCT-2 in Table 2.

<u>DST-2</u> It may be obtained by mean of the same involution σ on C_{2n} used for DCT-2, but now invoking (ii) in Proposition A1.0.9. In other

words, DCT-2 and DST-2 are obtained by restricting the DFT to the space of even and odd functions, respectively (odd and even with respect to a symmetry of C_n that fixes two edges).

The remaining discrete transforms may be obtained from DCT-1, DCT-2, DST-1, and DST-2 by applying Proposition A1.0.9 to the following involution on the path $P_n = \{0, 1, \ldots, n-1\}$

$$\sigma: \quad P_n \quad \to \quad P_n$$
$$k \quad \mapsto \quad n - 1 - k.$$

Clearly σ is the reflection with respect to the central point when n is odd; with respect to the central edge when n is even. In any case, we can take $X_1 = \{0, 1, \ldots, \left[\frac{n-2}{2}\right]\}$ and we have

$$X_0 = \begin{cases} \emptyset & \text{if } n \text{ is even} \\ \{\frac{n-1}{2}\} & \text{if } n \text{ is odd.} \end{cases}$$

For instance if we apply (i) of Proposition A1.0.9 to DCT-2 on P_{2n-1}, we get DCT-6, while if we apply (ii) of Proposition A1.0.9 to DCT-2 on P_{2n+1} we get DCT-8.

In other words, if we want to get a transform corresponding to $\beta_0, \beta_1, \beta_2, \beta_3$ in Table 1, we can take the transform corresponding to $\beta_0, \beta_1, \beta_1, \beta_0$ in the same table and then

- if $(\beta_2, \beta_3) = (2, 0)$ we can use the symmetry σ on P_{2n-1} and restrict to the even functions;
- if $(\beta_2, \beta_3) = (1, 0)$ we can use the symmetry σ on P_{2n+1} and restrict to the odd functions;
- if $(\beta_2, \beta_3) = (1, 1)$ we can use the symmetry σ on P_{2n} and restrict to the even functions;
- if $(\beta_2, \beta_3) = (1, -1)$ we can use the symmetry σ on P_{2n} and restrict to the odd functions.

Exercise A1.0.10 Use the well-known trigonometric formulæ $\cos \alpha \cos \beta = 2 \cos \frac{\alpha - \beta}{2} \cos \frac{\alpha + \beta}{2}$ and $\sin \alpha \sin \beta = 2 \cos \frac{\alpha - \beta}{2} \sin \frac{\alpha + \beta}{2}$ to derive the eigenvalues in Table 2 and Table 3 (but verify the boundary conditions!).

Exercise A1.0.11 Verify all the details of the derivation of the DTT's by means of Proposition A1.0.9. Show that no other transforms may be obtained following this procedure (if we continue to apply Proposition A1.0.9 to the DTT's no system (1.2) different from those in Table 1 can appear).

In the following table we report the value of $\pi(0), \pi(1), \ldots, \pi(n-1)$ for all the DTT's. We have used the formulæ for $\tilde{\pi}$ and $\overline{\pi}$ in, but in many cases we have divided the value by 2 or 4.

Table A1.4.

		$\pi(0)$	$\pi(k)$ $k = 1, \ldots, n-2$	$\pi(n-1)$
DCT-1		1	2	1
DCT-2	DST-1			
DCT-4	DST-2			
DCT-8	DST-4	1	1	1
DST-5	DST-6			
DST-7				
DCT-3	DCT-5	1	2	2
DCT-7				
DCT-6	DST-3	2	2	1
DST-8				

Now we give a table with the value of $\|S_h\|_\pi^2$ for each DTT.

Exercise A1.0.12 Prove the following trigonometric identities

$$\sum_{k=0}^{n-1} \cos^2[(k+\epsilon)\alpha] = \frac{n}{2} + \frac{\sin(n\alpha)\cos[(n+2\epsilon-1)\alpha]}{2\sin\alpha}$$

and

$$\sum_{k=0}^{n-1} \sin^2[(k+\epsilon)\alpha] = \frac{n}{2} - \frac{\sin(n\alpha)\cos[(n+2\epsilon-1)\alpha]}{2\sin\alpha}$$

for any $\epsilon \in \mathbb{R}$ and $\alpha \neq h\pi$.

Use (1) to prove the formulae in Table 5.

Exercise A1.0.13 Study the effect of transposition in Table 1 (for instance, if $\widetilde{S}_h(k)$ are the coefficients of DCT-3 and $\overline{S}_h(k)$ are the coefficients of DST-3 then $\widetilde{S}_h(n-1-k) = (-1)^{h+1}\overline{S}_h(k)$).

If X is the graph with edge set E, its adjacency operator is

$$[Af](x) = \sum_{y \sim x} f(y)$$

Table A1.5.

	$\|S_0\|_\pi^2$	$\|S_h\|_\pi^2$ $h = 1, \ldots, n-2$	$\|S_{n-1}\|_\pi^2$
DCT-1	$2n - 2$	$n - 1$	$2n - 2$
DCT-2	n	$\dfrac{n}{2}$	$\dfrac{n}{2}$
DCT-3 DST-3	n	n	n
DCT-4 DST-4	$\dfrac{n}{2}$	$\dfrac{n}{2}$	$\dfrac{n}{2}$
DCT-5 DCT-6	$2n - 1$	$\dfrac{2n - 1}{2}$	$\dfrac{2n - 1}{2}$
DCT-7 DST-8	$\dfrac{2n - 1}{2}$	$\dfrac{2n - 1}{2}$	$2n - 1$
DST-6 DCT-8 DST-7 DST-5	$\dfrac{2n + 1}{4}$	$\dfrac{2n + 1}{4}$	$\dfrac{2n + 1}{4}$
DST-1	$\dfrac{n + 1}{2}$	$\dfrac{n + 1}{2}$	$\dfrac{n + 1}{2}$
DST-2	$\dfrac{n}{2}$	$\dfrac{n}{2}$	n

for all $f \in L(X)$ and $x \in X$.

We recall that its Markov operator is

$$[Mf](x) = \frac{1}{\deg(x)} \sum_{y \sim x} f(y)$$

for all $f \in L(X)$ and $x \in X$. Clearly, if X is regular of degree d, we simply have $M = \frac{1}{d}A$. If X is not regular, then A and M give rise to two different notions of spectrum: the adjacency spectrum and the Markov spectrum respectively (the latter coincides with the spectrum of the simple random walk on X; see Section 1.8).

Exercise A1.0.14 Use DST-1 to show that the adjacency spectrum of the path P_n:

Figure A1.1.

is $\{\cos \frac{(h+1)\pi}{n+1} : h = 0, 1, \ldots, n-1\}$.

Compare with the Markov spectrum of P_{n+1} that was computed in Theorem 2.6.1.

Exercise A1.0.15

(1) Use DCT-2 to show that the spectrum of the simple random walk on the graph constituted by the path P_{n-1} with a loop in 0 and in $n-1$ is $\{\cos \frac{k\pi}{n} : h = 0, 1, \ldots, n-1\}$.

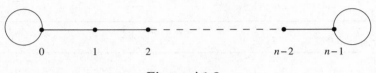

Figure A1.2.

(2) Use DCT-6 to show that the spectrum of the simple random walk on the graph constituted by the path P_{n-1} with a loop in 0 is $\{\cos \frac{h\pi}{n-\frac{1}{2}} : h = 0, 1, \ldots, n-1\}$.

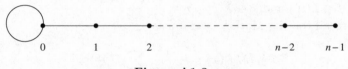

Figure A1.3.

(3) Show that the random walks in (1) and (2) can be obtained by lumping the simple random walk on the discrete circle.

For the solutions of these exercises see the books by Feller [87] and by Karlin and Taylor [136]. There is also a huge literature on the DFT's, the DTT's and other transforms and their algebraic theory and their numerical applications; see for instance [226], [31] and [72]. For the DTT's see also [121], [178] and [214]

In the last part of this appendix we use the DTT's to analyze a random walk on the homogeneous rooted tree $\mathbb{T}_{q,n}$ (of depth n and degree q). It is taken from [191]. See also [190]. Besides the basic notation and definitions introduced in Chapter 7, this part is essentially selfcontained. Indeed, although some results can be deduced from the theory developed in Chapters 7 and 8, the presentation below is independent and requires only elementary linear algebra.

Let Σ^j be the j-th level of the tree $\mathbb{T}_{q,n}$ and $L(\Sigma^j)$ be the corresponding space of complex valued functions on Σ^j. We define, for $j = 1, 2, \ldots, n$, an operator $d : L(\Sigma^j) \to L(\Sigma^{j-1})$ by setting, for $f \in L(\Sigma^j)$ and $x_1 x_2 \cdots x_{j-1} \in \Sigma^{j-1}$,

$$df(x_1, x_2, \ldots, x_{j-1}) = \sum_{x_j=0}^{q-1} f(x_1, x_2, \ldots, x_{j-1}, x_j).$$

Moreover we set $df \equiv 0$ if $f \in L(\Sigma^0) \equiv L(\emptyset)$.

The adjoint of the operator d is the operator $d^* : L(\Sigma^{j-1}) \to L(\Sigma^j)$ defined by setting, for $f \in L(\Sigma^{j-1})$ and $x_1 x_2 \cdots x_j \in \Sigma^j$,

$$d^* f(x_1, x_2, \ldots, x_j) = f(x_1, x_2, \ldots, x_{j-1}) \quad \text{for } j = 1, \ldots, n$$

and $d^* f = 0$ if $f \in L(\Sigma^n)$. Indeed, for $j = 1, 2, \ldots, n$, $f \in L(\Sigma^j)$ and $g \in L(\Sigma^{j-1})$ we have

$$\langle df, g \rangle = \sum_{x_1=0}^{q-1} \sum_{x_2=0}^{q-1} \cdots \sum_{x_{j-1}=0}^{q-1} df(x_1, x_2, \ldots, x_{j-1}) \overline{g(x_1, x_2, \ldots, x_{j-1})}$$

$$= \sum_{x_1=0}^{q-1} \sum_{x_2=0}^{q-1} \cdots \sum_{x_{j-1}=0}^{q-1} \left(\sum_{x_j=0}^{q-1} f(x_1, x_2, \ldots, x_j) \overline{g(x_1, x_2, \ldots, x_{j-1})} \right)$$

$$= \sum_{x_1=0}^{q-1} \sum_{x_2=0}^{q-1} \cdots \sum_{x_{j-1}=0}^{q-1} \left(\sum_{x_j=0}^{q-1} f(x_1, x_2, \ldots, x_j) \overline{d^* g(x_1, x_2, \ldots, x_{j-1}, x_j)} \right)$$

$$= \langle f, d^* g \rangle.$$

Moreover we have the obvious relation

$$dd^* = qI, \tag{1.4}$$

where I is the identity operator.

If we define $W_j = L(\Sigma^j) \cap \ker d$ (and $W_0 \equiv \mathbb{C}$), by the standard orthogonal decomposition $V = \ker T \oplus Im(T^*)$, where $T : V \to W$ is a linear map, we have

$$L(\Sigma^j) = W_j \oplus d^*(L(\Sigma^{j-1})) = W_j \oplus d^*(W_{j-1}) \oplus (d^*)^2(L(\Sigma^{j-2})) = \cdots =$$
$$= W_j \oplus d^*(W_{j-1}) \oplus \cdots \oplus (d^*)^{n-j} W_0.$$

Exercise A1.0.16 Prove that the above decomposition is orthogonal.

We now describe the random walk on the tree. The transition rule is the following: one passes from an internal vertex to its father with probability $\frac{1}{2}$ and with equal probability (i.e. $\frac{1}{2q}$) to one of its children; one passes from the root to one of its children with probability equal to $\frac{1}{q}$ and from a leaf to its father with probability 1. Looking only at the levels of the tree we have that one bounces back at the root and at the leaves while for the internal levels one goes up with probability $\frac{1}{2}$ and one goes down with the same probability.

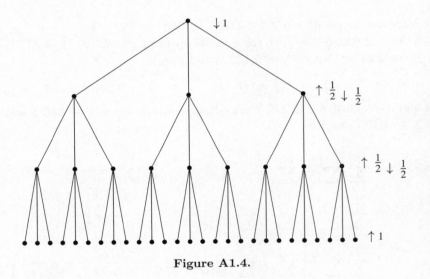

Figure A1.4.

The transition operator M, associated with the above Markov chain, can be expressed in terms of the operators d and d^*. More precisely, if, for $f \in L(\mathbb{T}_{q,n})$, we denote by f_j its restriction to $L(\Sigma^j)$, we have that

$$(Mf)_j = \frac{1}{2q}df_{j+1} + \frac{1}{2}d^* f_{j-1} \qquad \text{for } j = 1, 2, \ldots, n-1,$$

$$(Mf)_0 = \frac{1}{q}df_1$$

and finally

$$(Mf)_n = d^* f_{n-1}.$$

In virtue of (1.4) the space $U_k = W_k \oplus d^* W_k \oplus \cdots \oplus (d^*)^{n-k} W_k$ is clearly M-invariant. Set $f = \alpha_0^k g + \alpha_1^k d^* g + \cdots + \alpha_{n-k}^k (d^*)^{n-k} g$, with $g \in W_k$ and $\alpha_0^k, \alpha_1^k, \ldots \alpha_{n-k}^k$ undetermined coefficients. Note that $\alpha_j^k (d^*)^j g$ is the restriction of f to $L(\Sigma^{j+k})$. Then, for $k \geq 1$,

$$Mf = \frac{\alpha_0^k}{2q} dg + \frac{\alpha_1^k}{2q} dd^* g + \left(\frac{\alpha_1^k}{2q} d(d^*)^2 g + \frac{\alpha_0^k}{2} d^* g \right) + \cdots +$$
$$+ \left(\frac{\alpha_{n-k}^k}{2q} d(d^*)^{n-k} g + \frac{\alpha_{n-k-2}^k}{2} d^* (d^*)^{n-k-2} g \right)$$
$$+ \alpha_{n-k-1}^k d^* (d^*)^{n-k-1} g.$$

The jth term of this sum belongs to $L(\Sigma^{j+k})$ for $j = -1, 0, 1, \ldots, n-k$; moreover, for $k = 1$, the coefficient of dg in the first line is $\frac{\alpha_0^k}{q}$ (but this is irrelevant, as $dg = 0$). Thus, by (1.4), the above sum becomes

$$\frac{\alpha_1^k}{2} g + \left[\frac{\alpha_0^k}{2} + \frac{\alpha_2^k}{2} \right] d^* g + \cdots$$
$$+ \left[\frac{\alpha_{n-k}^k}{2} + \frac{\alpha_{n-k-2}^k}{2} \right] (d^*)^{n-k-1} g + \alpha_{n-k-1}^k (d^*)^{n-k} g.$$

Therefore f is an eigenvector for M if and only if $(\alpha_0^k, \alpha_1^k, \ldots, \alpha_{n-k}^k)$ is an eigenvector of the $(n-k+1) \times (n-k+1)$ matrix

$$\begin{pmatrix} 0 & \frac{1}{2} & 0 & & \\ \frac{1}{2} & 0 & \frac{1}{2} & & \\ 0 & \frac{1}{2} & 0 & \frac{1}{2} & \\ & \ddots & \ddots & \ddots & \ddots \\ & & & 1 & 0 \end{pmatrix}.$$

This is DST-3. Therefore the eigenvectors are given by

$$f = \sin\left(\frac{(h + \frac{1}{2})}{n - k + 1} \pi \right) g + \sin\left(2 \frac{(h + \frac{1}{2})}{n - k + 1} \pi \right) (d^*) g + \cdots +$$
$$+ \cdots + \sin\left((n - k) \frac{(n - k + \frac{1}{2})}{n - k + 1} \pi \right) (d^*)^{n-k} g$$

and the corresponding eigenvalues are $\lambda_h^k = \cos\left(\frac{h + \frac{1}{2}}{n - k + 1} \pi \right)$, where $g \in W_k$ and $h = 0, 1, \ldots, n - k$. Each eigenvalue has multiplicity equal to $\dim W_k = q^k - q^{k-1}$.

Similarly, if $k = 0$, $f = \alpha_0^0 g + \alpha_1^0 d^* g + \cdots + \alpha_n^0 (d^*)^n g$ is an eigenvector for M if and only if $(\alpha_0^0, \alpha_1^0, \ldots, \alpha_n^0)$ is an eigenvector of the $(n+1) \times (n+1)$ matrix

$$\begin{pmatrix} 0 & 1 & 0 & & & \\ \frac{1}{2} & 0 & \frac{1}{2} & & & \\ 0 & \frac{1}{2} & 0 & \frac{1}{2} & & \\ & \ddots & \ddots & \ddots & \ddots & \\ & & & & 1 & 0 \end{pmatrix}.$$

This is DCT-1. Therefore the eigenvectors are given by

$$f = \mathbf{1}_{\Sigma^0} + \cos\left(\frac{h}{n}\pi\right) \mathbf{1}_{\Sigma^1} + \cos\left(\frac{2h}{n}\pi\right) \mathbf{1}_{\Sigma^2} + \cdots + \cos(h\pi)\mathbf{1}_{\Sigma^n}$$

and the corresponding eigenvalues are $\lambda_h^0 = \cos\left(\frac{h}{n}\pi\right)$, where $h = 0, 1, \ldots, n$. Each eigenvalue has multiplicity equal to 1.

Appendix 2

Solutions of the exercises

1.4.3. We have $p^{(2m)}(x,y) \geq \frac{1}{2^{2m}}$ for all $x,y \in C_{2m+1}$. In fact, there always exists a path $(x_0 = x, x_1, \ldots, x_{2\ell-1}, x_{2\ell} = y)$ of even length $2\ell \leq 2m$ from x to y; if $\ell < m$ then the extended path

$$(x_0 = x, x_1, \ldots, x_{2\ell-1}, x_{2\ell}, x_{2\ell-1}, x_{2\ell}, \ldots, x_{2\ell-1}, x_{2\ell} = y)$$

(where the edge $(x_{2\ell-1}, x_{2\ell})$ appears $m - \ell$ times) fulfills our requirements. Therefore in the corollary we can take $n_0 = 2m$ and $\varepsilon = \frac{1}{2^{2m}}$, obtaining $A = \left(\frac{2^{2m}}{2^{2m}-1}\right)^2 \leq 2$ and $1 - c = \left(1 - \frac{1}{2^{2m}}\right)^{\frac{1}{2m}} \leq e^{-\frac{1}{m2^{2m+1}}}$.

2.5.4. *Hint.* One has $\|p^{(0)} - \pi\|_{TV} = \frac{n-1}{n}$ and $\|p^{(2)} - \pi\|_{TV} = \frac{1}{n(n-1)}$.

3.1.9. *Hint.* The stabilizer of any point (g, x) is trivial.

3.3.5. (1) Suppose that P is positive and take $\alpha \in \mathbb{C}$, $v, w \in V$. Then the quantity

$$\langle P(v + \alpha w), v + \alpha w \rangle = \langle Pv, v \rangle + \alpha \langle Pv, w \rangle + \overline{\alpha} \langle Pw, v \rangle + |\alpha|^2 \langle Pw, w \rangle$$

is positive. Therefore, in particular, $\alpha \langle Pv, w \rangle + \overline{\alpha} \langle Pw, v \rangle$ must be real and thus equal to its conjugate $\overline{\alpha} \langle P^*w, v \rangle + \alpha \langle P^*v, w \rangle$. As α is arbitrary, we obtain $\langle Pv, w \rangle = \langle P^*v, w \rangle$, and therefore $P^* = P$.

(2) Let $v \in V$ and $\lambda \in \mathbb{C}$. If $Av = \lambda v$, then $A(Bv) = BAv = \lambda Bv$. In other words, any eigensapce W of A is B-invariant. Then one may diagonalize the restriction $B|_W$ the restriction of B to W.

(3) If $P = \sum_{i=1}^{m} \lambda_i E_i$ is the spectral decomposition of P (E_i is the orthogonal projection onto the eigenspace corresponding to the eigenvalue λ_i), then the positivity yields $\lambda_i > 0$ for all $i = 1, 2, \ldots, m$. Then take $Q = \sum_{i=1}^{m} \sqrt{\lambda_i} E_i$. As far uniqueness is concerned, suppose that Q

is positive and $Q^2 = P$. Then $QP = PQ$ and, by (2), there exists a common basis of eigenvectors.

(4) Let P be the square root of A^*A, that is $P^2 = A^*A$. Set $U = AP^{-1}$. Then the linear operator U is unitary: $U^*U = AP^{-2}A^* = AA^{-1}(A^*)^{-1}A^* = I$. As for uniqueness we have that if $A = UP$ and $A = U_1P_1$ are two polar decompositions of A, then $U_2 := PP_1^{-1} = U^{-1}U_1$ is unitary and $U_2^*U_2 = I$ yields $P_1^{-1}P^2P_1^{-1} = I$, that is $P^2 = P_1^2$. But then the uniqueness of the square root in (3) applies forcing $P = P_1$ and $U = U_1$.

3.5.3. $\rho(g_1)\rho(g_2) = \rho(g_2)\rho(g_1), \forall g_1, g_2 \in G \Rightarrow \rho(g_2) = \lambda I, \forall g_2 \in G \Rightarrow$ every subspace is ρ-invariant.

3.7.3. (1) Let $n = |G|$ be the order of the group G. Then for any element $s \in G$ one has $s^n = 1_G$ and therefore

$$\rho(s)^n = \rho(s^n) = \rho(1_G) = I. \tag{2.1}$$

This implies that all the eigenvalues $\lambda_1, \ldots, \lambda_d$ of $\rho(s)$ are nth roots of unity.

(2) The eigenvalues $\lambda_1, \ldots, \lambda_d$ of $\rho(s)$ are complex numbers of modulus one. Therefore, $(\lambda_j)^{-1} = \overline{\lambda_j}$ for all j's.

3.7.10. (3) One has

$$\sum_{g \in S_n} (\chi_W(g))^2 = \sum_{a=0}^{n} (a-1)^2 \binom{n}{a} p(n-a)$$

$$= \sum_{a=0}^{n} a(a-1) \binom{n}{a} p(n-a) - a \sum_{a=0}^{n} \binom{n}{a} p(n-a) + n!$$

$$= n(n-1) \sum_{a=2}^{n} \binom{n-2}{a-2} p(n-a) - n \sum_{a=1}^{n} \binom{n-1}{a-1} p(n-a) + n!$$

$$= n(n-1) \sum_{a=0}^{n-2} \binom{n-2}{a} p(n-2-a)$$

$$- n \sum_{a=0}^{n-1} \binom{n-1}{a} p(n-1-a) + n!$$

$$= n! - n! + n! = n!.$$

3.8.3. (2) $\rho_{n-t}(g) = \rho_t(a)\rho_t(g)\rho_t(a)$ for all $g \in D_n$.

3.8.4. *Hint.* We have χ_1, χ_2 and ρ_t with $1 \leq t \leq \left[\frac{n}{2}\right]$.

3.9.11. (2) Suppose that (ρ, V) is irreducible. Consider $(\rho|_A, V)$ (the restriction of ρ to A) and let $W \leq V$ be a nontrivial irreducible $\rho|_A$-invariant subspace. By part (1) we have that W is one-dimensional. Set $H = \{g \in G : \rho(g)W \subseteq W\}$. Then $A \leq H$ and, if ℓ denotes the number of distinct subspaces among the $\rho(g)W$'s, with g varying in G, then clearly $\ell = |G/H|$. But $d_\rho \leq \ell$ as $\langle tW \rangle_{t \in T} = V$ where T is a set of representatives for the right cosets of H in G, that is $G = \coprod_{t \in T} tH$. As $|G/H| \leq |G/A|$ we conclude.

3.9.13. Formula (3.21) may be rewritten in the form

$$\sum_{t \in T} \frac{|C(t)|}{|G|} \chi_{\rho_1}(t) \overline{\chi_{\rho_2}(t)} = \delta_{\rho_1, \rho_2}$$

showing that the matrix $U = (U_{\rho,t})_{\rho \in \widehat{G}, t \in T}$ with $U_{\rho,t} = \sqrt{\frac{|C(t)|}{|G|}} \chi_\rho(t)$ is unitary (recall that $|T| = |\widehat{G}|$). Therefore

$$\sum_{\rho \in \widehat{G}} \sqrt{\frac{|C(t_1)|}{|G|}} \chi_\rho(t_1) \cdot \sqrt{\frac{|C(t_2)|}{|G|}} \overline{\chi_\rho(t_2)} = \delta_{t_1, t_2}$$

and we are done.

3.12.5. *Hint.* Show that $\ell(1_{S_5}) = 10$, $\ell(1,2) = 7$, $\ell(1,2)(3,4) = 6$, $\ell(1,2,3) = 4$, $\ell(1,2,3)(4,5) = 3$, $\ell(1,2,3,4) = 3$ and, finally, $\ell(1,2,3,4,5) = 2$. The computation yields 34 nonisomorphic graphs on five vertices.

3.13.8. (3) Suppose, by contradiction, that there exists an intermediate subgroup H such that $K < H < G$ and let $h \in H \setminus K$ and $g \in G \setminus H$. By double transitivity applied to $(K, hK), (K, gK) \in X \times X$, there exists $r \in G$ such that $rK = K$ and $rhK = gK$. But then $r \in K$, $rh \in H$ and $g \in H$, a contradiction.

3.14.1. *Hint.* (1) Use the fixed point character formula (3.22) to compute the characters of $M^{n-1,1}$, $M^{n-2,2}$ and $M^{n-2,1,1}$ and then use the decompositions (3.28), (3.29) and (3.32).

4.2.4. (1) Let $T \in Hom_G(L(X), L(X))$ be the G-invariant operator represented by the G-invariant matrix $r(x, y)$. As $r(x, y)$ is constant on the G-orbits on $X \times X$ there exist $\alpha_0, \alpha_1, \ldots, \alpha_N \in \mathbb{C}$ such that

$$r(x, y) = \alpha_0 r_0(x, y) + \alpha_1 r_1(x, y) + \cdots + \alpha_N r_N(x, y) \qquad (2.2)$$

where

$$r_i(x,y) = \begin{cases} 1 & \text{if } (x,y) \in \Theta_i \\ 0 & \text{otherwise.} \end{cases}$$

But $r_i(x,y)$ is the matrix that represents the operator Δ_i and this proves 1.

(2) It is immediate to check that, for $g \in G$ and $(x,y) \in \Theta_s$, $\Xi_{i,j}(gx,gy) = g\Xi_{i,j}(x,y)$ and thus, in particular, $|\Xi_{i,j}(gx,gy)| = |\Xi_{i,j}(x,y)| = \xi_{i,j}(s)$ depends only on the G-orbit Θ_s.

(3) Denote by $r = r_i \cdot r_j$ the matrix that represents the operator $\Delta_i\Delta_j$. For $(x,y) \in \Theta_s$ we have

$$r(x,y) = \sum_{z \in X} r_i(x,z)r_j(z,y) = |\{z : r_i(x,z) = 1, r_j(z,y) = 1\}|$$
$$= |\{z : (x,z) \in \Theta_i, (z,y) \in \Theta_j\}| = \xi_{i,j}(s),$$

showing that the coefficient α_s in (2.2) equals $\xi_{i,j}(s)$.

(4) It is clear that the set $\{(y,x) : (x,y) \in \Theta_i\}$ is a G-orbit, say $\Theta_{i'}$. Then, for $f, g \in L(X)$, we have

$$\langle \Delta_i f, g \rangle = \sum_{x \in X} \Delta_i f(x)\overline{g(x)}$$
$$= \sum_{x \in X} \sum_{y \in X:(x,y)\in\Theta_i} f(y)\overline{g(x)}$$
$$= \sum_{y \in X} f(y) \sum_{x \in X:(y,x)\in\Theta_{i'}} \overline{g(x)}$$
$$= \langle f, \Delta_{i'} g \rangle.$$

4.8.4. The automorphism $\tau : A_4 \to A_4$ given by $\tau(123) = (132)$ and $\tau(124) = (142)$ satisfies the condition of Gelfand's lemma (Exercise 4.3.3).

4.9.3. *Hint.* If P is in detailed balance with a probability measure π, it is also in detailed balance with $\tilde{\pi}(x) = \frac{1}{|G|}\sum_{g\in G} \pi(gx)$. Note that this does not require the transitivity of the G-action on X.

4.9.5. (1) The function $^K f(x) = \frac{1}{|K|}\sum_{k\in K} f(kx)$ is the projection of f on the space of K-invariant functions on X; therefore $^K f \equiv 0$. But $f(x_0) = {}^K f(x_0)$.

(2) In each V_i use an orthonormal basis $f_{i,1}, f_{i,2}, \ldots, f_{i,d_i}$ with $f_{i,1} = \sqrt{d_i}\phi_i$.

5.1.17. *Hint.* Show that any simple, connected graph X with no loops and $|X| \leq 5$ has a nontrivial automorphism. See Examples 3.12.3, 3.12.4 and Exercise 3.12.5.

7.2.1. *Hint.* (2) The isomorphism is given by

$$Aut(\mathbb{T}_{q,2}) \ni g \mapsto (g_{g_\emptyset^{-1}(1)}, g_{g_\emptyset^{-1}(2)}, \ldots, g_{g_\emptyset^{-1}(q)}; g_\emptyset).$$

8.2.9. *Hint.* Any automorphism of $\Theta_{n,m+1}$ is also an isometry of the tree in Example 8.1.11 and an isometry of $\{\phi \in \Theta_{n,m+1} : |\mathrm{dom}(\phi)| = n\}$ which coincides with the Hamming scheme.

8.2.10. Solution: $\Omega_{3,2}$, $\Omega_{3,1}$ and $\Omega_{2,1}$ are empty.

8.3.7. *Hint.* See Exercise 8.8.14.

8.3.8. *Hint.* We have $a_{n,0} = 1$, $a_{n,j} = 0$ for $j \geq 1$, $b_{n,0} = q - 1$, $b_{n,j} = q$ for $j = 1, 2, \ldots, n$ and

$$C_{i,k}^n = \begin{cases} 1 & \text{if } n > N - k, \ i = n - N + k \\ 1 & \text{if } n \leq N - k, i = 0 \\ 0 & \text{otherwise.} \end{cases}$$

8.6.7. For $0 \leq h, m \leq n$, the orbits of $GL(n, q)$ on $\mathcal{G}_h \times \mathcal{G}_m$ are the sets

$$\Omega_r = \{(V, W) \in \mathcal{G}_h \times \mathcal{G}_m : \dim(V \cap W) = r\},$$

for $\max\{0, h + m - n\} \leq r \leq \min\{h.m\}$. Then we can apply Exercise 3.13.5.

8.7.2. *Hint.* Use the map $\begin{pmatrix} A & 0 \\ C & B \end{pmatrix} \mapsto (A, B, CA^{-1})$.

8.7.12. Clearly $\delta(\phi, \phi_0) = k$ if and only if $\dim(\ker \phi) = n - k$ if and only if $\dim(Im\phi) = k$. We have $\binom{n}{k}_q$ choices for $\ker \phi$. Once we have chosen $\ker \phi$, we can take a basis $v_1, v_2, \ldots, v_n$ for $\mathbb{F}_q^n$ such that $v_{k+1}, v_{k+2}, \ldots, v_n \in \ker \phi$. Then ϕ is an arbitrary linear map sending $v_1, v_2, \ldots, v_k$ onto a set $w_1, w_2, \ldots, w_k$ of linear independent vectors in $\mathbb{F}_q^m$; this can be done in $(q^m - 1)(q^m - q) \cdots (q^m - q^{k-1})$ different ways.

8.7.14. It is clear that χ_ψ is a character of $\mathcal{X}_{n,m}^q$. Moreover $\chi_{\psi_1 + \psi_2} = \chi_{\psi_1} \cdot \chi_{\psi_2}$ and therefore $\psi \mapsto \chi_\psi$ is a group homomorphism. Since χ_ψ is trivial if and only if $\psi \equiv 0$, the homomorphism is injective. Since

$|\widehat{\mathcal{X}^q_{n,m}}| = |\mathcal{X}^q_{n,m}|$, it is also surjective. Note that we have only used the *existence* of χ and the fact that $\widehat{\mathcal{X}^q_{n,m}} \cong \mathcal{X}^q_{n,m}$. For more on character theory of $\mathbb{F}^n_q$ we refer to [155].

8.7.15. It suffices to show that $\overline{V_k}$ is $G^q_{n,m}$-invariant: if $(\alpha, \beta, \gamma) \in G^q_{n,m}$, $\psi \in \mathcal{X}^q_{m,n}$ and $\phi \in \mathcal{X}^q_{n,m}$, then

$$
\begin{aligned}
[(\alpha, \beta, \gamma)\chi_\psi](\phi) &= \chi_\psi[(\alpha, \beta, \gamma)^{-1}\phi] \\
&= \chi_\psi(\beta^{-1}\phi\alpha - \beta^{-1}\gamma\alpha) \\
&= \overline{\chi_\psi(\beta^{-1}\gamma\alpha)} \cdot \chi[Tr(\psi\beta^{-1}\phi\alpha)] \\
&= \overline{\chi_\psi(\beta^{-1}\gamma\alpha)} \cdot \chi_{\alpha\psi\beta^{-1}}(\phi)
\end{aligned}
$$

and clearly $rk(\alpha\psi\beta^{-1}) = rk(\psi)$.

8.7.16. We denote by (V, v, w) a triple with $V \subseteq \mathbb{F}^n_q$ a subspace with $\dim(V) = n - 1$, $v \in \mathbb{F}^n_q \setminus V$ and $w \in \mathbb{F}^n_q \setminus \{0\}$. Also denote by $\phi_{V,v,w}$ the linear map in $\mathcal{X}^q_{n,m}$ such that $\ker \phi_{V,v,w} = V$ and $\phi_{V,v,w}(v) = w$. For $F \in L(\mathcal{X}^q_{n,m})$ and $\phi \in \mathcal{X}^q_{n,m}$ we have

$$
(\Delta F)(\phi) = \frac{1}{q^n - q^{n-1}} \sum_{(V,v,w)} F(\phi + \phi_{V,v,w}). \tag{2.3}
$$

Indeed, for a fixed $\phi_{V,v,w}$ one has $|\{(V, v', w') : \phi_{V,v',w'} = \phi_{V,v,w}\}| = q^n - q^{n-1}$. Moreover, if $\psi \in \mathcal{X}^q_{n,m}$ and $\psi(w) = av + v_1$ with $a \in \mathbb{F}_q$ and $v_1 \in V$, then

$$
\chi_\psi(\phi_{V,v,w}) = \chi[Tr(\psi\phi_{V,v,w})] = \chi(a). \tag{2.4}
$$

Note also that

$$
\chi\left(\sum_{a \in \mathbb{F}_q \setminus \{0\}} a\right) = -1 \tag{2.5}
$$

since χ is orthogonal to the trivial character.

We need a few more formulae. If $Im\psi \subseteq V$ then

$$
\sum_{v,w} \chi_\psi(\phi_{V,v,w}) = (q^m - 1)(q^n - q^{n-1}) \tag{2.6}
$$

(this is the number of (v, w)'s for a fixed V: for, if $Im\psi \subseteq V$ then $Tr[\psi\phi_{V,v,w}] = 0$ and $\chi_\psi(\phi_{V,v,w}) = \chi(0) = 1$).

If $Im\psi \subsetneq V$ then

$$
\sum_{\substack{v,w: \\ \psi(w) \in V}} \chi_\psi(\phi_{V,v,w}) = (q^{m-1} - 1)(q^n - q^{n-1}) \tag{2.7}
$$

(the proof is the same as for (2.6)) and

$$\sum_{\substack{v,w: \\ \psi(w) \notin V}} \chi_\psi(\phi_{V,v,w}) = -q^{m-1}(q^n - q^{n-1}) \qquad (2.8)$$

(if $\psi(w) \notin V$, then we can apply (2.4) and (2.5); the right hand side of (2.8) is just $-|\{(v,w) : \psi(w) \notin V\}| \cdot \frac{1}{q-1})$.

Finally, if $rk(\psi) = k$ then

$$\Delta\chi_\psi = \chi_\psi \cdot \frac{1}{q^n - q^{n-1}} \sum_{(V,v,w)} \chi_\psi(\phi_{V,v,w})$$

$$(\text{by } (2.3)) = \chi_\psi \cdot \frac{1}{q^n - q^{n-1}} \left[\sum_{V: Im\psi \subseteq V} \sum_{v,w} \chi_\psi(\phi_{V,v,w}) \right.$$

$$+ \sum_{V: Im\psi \subsetneq V} \sum_{\substack{v,w \\ \psi(w) \in V}} \chi_\psi(\phi_{V,v,w})$$

$$+ \left. \sum_{V: Im\psi \subsetneq V} \sum_{\substack{v,w \\ \psi(w) \notin V}} \chi_\psi(\phi_{V,v,w}) \right]$$

$$= \chi_\psi \cdot \frac{1}{q^n - q^{n-1}} \left[\binom{n-k}{1}_q (q^m - 1)(q^n - q^{n-1}) \right.$$

$$+ \binom{k}{1}_q q^{n-k}(q^n - q^{n-1})(q^{m-1} - 1)$$

$$- \left. \binom{k}{1}_q q^{n-k}(q^n - q^{n-1})q^{m-1} \right]$$

$$= \frac{(q^m - 1)(q^{n-k} - 1) - q^{n-k}(q^k - 1)}{q - 1} \cdot \chi_\psi$$

(we have used (2.6), (2.7) and (2.8) and Corollaries 8.5.5 and 8.5.6 to count the number of V's satisfying $Im\psi \subsetneq V$ and $Im\psi \subseteq V$, respectively).

8.8.12. Hint. The inverse map T^{-1} is given by

$$T^{-1}(k,\ell) = \begin{cases} (k - n + 2h, \ell) & \text{if } n - h < h - \ell \\ (k + \ell, \ell) & \text{if } n - h \geq h - \ell. \end{cases}$$

8.8.16. Hint. The $\psi(h,\ell,u;s)$'s are essentially the spherical functions of the Hamming scheme (see Theorem 5.3.2).

9.1.1. $(1)\theta$ may be defined on the elements of a basis $\{e_i \otimes f_j\}$ of $V \bigotimes W$ by setting $\theta(e_i \otimes f_j) = \psi(e_i, f_j)$.

(2) Applying (1) with U in place of Z we get the existence of a linear map $\varepsilon : V \bigotimes W \to U$ such that $(*)$ $\psi = \varepsilon \circ \phi$. In virtue of property (b) of ψ with $V \bigotimes W$ in place of Z, we have a linear map $\xi : U \to V \bigotimes W$ such that $(**)$ $\phi = \xi \circ \psi$.

Then, from $(*)$ and $(**)$ it follows that $\phi = \xi \circ \varepsilon \circ \phi$ and since ϕ is surjective, $\xi \circ \varepsilon = id_{V \bigotimes W}$; in particular ε is injective.

From (a) we know that $\psi(V \times W)$ generates U and therefore, from $(*)$ we get that ε is also surjective and therefore it is the desired isomorphism.

9.4.7. (1) The diagonal matrices form a maximal abelian subalgebra in the full matrix algebra $M_{n,n}(\mathbb{C})$.

(2) Extend each basis of V_ρ^K to an orthonormal basis $\{v_1^\rho, v_2^\rho, \ldots, v_{d_\rho}^\rho\}$ for V_ρ for all $\rho \in I$. Then, any function $f \in T_{v_k}^\sigma V_\sigma$ is of the form

$$f(hx_0) = \sqrt{\frac{d_\sigma}{|X|}} \sum_{j=1}^{d_\sigma} a_j \phi_{j,k}^\sigma(h)$$

(f is the T_{v_k}-image of $\sum_{j=1}^{d_\sigma} a_j v_j^\sigma$) and therefore

$$(E_i^\rho)(gx_0) = \frac{d_\rho}{|G|} \sum_{h \in G} f(hx_0) \overline{\phi_{i,i}^\rho(g^{-1}h)}$$

$$= \frac{d_\rho}{|G|} \sqrt{\frac{d_\sigma}{|X|}} \sum_{j=1}^{d_\sigma} a_j [\phi_{j,k}^\sigma * \phi_{i,i}^\rho](g)$$

$$= \sqrt{\frac{d_\sigma}{|X|}} \sum_{j=1}^{d_\sigma} a_j \delta_{\sigma,\rho} \delta_{k,i} \phi_{j,i}^\sigma(g)$$

$$= \begin{cases} f(gx_0) & \text{if } \rho = \sigma \text{ and } i = k \\ 0 & \text{otherwise.} \end{cases}$$

9.4.9. *Hint.* Use Wielandt's lemma to prove that the multiplicity of $S^{n-1,1}$ in $L(S_{a+b+c}/(S_a \times S_b \times S_c))$ is 2 when $\max\{a, b, c\} > 0$.

9.5.8. Take an orthonormal basis $v_1, v_2, \ldots, v_d$ in V such that v_1 is K-invariant. Use the fact that $\frac{1}{|K|} \sum_{k \in K} \rho(k)$ is the projection from

V onto the space $V^K = span_{\mathbb{C}}\langle v_1 \rangle$ of K-invariant vectors. Note also that

$$\frac{1}{|K|} \sum_{k \in K} \chi(gk) = \frac{1}{|K|^2} \sum_{k,k_2 \in K} \chi(k_2^{-1} g k k_2) = \frac{1}{|K|^2} \sum_{k_1,k_2 \in K} \chi(k_2^{-1} g k_1).$$

9.5.13. Fix $\rho \in \widehat{G}$ and let $\{v_1, v_2, \ldots, v_{d_\rho}\}$ be an orthonormal basis in V_ρ. Then, for $v \in V_\rho$ and $f \in V_\rho'$ one has

$$\langle F, \mathcal{T}(v \otimes f) \rangle_{L(G)} = \frac{d_\rho}{|G|} \sum_{g \in G} F(g) \overline{f[\rho(g^{-1})v]}$$

$$= \frac{d_\rho}{|G|} \sum_{g \in G} F(g) \overline{f[\sum_{i=1}^{d_\rho} \langle \rho(g^{-1})v, v_i \rangle_{V_\rho} v_i]}$$

$$= \frac{d_\rho}{|G|} \sum_{i=1}^{d_\rho} \overline{f(v_i)} \sum_{g \in G} F(g) \langle \rho(g)v_i, v \rangle_{V_\rho}$$

$$= \frac{d_\rho}{|G|} \sum_{i=1}^{d_\rho} \overline{f(v_i)} \langle \widehat{F}(\rho)v_i, v \rangle_{V_\rho}$$

$$= \frac{d_\rho}{|G|} \sum_{i=1}^{d_\rho} \langle \widehat{F}(\rho)v_i, [v \otimes f](v_i) \rangle_{V_\rho}$$

$$= \langle \widehat{F}, v \otimes f \rangle_{HS,G}.$$

9.5.14. (1) Suppose that $\theta : W' \to W$ is the Riesz map. If $S : U \to W \otimes V$ is a linear map, $u \in U, w \in W$ and $v \in V$, denote by $[Su](w,v)$ the value of the bi-antilinear map Su on the pair (w,v). Then the explicit isomorphism is given by:

$$\begin{array}{ccc} \mathrm{Hom}_G(\pi' \otimes \eta, \sigma) & \longrightarrow & \mathrm{Hom}_G(\eta, \pi \otimes \sigma) \\ T & \longmapsto & \widetilde{T} \end{array}$$

where $\widetilde{T}$ is defined by setting:

$$[\widetilde{T}u](w,v) = \langle T[\theta^{-1}(w) \otimes u], v \rangle_V$$

Indeed, the expression $\langle T[\theta^{-1}(w) \otimes u], v \rangle_V$ is linear in T and in u, bi-antilinear in (w,v). Moreover, $\widetilde{T}$ commutes with the action of G: for any $g \in G$ we have

$$
\begin{aligned}
[\widetilde{T}\eta(g)u](w,v) &= \langle T[\theta^{-1}(w) \otimes \eta(g)u], v \rangle_V \\
&= \langle T\{(\pi'(g) \otimes \eta(g))[\pi'(g)^* \theta^{-1}(w) \otimes u]\}, v \rangle_V \\
&= \langle \sigma(g)T[\theta^{-1}(\pi(g)^*w) \otimes u], v \rangle_V \\
&= \langle T[\theta^{-1}(\pi(g)^*w) \otimes u], \sigma(g)^*v \rangle_V \\
&= [\widetilde{T}u](\pi(g)^*w, \sigma(g)^*v) \\
&= [(\pi(g) \otimes \sigma(g))\widetilde{T}u](w,v).
\end{aligned}
$$

Finally, the inverse of $T \mapsto \widetilde{T}$ is the map $T \mapsto \widehat{T}$ given by the following rule: if $T \in \mathrm{Hom}_G(\eta, \pi \otimes \sigma)$ and $\omega \otimes u \in W' \otimes U$ then $\widehat{T}(\omega \otimes u) = v_0$, where v_0 is the unique vector in V such that

$$
[Tu](\theta(\omega), v) = \langle v_0, v \rangle_V
$$

for all $v \in V$.

(2) This is simple:

$$
\begin{aligned}
\langle \chi_{\pi' \otimes \eta}, \chi_\sigma \rangle_{L(G)} &= \langle \overline{\chi_\pi} \cdot \chi_\eta, \chi_\sigma \rangle_{L(G)} = \langle \chi_\eta, \chi_\pi \cdot \chi_\sigma \rangle_{L(G)} \rangle_{L(G)} \\
&= \langle \chi_\eta, \chi_{\pi \otimes \sigma} \rangle_{L(G)}.
\end{aligned}
$$

9.7.4. $C_2 \times C_2$ is an homomorphic image of Q: if $C_2 \times C_2 = \langle a \rangle \times \langle b \rangle$ then $q(-1) = 1, q(\pm i) = a, q(\pm j) = b$ and $q(\pm k) = ab$ gives such an epimorphism. If χ_1, χ_2, χ_3 and χ_4 are the characters (=irreducible representations) of $C_2 \times C_2$, then the compositions $\chi_i \circ q$ yields four one-dimensional (irreducible) representations of Q. By the Peter–Weyl theorem there is only left one of dimension two: this is ρ from (9.22).

9.7.5. (i) Introduce a (complex) scalar product $((\cdot, \cdot))$ in V by setting $((v_i, v_j)) = \delta_{i,j}$. Then, by Proposition 3.3.1, the scalar product $(\cdot, \cdot)$ defined by $(v, w) = \sum_{g \in G}((\rho(g)v, \rho(g)w))$ for all $v, w \in V$ is ρ-invariant. Using the Gram–Schmidt orthonormalization process transform the v_i's into a $(\cdot, \cdot)$-orthonormal basis $\{w_1, w_2, \ldots, w_d\}$. Note that the coefficients of the v_i's in the expression of the w_j's are all real. Indeed for all $i, j = 1, 2, \ldots, d$ we have

$$
(v_i, v_j) = \sum_{g \in G}((\rho(g)v_i, \rho(g)v_j)) = \sum_{g \in G} \sum_{k=1}^{d}((\rho(g)v_i, v_k))((\rho(g)v_j, v_k))
$$

is real as each summand in the right hand side is real by hypothesis.

(ii) Denote by $\overline{\rho}(g)$ the matrix representing $\rho(g)$ with respect to the basis $\{w_1, w_2, \ldots, w_d\}$. Then $\widetilde{\rho}$ and $\overline{\rho}$ are equivalent and therefore, by

Proposition 3.3.2, unitarily equivalent. Therefore there exist a unitary matrix U such that $U\widehat{\rho}U^* = \overline{\rho}$ is real.

9.7.11. (1) We have $gug^{-1} = u^{-1} \Leftrightarrow g^2 = (gu^{-1})^2$ and the number of u for which the right hand side identity holds true equals the number of $v(\equiv gu^{-1})$ for which $v^2 = g^2$.

(2)$\sum_{g \in G} p(g) = \sum_{g \in G} \rho(g^2) = \sum_{g \in G} \rho(t)|\{g \in G : g^2 = t\}| = \sum_{t \in G} \rho(t)^2$.

(3) Denote by $\mathcal{A}$ the ambivalence classes of G so that $|\mathcal{A}| = h$. Then

$$\sum_{g \in G} p(g) = |\{(g, u) \in G \times G : gug^{-1} = u^{-1}\}|$$

$$= \sum_{u \in G} |\{g \in G : gug^{-1} = u^{-1}\}|$$

$$= \sum_{C \in \mathcal{A}} \sum_{u \in C} |\{g \in G : gug^{-1} = u^{-1}\}|$$

$$= \sum_{C \in \mathcal{A}} \sum_{u \in C} \frac{|G|}{|C|}$$

$$= h|G|.$$

9.7.15. (1) Take an orthonormal basis $v_1, v_2, \ldots, v_{d_i}$ in V_i such that v_1 is K-invariant (see also the Solution of Exercise 9.5.8). Then proceed as in the proof of Theorem 9.7.7.

(2) Proceed as in the proof of Corollary 9.7.8, but using the spherical Fourier transform.

(3) Proceed as in (i) of Theorem 9.7.10.

(4) Denote by $\phi_i(j)$ the value of ϕ_i on Ω_j. First prove that $\phi_i(j) = \overline{\phi_i(\theta(j))}$. Then proceed as in (ii) of Theorem 9.7.10 using the dual orthogonality relations for the spherical functions (see, for instance, Exercise 5.1.11).

(5) Use the projection formula (4.21).

9.8.14. *Hint.* Use Exercises 9.8.6, 9.5.8 and the orthogonality relations for the spherical functions.

9.8.15. *Hint.* Define an equivalence relation on G by setting $g_1 \sim g_2$ when there exist $h \in H$ and $k \in K$ such that $g_2 = hkg_1h^{-1}$.

9.9.3. Let $G = S_3$ be the symmetric group of degree three and set $\Delta = \{(1, 2), (1, 3, 2)\}$. Then $\Delta\Delta^{-1} = \{1_G, (2, 3)\}$ is a subgroup and coincides with $G(\Delta\Delta^{-1})$. However the latter does not contain the element $(1,3) = (1,2)(1,3,2) = (1,2)^{-1}(1,3,2) \in G_\Delta$. This shows the statement for the first inclusion.

Taking Δ any subgroup of G which is not normal yields and example where the second inclusion fails to be an equality.

9.9.4. We show it by recurrence on k. For $k = 1$ the statement is obvious. Suppose the statement is true for $k - 1$. We first show that $\Delta^k \subseteq \text{supp}\mu^{*k}$. If $g = \delta\delta_1$ with $\delta \in \Delta^{k-1}$ and $\delta_1 \in \Delta$, then

$$\mu^{*k}(g) = \sum_{h \in G} \mu^{*(k-1)}(\delta\delta_1 h)\mu(h^{-1}) \geq \mu^{*(k-1)}(\delta)\mu(\delta_1) > 0$$

where in the first inequality follows by taking $h = (\delta_1)^{-1}$ and the last one follows by the inductive step. Conversely, take $g \in \text{supp}\mu^{*k}$, then there exists $t \in G$ such that $\mu^{*(k-1)}(gt) > 0$ and $\mu(t^{-1}) > 0$. This implies that $gt \in \Delta^{k-1}$ (by the inductive step) and that $t^{-1} \in \Delta$. Therefore $g = (gt)t^{-1} \in \Delta^k$.

10.1.1 The strategy is as follows. Denote by $\sigma(i)$ the number inside box i, $i = 1, 2, \ldots, 2n$. Then, prisoner with registration number i opens box i. If he finds his own number he stops. Otherwise he opens box $\sigma(i)$. If he finds his own number he stops. Otherwise he opens box $\sigma(\sigma(i))$. And so on. The prisoners are (all) safe if and only if the cyclic decomposition of the permutation $\sigma \in S_{2n}$ has no cycle of length k with $k \geq n + 1$.

Now, a permutation $\sigma \in S_{2n}$ may contain at most one single k-cycle with $n + 1 \leq k \leq 2n$. Moreover, there are exactly $\frac{(2n)!}{k}$ distinct permutations containing a cycle of length $k \geq n + 1$. Indeed, there are $\binom{2n}{k}$ distinct ways to choose the k numbers for the cycle, $(k-1)!$ ways to order them and $(2n - k)!$ ways for permuting the remaining $2n - k$ numbers. Thus, the number of permutations in S_{2n} that contain a cycle of length $\geq n + 1$ is exactly $\frac{(2n)!}{n+1} + \frac{(2n)!}{n+2} + \cdots + \frac{(2n)!}{2n}$. Thus, the probability that a random permutation $\sigma \in S_{2n}$ contains such a long cycle is $\frac{1}{n+1} + \frac{1}{n+2} + \cdots + \frac{1}{2n}$; equivalently, the probability that σ does not contain a cycle of length exceeding n (so that the strategy applies succesfully) is

$$1 - \frac{1}{n+1} - \frac{1}{n+2} - \cdots - \frac{1}{2n} \approx 1 - \ln 2n + \ln n = 1 - \ln 2 \approx 0.3.$$

10.1.7. *Hint.* (1) Use $(i_1, i_2, \ldots, i_k) = (i_1, i_k)(i_1, i_{k-1}) \cdots (i_1, i_2)$.
(2) Use $(i+1, i+2)(i, i+1)(i+1, i+2) = (i, i+2)$.

10.3.18. *Hint.* From the proof of Proposition 10.3.16, for all $F \in Hom_{S_n}(S^\lambda, M^\lambda)$ one has $F = c1_{S^\lambda}$.

10.5.16. *Hint.* If $\{i_1, i_2, \ldots, i_k\} \mapsto \{j_1, j_2, \ldots, j_{k-1}\}$ and h is as above, then $j_h \equiv i_{h+1} \geq 2h$ and $h = \min\{t : j_t \geq 2t\}$.

10.6.7. (1) The map is well defined because if $\pi e_{\{t\}} = \eta e_{\{t\}}$, then $\pi\{t\} = \eta\{t\}$ and $\pi\eta^{-1} \in R_{\pi t} \equiv R_{\eta t}$ and therefore

$$\eta B_t = B_{\eta t}\eta = B_{\pi t}(\pi\eta^{-1})\eta = B_{\pi t}\pi = \pi B_t.$$

(2) From (1) we get that $A_t B_t = \sum_{\sigma \in C_t} \sum_{\gamma \in R_t} \text{sign}(\sigma)\sigma\gamma$ belongs to the two-sided ideal of $L(S_n)$ generated by the matrix coefficients of S^λ. Then we can apply the operator $P : L(S_n) \to L(S_n)$ defined by $Pf = \frac{1}{n!}\sum_{\pi \in S_n} \pi f \pi^{-1}$, $f \in L(S_n)$, that projects onto the subalgebra of central functions of S_n. Note that the value of $A_t B_t$ at the identity of S_n is equal to 1.

(3) Apply (2) and note that, for $\sigma \in C_t$ and $\theta \in \mathcal{C}_\mu$, one has

$$|\mathcal{C}_\mu| \cdot |\{(\gamma, \pi) \in R_t \times S_n : \pi\sigma\gamma\pi^{-1} = \theta\}| = n!|\{\gamma \in R_t : \sigma\gamma \in \mathcal{C}_\mu\}|.$$

(4) Use the following facts: if $\sigma \in C_t$, $\gamma \in R_t$ and $\sigma(\gamma(j)) = j$, then $\gamma(j) = j$ and $\sigma(j) = j$. Also recall that $\sum_{j=1}^m j^2 = \frac{m(m+1)(2m+1)}{6}$ and analyze the pairs (i, j) of numbers belonging to different rows and columns in t.

10.7.3. *Hint.* For (1) use integration by parts and an inductive argument.

10.7.4. *Hints.* (1) From $\log \frac{1+x}{1-x} = 2\left(x + \frac{x^3}{3} + \frac{x^5}{5} + \cdots +\right)$, and setting $x = \frac{1}{2n+1}$ deduce that

$$\frac{2}{2n+1} < \log\frac{n+1}{n}$$
$$< \frac{2}{2n+1}\left\{1 + \frac{1}{3}\left[\frac{1}{(2n+1)^2} + \frac{1}{(2n+1)^4} + \cdots\right]\right\}$$
$$= \frac{2}{2n+1}\left[1 + \frac{1}{12n(n+1)}\right].$$

(2) Use (1) to prove that $\frac{a_{n+1}}{a_n} < 1$ and that $\frac{b_{n+1}}{b_n} > 1$.

(3) Write Wallis's formula in the form $\sqrt{\pi} = \lim_{n\to\infty} \frac{2^{2n}(n!)^2}{\sqrt{n}(2n)!}$ and express $n!$ and $(2n)!$ in terms of a_n.

(4) Use $e^{1/12n} < 1 + \frac{1}{12n-1}$.

A1.0.12. (1) Using $\cos\theta = \frac{e^{i\theta}+e^{-i\theta}}{2}$ and $\sin\theta = \frac{e^{i\theta}-e^{-i\theta}}{2i}$ we have

$$\sum_{k=0}^{n-1}\cos^2[(k+\epsilon)\alpha] = \frac{1}{4}\cdot\frac{1-e^{2n\alpha i}}{1-e^{2\alpha i}}e^{2\epsilon\alpha i}$$

$$+ \frac{1}{4}\cdot\frac{1-e^{2n\alpha i}}{1-e^{2\alpha i}}e^{2\epsilon\alpha i} + \frac{n}{2}$$

$$= \frac{n}{2} + \frac{1}{4}\cdot\frac{\sin(2n-1+2\epsilon)\alpha] - \sin[(2\epsilon-1)\alpha]}{\sin\alpha}$$

$$= \frac{n}{2} + \frac{\sin(n\alpha)\cos[(n+2\epsilon-1)\alpha]}{2\sin\alpha}.$$

Bibliography

[1] G.M. Adelson-Velskii, B. Ju. Veisfeiler, A.A. Leman and I.A. Faradžev, An example of a graph which has no transitive group of automorphisms. (Russian) *Dokl. Akad. Nauk SSSR* **185** (1969), 975–976.

[2] D. Aldous and J. Fill, *Reversible Markov Chains and Random Walks on Graphs*, in preparation (http://stat-www.berkeley.edu/users/aldous/RWG/book.html).

[3] M. Aigner, *Combinatorial Theory*. Reprint of the 1979 original. Classics in Mathematics. Springer-Verlag, Berlin, 1997.

[4] J.L. Alperin and R.B. Bell, *Groups and Representations*, Graduate Texts in Mathematics, 162. Springer-Verlag, New York, 1995.

[5] H. Akazawa and H. Mizukawa, Orthogonal polynomials arising from the wreath products of a dihedral group with a symmetric group, *J. Combin. Theory Ser. A* **104** (2003), 371–380.

[6] M. Artin, *Algebra*. Prentice Hall, Englewood Cliffs, NJ, 1991.

[7] R.A. Bailey, *Association Schemes, Designed Experiments, Algebra and Combinatorics*. Cambridge Studies in Advanced Mathematics **84**, Cambridge University Press, 2004.

[8] R.A. Bailey, Association Schemes: A Short Course, note del seminario di Geometria Combinatoria, Roma (2005).
http://www.mat.uniroma1.it/~combinat/quaderni/quad18E/18E_bailey.html.

[9] R.A. Bailey, Ch.E. Praeger, C.A. Rowley and T.P Speed, Generalized wreath products of permutation groups. *Proc. London Math. Soc.* **47** (1983), 69–82.

[10] E. Bannai, Orthogonal polynomials in coding theory and algebraic combinatorics, in *Orthogonal Polynomials* (P. Nevai Ed.), 25–53, Kluwer Academic, Dordrecht 1990.

[11] E. Bannai, An introduction to association schemes. Methods of discrete mathematics (Braunschweig, 1999), 1–70, *Quad. Mat.*, **5**, Dept. Math., Seconda Univ. Napoli, Caserta, 1999.

[12] E. Bannai and T. Ito, *Algebraic Combinatorics* Benjamin, Menlo Park, CA, 1984.

[13] E. Bannai, N. Kawanaka and S.-Y. Song, The character table of the Hecke algebra $H(\mathrm{GL}_{2n}(F_q), \mathrm{Sp}_{2n}(F_q))$. *J. Algebra* **129** (1990), 320–366.

[14] E. Bannai and H. Tanaka, The decomposition of the permutation character $1_{\mathrm{GL}(n,q^2)}^{\mathrm{GL}(2n,q)}$. *J. Algebra* **265** (2003), 496–512.

[15] L. Bartholdi and R.I. Grigorchuk. On parabolic subgroups and Hecke algebras of some fractal groups. *Serdica Math. J.* **28** (2002), 47–90.

[16] H. Bass, M.V. Otero-Espinar, D. Rockmore and Ch. Tresser, *Cyclic Renormalization and Automorphism Groups of Rooted Trees.* Lecture Notes in Mathematics, **1621**. Springer-Verlag, Berlin, 1996.

[17] R. Beaumont and R. Peterson, Set transitive permutation groups, *Canad. J. Math.* **7** (1955), 35–42.

[18] E. Behrends, *Introduction to Markov Chains. With Special Emphasis on Rapid Mixing.* Advanced Lectures in Mathematics. Friedr. Vieweg & Sohn, Braunschweig, 2000.

[19] M.B. Bekka, P. de la Harpe, Irreducibility of unitary group representations and reproducing kernels Hilbert spaces. Appendix by the authors in collaboration with Rostislav Grigorchuk. *Expo. Math.* **21** (2003), 115–149.

[20] M.B. Bekka, P. de la Harpe and A. Valette, *Kazhdan's Property (T).* Preprint (2006).
http://www.unige.ch/math/biblio/preprint/2006/
KazhdansPropertyT.pdf

[21] R. Bellman, *Introduction to Matrix Analysis.* Reprint of the second (1970) edition. With a foreword by Gene Golub. Classics in Applied Mathematics, 19. Society for Industrial and Applied Mathematics (SIAM), Philadelphia, PA, 1997.

[22] E.R. Belsley, Rates of convergence of random walk on distance regular graphs. *Probab. Theory Related Fields* **112** (1998), 493–533.

[23] F.A.Berezin and I.M. Gelfand, Some remarks on the theory of spherical functions on symmetric Riemannian manifolds, *Tr. Mosk. Math. O.-va* **5** (1956) 311–351; [Gelfand's collected papers, Vol. II, Springer (1988) 275–320].

[24] N. Bergeron and A. Garsia; Zonal polynomials and domino tableaux, *Discrete Math.* **88** (1992) 3–15.

[25] N. Biggs, *Algebraic Graph Theory.* Second edition. Cambridge Mathematical Library. Cambridge University Press, Cambridge, 1993.

[26] L.J. Billera, S.P. Holmes and K. Vogtmann, Geometry of the space of phylogenetic trees. *Adv. in Appl. Math.* **27** (2001), 733–767 [Erratum *Adv. in Appl. Math.* **29** (2002), no. 1, 136].

[27] P. Billingsley, *Probability and Measure.* Third edition. Wiley Series in Probability and Mathematical Statistics. John Wiley, New York, 1995.

[28] K.P. Bogart, An obvious proof of Burnside's lemma, *Amer. Math. Monthly* **98** (1991), no. 10, 927–928.

[29] E. Bolker, The finite Radon transform, *Contemporary Math.* **63** (1987), 27–50.

[30] Ph. Bougerol, Théorème central limite local sur certains groupes de Lie. *Ann. Sci. École Norm. Sup.* (4) **14** (1981), 403–432.

[31] E. Bozzo and C. Di Fiore, On the use of certain matrix algebras associated with discrete trigonometric transforms in matrix displacement decomposition. *SIAM J. Matrix Anal. Appl.* **16** (1995), 312–326.

[32] P. Bremaud, *Markov Chains. Gibbs Fields, Monte Carlo Simulation, and Queues.* Texts in Applied Mathematics, 31. Springer-Verlag, New York, 1999.

[33] M. Brender, Spherical functions on the symmetric groups. *J. Algebra* **42** (1976), 302–314.

[34] M. Brender, A class of Schur algebras. *Trans. Amer. Math. Soc.* **248** (1979), 435–444.

[35] W.L. Briggs and V.E. Henson, *The DFT. An Owner's Manual for the Discrete Fourier Transform.* Society for Industrial and Applied Mathematics (SIAM), Philadelphia, PA, 1995.

[36] A.E. Brouwer, A.M. Cohen and A. Neumaier, *Distance-Regular Graphs.* Ergebnisse der Mathematik und ihrer Grenzgebiete (3) [Results in Mathematics and Related Areas (3)], 18. Springer-Verlag, Berlin, 1989.

[37] D. Bump, *Lie Groups.* Graduate Texts in Mathematics, 225. Springer-Verlag, New York, 2004.

[38] D. Bump and D. Ginzburg, Frobenius–Schur numbers. *J. Algebra* **278** (2004), 294–313.

[39] P.J. Cameron, *Permutation Groups.* London Mathematical Society Student Texts, 45. Cambridge University Press, Cambridge, 1999.

[40] P.J. Cameron and J.H. van Lint, *Designs, Graphs, Codes and their Links.* London Mathematical Society Student Texts, 22. Cambridge University Press, Cambridge, 1991.

[41] L. Carlitz, Representations by quadratic forms in a finite field. *Duke Math. J.* **21** (1954), 123–137.

[42] T. Ceccherini-Silberstein, Yu. Leonov, F. Scarabotti and F. Tolli, Generalized Kaloujnine groups, uniseriality and height of automorphisms. *Internat. J. Algebra Comput.*, **15**, (2005), 503–527.

[43] T. Ceccherini-Silberstein, F. Scarabotti and F. Tolli, Finite Gelfand pairs and their applications to probability and statistics, *J. Math. Sci. (N.Y.)*, **141** (2007), 1182–1229.

[44] T. Ceccherini-Silberstein, F. Scarabotti and F. Tolli, Trees, wreath products and finite Gelfand pairs, *Adv. Math.*, **206** (2006), 503–537.

[45] T. Ceccherini-Silberstein, A. Machì, F. Scarabotti and F. Tolli, *Induced Representations and Mackey theory*, submitted.

[46] F.R.K. Chung and R.L. Graham, Stratified random walks on the n-cube. *Random Structures Algorithms* **11** (1997), no. 3, 199–222.

[47] H.S.M. Coxeter, *Introduction to Geometry.* Reprint of the 1969 edition. Wiley Classics Library. John Wiley, New York, 1989.

[48] Ch.W. Curtis and I. Reiner, *Methods of Representation Theory. With Applications to Finite Groups and Orders.* Vol. I and II. Pure and Applied Mathematics (New York). A Wiley-Interscience Publication. John Wiley, New York, 1981 and 1987.

[49] G. Davidoff, P. Sarnak and A. Valette, *Elementary Number Theory, Group Theory, and Ramanujan Graphs*, London Mathematical Society Student Texts, 55. Cambridge University Press, Cambridge, 2003.

[50] Ph.J. Davis, *Circulant Matrices.* John Wiley, New York-, 1979.

[51] Ph. Delsarte, An algebraic approach to the association schemes of coding theory, *Philips Res. Rep. Suppl.* **10** (1973). Available at: http://users.wpi.edu/~martin/RESEARCH/philips.pdf

[52] Ph. Delsarte, Association schemes and t-designs in regular semilattices. *J. Combin Theory Ser. A* **20** (1976), 230–243.

[53] Ph. Delsarte, Hahn polynomials, discrete harmonics and t-designs, *SIAM J. Appl. Math.* **34** (1978), 154–166.

[54] Ph. Delsarte and V.I. Levenshtein, Association schemes and coding theory. Information theory: 1948–1998. *IEEE Trans. Inform. Theory* **44** (1998), 2477–2504.

[55] P. Diaconis, *Group Representations in Probability and Statistics*, IMS Hayward, CA. 1988.

[56] P. Diaconis, A generalization of spectral analysis with application to ranked data. *Ann. Statist.* **17** (1989), no. 3, 949–979.

[57] P. Diaconis, Patterned matrices. Matrix theory and applications (Phoenix, AZ, 1989), 37–58, *Proc. Sympos. Appl. Math.*, **40**, Amer. Math. Soc., Providence, RI, 1990.

[58] P. Diaconis, The cutoff phenomenon in finite Markov chains, *Proc. Natl. Acad. Sci. USA*, **93** (1996), 1659–1664.

[59] P. Diaconis, Random matching: an application of zonal polynomials to a (relatively) natural problem (unpublished).

[60] P. Diaconis, Random walks on groups: characters and geometry. Groups St. Andrews 2001 in Oxford. Vol. I, 120–142, London Math. Soc. Lecture Note Ser., **304**, Cambridge Univ. Press, Cambridge, 2003.

[61] P.Diaconis and D. Freedman, An elementary proof of Stirling's formula. *Amer. Math. Monthly* **93** (1986), no. 2, 123–125.

[62] P. Diaconis, R.L. Graham and J.A. Morrison, Asymptotic analysis of a random walk on a hypercube with many dimensions. *Random Struct Algor* **1** (1990), 51–72.

[63] P. Diaconis and S.P. Holmes, Matchings and phylogenetic trees. *Proc. Natl. Acad. Sci. USA* **95** (1998), 14600–14602.

[64] P. Diaconis and S.P. Holmes, Random walks on trees and matchings. *Electron. J. Probab.* **7** (2002), 17 pp. (electronic)

[65] P. Diaconis and E. Lander, Some formulas for zonal polynomials, unpublished (1986).

[66] P. Diaconis and D. Rockmore, Efficient computation of the Fourier transform on finite groups. *J. Amer. Math. Soc.* **3** (1990), 297–332.

[67] P. Diaconis and D. Rockmore, Efficient computation of isotypic projections for the symmetric group. Groups and computation (New Brunswick, NJ, 1991), 87–104, *DIMACS Ser. Discrete Math. Theoret. Comput. Sci.*, **11**, Amer. Math. Soc., Providence, RI, 1993.

[68] P. Diaconis and L. Saloff-Coste, Separation cutoffs for death and birth chains, Preprint (2005).

[69] P. Diaconis and M. Shahshahani, Generating a random permutation with random transpositions, *Z. Wahrsch. Verw. Geb.*, **57** (1981), 159–179.

[70] P. Diaconis and M. Shahshahani, Time to reach stationarity in the Bernoulli–Laplace diffusion model, *SIAM J. Math. Anal.* **18** (1987), 208–218.

[71] J. Dieudonné, *Treatise on Analysis*, Vol. VI. Translated from the French by I.G. Macdonald. Pure and Applied Mathematics, 10-VI. Academic Press, New York, 1978.

[72] C. Di Fiore, Matrix algebras and displacement decompositions. *SIAM J. Matrix Anal. Appl.* **21** (1999), 646–667 (electronic).

[73] Ch.F. Dunkl, A Krawtchouk polynomial addition theorem and wreath products of symmetric groups, *Indiana Univ. Math. J.* **25** (1976), 335–358.

[74] Ch.F. Dunkl, An addition theorem for some *q*-Hahn polynomials. *Monatsh. Math.* **85** (1978), 5–37.

[75] Ch.F. Dunkl, An addition theorem for Hahn polynomials: the spherical functions, *SIAM J. Math. Anal.* **9** (1978), 627–637.

[76] Ch.F. Dunkl, Spherical functions on compact groups and applications to specical functions, *Symposia Mathematica* **22** (1979), 145–161.

[77] Ch.F. Dunkl, Orthogonal functions on some permutation groups, *Proc. Symp. Pure Math.* **34**, Amer. Math. Soc., Providence, RI, (1979), 129–147.

[78] Ch.F. Dunkl, A difference equation and Hahn polynomials in two variables, *Pacific J. Math.*, **92** (1981), 57–71.

[79] Ch.F. Dunkl and D.E. Ramirez, Krawtchouk polynomials and the symmetrization of hypergroups, *SIAM J. Math. Anal.* **5** (1974), 351–366.

[80] J.R. Durbin, Spherical functions on wreath products, *J. Pure Appl. Algebra* **10** (1977/78), 127–133.

[81] R. Durrett, *Probability: Theory and Examples.* Second edition. Duxbury Press, Belmont, CA, 1996.

[82] H. Dym and H.P. McKean, *Fourier Series and Integrals.* Probability and Mathematical Statistics, No. 14. Academic Press, New York-London, 1972.

[83] J.D. Farley, Quasi-differential posets and cover functions of distributive lattices. I. A conjecture of Stanley. *J. Combin. Theory Ser. A* **90** (2000), 123–147.

[84] J.D. Farley, Quasi-differential posets and cover functions of distributive lattices. II. A problem in Stanley's enumerative combinatorics. *Graphs Combin.* **19** (2003), 475–491.

[85] J.D. Farley and K. Sungsoon, The automorphism group of the Fibonacci poset: a "not too difficult" problem of Stanley from 1988. *J. Algebraic Combin.* **19** (2004), no. 2, 197–204.

[86] J. Faraut, Analyse Harmonique sur les paires de Gelfand et les espaces hyperboliques, in *Analyse harmonique*, J.L. Clerc, P. Eymard, J. Faraut, M. Raés, R. Takahasi, Eds. (École d'été d'analyse harmonique, Université de Nancy I, Septembre 15 au Octobre 3, 1980). C.I.M.P.A. V, 1983.

[87] W. Feller, *An Introduction to Probability Theory and its Applications*, Vol. I. Second edition John Wiley, New York, 1971.

[88] A. Figà-Talamanca, Note del Seminario di Analisi Armonica, A.A. 1990-91, Università di Roma "La Sapienza".

[89] A. Figà-Talamanca, Local fields and trees. Harmonic functions on trees and buildings (New York, 1995), 3–16, *Contemp. Math.*, **206**, Amer. Math. Soc., Providence, RI, 1997.

[90] A. Figà-Talamanca, An application of Gelfand pairs to a problem of diffusion in compact ultrametric spaces, in *Topics in Probability and Lie Groups: Boundary Theory*, 51–67, CRM Proc. Lecture Notes, **28**, Amer. Math. Soc., Providence, RI, 2001.

[91] A. Figà-Talamanca and C. Nebbia, *Harmonic Analysis and Representation Theory for Groups Acting on Homogeneous Trees.* London Mathematical Society Lecture Note Series, 162. Cambridge University Press, Cambridge, 1991.

[92] S. Fomin, Duality of graded graphs. *J. Algebraic Combin.* **3** (1994), 357–404.

[93] S. Fomin, Schensted algorithms for dual graded graphs. *J. Algebraic Combin.* **4** (1995), 5–45.

[94] W. Fulton, *Young Tableaux. With Applications to Representation Theory and Geometry.* London Mathematical Society Student Texts, 35. Cambridge University Press, Cambridge, 1997.

[95] W. Fulton and J. Harris, *Representation Theory. A First Course.* Springer-Verlag, New York, 1991.

[96] F.R. Gantmacher, *The Theory of matrices.* Vols. 1, 2. Translated by K.A. Hirsch, Chelsea Publishing, New York 1959.

[97] A. Garsia, Gelfand pairs in finite groups, unpublished MIT manuscript (1985).

[98] I.M. Gelfand, Spherical functions on symmetric Riemannian spaces, *Dokl. Akad. Nauk. SSSR* **70** (1959), 5–8; [Collected papers, Vol. II, Springer (1988) 31–35].

[99] B.V. Gnedenko, *The Theory of Probability.* "Mir", Moscow, 1978 and 1982, or Chelsea Publishing, New York 1967.

[100] R. Godement, A theory of spherical functions. I. *Trans. Amer. Math. Soc.* **73**, (1952), 496–556.

[101] R. Gow, Two multiplicity-free permutation representations of the general linear group $GL(n, q^2)$. *Math. Z.* **188** (1984), 45–54.

[102] C.D. Godsil, *Algebraic Combinatorics.* Chapman & Hall, New York, 1993.

[103] P. Graczyk, G. Letac and H. Manam, The hyperoctaedral group, symmetric group representations and the moments of the real Wishart distribution, *J. Theoret. Probab.* **18** (2005), 1–42.

[104] R.L. Graham, D.E. Knuth and O. Patashnik, *Concrete Mathematics*, Addison-Wesley, Reading, MA, 1989.

[105] R.L. Graham, R. Li and W. Li, On the structure of t-designs, *SIAM J. Alg. and Disc. Meth.* **1** (1980), 8–14.

[106] A.S. Greenhalgh, Measure on groups with subgroups invariance properties, *Technical report No. 321*, Department of Statistics, Stanford University, 1989.

[107] R.I. Grigorchuk, Just infinite branch groups. New horizons in pro-*p* groups, 121–179, Progr. Math., **184**, Birkhäuser Boston, Boston, MA, 2000.

[108] L.C. Grove, Spherical functions on finite split extensions. *Colloq. Math.* **44** (1981), 185–192.

[109] L.C. Grove, *Groups and Characters*, Pure and Applied Mathematics (New York). John Wiley, New York, 1997.

[110] L.C. Grove, *Classical Groups and Geometric Algebra*, Graduate Studies in Mathematics, 39. American Mathematical Society, Providence, RI, 2002.

[111] A. Hanaki, and K. Hirotsuka, Irreducible representations of wreath products of association schemes. *J. Algebraic Combin.* **18** (2003), 47–52.

[112] F. Harary, Exponentiation of permutation groups, *Amer. Math. Monthly* **66** (1959), 572–575.

[113] F. Harary and E.M. Palmer, *Graphical Enumeration.* Academic Press, New York, 1973.

[114] P. de la Harpe, *Topics in Geometric Group Theory*, University of Chicago Press, 2000.

[115] P. de la Harpe, *Représentsations des Groupes Finis*, Preprint 2007.

[116] O. Häggström, *Finite Markov Chains and Algorithmic Applications.* London Mathematical Society Student Texts, 52. Cambridge University Press, Cambridge, 2002.

[117] S. Helgason, *Differential Geometry, Lie Groups, and Symmetric Spaces.* Corrected reprint of the 1978 original. Graduate Studies in Mathematics, 34. American Mathematical Society, Providence, RI, 2001.

[118] S. Helgason, *Groups and Geometric Analysis. Integral Geometry, Invariant Differential Operators, and Spherical Functions.* Corrected reprint of the 1984 original. Mathematical Surveys and Monographs, 83. American Mathematical Society, Providence, RI, 2000.

[119] A. Henderson, Spherical functions of the symmetric space $G(\mathbb{F}_{q^2})/G(\mathbb{F}_q)$. *Represent. Theory* **5** (2001), 581–614 (electronic).

[120] I.N. Herstein, *Topics in Algebra.* Second edition. Xerox College Publishing, Lexington, Mass., 1975.

[121] F.B. Hildebrand, *Finite-Difference Equations and Simulations.* Prentice-Hall, Englewood Cliffs, NJ, 1968.

[122] D.G. Higman, Coherent configurations. II. Weights. *Geometriae Dedicata* **5** (1976), no. 4, 413–424.

[123] T.W. Hungerford, *Algebra.* Reprint of the 1974 original. Graduate Texts in Mathematics, 73. Springer-Verlag, New York, 1980.

[124] L.-K. Hua, On the Automorphisms of the symplectic group over any field. *Annals of Math.* **49** (1948), 739–759.

[125] B. Huppert, *Character Theory of Finite Groups*, De Gruyter Expositions in Mathematics, 25, Walter de Gruyter, 1998.

[126] N.F.J. Inglis, M.W. Liebeck and J. Saxl, Multiplicity-free permutation representations of finite linear groups. *Math. Z.* **192** (1986), 329–337.

[127] N.F.J. Inglis and J. Saxl, An explicit model for the complex representations of the finite general linear groups. *Arch. Math.* (Basel) **57** (1991), 424–431.

[128] R.E. Ingram, Some characters of the symmetric group. *Proc. Amer. Math. Soc.* **1** (1950), 358–369.

[129] I.M. Isaacs, *Character Theory of Finite Groups*, Corrected reprint of the 1976 original [Academic Press, New York]. Dover Publications, New York, 1994.

[130] A.T. James, Zonal polynomials of the real positive definite symmetric matrices, *Ann. Math.* **74** (1961), 456–469.

[131] G.D. James, A characteristic free approach to the representation theory of S_n, *J. Algebra* **46** (1977), 430–450.

[132] G.D. James, *The Representation Theory of the Symmetric Group*, Lecture Notes in Math., **682**, Springer-Verlag, Berlin, 1978.

[133] G.D. James, *Representations of General Linear Groups.* London Mathematical Society Lecture Note Series, 94. Cambridge University Press, Cambridge, 1984.

[134] G.D. James, and A. Kerber, *The Representation Theory of the Symmetric Group*, Encyclopedia of Mathematics and its Applications, **16**, Addison-Wesley, Reading, MA, 1981.

[135] S. Karlin and J. McGregor, The Hahn polynomials, formulas and an application, *Scripta Math.* **26** (1961), 33–46.

[136] S. Karlin and H.M. Taylor, *An Introduction to Stochastic Modeling.* Third edition. Academic Press, San Diego, CA, 1998.

[137] J. Karlof, The subclass algebra associated with a finite group and subgroup. *Trans. Amer. Math. Soc.* **207** (1975), 329–341.

[138] L.K. Kates, Zonal polynomials, PhD thesis, Princeton University, 1980.

[139] Y. Katznelson, *An Introduction to Harmonic Analysis*, Third edition. Cambridge Mathematical Library. Cambridge University Press, Cambridge, 2004.

[140] N. Kawanaka, On subfield symmetric spaces over a finite field. *Osaka J. Math.* **28** (1991), 759–791.

[141] N. Kawanaka and H. Matsuyama, A twisted version of the Frobenius–Schur indicator and multiplicity-free permutation representations, *Hokkaido Math. J.* **19** (1990), 495–508.

[142] J.G. Kemeny and J.L. Snell, *Finite Markov Chains*, Reprinting of the 1960 original. Undergraduate Texts in Mathematics. Springer-Verlag, New York, 1976.

[143] A. Kerber, *Applied Finite Group Actions*. Second edition. Algorithms and Combinatorics, 19. Springer-Verlag, Berlin, 1999.

[144] A. Kleshchev, *Linear and Projective Representations of Symmetric Groups*. Cambridge Tracts in Mathematics, 163. Cambridge University Press, Cambridge, 2005.

[145] A.A. Klyachko, Models for complex representations of groups $GL(n, q)$. (Russian) *Mat. Sb. (N.S.)* **120**(162) (1983), 371–386.

[146] A.U. Klimyk and N. Ja. Vilenkin, *Representation of Lie Groups and Special Functions. Volume 2.* Mathematics and its Applications **81**. Kluwer Academic, Dordrecht, 1993.

[147] T.H. Koornwinder, A note on the multiplicity free reduction of certain orthogonal and unitary groups. *Nederl. Akad. Wetensch. Indag. Math.* **44** (1982), 215–218.

[148] A. Krieg, Hecke algebras. *Mem. Amer. Math. Soc.* **87** (1990), no. 435.

[149] S. Lang, $SL_2(R)$. Reprint of the 1975 edition. Graduate Texts in Mathematics, 105. Springer-Verlag, New York, 1985.

[150] S. Lang, *Linear Algebra*. Reprint of the third edition. Undergraduate Texts in Mathematics. Springer-Verlag, New York, 1989.

[151] S. Lang, *Algebra*. Revised Third Edition. Graduate Texts in Mathematics, 211. Springer-Verlag, New York, 2002.

[152] G. Letac, Problèmes classiques de probabilité sur un couple de Gelfand in *Analytical problems in probability*, p. 93–120, Lecture Notes in Math., **861**, Springer Verlag, New York, 1981.

[153] G. Letac, Les fonctions sphériques d'un couple de Gelfand symétrique et les chaînes de Markov. *Advances Appl. Prob.*, **14** (1982), 272–294.

[154] G. Letac and L. Takács, Random walks on an m-dimensional cube. *J. Reine Angew. Math.* **310** (1979), 187–195.

[155] W.C.W. Li, *Number Theory with Applications*, Series on University Mathematics, 7. World Scientific Publishing, River Edge, NJ, 1996.

[156] J.H van Lint and R.M. Wilson, *A Course in Combinatorics*. Second edition. Cambridge University Press, Cambridge, 2001.

[157] L.H. Loomis, *An Introduction to Abstract Harmonic Analysis*, Van Nostrand, Toronto, 1953.

[158] G. Lusztig, Symmetric spaces over a finite field. The Grothendieck Festschrift, Vol. III, 57–81, Progr. Math., 88, Birkhduser Boston, Boston, MA, 1990.

[159] F.J. MacWilliams and N.J.A. Sloane, *The Theory of Error-Correcting Codes* I and II. North-Holland Mathematical Library, Vol. 16. North-Holland, Amsterdam, 1977.

[160] A. Machì , *Gruppi. Una introduzione a idee e metodi della Teoria dei Gruppi*, Springer, Berlin, 2007.

[161] A. Markoe, *Analytic Tomography*, Encyclopedia of Mathematics and its Applications, **106** Cambridge University Press, 2006.

[162] G. Mauceri, Square integrable representations and the Fourier algebra of a unimodular group. *Pacific J. Math.* **73** (1977), 143–154.

[163] H. Mizukawa, Zonal spherical functions on the complex reflection groups and $(n + 1, m + 1)$-hypergeometric functions, *Adv. Math* **184** (2004), 1–17.

[164] H. Mizukawa and H. Tanaka, $(n + 1, m + 1)$-hypergeometric functions associated to character algebras, *Proc. Amer. Math. Soc.* **132** (2004), 2613–2618.

[165] I.G. Macdonald, *Symmetric Functions and Hall Polynomials*, second edition, Oxford Univeristy Press 1995.

[166] G.W. Mackey, Multiplicity free representations of finite groups. *Pacific J. Math.* **8** (1958) 503–510.

[167] P. Matthews, Mixing rates for a random walk on the cube. *SIAM J. Algebraic Discrete Methods* **8** (1987), 746–752.

[168] M.A. Naimark and A.I. Stern, *Theory of Group Representations*, Springer-Verlag, New York, 1982.

[169] M.B. Nathanson, *Elementary Methods in Number Theory*. Graduate Texts in Mathematics, 195. Springer-Verlag, New York, 2000.

[170] C. Nebbia, Classification of all irreducible unitary representations of the stabilizer of the horicycles [horocycles] of a tree. *Israel J. Math.* **70** (1990), 343–351.

[171] P.M. Neumann, A lemma that is not Burnside, *Math. Sci.* **4** (1979) 133–141.

[172] J.R. Norris, *Markov Chains*. Reprint of 1997 original. Cambridge Series in Statistical and Probabilistic Mathematics, 2. Cambridge University Press, Cambridge, 1998.

[173] A. Okounkov and A.M. Vershik, A new approach to representation theory of symmetric groups. *Selecta Math. (N.S.)* **2** (1996), no. 4, 581–605.

[174] A. Okounkov and A.M. Vershik, A new approach to representation theory of symmetric groups. II. (Russian) *Zap. Nauchn. Sem. S.-Peterburg. Otdel. Mat. Inst. Steklov. (POMI)* **307** (2004), Teor. Predst. Din. Sist. Komb. i Algoritm. Metody. 10, 57–98, 281; translation in *J. Math. Sci. (N. Y.)* **131** (2005), no. 2, 5471–5494.

[175] M.E. Orrison, Radon transforms and the finite general linear groups. *Forum Math.* **16** (2004), no. 1, 97–107.

[176] M.H. Peel, Specht modules and the symmetric groups, *J. Algebra* **36** (1975), 88–97.

[177] K. Petersen, *Ergodic Theory*. Corrected reprint of the 1983 original. Cambridge Studies in Advanced Mathematics, 2. Cambridge University Press, Cambridge, 1989.

[178] M. Puschel, J.M.F. Moura, The algebraic approach to the discrete cosine and sine transforms and their fast algorithms. *SIAM J. Comput.* **32** (2003), 1280–1316.

[179] M. Reed and B. Simon, *Methods of Modern Mathematical Physics. I. Functional Analysis*, Second edition. Academic Press, New York, 1980.

[180] F. Ricci, *Analisi di Fourier non commutativa*, notes of a course given at the Scuola Normale Superiore di Pisa in the academic year 2001–2002 [http://math.sns.it/HomePages/Ricci/corsi.html].

[181] W. Rudin, *Fourier Analysis on Groups* Reprint of the 1962 original. Wiley Classics Library. John Wiley, New York, 1990.

[182] B.E. Sagan, *The Symmetric Group. Representations, Combinatorial Algorithms, and Symmetric Functions*. Second edition. Graduate Texts in Mathematics, 203. Springer-Verlag, New York, 2001.

[183] J. Saxl, On multiplicity-free permutation representations, in *Finite Geometries and Designs*, p. 337–353, London Math. Soc. Lecture Notes Series, **48**, Cambridge University Press, 1981.

[184] L. Saloff-Coste, *Lectures on Finite Markov Chains*, in: E. Giné, G.R. Grimmett and L. Saloff-Coste, *Lectures on probability theory and statistics*. Lectures from the 26th Summer School on Probability Theory held in Saint-Flour, August 19–September 4, 1996. Edited by P. Bernard. Lecture Notes in Mathematics, 1665. Springer-Verlag, Berlin, 1997.

[185] L. Saloff-Coste, Total variation lower bounds for finite Markov chains: Wilson's lemma. Random walks and geometry, 515–532, Walter de Gruyter, Berlin, 2004.

[186] F. Scarabotti, Time to reach stationarity in the Bernoulli–Laplace diffusion model with many urns, *Adv. in Appl. Math.* **18** (1997), 351–371.

[187] F. Scarabotti, Radon transforms on the symmetric group and harmonic analysis of a class of invariant Laplacians. *Forum Math.* **10** (1998), 407–411.

[188] F. Scarabotti, Fourier analysis of a class of finite Radon transforms, *Siam J. Discrete Math.* **16** (2003), 545–554.

[189] F. Scarabotti, Harmonic analysis of the space of $S_a \times S_b \times S_c$-invariant vectors in the irreducible representations of the symmetric group. *Adv. in Appl. Math.* **35** (2005), 71–96.

[190] F. Scarabotti, The discrete sine transform and the spectrum of the finite q-ary tree, *SIAM J. Discrete Math.* **19** (2005), 1004–1010.

[191] F. Scarabotti and F. Tolli, Spectral analysis of finite Markov chains with spherical symmetries, *Adv. in Appl. Math.*, **38** (2007), no. 4, 445–481.

[192] F. Scarabotti and F. Tolli, Harmonic analysis of finite lamplighter random walks, *J. Dynam. Control. Syst.*, to appear.

[193] F. Scarabotti and F. Tolli, Harmonic analysis on a finite homogeneous space, preprint 2007.

[194] E. Schröder, Vier combinatorische Probleme, *Zeit. f. Math. Phys.*, **15** (1870), 361–376.

[195] E. Seneta, *Non-negative Matrices and Markov Chains*, Springer Series in Statistics, second rev. edition, Springer-Verlag Berlin 2006.

[196] J.P. Serre, *Linear Representations of Finite Groups*, Graduate Texts in Mathematics, Vol. 42. Springer-Verlag, New York, 1977.

[197] G.E. Shilov, *Linear Algebra*. Revised English edition. Translated from the Russian and edited by Richard A. Silverman. Dover Publications, Inc., New York, 1977.

[198] B. Simon, *Representations of Finite and Compact Groups*, American Math. Soc., 1996.

[199] A.N. Shiryaev, *Probability* Translated from the first (1980) Russian edition by R.P. Boas. Second edition. Graduate Texts in Mathematics, 95. Springer-Verlag, New York, 1996.

[200] N.J.A. Sloane, Recent bounds for codes, sphere packings and related problems obtained by linear programming and other methods. *Contemp. Math.* **9** (1982), 153–185.

[201] R.P. Stanley, Differential posets. *J. Amer. Math. Soc.* **1** (1988), 919–961.

[202] R.P. Stanley, Some combinatorial properties of Jack symmetric functions, *Adv. in Math.* **77** (1989), 76–115.

[203] R.P. Stanley, Variations on differential posets. Invariant theory and tableaux (Minneapolis, MN, 1988), 145–165, *IMA Vol. Math. Appl.*, **19**, Springer, New York, 1990.

[204] R.P. Stanley, *Enumerative Combinatorics,* Vol. 1, Cambridge University Press, 1997.

[205] R.P. Stanley, *Enumerative combinatorics.* Vol. 2. With a foreword by Gian-Carlo Rota and Appendix 1 by Sergey Fomin. Cambridge Studies in Advanced Mathematics, 62. Cambridge University Press, Cambridge, 1999.

[206] D. Stanton, Three addition theorems for some q-Krawtchouk polynomials. *Geom. Dedicata* **10** (1981), 403–425.

[207] D. Stanton, A partially ordered set and q-Krawtchouk polynomials. *J. Combin. Theory Ser. A* **30** (1981), 276–284.

[208] D. Stanton, Generalized n-gons and Chebychev polynomials. *J. Combin. Theory Ser. A* **34** (1983), 15–27.

[209] D. Stanton, Orthogonal polynomials and Chevalley groups, in *Special Functions: Group Theoretical Aspects and Applications* (R. Askey et al., Eds.) 87–128, Dordrecht, Boston, 1984.

[210] D. Stanton, Harmonics on Posets. *J. Comb. Theory Ser. A* **40** (1985), 136–149.

[211] D. Stanton, An introduction to group representations and orthogonal polynomials, in *Orthogonal Polynomials* (P. Nevai Ed.), 419–433, Kluwer Academic Dordrecht, 1990.

[212] S. Sternberg, *Group Theory and Physics*, Cambridge University Press, Cambridge, 1994.

[213] G. Strang, *Linear Algebra and its Applications.* Second edition. Academic Press New York, 1980.

[214] G. Strang, The discrete cosine transform. *SIAM Rev.* **41** (1999), 135–147.

[215] D.W. Stroock, *An Introduction to Markov Processes.* Graduate Texts in Mathematics, 230. Springer-Verlag, Berlin, 2005.

[216] L. Takacs, Harmonic analysis on Schur algebras and its applications in the theory of probability, in *Probability Theory and Harmonic Analysis* (Cleveland, Ohio, 1983), 227–283. Monogr. Textbooks Pure Appl. Math., 98, Dekker, New York, 1986.

[217] O. Tamaschke, On Schur-rings which define a proper character theory on finite groups. *Math. Z.* **117** 1970 340–360.

[218] O. Tamaschke, *Schur-Ringe.* Vorlesungen an der Universitdt T—bingen im Sommersemester 1969. B. I-Hochschulskripten, 735a*. Bibliographisches Institut, Mannheim, 1970.

[219] H. Tarnanen, M. Aaltonen and J.-M. Goethals, On the nonbinary Johnson scheme, *European J. Combin.* **6** (1985), 279–285.

[220] A. Terras, *Fourier Analysis on Finite Groups and Applications.* London Mathematical Society Student Texts, 43. Cambridge University Press, Cambridge, 1999.

[221] J.G. Thompson, Fixed point free involutions and finite projective planes, in *Finite Simple Groups II* (M-J.Collins Ed.), New York 1980.

[222] R.M. Thrall, On symmetrized Kronecker powers and the structure of the free Lie ring, *Amer. J. Math.*, **64** (1942), 371–388.

[223] D. Travis, Spherical functions on finite groups. *J. Algebra* **29** (1974), 65–76.

[224] D. Vere-Jones, Finite bivariate distributions and semigroups of non-negative matrices, *Quart. J. Math. Oxford Ser.* (2) **22** (1971), 247–270.

[225] E. Vinberg, private communication.

[226] Z. Wang, Fast algorithms for the discrete W transform and for the discrete Fourier transform. *IEEE Trans. Acoust. Speech Signal Process.* **32** (1984), 803–816.

[227] A. Wawrzyńczyk, *Group Representations and Special Functions. Examples and Problems Prepared by Aleksander Strasburger.* Translated from the Polish by Bogdan Ziemian. Mathematics and its Applications (East European Series). D. Reidel Publishing Co., Dordrecht; PWN—Polish Scientific Publishers, Warsaw, 1984.

[228] H. Wielandt, *Finite Permutation Groups*, Academic Press, New York 1964.

[229] E. Wigner, On representations of certain finite groups, *Amer. J. Math.* **63** (1941), 57–63.

[230] P. Winkler, *Mathematical Puzzles that Boggle the Mind*, AK Peters Ltd. 2007.

[231] W. Woess, *Catene di Markov e Teoria del Potenziale nel Discreto*, Quaderni dell'Unione Matematica Italiana, 41, Pitagora Editrice, Bologna, 1996.

[232] W. Woess, *Random Walks on Infinite Graphs and Groups*, Cambridge Tracts in Mathematics, 138. Cambridge University Press, Cambridge, 2000.

[233] E.M. Wright, Burnside's lemma: a historical note, *J. Comb. Theory (B)*, **30** (1981), 89–90.

[234] P.-H. Zieschang, *Theory of Association Schemes.* Springer Monographs in Mathematics. Springer-Verlag, Berlin, 2005.

Index